HANDBOOK of Microwave Techniques and Equipment

HANDBOOK

of Microwave Techniques and Equipment

Harry E. Thomas

CONSULTING ENGINEER

PRENTICE-HALL, INC.

ENGLEWOOD CLIFFS, N.J.

PRENTICE-HALL INTERNATIONAL, INC., *London*
PRENTICE-HALL OF AUSTRALIA, PTY. LTD., *Sydney*
PRENTICE-HALL OF CANADA, LTD., *Toronto*
PRENTICE-HALL OF INDIA PRIVATE LTD., *New Delhi*
PRENTICE-HALL OF JAPAN, INC., *Tokyo*

Handbook of
Microwave Techniques
and Equipment

Harry E. Thomas

ISBN: 0-13-380329-5

10 9 8 7 6 5 4 3 2

Library of Congress Catalog Card Number: 72-2347

Printed in the United States of America

Contents

Appendices, 263

Index, 313

Preface

A microwave engineering handbook, in order to be completely inclusive of supporting theory, operation, components, and hardware design, must be either voluminous or severely condensed. This book is intended to establish a balance between these two extremes.

Subjectwise, its approach is 5-pronged, the first dealing with concepts, transmission lines, wave propagation, and systems; the material then branches in the following four directions:

(1) Theory and operation of passive and active components. This of course details the entire field of electronic component types plus transmission lines, waveguides, cavities, filters, switches, etc.

(2) Measurement factors, instruments, and supporting items such as attenuators, termination, stubs, tuners, etc.

(3) Specialized areas of millimeter waves plus the development, transition, and operating equipment using integrated solid-state circuits. Beyond MIC's, specialized attention is given to masers and parametric amplifiers.

(4) Antenna theory design and construction. This material covers conventional approaches plus final arrival at IC-oriented phased array construction.

The material is arranged and slanted toward providing factual and procedural tools for the working engineer, the circuit designer, and the laboratory technician. The approach consists of expressing in condensed form the basic facts and theory concerning a subject and then supporting the basic material with data, summaries, illustrations, and physical facts. Once the above working personnel has been oriented or refreshed by the condensed background and operational theories, he finds circuits, hardware designs, and equipment illustrated which provide usable information for productive action. Tables, physical data, condensed procedures, instrument photographs, etc., constitute most of this information.

HARRY E. THOMAS

HANDBOOK of Microwave Techniques and Equipment

Basic Microwave Concepts and Operation 1

BASIC

Self-descriptively, microwave energy concerns extremely high frequencies or wavelengths in microwave dimension; in descriptive nomenclature microwave transmission (mostly through waveguides) is a wave-propagation phenomenon—thus we have the appellation "microwaves."

The frequency and wavelengths in the electromagnetic domain are related by a physical constant—the speed of light in space:

$$\begin{aligned} c &= 984{,}000 \text{ ft/sec} \\ &= 1.18 \times 10^{10} \text{ in./sec} \\ &= 3 \times 10^{8} \text{ m/sec} \end{aligned}$$

The relationship is a product,

$$\lambda \times f = c$$

where λ = wavelength (meters)

f = frequency (Hz)

c = speed of light (meters/second)

When magnetic waves pass through a dielectric they are slowed down; the amount is proportional to the square root of the dielectric constant. In Teflon, for instance, with a dielectric constant of 2.1, the speed of light (c_t) now becomes:

$$c_t = 3/\sqrt{2.1} \times 10^{8} \text{ m/sec.}$$

MICROWAVE SPECTRUM

The frequency and wavelength range of microwaves within the electromagnetic spectrum in hertz and kilomegahertz is from

$$0.3 \text{ kMHz} = 300 \text{ MHz} - \lambda = 100 \text{ cm}$$

to

$$300 \text{ kMHz} = 300{,}000 \text{ MHz} - \lambda = 1 \text{ mm}$$

Table 1-1 shows wavelengths and frequencies for the entire electromagnetic spectrum; this entirety is in turn broken down into the radio-communication spectrum and the various microwave bands used also for communication and for radar, plus the range of light-related emissions.

Note that as we progress through the entire electromagnetic spectrum, the entire emission nomenclature changes from

1. Discrete frequencies—used mostly in the audio and speech range.
2. Conventional wavelengths—from very low frequency up through microwave lengths.
3. So-called rays—heat-, light-, molecular-electron-related emissions. All such emissions, wherever they exist, are also frequency- (and wavelength-)oriented phenomena.

Table 1-2 tabulates these wavelengths and frequencies for convenience in conversion; note that the wavelength part of the tabulation extends into micron and angstrom quantities, the former units used when referring to minute linear dimensions in Integrated Circuit construction and the latter universally used when referring to the wavelength of light and special rays. An angstrom (Å) = 10^{-10} m = 10^{-7} mm.

TABLE 1-1 MICROWAVE AND ELECTROMAGNETIC SPECTRUMS

	Wavelength Limits (cm)		*Frequency Limits (GHz)*		
Region	*Maximum*	*Minimum*	*Minimum*	*Maximum*	*Remarks*
Radio	Infinite	0.1	D.C.	300	VLF/LF/MF/ HF/VHF/UHF/SHF/Exp.
Infra-red	0.1	0.00008	300	375,000	Heat & Black Light
Light (visible)	0.00008	0.000038	375,000	790,000	Starts with Red, progresses through orange, yellow, green, blue, and violet
Ultra-violet	0.000038	0.0000012	790,000	22,500,000	Chemical and invisible
X-rays	0.0000012	0.0000000006	22,500,000	45,000,000,000	–
Gamma rays	0.000000014	0.0000000001	45,000,000,000	270,000,000,000	Radioactivity
Cosmic rays	0.0000000001	Indefinite	270,000,000,000	Indefinite	Little known

Wavelength band (m)	*Frequency band*	*Approximate number of KHz per meter change in wavelength*	*Approximate number of meters per KHz change in frequency*	*Official FCC band abbreviation (frequency)*
Very long waves (infinity to 10,000)	0 to 30 KHz/sec	Below 0.01	Over 333	Very low (VLF)
Long waves (10,000 to 1000)	30 to 300 KHz/sec	0.05	20	Low (LF)
Medium waves (1000 to 100)	300 to 3000 KHz/sec	5	0.2	Medium (MF)
Short waves (100 to 10)	3 to 30 MHz/sec	500	0.002	High (HF)
Very short waves (10 to 1)	30 to 300 MHz/sec	5×10^4	0.2×10^{-4}	Very high (VHF)
Ultra short waves (1 to 0.1) (microwaves)	0.3 to 3 GHz/sec	5×10^6	0.2×10^{-6}	Ultra high (UHF)
Super short waves (10 to 1 cm) (microwaves)	3 to 30 GHz/sec	5×10^8	0.2×10^{-8}	Super high (SHF)
Extremely short waves (10 to 1 mm) (millimeter waves)	30 to 300 GHz/sec	5×10^{10}	0.2×10^{-10}	Extremely high (EHF)
Quasi-optical (1 to 0.1 mm)	300 to 3000 GHz/sec	5×10^{12}	0.2×10^{-12}	

TABLE 1-2 FREQUENCY AND WAVELENGTH CONVERSION

Wavelength unit	*Equivalence (in.)*	*Equivalence (GHz)*
1 meter	39.37	0.3
1 decimeter	3.937	3
1 centimeter	0.3937	30
1 millimeter	0.03937	300
1 micron	0.00003937	300,000
1 millimicron	0.00000003937	300,000,000
1 angstrom unit	0.000000003937	3,000,000,000

(a) WAVELENGTH-FREQUENCY

Meter	*Decimeter*	*Centimeter*	*Millimeter*	*Micron*	*Millimicron*	*Angstrom unit*
1 Meter	10	100	1,000	1,000,000	1,000,000,000	10,000,000,000
0.1	1 Decimeter	10	100	100,000	100,000,000	1,000,000,000
0.01	0.1	1 Centimeter	10	10,000	10,000,000	100,000,000
0.001	0.01	0.1	1 Millimeter	1,000	1,000,000	10,000,000
0.000001	0.00001	0.0001	0.001	1 Micron	1,000	10,000
0.000000001	0.00000001	0.0000001	0.000001	0.001	1 Millimicron	10
0.0000000001	0.000000001	0.00000001	0.0000001	0.0001	0.1	1 Angstrom unit

(b) METERS-MICRONS-ANGSTROMS

BASIC WHYS OF MICROWAVE USAGE

Microwaves offer special advantages in four specific fields:

1. In communication they consume a proportionately smaller percentage of the total bandwidth of information channels where modulation rides upon the main carrier wave; this inherently permits the use of more channels than at lower carrier frequencies. We capitalize on this qualification in space communication and telemetry.

2. In navigation and radar, microwave antennas give extreme directivity because the short wavelengths allow focusing and directivity much the same way as lenses focus and direct light rays.

3. For industrial heating, drying, cooking, and physical diathermy, microwave energy is more easily directed, controlled, and concentrated than low-frequency energy (see applications in Chapter 11).

4. Useful molecular resonances exist at microwave frequencies in diodes of certain crystal materials; in these crystals we can generate or convert energy by means of atomic oscillations in and around the avalanche condition. The Gunn, Read, and Impatt diodes embodying this effect will be discussed in Chapter 12.

All in all, these characteristics of microwave operation constitute a useful and considerable part of the electronic regime. We have but to reconcile the electromagnetic phenomena underlying some of the above applications to the physical operating equipment to textually document the microwave technology. Such documentation constitutes the contents of this volume.

ELEMENTS OF THE MICROWAVE REGIME

As in the other branches of electronic technology, we are concerned with specific details related to physical equipment. Microwave technology may thus be divided into four main areas:

1. Generation and processing (Chapters 7 and 11).
2. Transmission (Chapters 2, 3, 4, 5 and 6).
3. Measurement (Chapters 8, 9, 10 and 11).
4. Application: Communication, navigational, industrial (Chapters 10 and 11).

Relating these to the final text, the following preview of our ultimate treatment of items 1, 2, and 3 will constitute our approach to the main material in the following manner:

1. How to generate or amplify moderate to large amounts of microwave energy for signal emission, particularly for use in communication and reception in space, or for the application of this power as applied to the processing of material and industrial substances. Here we are concerned primarily with oscillation and amplification in our principal devices.
2. How microwaves must be transmitted for communication and, secondarily, through specialized conductors for short-range and direct application in industrial equipment. This, of course, includes antennas, transmission lines, and waveguides.
3. How to detect, evaluate, and interpret energy so that microwaves may be measured in small amounts of power, a continuing and intimate process that constitutes most of the material in Chapters 8 through 11. A portion of this material centers on receiver equipment and communication systems.

COMPONENT-HARDWARE SUMMARY

Table 1-3 alphabetically summarizes the field of microwave components as they are used in most fields and applications. In a few cases the name of the component self-evidently indicates whether it concerns power or measurement (modulators, slotted lines, etc.). Each of these items will have its structure and design theory of operation studied within the appropriate chapters to come.

Table 1-4 lists the direct areas of commercial microwave application and the included equipment or specific required instruments. This listing aims to give the reader a practical preview of the ultimate usage of microwave equipment. A number of subcategorizations appear in their appropriate places (microwave tubes, IC microwave assemblies, etc.). Of particular interest here are the listings of microwave measuring instruments (Table 10-1). The microwave glossary of terms and definitions adds a number of other categorizations applicable to special molecular devices, diodes, switching modes, etc.

WAVE MOTION—MICROWAVE TRANSMISSION

Microwave technology (except in the case of specially constructed microwave vacuum tubes) centers about the transmission of microwaves through shielded cables or waveguides, the primary object being transfer of power with a minimum loss. High-frequency energy travels better through the interior space of a hollow conductor than through the conductor itself. As we shall see, it "bounces" its way across and along the interior surfaces.

This entire phenomenon is basically linked to wave motion, because very high frequencies do not remain localized or concentrated in a pair of conductors as do dc or low-frequency energy. Microwave energy, accordingly, cannot be applied to, or operate upon, lumped surface constants—in effect, it "sifts" through R's, L's, and C's and must be handled, detected, or measured only in terms of physical distances related to the wave motion. And wave motion is described in terms of the height of and the distance

TABLE 1-3 MICROWAVE COMPONENT HARDWARE

Attenuators
- Coaxial
- Diode
- Ferrite
- Solid-state
- Step
- Stripline
- Turrent
- Waveguide

Bolometers
- Barretters
- Coaxial
- Thermistors
- Waveguide

Cavities

Circulators
- Coaxial
- Ferrite
- Miniature
- Paramp
- Stripline
- Thick-film
- Waveguide

Couplers
- Bi-Directional
- Directional, coaxial
- Directional, stripline
- Directional, waveguide
- Dual
- Multi-hole
- Rotary-joint

Detectors
- Bolometer
- Coaxial
- Crystal
- Diode
- Thermistor
- Thick-film
- Waveguide

Filters
- Cavity
- Coaxial
- Crystal
- Ferrite
- Frequency multiplexers
- Interdigital
- Solid-state
- Tubular
- Waveguide
- YIG

Hybrids
- Coaxial
- Miniature
- Stripline
- Waveguide

Isolators
- Active
- Coaxial
- Ferrite
- Miniature
- Thick-film
- Waveguide

Junctions
- Coaxial
- Stripline
- Waveguide

Mismatches
- Coaxial
- Miniature
- Waveguide

Mixers
- Balanced
- Coaxial
- Crystal
- Diode
- Harmonic
- Miniature
- Stripline
- Waveguide
- YIG-tuned

Modulators
- Diode
- Ferrite
- Laser
- Magnetron
- Phase
- PIN
- Pulse
- Single-sideband
- Solid-state

Multipliers
- Coaxial-waveband
- Crystal
- Step-recovery
- Thin-film
- Varactor
- Waveguide-waveguide

Power Dividers
- Coaxial
- Stripline
- Waveguide

Probes and Samplers
- Broadband
- Tuned
- Untuned
- Variable

Slotted Lines
- Coaxial
- Waveguide

Terminations
- Coaxial
- Loads
- Metalized Glass
- Oil-filled
- Stripline
- Thermopile
- Transformers
- Transitions
- Waveguides

Tuners
- Coaxial
- Line-Stretchers
- Phase Shifters
- Shorts
- Stubs
- Waveguide

TABLE 1-4 COMMERCIAL MICROWAVE APPLICATIONS

Communication	
Radar	Varactor multipliers
Tracking missiles; satellites	Parametric amplifiers
Switch drivers	PIN modulators
Instrumentation	
Microstrip transmission	
Thin-film hybrids; couplers	
Laser measurement; therapy	
Industrial	
Thickness gauging	Structural flaw detection
Flow detection	Irradiation processing
Ultrasonic cleaning	

between crests and valleys (or nodal voltage points) of the waves as they pass through or across distributed amounts of the R's, L's, and C's making up the element of a component.

Even in generator tubes, detectors, or other nontransmission devices, the parts or elements of these devices must be waveguide-like or transmission line-like pieces of metal possessing distributed electrical characteristics.

SKIN EFFECT

Skin effect is the second key fact in microwave-motion energy transmission. Skin effect centers around the low-power losses that reside upon the inside surface contour and composition of waveguide structures; as a waveform passes over, or is contained in, a rectangular or hollow tube, its energy does not penetrate the metal to any significant depth; it passes through and onward with minimum loss. This loss, analogous to I^2R in a conducting circuit, is usually listed in dB (decibel) power loss per 100 ft.

To illustrate, we find, conversely, that at high frequencies the currents in a solid round conductor are not uniformly distributed but "crowd" or concentrate with higher density near the outside surface; so a hollow conductor of equal diameter will carry just as much current and produce no more I^2R losses.

In a solid conductor this is due to magnetic flux lines surrounding it, which encircle part but not all of the cross-sectional area. Those parts of the cross section circled by the largest number of flux lines (the center area) have higher inductance and reactance than those located toward the exterior and, hence, cause the currents to redistribute or crowd toward the outside. To summarize, in a solid round wire this effect produces a maximum density at the surface and a minimum at the center; with a square bar the greatest concentration is at the corners, and with a flat strip the current density is greatest at the edges. In effect, the skin of a conductor carries practically all the high-frequency energy, with penetration of current beneath the skin decreasing as ac frequency increases.

The exponential falloff of current density with frequency is defined quantitatively as *skin depth*, a distance at which the current is 37 per cent ($1/e$) of its surface density. Its value for most microwaves is in microns and varies with the resistivity of the metal being used to plate the interior of the waveguide. For instance, in copper, in mils,

$$\begin{aligned}\delta &= \frac{2.61}{\sqrt{f}} \qquad \text{at } f = 100 \text{ MHz}\\ &= 2.61 \times 10^{-4}\\ &= 0.000261 \text{ mil}\\ &= 0.261 \text{ micron}\end{aligned}$$

Also, the crowding of currents into increasingly smaller skin depths means that the longitudinal resistivity of the current path has been increased so that surface path losses are increased with frequency.

These effects are pictured in Fig. 1-1, where, compositely, we show how the depth of penetration (in microns) decreases and the resistivity increases with frequency.

Waveguides are thus plated (particularly on their interior surfaces) mostly with silver or copper, since a few thousandths of an inch plating is more than sufficient to contain all the current on the inside of a waveguide. Where it is impractical to uniformly plate interior surfaces of a waveguide, solid silver construction is used.

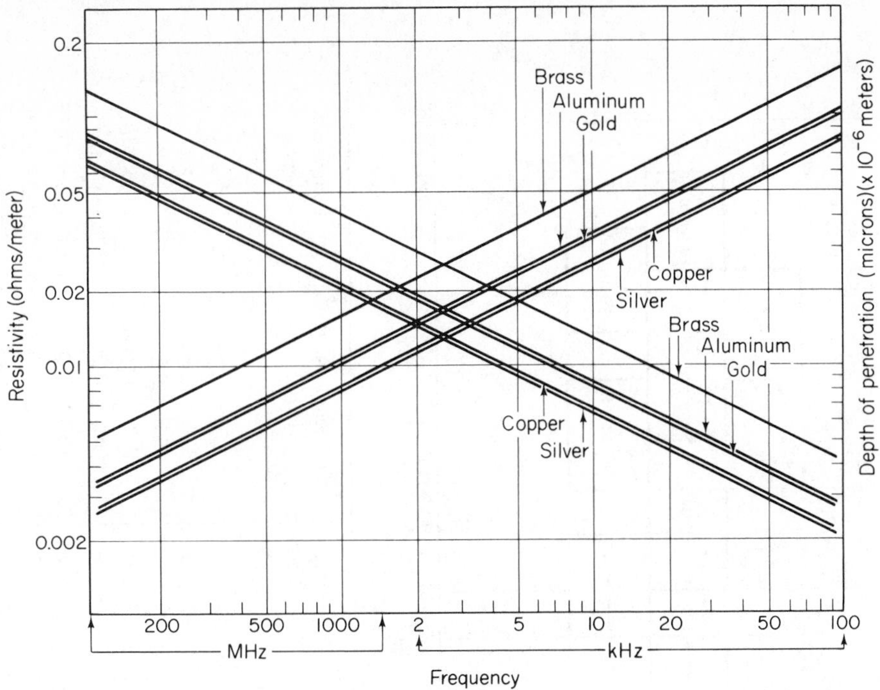

Fig. 1-1 RESISTIVITY–PENETRATION RELATIONSHIPS

WAVEGUIDES

After accounting for skin effects, wave motion at microwave frequencies requires physical and electrical guidance of signal energies in hollow waveguide structures. Besides minimizing losses, these structures, if properly designed, promote and aid the propagation of energy. This is done by accounting for "mode" operation or wave-motion characteristics at all frequencies and tailoring them to the size and dimensions of the hollow, signal-carrying pipes. This design and tailoring extend from roughly 1 to 4 kMHz; over this range there are allocated 8 to 10 frequency bands. Fig. 1-2 diagrammatically gives a layout of the bands plus the approximate rectangular waveguide dimensions for corresponding hardware.

ELECTRON TUBES, HARDWARE, AND SOLID-STATE DEVICES

Electron vacuum tubes, solid-state components, and specialized hardware constitute the main structure of wave-motion devices. Each of the first two utilize special electron characteristics and are not always directly related to waveguide structures. They are used mostly in power-generating and transmission devices.

Table 1-5 lists the main categories of electron vacuum tubes, divided into power and small signal-amplifying devices. Their special electron-related characteristics are discussed in Chapter 7. Solid-state microwave accessories appear in Chapter 12.

Contributory to and compatible with the above, we have measurement equipment utilizing many wave-motion structures, the elements often being arranged in special geometric manner and tailored to give controlled attenuation, power detection, or phase manipulation.

Finally, the hardware divides itself into transfer, attenuation, signal-detection, impedance, or power-measuring devices. Specifically, they are covered by the following categories:

1. Couplers, tees, isolators, cavities (Chapter 5).

2. Junctions, switches, or duplexers (Chapter 6).

3. Impedance matching (Chapter 8).

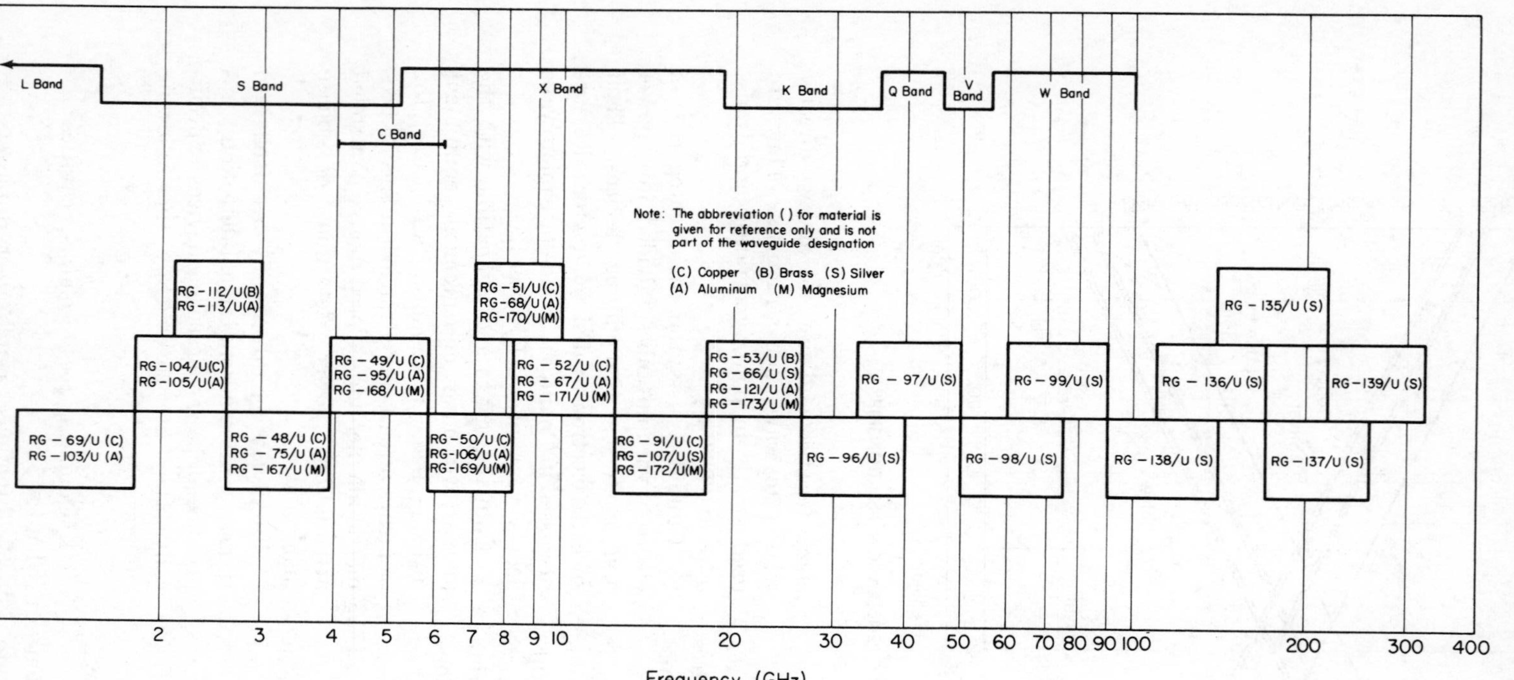

Standard waveguides

Fig. 1-2 FREQUENCY BANDS VERSUS WAVEGUIDE DIMENSIONS

Band Letter	Frequency range (GHz)	Dimensions (inches)
S	2.6 to 3.95	$3 \times 1\frac{1}{2}$
G	3.95 to 5.85	2×1
J	5.3 to 8.2	$1\frac{1}{2} \times 3/4$
H	7.05 to 10.0	$1\frac{1}{4} \times 5/8$
X	8.2 to 12.4	$1 \times 1/2$
M	10.0 to 15.0	0.850×0.475
P	12.4 to 18.0	0.702×0.391
K	18.0 to 26.5	0.500×0.250
R	26.5 to 40	0.360×0.220

Waveguide dimensions

Finally, we see that the field of microwaves is a specialized extension of the broader electronic regime having particular merit in its directive, reflective, and resonant qualifications, so it has become useful in communication, molecular, and quasi-optical applications. Its transmission, generative, and hardware functions are specialized, waveguide-oriented structures relying and capitalizing in size and shape upon the facts of low skin-effect losses and wave motion in hollow conductors.

TABLE 1-5 ELECTRON TUBE CATEGORIES

Device	Features	Frequency power availability	Gain	Efficiency	Amplifier bandwidth	Principal applications
Triode (ceramic) (planar) (pencil)	Simple and reliable Low cost Efficient Stable	To 4 Gc 1 kW – 5 MW 0.01 W – 200 kW	8–20 dB	30–60%	0.5–2%	TV and radar transmitters Receivers Relays
Magnetron	Compact Low cost Simple and reliable	0.3–100 Gc To 10 MW To 10 kW		30–55%		Radar transmitters ECM and ECCM transmitters
VTM (Voltage-tuned magnetron)	Voltage tuning Near-octave tuning Compact Lightweight	0.2–10 Gc 3 MW – 50 W		1–25%		Signal generators Test equipment Master oscillator Local oscillator
Reflex klystron	Low cost Wide tuning Voltage tuning Long life	1–130 Gc 1–1000 MW		0.5–5%		Local oscillator Relays Signal generators Paramp pumps. Telemetry
Two-cavity klystron	Higher power than reflex klystron	8–20 Gc 1–20 W		0.5–5%		Paramp pumps Radar chain driver Signal generators. Telemetry
Drift-tube klystron	Higher power than reflex klystron (millimeter frequencies)	20–80 Gc 0.5–50 W		0.5–5%		Same as two-cavity klystron
Klystron amplifier	High pulse and CW power High gain Flexibility Good bandwidths	0.1–10 Gc 20 kW – 30 MW 0.1–300 kW	15–90 dB	25–45%	2–15%	Radar transmitters Communication transmitters
Amplitron	Efficient. High power Low cost. Compact Simple and reliable	0.5–35 Gc 0.1–35 MW 1 W – 200 kW	8–20 dB	50–80%	8–12%	Radar transmitters
Stabilotron	Highly stable oscillator (any power)	0.5–10 Gc 50 – 600 kW 1–1000 W		40–50%		Radar transmitters
O-BWO	Octave electronic tuning All frequencies Reliable and rugged	1–400 Gc 1–100 kW 1 MW – 1 kW		1–10%		Signal generators Local oscillators
M-BWO	Octave electronic tuning Reliable and rugged High CW power over octave Good efficiency	1–35 Gc 25–350 W		30–50%		ECM transmitters Master oscillator-driver
TWT	Octave bandwidths High gain Long life. Low noise	1–180 Gc 100 W – 3 MW 1 MW – 1 kW	10–50 dB	10–30%	10–100%	Radar transmitters. Relays Test equipment ECM and ECCM. Receivers
CFA (Crossed field amplifier)	Compact Efficient Light Phase-stable	1–10 Gc 50 – 250 kW 200 – 2000 W	20 – 30 dB	30 – 50%	10 – 30%	Same as TWT

Transmission Lines 2

STRUCTURE—DEFINITIONS

Transmission lines are parallel, two-conductor, energy-transfer structures for continuously passing quantities of electronic power throughout their length from one point to another. Twisted two-wire cables (shielded or unshielded) or parallel spaced open-wire lines are the most common form, either type being used to transfer power or signal energy from source to load [Fig. 2-1(a)].

Other structures use one outer conductor to shield the inner conductor (coaxial), using spaced washer insulators or solid insulation [Fig. 2-1(b)]. We see open-wire spacing of conductors in old-style telegraph lines, sometimes made with spaced bar insulators or held apart by continuous plastic molding [Fig. 2-1(c)]. This latter type is the common TV lead-in conductor most commonly known as a 300-Ω line.

Each of these structures possesses electrical characteristics bearing upon the propagation of alternating current along them. Although used mostly at frequencies below the microwave region (1000 MHz), some coaxial lines are used for passing microwave energy.

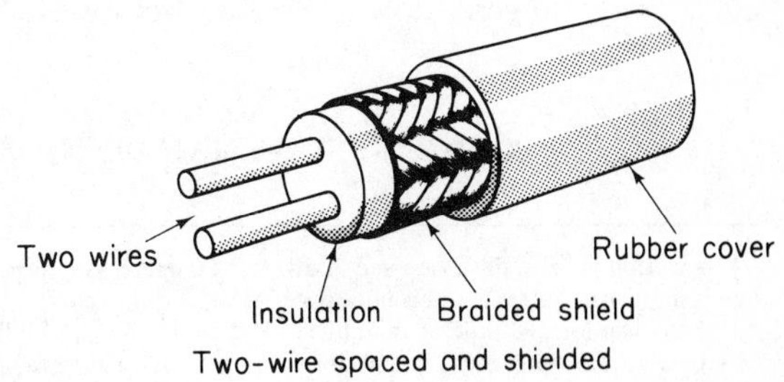

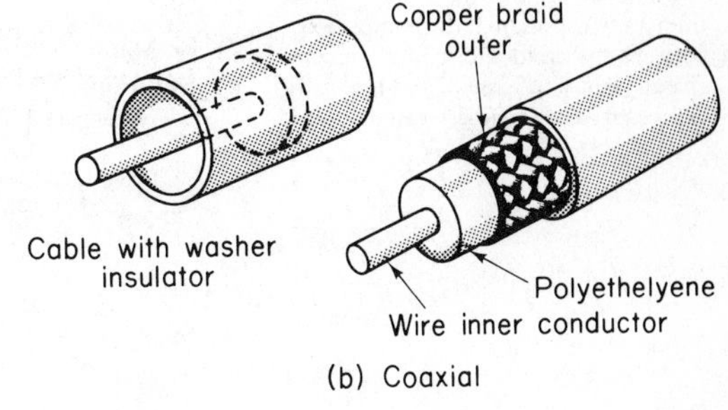

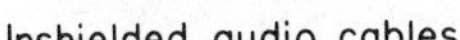

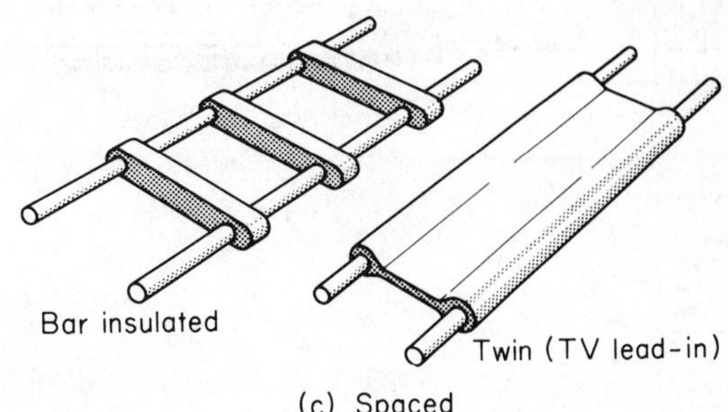

Fig. 2-1 TWO-WIRE AND SPACED TRANSMISSION LINES

Common electrical characteristics and constants are given in Table 2-1.

The conductors in all transmission lines possess resistance, usually stated in ohms per foot, linear self-inductance, and distributed capacitance between the lines, respectively specified commercially in microhenries per foot and micromicrofarads per foot. These reactances at low frequencies (say 60 and 400 Hz) are negligible with respect to resistance.

As frequency increases these reactances increase, until above 50 kHz they retard the transfer of energy so that a complete source cycle may be sent down the line before the start of the cycle reaches the load. Now, a 400-Hz line would have to be over 400 miles long to have this happen, so in commercial practice, say in an aircraft using 400-Hz power, a 200-ft length would be a very short line. However, at 10 kHz (say in a radar system) a 2-in. line would be relatively long.

It is convenient, as we shall see later, to talk of the physical dimensions of lines in terms of quarter wavelengths. Thus, a transmission line is considered to be electrically short when it is overall physically small compared to a quarter wavelength of the signal frequency that it is carrying. Radio-frequency lines are thus long lines, as we shall consider them in this chapter.

TABLE 2-1 COAXIAL LINE DEFINITIONS AND CONSTANTS

Attenuation is generally expressed in dB per unit, usually 100 ft, and is indicative of the power loss. A "rule of thumb" is that each 3 dB represents a loss of one-half of the power.

$$A = 4.35\frac{R_t}{Z_0} + 2.78\sqrt{\epsilon}pF$$

Capacitance permits the storage of electricity when potential differences exist between the conductors. In the case of coaxial cables, it is small and is usually expressed in micro-micro-farads.

$$C = 1016\frac{\sqrt{\epsilon}}{Z_0}$$

Characteristic impedance

$$Z = \sqrt{\frac{L}{C}} \text{ ohms where:}$$

L = Unit distributed inductance
C = Unit distributed capacitance

There are three main impedance groups in coaxial cable, namely, 50, 70, and 93 ohms.

Resistance – the energy loss conversion factor expressed in ohms.

$$R_t = 0.1\left(\frac{1}{d} + \frac{1}{D}\right)\sqrt{F} \text{ for coaxial copper line}$$

$$R_t = \frac{0.2}{p}\sqrt{F} \text{ for open 2-wire copper line}$$

Inductance is the property of an electric circuit which determines, for a given rate of change of current in the circuit, the electromotive force induced in the same circuit.

Usually, in coaxial cables, the induced values are so low that they are normally not considered in calculations except where inductance might be a factor in associated equipment

$$L = 1016\sqrt{\epsilon}Z_0$$

Velocity of propagation is the ratio of the dielectric constant of air to the square root of the dielectric constant of the inculator. It is usually expressed as a percentage.

$$\frac{V_L}{V_S} = \frac{1}{\sqrt{\epsilon}}$$

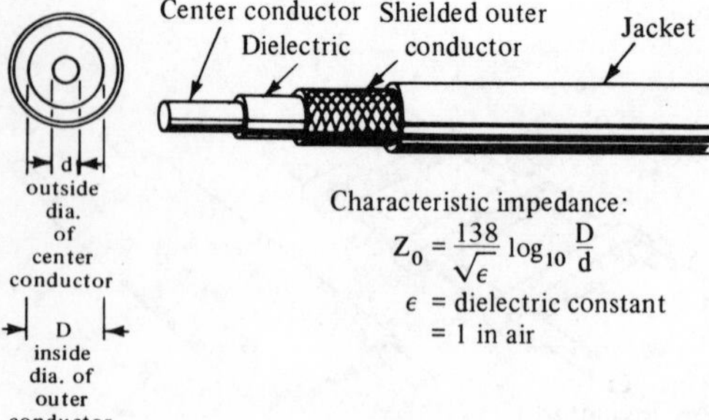

Characteristic impedance:

$$Z_0 = \frac{138}{\sqrt{\epsilon}}\log_{10}\frac{D}{d}$$

ϵ = dielectric constant = 1 in air

Symbols

R_t = total line resistance in ohms per 1000 ft
d = diameter of conductors (center conductor for coaxial line) in inches
D = diameter of inner surface of outer coaxial conductor in inches
F = frequency in megacycles
L = inductance of line in micromicro henries per foot
Z_0 = surge impedance of transmission line in ohms
ϵ = dielectric constant of transmission line insulation
C = capacitance of line in micromicrofarads per foot
V_L = velocity of propagation in transmission line } same unit
V_S = velocity of propagation in free space } same unit
A = attenuation in decibels per 100 feet
p = power factor of dielectric medium

STRUCTURE—THE LINE AS A CIRCUIT

A transmission line (even though it carries microwave energy) differs from a waveguide in that it has discrete measurable electrical constants distributed in each conductor throughout its length. These constants are its self-inductance, capacitance, and resistance per unit length, spread out in each leg of the line throughout its linear dimensions.

Circuitwise, even while considering the fact of their distributed nature, we can better understand dynamic line operation by arranging these sectional-unit linear quantities as lumped constants. We can then multiply each one by the total number of sections in the line to obtain overall line quantities.

First, taking the whole line as a replica of the distributed sectional units (L, C, and R per foot), we see this analysis in Fig. 2-2(a), where each line is broken down into its four main electrical constants.

1. R, the unit-linear-series dc resistance of each conductor.
2. L, the unit-linear-series self-inductance of each conductor. This constant is augmented slightly by some mutual inductance caused by interaction of adjacency currents in the two wires.
3. C, the unit-linear cross capacitance between the wires.
4. G, the unit-linear conductance between one line and the other through the resistance of the insulator material.

(a) Equivalent circuit breakdown

(b) Symmetrical circuit arrangement

(c) Series impedance — "T" arrangement

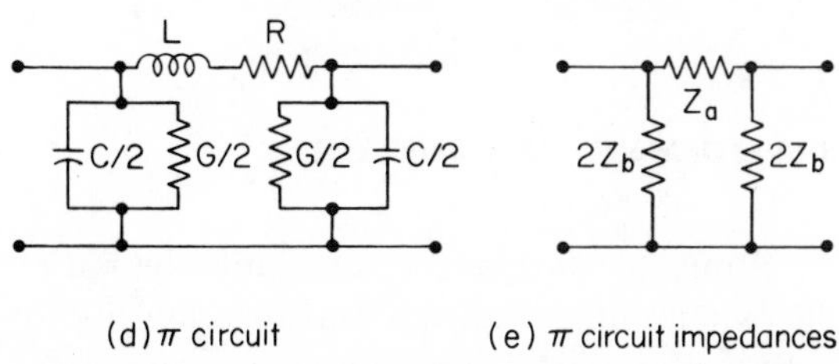

(d) π circuit (e) π circuit impedances

Fig. 2-2 *L-C-R* **TRANSMISSION LINE CIRCUIT ANALOGY**

We must, however, rearrange these constants so that the line will present an identical and symmetrical electrical impedance when viewed from either end. This is seen in Fig. 2-2(b), where we split the R's and L's and place them on each side of the parallel capacitance and conductance. When mathematically labeling these constants as impedances, z_1 for the series constants and z_2 for the parallel ones, the split-series arrangement gives us the schematic of Fig. 2-2(c). Another parallel, or π arrangement, halving the shunt constants and impedances, is seen in Fig. 2-2(d).

CHARACTERISTIC IMPEDANCE

Next, using the same circuit configurations but arranging them as successive lumped-constant sections, the line (when considered infinitely long) becomes somewhat closer to an actual operating structure, as seen in Fig. 2-3(a). So when we apply a signal, say switch a battery across the line, C_1 starts to charge, being retarded somewhat by L_1. But the voltage rise passes on to C_2, attempting to charge it (through the retarding effects of L_2), and so on continuously down through all the L's and C's making up the line.

The process is self-limiting, however, and ultimately stabilizes because the inductor retarding effects eventually oppose the build-

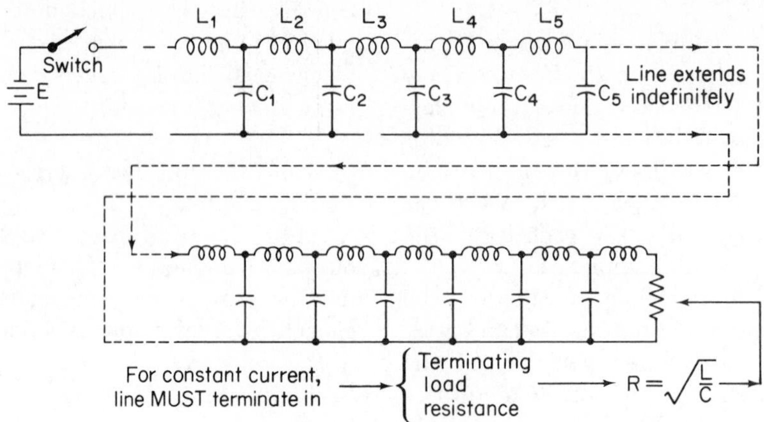

Fig. 2-3 LUMPED CONSTANT-LINE STRUCTURE

up of capacitor-charge voltages (which decrease as the source voltage proceeds down the line), and we finally wind up with the whole line acting as if it were a single finite resistance, known as its *characteristic impedance.*

Viewing it another way, the "mix" or ratio of these L's and C's determines how much current will flow when voltage is applied to the line, the same as any load resistor. This ratio-determined impedance is, numerically,

$$Z_0 = \sqrt{\frac{L}{C}}$$

Practically, it takes the form of a resistor which is connected at the end of the line or across any intermediate point (or even across the terminals of the first section) and which controls the current drawn and the voltages developed, as they would be when the input switch is thrown or signal is applied.

For instance, if 100 ft of a twisted pair line has an inductance of 0.25 μH and a total capacitance of 1,000 mmF, its impedance

$$Z = \sqrt{\frac{L}{C}} = \sqrt{\frac{0.25 \times 10^{-3}}{10^{-9}}} = 500\ \Omega$$

These L's and C's are actually measurable and tell us that no matter where we cut or terminate this line and connect a load across it, it should always be executed with a 500-Ω resistor.

Or, another way, if we found that this line had 0.0025-μH inductance and 10-mmF capacitance per running foot, the result would be the same.

For typical coaxial and twisted pair constants and impedances, see Table 2-1.

This constant, perhaps more than any other, tells us many things.

1. It is the proper termination of a section of line that ensures maximum power transfer or signal passage. Thus it electrically matches the load and the generating source.

2. Correct characteristic-impedance termination ensures a minimum of reflections when applying power. These unwanted, and transient voltage and current gyrations occur when applied voltage meets an open-ended or misterminated transmission-line junction; they cause power losses and transmission inefficiency.

3. It tells the speed at which voltage change travels along the line.

DEVELOPMENT OF CONSTANTS

Study of the passage of input voltage or current impulses down a transmission line in conjunction with a few simple electrical relationships establishes the technology for measuring, designing, and operating these units.

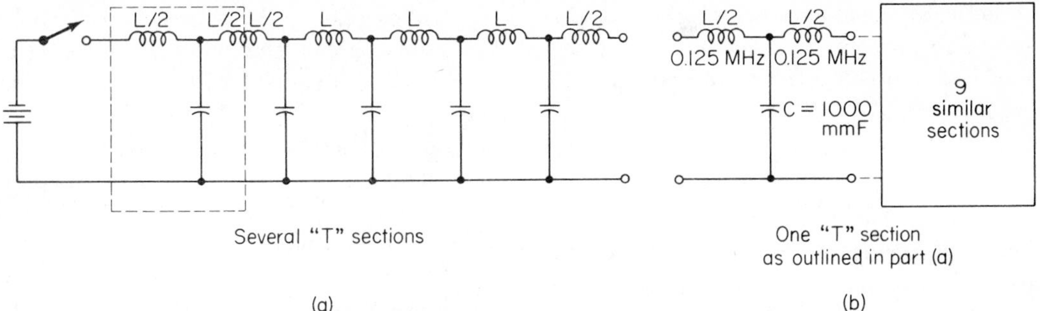

Fig. 2-4 T SECTION LINE CONSTANTS

Such relationships enable us to derive the formulas for characteristic impedance, impulse or wavefront transmission time, and standing-wave or nodal-voltage-point amplitudes.

To do this, refer to the sectional lumped-constant inductance and capacitance shown in the "T" section of the line in Fig. 2-4(a).

Let us recall a few fundamental relationships:

$$Q = IT \qquad (1)$$

where the quantity of electricity (Q), in coulombs, is equal to the current multiplied by the time it flows. These are the electrons that have flowed from the battery.

$$Q = CE \qquad (2)$$

where the Q in a capacitor is equal to the capacitance times the voltage across it. This is the number of electrons that have passed into the line.

Now, since the energy from the battery must equal that contained by the line,

$$Q = IT = CE \qquad (3)$$

and, since the capacitor charging is a matter of changing voltages, the rates of change must be taken into account and dynamically we know that

$$E = L\frac{dI}{dT} \qquad (4)$$

or at any instant the capacitor voltage must be proportional to the inductor feeding it multiplied by the rate of change (with time) of the current through it or, transposing (4),

$$E\,dT = L\,dI \qquad (5)$$

so finally, after things have stabilized,

$$ET = LI \qquad (6)$$

So, now, multiplying each side of equations (3) and (6) and equating,

$$IT \times ET = CE \times LI$$

and

$$EIT^2 = LCEI$$

With simplification,

$$T = \sqrt{LC} \qquad (7)$$

which gives us in seconds the time it takes voltage change to travel along a single section of L and C.

Now, since we are also interested in the ratio of E and I (the impedance), we can simplify (6) and (3), this time by dividing one by the other

$$\frac{ET}{IT} = \frac{LI}{CE} \qquad (8)$$

and then simplifying by multiplying each side by E/I and canceling the T's

$$\frac{E^2T}{I^2T} = \frac{L}{C} \qquad \frac{E}{I} = \frac{L}{C}$$

finally taking the square root,

$$Z_0 = \sqrt{\frac{L}{C}} \qquad (9)$$

Practically, we can approach a transmission-line problem and say our 100-ft section of 1000-ft line by referring again to page where:

$$L = 0.25\ \mu\text{H}$$

$$C = 1000\ \text{mmF}$$

Note that even in transforming the line into a "T" section where L was split into 0.125-mH sections, our fundamental impedance remains the same. Thus

$$Z = \sqrt{\frac{L}{C}} = \sqrt{\frac{0.025 \times 10^{-3}}{10^{-9}}} = 0.5 \times 10^{-3}$$
$$= 500\ \Omega$$

and

$$T = \sqrt{LC} = \sqrt{0.25 \times 10^{-3} \times 10^{-9}}$$
$$= 0.5 \times 10^{-6}$$
$$= 0.5\ \mu\text{sec}$$

VOLTAGES ON A TRANSMISSION LINE

Sine-wave inputs best illustrate how voltages vary as they pass along the lumped constants of a transmission line. Figure 2-5 shows the sine-wave input currents and voltages existing at three points along a lumped-constant line where voltage amplitudes will be studied. Figure 2-5(a) is the schematic of the line properly terminated. Figure 2-5(b) is the applied sine-wave voltage, Fig. 2-5(c) is the wave existing at point A, Fig. 2-5(d) the voltage at point B, and Fig. 2-5(e) the voltage existing at point C. To repeat, the delayed buildup of voltage at various points on the line is caused by the time taken in charging the lumped parallel capacitances plus the delaying effect of the series inductances.

More specifically, in the above example we should note that at times T_3, T_5, and T_7, current is just starting to flow in the lumped capacitances C_3, C_5 and C_7, and that by the time C_7 is charging, the input voltage has gone through three quarters of a cycle of voltage variation.

We can show this variation, or how a voltage wave travels down a line, by placing an oscilloscope at the source and at each of the points along the line, or better still a multibeam instrument with each single channel triggered individually by the maximum voltage at the observation points. The display would give us sine-wave plots displaced at quarter-cycle time intervals, essentially the same as in Fig. 2-5.

If the scope traces were all triggered by the input sine-waves, it would be necessary to insert delay networks (artificial adjustable transmission lines) in the signal-trace amplifier channels to displace the individual voltage points properly for correct observation. The size of the delay-line network also tells us the speed of voltage travel along the line.

Looking at the variations another way, Fig. 2-6 plots the voltages at different time intervals—T_6, T_7, and T_8—showing how the wavefront (say, the maximum positive sine-wave peak) passes on and forward along the line.

VELOCITY AND WAVELENGTH

A voltage waveform that is initiated and impinged upon a transmission line does not reach a specified point until some time later. It is delayed by capacitor charging plus inductive choking of current, so, as noted above, we can physically measure the time delay of any portion of it—say the peak of a sine wave or the wavefront of a rectangular pulse caused by switching. The time consumed in a wave's travel over a single section we saw was $T = \sqrt{L \times C}$. Taking a single section that is 1000 ft long and using the T (in our previous example) of 0.5-μsec velocity,

$$\frac{\text{distance}}{\text{time}} = V$$

$$= \frac{1000}{0.5 \times 10^{-6}} \begin{aligned} &= 2000 \times 10^{+6} \\ &= 2000 \text{ million ft/sec} \\ &= 300{,}000 \text{ miles/sec} \end{aligned}$$

But if the input voltage was a sine wave of 2 MHz, the entire first cycle applied to the

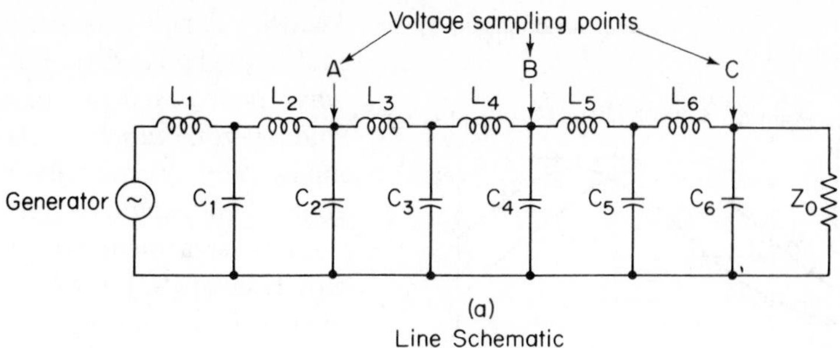

(a)
Line Schematic

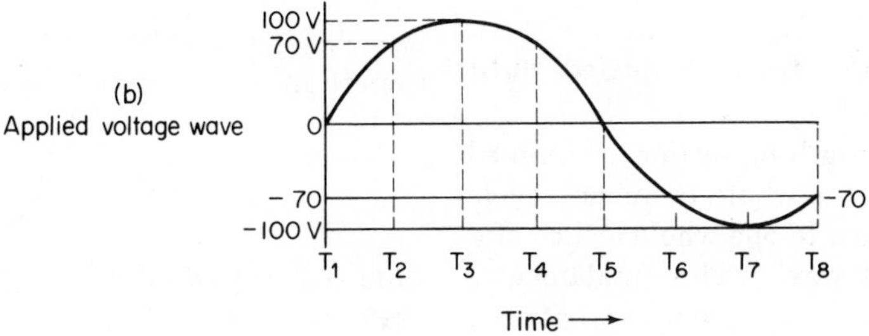

(b)
Applied voltage wave

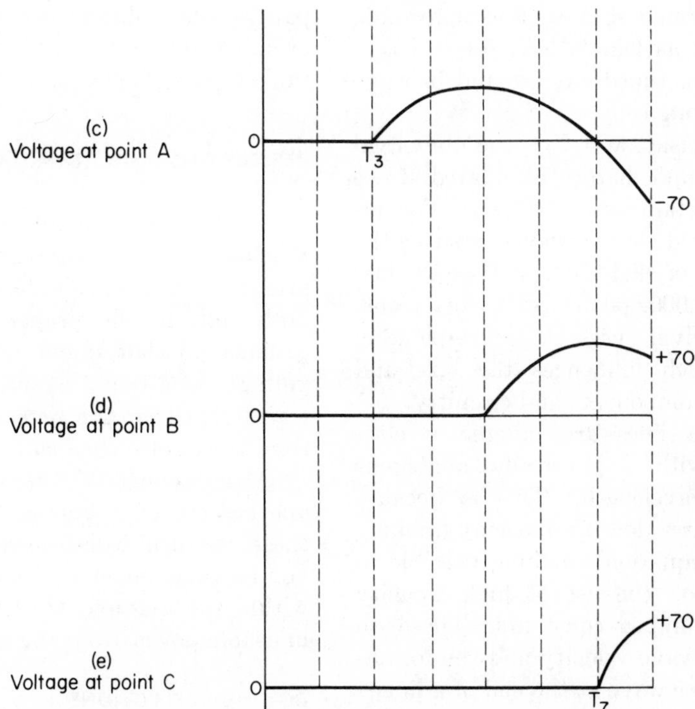

(c)
Voltage at point A

(d)
Voltage at point B

(e)
Voltage at point C

Fig. 2-5 SINEWAVE VOLTAGE TRAVEL ON TRANSMISSION LINE

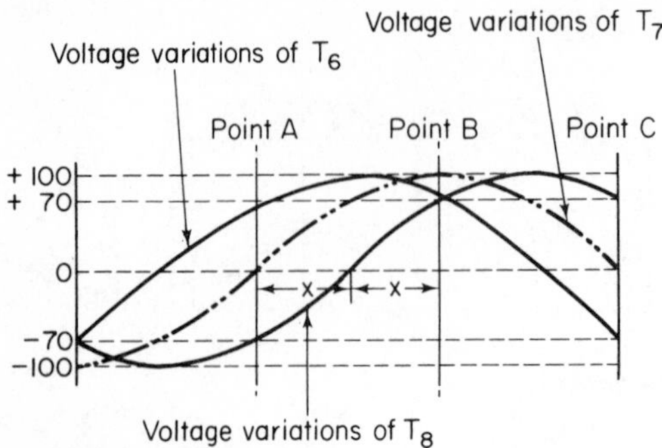

Fig. 2-6 SUCCESSIVE TIME INTERVAL LINE VOLTAGE VARIATIONS

line would not have been completed until $\frac{1}{2}$ μsec had elapsed.

So, a 1000-ft length or section of lumped constants absorbs one whole frequency cycle or is physically equal to one whole length of a voltage or current wave. This fundamental condition establishes a relationship between frequency and physical dimension known as *wavelength*. This equation is a fundamental relationship used widely throughout all electrical and electronic studies; it is a formula almost as basic as Ohm's law, for it has a direct analogy in tuned circuits and in wave propagation through space.

Thus, when using $\lambda = V/f$, the numerical wavelength is immediately calculated if we know or can obtain either V or f. For instance, in radiated electric waves in space the speed is that of light ($E = V = 186{,}000$ miles/sec $= 300{,}000{,}000$ m/sec), so any transmitter or receiver uses a corresponding wavelength which influences the size and nature of their components and circuitry.

Transmission lines are inherently electrically long with a physically analogous long-distance wavelength. This is because they usually carry low frequencies (making small f in our equation), making possible in their construction the use of high-capacity and inductive lumped constants. These, in turn, delay the wave velocity and the formation of a complete wave at the end of a line—again making for a longer wavelength.

Throughout design, measurement, and construction of transmission lines it is more convenient to use the electrical and physical constants that accompany a quarter-wavelength. This is because the voltage, current, and impedance distribution at the quarter-wave point are a maximum (as in a complete sine wave), which makes tailoring, adjustment, and measurement more convenient than when using a complete line.

This treatment, of course, is allied to circuit resonance, for we calculate resonant frequency in a tuned circuit from its total inductance and capacitance:

$$f = \frac{1}{2\pi\sqrt{LC}}$$

which is somewhat analogous to our wavelength formula. Since $T = \sqrt{LC}$,

$$f = \frac{1}{2\pi T}$$

and if $f = V/\lambda$, $V/\lambda = 1/2\pi T$, and $\lambda/V = 2\pi T$, then

$$h = 2\pi TV$$

but $V = D/T$, so $\lambda = 2\pi D$, which is a compatible dimension, since when an inductor exists in free space it radiates as a dipole or a full-wavelength element.

REFLECTIONS AND TERMINATION

In commercial, practical usage most transmission lines carry alternating current. Operators and designers aim to correctly terminate both ends in the proper characteristic impedance so that under normal current and voltage conditions maximum power will be transferred from the generator to the line and from the line to the load.

Mismatching of the impedances causes another set of losses because of reflections; these in turn cause unwanted, out-of-phase voltages and currents to course back and forth within the line and to dissipate or subtract unusable power from the source.

DC-LINE REFLECTIONS

Dc-line reflections are first studied by the action of dc wavefronts meeting the termina-

tion presented by an open or shorted end connection.

In the case of an open circuit at the receiving end when a negative wavefront (Fig. 2-6) passes down the line, both E and I wavefronts pass along in phase, producing successive charging of each capacitor and its preceding inductor. However, when the last capacitor is charged and when there remains no voltage across the last inductor, the field collapses and extra or abnormal current flows in the same direction into the last capacitor. Since there is no succeeding L, C, or terminating impedance, there is no place for the stored inductor energy to go and a *double* positive voltage appears across it. Thus the capacitor charging current in the last inductor must drop to zero, and since there is no place for the current to flow from the next-to-last inductor, its field also collapses and doubles the voltage across the next-to-last capacitor.

This change of voltage (moving backward) is the same as though a voltage wave upon arriving at the end of the line found no place to go and was forced to travel back in the same polarity. Such action is called *reflection*.

For the action of reflected currents, the process is repeated with the collapsing field

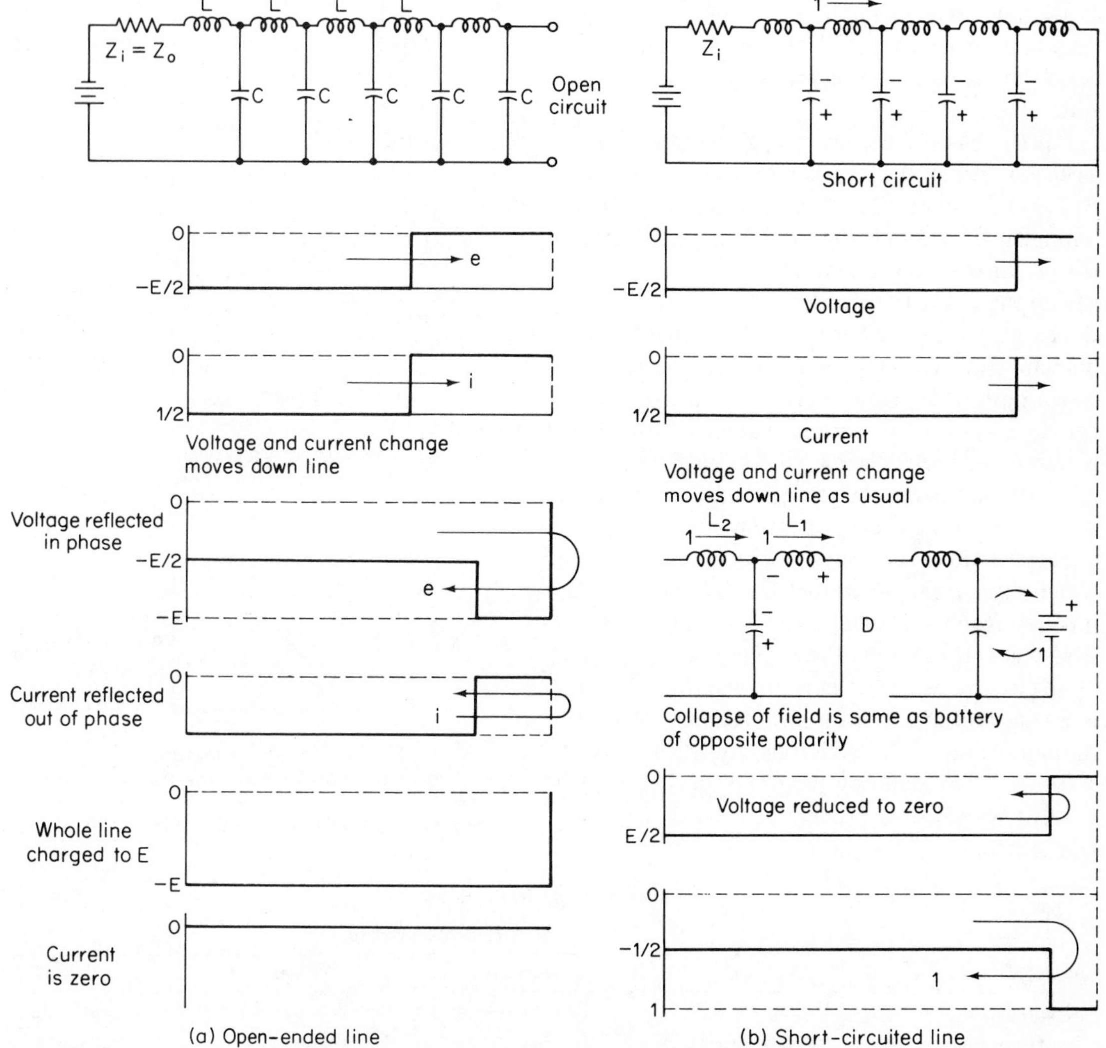

Fig. 2-7 DC AND VOLTAGE REFLECTION SEQUENCES

around each coil resulting in zero current. To repeat, as each capacitor is charged the current drops to zero, effectively reflecting the current changes in the opposite polarity. Thus a negative-polarity current wave proceeds backward along the line.

Note that upon arriving back at the sending end, the positive reflected voltage wave has now produced another $E/2$ voltage, so the line has ideally had its input voltage doubled, even though no power transfer has been accomplished. In actual circuits the reflection phenonenon performs much like a self-oscillatory circuit within the whole internal system, which supports the back-and-forth movement of sending and reflected voltages and currents until losses within the line cause the excursion to eventually die away to zero.

In the case of a short-circuit termination, as shown in Fig. 2-7, we see the usual negative $E/2$ and $I/2$ waves passing as usual down the line until they reach the last inductor. Here, with a short circuit across it, the last inductor has no capacitor to charge, and energy current stored in it must pass right back through the bottom capacitor terminal and immediately discharges it to zero. This is the same as producing an $E/2$ voltage wavefront of opposite (positive) polarity to the initial waveform (it has decreased from $-E/2$ to zero). The effect is again a reflection producing a backward-traveling, positive-polarity wave which discharges the inductors and capacitors until it reaches the input impedance, where it also exactly cancels the applied voltage.

Now, the reflected current wave produces current doubling because the energy of the inductors add to those charged by the original wave, so a like-polarity (negative in this case) current wave $(+I)$ travels backward toward the source.

AC-LINE REFLECTIONS

Sine-wave voltages applied to a transmission line produce continuous, regularly recurring reflections under open- and short-circuit termination identical in nature to

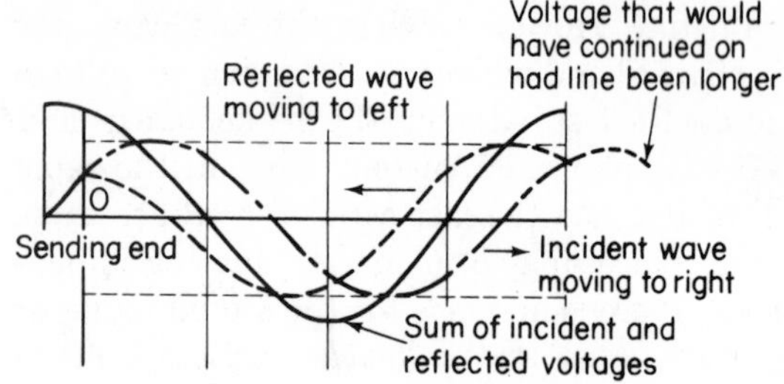

(a) Voltage excursions

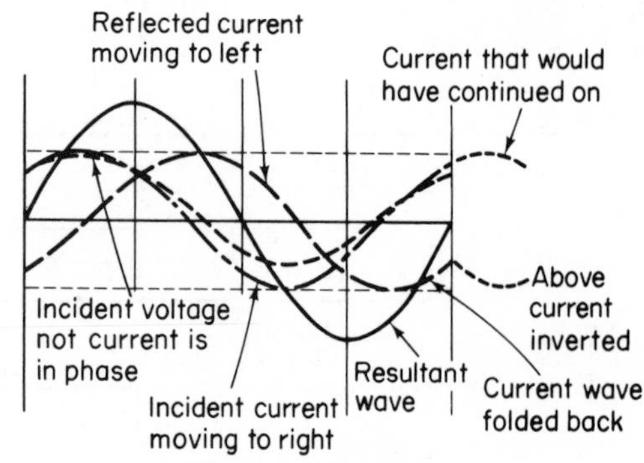

(b) Current excursions

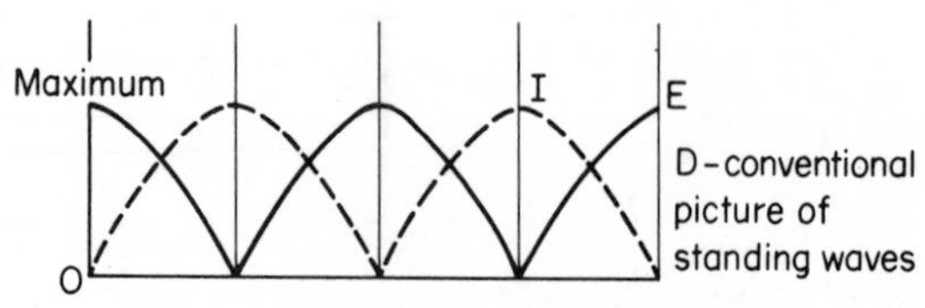

(c) Standing-wave summation of current and voltage waveforms

Fig. 2-8 AC VOLTAGE AND CURRENT LINE REFLECTIONS

Summary of dc Reflections

	Reflected waves			
Line termination	Voltage	Current	Voltage	Current
Open-circuit	In phase	Reverse phase	2 × ampl.	Same ampl.
Short-circuit	Reverse phase	In phase	Same ampl.	2 × ampl.

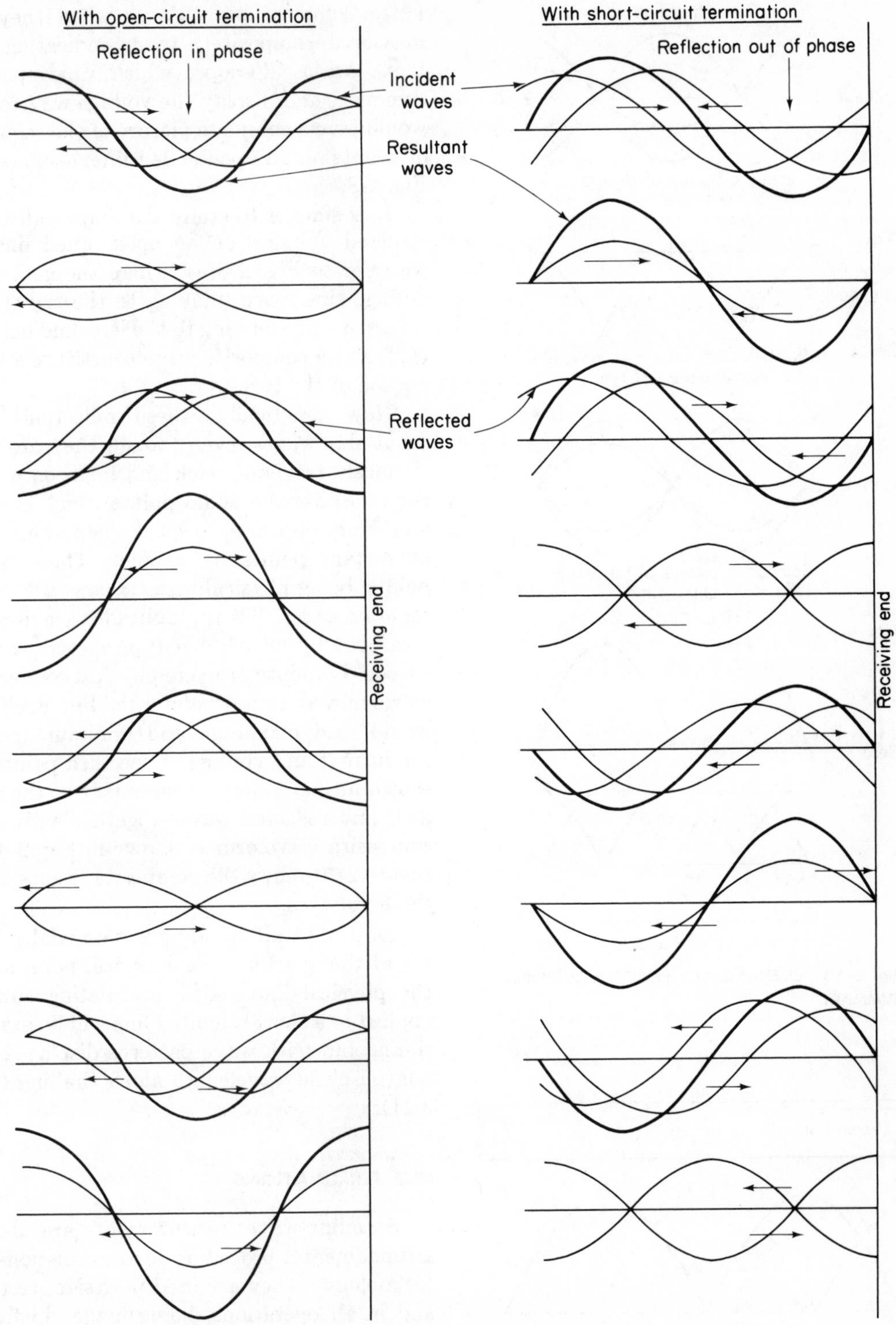

Fig. 2-9 REFLECTED AND INCIDENT TRANSMISSION LINE WAVEFORMS

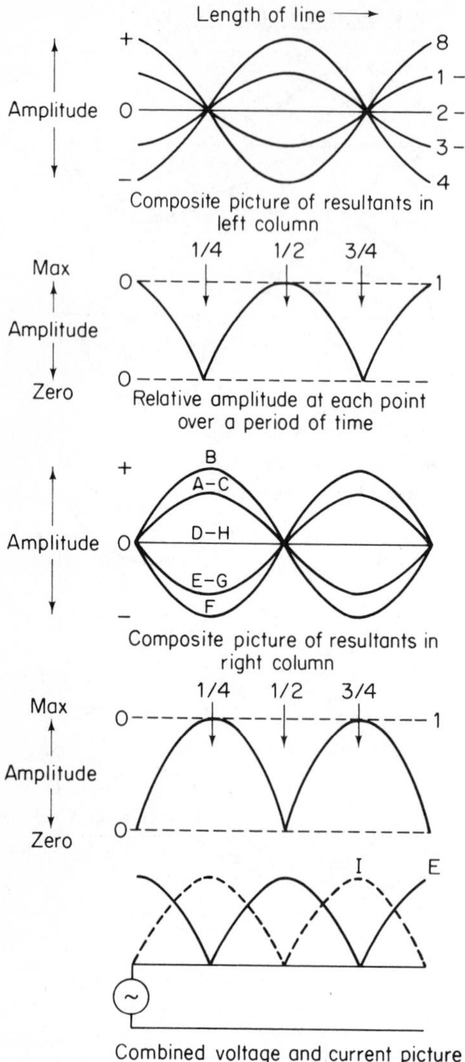

Fig. 2-10 OPEN-LINE COMPOSITE WAVEFORM SUMMARY

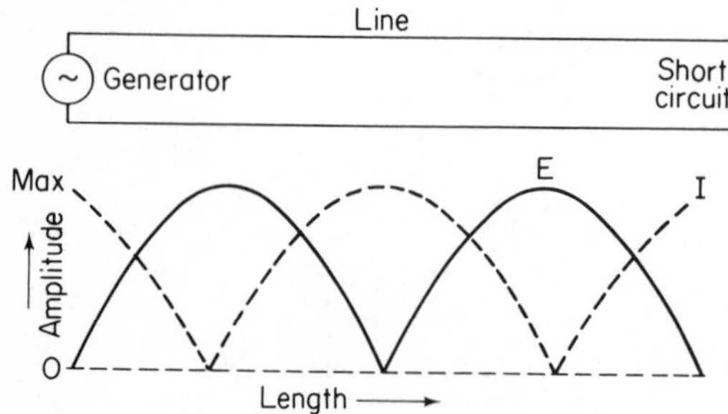

Fig. 2-11 SHORTED-LINE COMPOSITE WAVEFORMS

those described above. The incident (applied) voltages plus the reflected voltages deliver vector sums of both waves when they are measured compositely by ac-indicating meters. An oscilloscope, which can separate, summate, and display the voltage waveforms, would give a complete picture of the sequence of events on an open-ended line, as shown in Fig. 2-8(a).

It is simpler to study the graphically constructed voltages on an open-ended line, so we refer to Fig. 2-8(a), where the generated voltage sine wave (moving to the right), the reflected wave moving to the left (and delayed 90°), and a composite wave constitute a summation of the two.

Now practically, these reflections and input sine waves, even though they are continuously traveling back and forth on a line, combine to make nodal points which become stationary or are, so to speak, "standing still" at certain points on a line. These nodal points, being physically stationary, are called *standing waves*. Their amplitudes are directly measured by ac rectifier-type meters, and if we could tap into or make physical connection as we moved a meter along the line itself, we would read maximum and minimum points for current and voltage at specified points, as shown in Fig. 2-8(c). A summary of the incident and reflected waves together with their composite waveform is shown in Fig. 2-9(b). Figure 2-10 shows the composite results of all these voltages.

Note that all standing-wave nodal points are at the quarter-cycle electrical point along the physical line. Also, alternating current applied to a short-circuited line will be exactly similar but with wave patterns displaced one quarter-cycle wavelength along the line (Fig. 2-11).

LINE TERMINATIONS

Standing-wave measurements are almost a fundamental procedure in transmission-line technology. They are used in design, testing, and in all operations, being made chiefly to check or to make adjustments of line terminations. The types of terminations may be

1. Normal. Z_0 characteristic impedance; with a perfect matched termination, the standing wave should be zero [Fig. 2-12(a)].
2. Open-circuit [Fig. 2-12(b)].
3. Closed-circuit [Fig. 2-12(c)].
4. R's less or greater than Z_0 [Fig. 2-12(d)].
5. X's or C's less or greater than Z_0 [Fig. 2-12(e)].

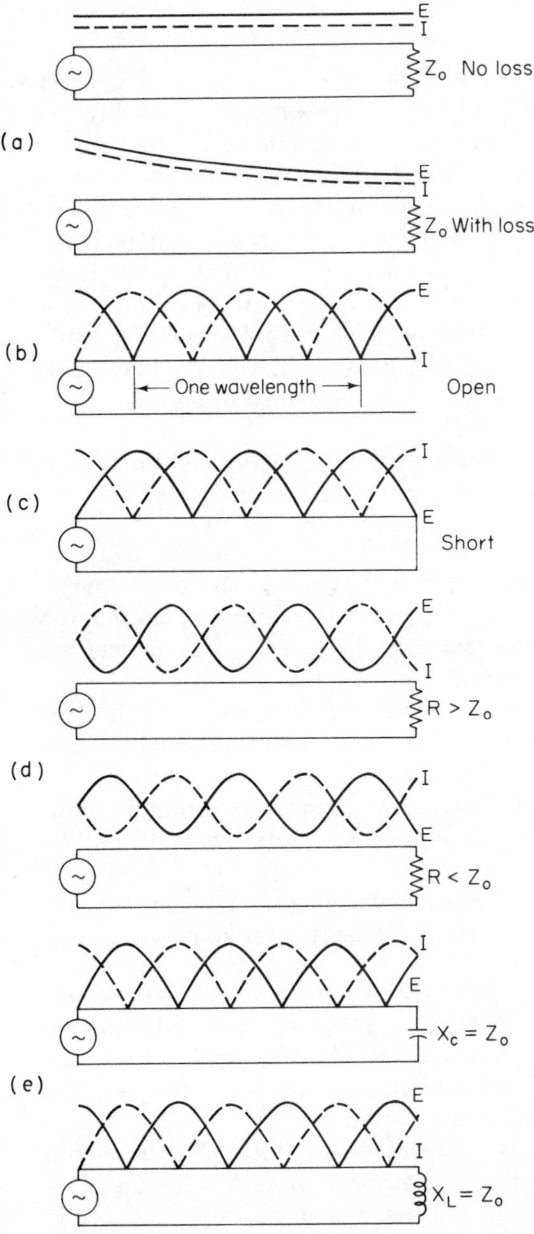

Fig. 2-12 STANDING WAVE SUMMARY WITH VARIOUS TERMINATIONS

The standing E and I cyclic measurements occurring under each of the above conditions are pictured against each of the termination configurations (assuming lossless transmission). Note that these are amplitude measurements and are not to be considered waveforms, although they are generated from signal-input sine-wave voltages.

In type 1 with termination using Z_0, there will be a constant reading on an ac meter moved along the line. With line losses, meter current and voltage amplitudes will slope downward toward the receiving end.

On an open-circuited line (type 2) voltage will be a maximum at the end, with the current here at a minimum. If the line is electrically long, the distance between maximum points or between zero points will be one half-wavelength, and between alternate points the distance will be one full wavelength.

Conditions on a short-circuited line (type 3) will be the conjugate to those on an open line.

Whenever the termination is resistive (type 4) and of any other value than Z_0, there will be reflections; for instance, terminating resistance lower than Z_0 approaches the short-circuit conditions. R_1 higher than Z_0 makes the line appear as an open circuit with standing voltage waves greatest at the end of the line and current waves a minimum but not zero.

We may calculate the amount of reflected voltage by the equation

$$E_r = E_t \frac{R_l - Z_0}{R_l - Z_0}$$

where E_i = incident voltage

E_r = reflected voltage

R_l = termination R

Z_0 = characteristic impedance

With reactance termination (X_l or X_c, type 5), both current and voltage waves are shifted in phase upon arriving at the end of the line. In either case, energy is not absorbed

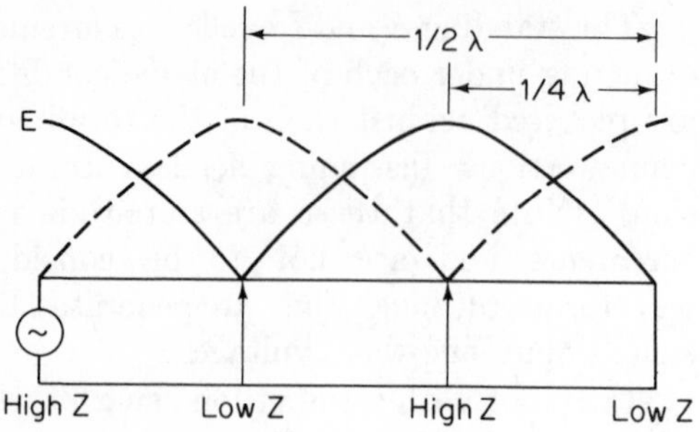

(a) Short-circuit termination

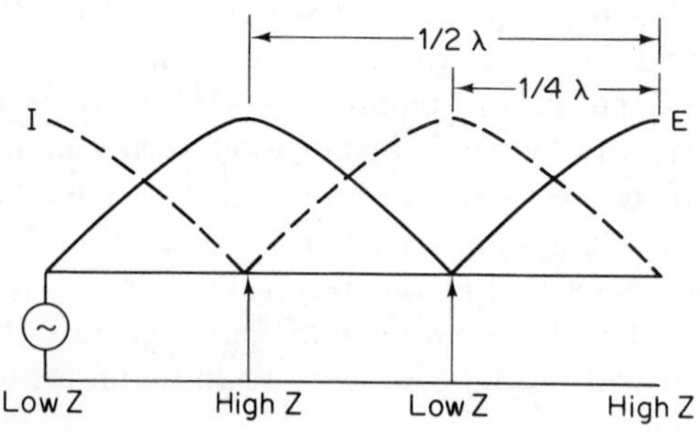

(b) Open-circuit termination

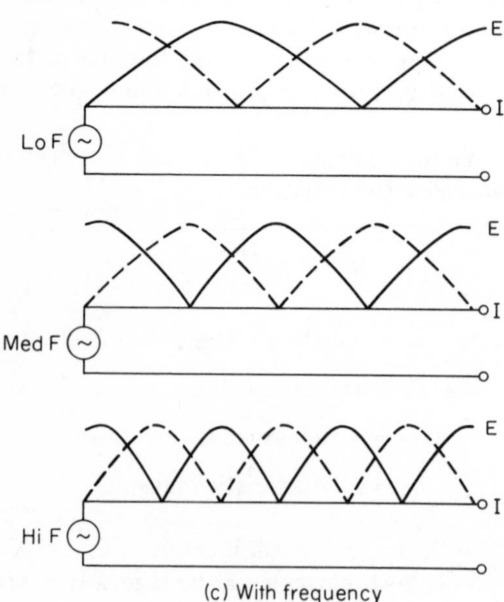

(c) With frequency

Fig. 2-13 LINE IMPEDANCE VARIATIONS

but returned to the line in the form of reflections which have a phase component (not present in resistive terminations) which is determined by how the magnitude of the reactance compares to Z_0.

The diagrammed standing waves show the most favorable case, or where X_c or X_l are equal to Z_0; note, however, that the phase differences still exist.

IMPEDANCE

Transmission-line impedance is a key factor in design and operation. Probably the best example is the attainment of best energy transfer from a generator, for it is here that it is evident that a source impedance should exactly equal the characteristic impedance. Particularly, we can note that excess losses occur when the far end of the line, being stimulated by a generator mismatch, has a termination that is not Z_0; thus double trouble results from line and generator both not being equal to Z_0.

Other design and operational conditions are also sensitive to impedance variations:

1. Tap-offs along a transmission line may be necessary for measurements or for power transfer; for correct, efficient operation the impedance must be known at these selected points.
2. Transformation to larger or smaller impedances necessitates matching (and measurement) between sections.
3. Frequency changes resulting in electrical-line-length variations require impedance matching.
4. Impedance inversion may be used after measurement for certain design reasons.

Figure 2-13 summarizes the impedance variations in open-ended and short-circuit-terminated lines. At the points of voltage maxima and minima on both, the line impedance is resistive.

On a short-circuited line, points of an odd number of quarter-wavelengths from the receiving end have a high impedance [Fig. 2-13(a)]. At points that are an even number of quarter-wavelengths from the short circuit, the impedance is low. This is the point usually employed for insertion of a series signal source or generator.

The impedance variations for an open-circuited line are the conjugate of those for the short-circuited conditions, low impedance for high impedance, and high Z_0 for low.

Now, if applied frequency is increased or decreased, the electrical length of the line is changed, a lower frequency causing the electrical wavelength to become greater and the nodal points to become separated. We see this effect upon the E–I variations of Fig. 2-14(b), where a generator feeds a full wave line. Decreasing the generator frequency lowers the electrical length to three-fourths of a wavelength; doubling the frequency applied to (a) makes the line shorter, accommodating one and a half wavelengths.

LINES AS CIRCUIT ELEMENTS

In moderately high frequencies, quarter-wavelength and one-eighth-wavelength sections of transmission line behave like and are used as resonant circuits. For instance, a section of line less than a quarter-wavelength appears inductive to a generator; if it is exactly one-eighth wavelength, the inductive reactance is equal to Z_0 of the line [Fig. 2-14(a)]. An open section of line less than a quarter-wavelength appears capacitive and at one-eighth wavelength is equal to Z_0 of the line [Fig. 2-14(b)].

Likewise, a quarter-wavelength-line section acts like a resonant circuit, that is, open circuited [Fig. 2-14(c)] as a parallel resonant connection and short circuited [Fig. 2-14(d)] as a series resonant array. From the voltage and current variations we see that a quarter-wavelength section connected correctly can act as an impedance transformer. Thus, when a quarter-wave section is terminated in a resistance greater than Z_0, the line will invert impedance to cause it to resemble a resistance less than Z_0.

This inversion can be numerically expressed as

$$\frac{Z_R}{Z_0} = \frac{Z_0}{Z_S}$$

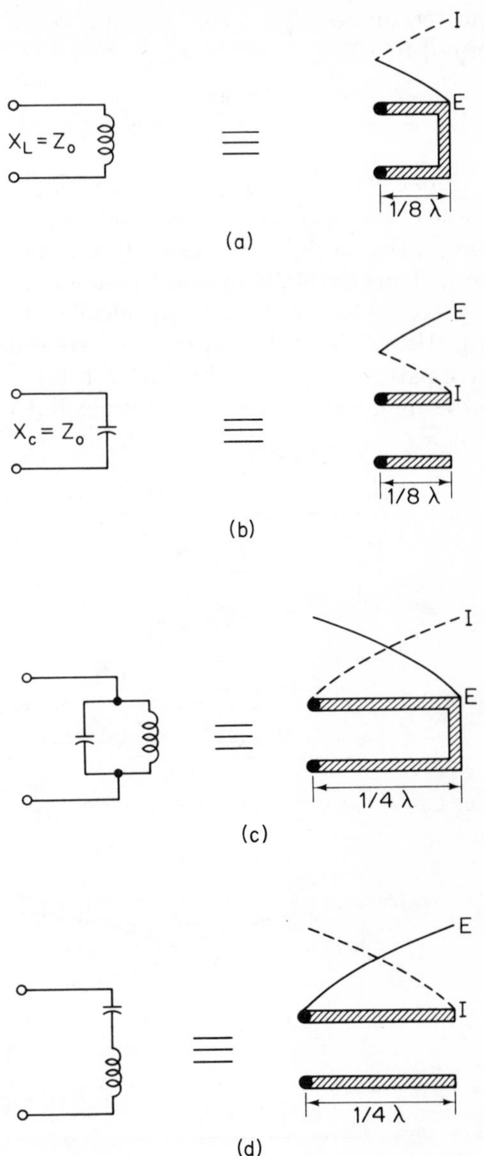

Fig. 2-14 LINE VERSUS RESONANT CIRCUIT ANALOGIES

where Z_R = receiving-end impedance (1)

Z_S = sending-end impedance

Z_0 = characteristic impedance

$$Z_S = \frac{Z_0^2}{Z_R}$$

EXAMPLE. Z_S at the sending end of a 50-Ω

line, terminated with a 25-Ω resistor, is transformed to 100 Ω:

$$Z_S = \frac{Z_0^2}{Z_R} = \frac{50^2}{25} = 100\ \Omega$$

From equation (1) we calculate that the reflected voltage from this termination is one-third of the incident voltage (with a negative sign). Thus the standing-wave voltage at the end is equal to two-thirds of the incident voltage. Using this information, this standing-wave pattern appears as in Fig. 2-15(a). At the sending end the voltage is higher, but the current ratio brings the sending-end impedance to 100 Ω.

With a terminating resistor given for Z_0 we would have, with the line terminated in 250 Ω an input impedance of 10 Ω:

$$Z_S = \frac{Z_0^2}{Z_R} = \frac{2500}{250} = 10\ \Omega$$

A quarter-wave line also inverts reactances; if a section is terminated in a pure capacitance, its input will appear inductively reactive; the conjugate (capacitance reactance) occurs when terminating an inductively reactive line.

From our background on one-eighth- and quarter-wavelength lines we can simply predict transformations using a $\frac{3}{8}\lambda$ line; here we merely consider each section by itself, a $\lambda/8$ section converted directly to a $\frac{1}{4}\lambda$ section. If the $\lambda/8$ wave section is shorted, it displays inductive reactance to the $\lambda/4$ section, which in turn inverts the impedance so that it appears capacitative and equal to Z_0. An open $\lambda/8$ section produces inductive reactance at the end of the $\lambda/4$ section.

Now, we consider half-wave sections as a composite of two quarter-wavelength sections

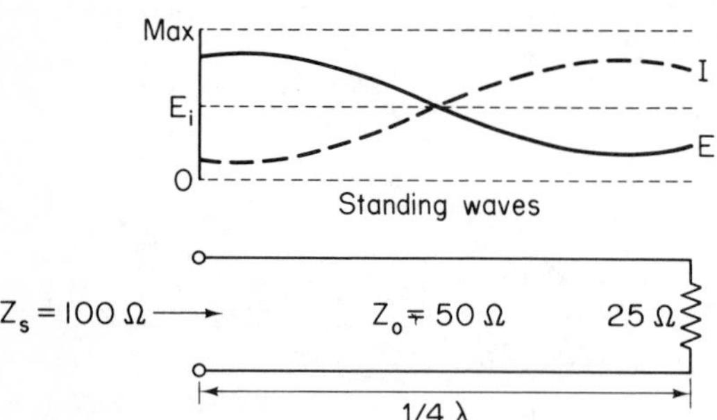

(a) Low R termination looks like high R at input

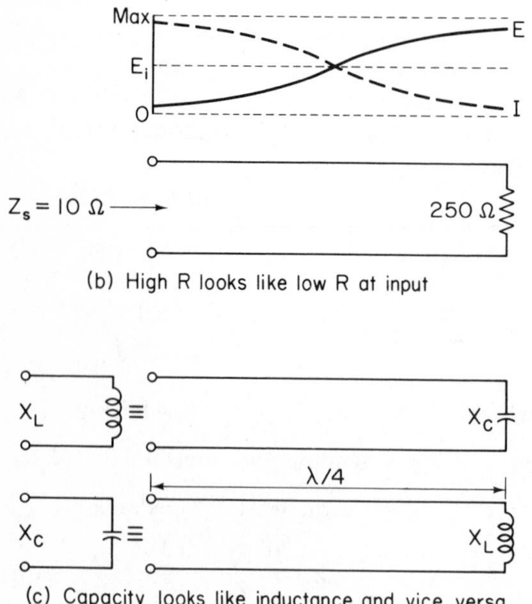

(b) High R looks like low R at input

(c) Capacity looks like inductance and vice versa

Fig. 2-15 EXAMPLE OF IMPEDANCE TRANSFORMATION

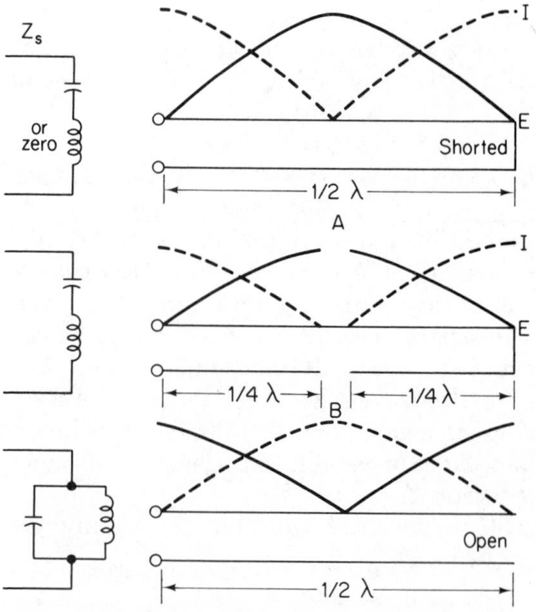

Fig. 2-16 QUARTER-WAVE INVERSION

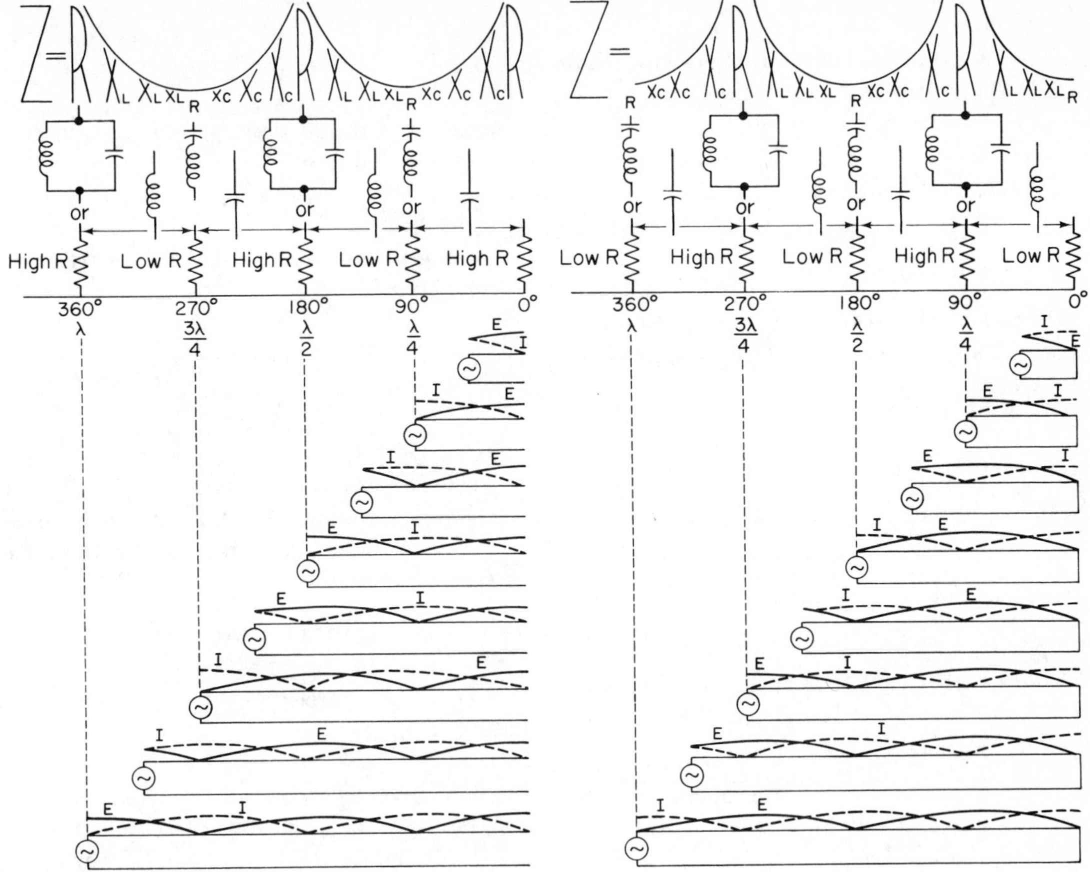

Fig. 2-17 LINE SECTION IMPEDANCE DIAGRAM

in performing their inverting function. Figure 2-16(a) shows an end-shorting quarter-wave section which inverts a low terminal input impedance (to open circuit) while the second inverts the open or high Z_0 impedance again to present the ultimate shorted end section.

LINE-SECTION IMPEDANCE CHART

Figure 2-17 is a diagrammatical illustration, in open and shorted line sections of:

1. Impedance.
2. Standing-wave voltage variations.
3. Equivalent circuit configurations.

The standing-wave variations are referenced to a generator positioned at the left-hand end of each standing-wave display with its equivalent impedance and circuit referenced directly above each generator position. The generalized impedance variation is shown above the positional references.

IMPEDANCE FROM PHYSICAL CHARACTERISTICS

Calculation of characteristic impedance is readily accomplished in two-wire open and coaxial lines.

In a two-wire line

$$Z_0 = 276 \log_{10} \times \frac{2D}{d}$$

where D = distance between wires

d = diameter of wire

The conventional 600-Ω open-wire line readily illustrates this calculation. If $D = 6$ in. and $d = 0.08$ in. (#12 wire),

$$Z_0 = 276 \log_{10} \times \frac{12}{0.08}$$
$$= 276 \times \log 150$$
$$= 276 \times 2.176 = 600\ \Omega$$

Impedances of commonly spaced two-wire lines in terms of the ratio D/d are given in Fig. 2-18(a).

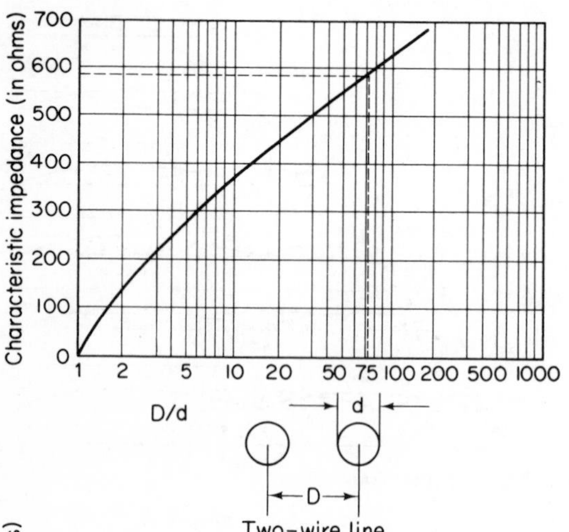

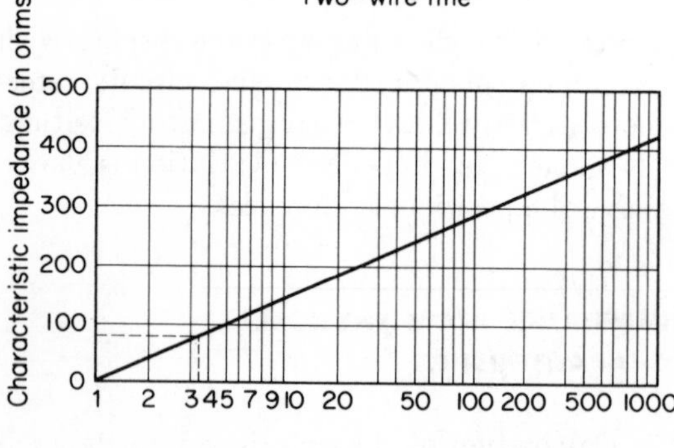

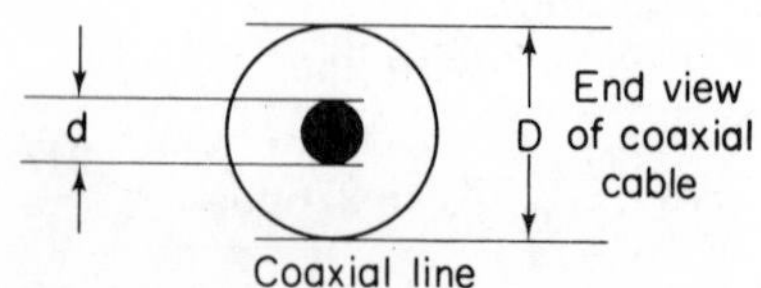

Fig. 2-18 IMPEDANCE CHARACTERISTIC OF TWO-WIRE AND COAXIAL LINES

In a coaxial line

$$Z_0 = 138 \log_{10} \frac{D}{d}$$

where D = inside diameter of outside conductor

d = diameter of inner conductor

Using a $\frac{1}{4}$-in. inner rod within a 0.875-in. copper tubing,

$$Z_0 = 138 \log_{10} \frac{0.875}{0.25}$$
$$= 138 \log_{10} 3.5$$
$$= 75\ \Omega$$

This cable is commonly used as TV lead-in wire, and common variations are shown in Fig. 2-18(b) with Z_0 plotted against the ratio of D/d.

LOSSES IN RADIOFREQUENCY LINES

There are three varieties of losses in rf lines.

1. Copper loss, manifested as heat dissipated by electron flow in the conductor.
2. Dielectric losses, which also appear as heat originating in the insulators.
3. Radiation or induction losses, which represent power transferred from the lines to adjacent circuits or into free space.

Copper losses are composed of I^2R heating effect, a greater proportion of which is due to surface currents existing as a result of skin effects (see Chapter 1).

Dielectric losses range from negligible amounts in open-wire lines to coaxial lines using solid or beaded insulating material. A number of insulating materials, arranged according to increasing loss characteristics and insulating properties, appear in Table 2-2. A number of other factors enter into the design and operation of coaxial lines—voltage breakdown, flexibility, constructional factors such as ease of providing terminations and support, durability, etc. A number of these are covered under measurement equipment (Chapter 10) and in equipment applications (Chapter 11).

TABLE 2-2 COAXIAL LINE DESIGN DATA

▼ START HERE TO SELECT BY TYPE NUMBER ▼ START HERE TO SELECT BY CHARACTERISTIC IMPEDANCE

RG cable type	Alpha wire no.	Inner conductor			PE di-elec-tric	Shields			Jacket		Armor (0.0126 aluminum wire)	Oper. (V) rms	Lbs. per M ft	Nominal impedance (ohms)		Cap. (μμf) ft	Attenuation (dB/100 ft) frequency in MHz								VP %	RG cable type
		MAT.	STRAND	O.D.	O.D.	INNER	OUTER	O.D.	MAT.	O.D.							10	50	100	200	400	600	1000	3000		
5B/U	800	SC	1	0.051	0.185	SC	SC	0.260	NCV	0.335	–	3000	83	50	△	29.5	0.65	1.6	2.4	3.6	5.2	6.6	8.8	16.7	65.9	5B/U
6A/U	802	CW	1	0.0285	0.189	SC	C	0.264	NCV	0.336	–	2700	74	75	□	20	0.70	1.8	2.9	4.3	6.5	8.3	11.2	22	65.9	6A/U
8/U	803	C	7	0.086	0.295	C	–	0.340	V	0.415	–	4000	99	52	△	29.5	0.56	1.35	2.1	3.1	5.0	6.5	8.8	17.5	65.9	8/U
8/AU	804	C	7	0.0285	0.295	C	–	0.340	NCV	0.415	–	4000	99	52	△	29.5	0.56	1.35	2.1	3.1	5.0	6.5	8.8	17.5	65.9	8A/U
9A/U	842	SC	7	0.086	0.285	SC	SC	0.355	NCV	0.430	–	4000	126	52	▲	29.5	0.45	1.26	2.3	3.4	5.2	6.5	9.0	17	65.9	9A/U
9B/U	805	SC	7	0.086	0.285	SC	SC	0.355	NCV	0.430	–	4000	126	50	△	30	0.45	1.26	2.3	3.4	5.2	6.5	9.0	17	65.9	9B/U
10A/U	806	C	7	0.086	0.295	C	–	0.340	NCV	0.415	0.475	4000	121	52	△	29.5	0.56	1.35	2.1	3.1	5.0	6.5	8.8	17.5	65.9	10A/U
11/U	807	TC	7	0.048	0.295	C	–	0.340	V	0.415	–	4000	89	75	□	20.5	0.65	1.5	2.15	3.2	4.7	6.0	8.2	18	65.9	11/U
11A/U	808	TC	7	0.048	0.292	C	–	0.340	NCV	0.412	–	4000	89	75	□	20.5	0.65	1.5	2.15	3.2	4.7	6.0	8.2	18	65.9	11A/U
12A/U	809	TC	7	0.048	0.292	C	–	0.340	NCV	0.412	0.475	4000	113	75	□	20.5	0.65	1.5	2.15	3.2	4.7	6.0	8.2	18	65.9	12A/U
13A/U	810	TC	7	0.048	0.290	C	C	0.355	NCV	0.430	–	4000	114	74	□	20.5	0.65	1.5	2.15	3.2	4.7	6.0	8.2	18	65.9	13A/U
14A/U	811	C	1	0.102	0.383	C	C	0.463	NCV	0.558	–	5500	201	52	△	29.5	0.28	0.85	1.5	2.3	3.5	4.4	6.0	11.7	65.9	14A/U
17A/U	812	C	1	0.188	0.695	C	–	0.760	NCV	0.885	–	11000	446	52	△	29.5	0.23	0.60	0.95	1.5	2.4	3.2	4.5	9.5	65.9	17A/U
18A/U	813	C	1	0.188	0.695	C	–	0.760	NCV	0.885	0.945	11000	496	52	△	29.5	0.23	0.60	0.95	1.5	2.4	3.2	4.5	9.5	65.9	18A/U
19A/U	843	C	1	0.250	0.925	C	–	0.990	NCV	1.135	–	14000	720	52	△	29.5	0.14	0.42	0.69	1.1	1.8	2.45	3.5	7.7	65.9	19A/U
20A/U	844	C	1	0.250	0.925	C	–	0.990	NCV	1.135	1.195	14000	786	52	△	29.5	0.14	0.42	0.69	1.1	1.8	2.45	3.5	7.7	65.9	20A/U
34B/U	814	C	7	0.075	0.470	C	–	0.535	NCV	0.640	–	5200	195	75	□	20	0.29	0.85	1.3	2.1	3.3	4.5	6.0	12.5	65.9	34B/U
35B/U	815	C	1	0.1045	0.690	C	–	0.760	NCV	0.880	0.945	10000	425	75	□	20.5	0.23	0.61	0.85	1.25	1.95	2.47	3.5	8.6	65.9	35B/U
55/U	845	C	1	0.032	0.121	TC	TC	0.176	PE	0.206	–	1900	31	53.5	△	28.5	1.3	3.2	4.8	7.0	10.5	13.0	17	32	65.9	55/U
55A/U	816	SC	1	0.035	0.121	SC	SC	0.176	NCV	0.216	–	1900	36	50	△	29.5	1.3	3.2	4.8	7.0	10.5	13.0	17	32	65.9	55A/U
55B/U	817	SC	1	0.032	0.121	TC	TC	0.176	PE	0.206	–	1900	32	53.5	△	28.5	1.3	3.2	4.8	7.0	10.5	13.0	17	32	65.9	55B/U
58/U	818	C	1	0.032	0.121	C	–	0.150	V	0.200	–	1900	24	53.5	△	28.5	1.4	3.5	5.3	8.3	11.5	17.8	20	40	65.9	58/U
58A/U	819	TC	19	0.0375	0.120	TC	–	0.150	V	0.199	–	1900	25	50	△	29.5	1.6	4.1	6.2	9.2	14.0	17.5	23.5	45	65.9	58A/U
58C/U	820	TC	19	0.0375	0.120	TC	–	0.150	NCV	0.199	–	1900	25	50	△	29.5	1.6	4.1	6.2	9.2	14.0	17.5	23.5	45	65.9	58C/U
59/U	821	CW	1	0.0253	0.150	C	–	0.191	V	0.250	–	2300	36	73	□	21	1.1	2.7	4.0	5.7	8.5	10.8	14.0	26	65.9	59/U
59B/U	823	CW	1	0.023	0.150	C	–	0.191	NCV	0.246	–	2300	36	75	□	20.5	1.1	2.7	4.0	5.7	8.5	10.8	14.0	26	65.9	59B/U
62/U	824	CW	1	0.025	0.151	C	–	0.191	V	0.250	–	750	34	93	▲	13.5	0.82	1.9	2.7	3.9	5.8	7.0	9.0	17	84	62/U
62A/U	846	CW	1	0.025	0.151	C	–	0.191	NCV	0.249	–	750	34	93	▲	13.5	0.82	1.9	2.7	3.9	5.8	7.0	9.0	17	84	62A/U
63B/U	847	CW	1	0.0253	0.295	C	–	0.340	NCV	0.415	–	1000	78	125	●	10	0.60	1.4	2.0	2.9	4.1	5.1	6.5	11.3	84	63B/U
71/U	839	CW	1	0.025	0.151	C	TC	0.198	PE	0.259	–	750	42	93	▲	13.5	0.82	1.9	2.7	3.9	5.8	7.0	9.0	17	84	71/U
71A/U	840	CW	1	0.025	0.151	TC	TC	0.198	V	0.245	–	750	42	93	▲	13.5	0.82	1.9	2.7	3.9	5.8	7.0	9.0	17	84	71A/U
71B/U	841	CW	1	0.025	0.151	C	TC	0.208	PE	0.250	–	750	42	93	▲	13.5	0.82	1.9	2.7	3.9	5.8	7.0	9.0	17	84	71B/U
74A/U	825	C	1	0.102	0.383	C	C	0.564	PE	0.558	0.615	5500	230	52	△	29.5	0.28	0.85	1.5	2.3	3.5	4.4	6.0	11.7	65.9	74A/U
79B/U	848	CW	1	0.025	0.295	C	–	0.340	NCV	0.415	0.475	1000	122	125	●	10	0.60	1.4	2.0	2.9	4.1	5.1	6.5	11.3	84	79B/U
164/U	826	C	1	0.1045	0.690	C	–	0.760	NCV	0.890	–	10000	392	75	□	20.5	0.23	0.61	0.85	1.25	1.95	2.47	3.5	8.6	65.9	164/U
174/U	827	CW	7	0.019	0.060	TC	–	0.069	V	0.105	–	–	–	50	△	30	–	–	–	–	2.0	–	–	–	65.9	174/U
177/U	828	C	1	0.195	0.690	SC	SC	0.760	NCV	0.910	–	14000	465	50	△	30	0.23	0.60	0.95	1.5	2.4	3.2	4.5	9.5	65.9	177/U
212/U	829	SC	1	0.056	0.189	SC	SC	0.265	NCV	0.336	–	3000	85	50	△	29.5	0.65	1.6	2.4	3.6	5.2	6.6	8.8	16.7	65.9	212/U
213/U	830	C	7	0.090	0.292	C	–	0.340	NCV	0.412	–	4000	100	50	△	30.5	0.56	1.35	2.1	3.1	5.0	6.5	8.8	17.5	65.9	213/U
214/U	831	SC	7	0.090	0.292	SC	SC	0.360	NCV	0.432	–	4000	129	50	△	30.5	0.45	1.26	2.3	3.4	5.2	6.5	9.0	17	65.9	214/U
215/U	832	C	7	0.090	0.292	C	–	0.340	NCV	0.412	–	4000	122	50	△	30.5	0.56	1.35	2.1	3.1	5.0	6.5	8.8	16.7	65.9	215/U
216/U	833	TC	7	0.048	0.292	C	C	0.360	NCV	0.432	–	4000	115	75	□	20.5	0.65	1.5	2.15	3.2	4.7	6.0	8.2	18	65.9	216/U
217/U	834	C	1	0.106	0.380	C	C	0.463	NCV	0.555	–	5500	202	50	△	30	0.28	0.85	1.5	2.3	3.5	4.4	6.0	11.7	65.9	217/U
218/U	835	C	1	0.195	0.690	C	–	0.760	NCV	0.880	–	11000	457	50	△	30	0.225	0.60	0.95	1.5	2.4	3.2	4.5	9.5	65.9	218/U
219/U	836	C	1	0.195	0.690	C	–	0.760	NCV	0.880	0.945	11000	507	50	△	30	0.225	0.60	0.95	1.5	2.4	3.2	4.5	9.5	65.9	219/U
220/U	850	C	1	0.260	0.910	C	–	0.990	NCV	1.120	–	14000	725	50	△	29.5	0.17	–	0.69	1.12	1.85	–	3.6	7.7	–	220/U
221/U	851	C	1	0.260	0.910	C	–	0.990	NCV	1.120	1.195	14000	790	50	△	29.5	0.17	–	0.69	1.12	1.85	–	3.6	7.7	–	221/U
223/U	837	SC	1	0.036	0.120	SC	SC	0.176	NCV	0.216	–	1900	36	50	△	30	1.3	3.2	4.8	7.0	10.5	13.0	17.0	32	65.9	223/U
224/U	838	C	1	0.106	0.380	C	C	0.463	NCV	0.555	0.615	5500	232	50	△	30	0.28	0.85	1.5	2.3	3.5	4.4	6.0	11.7	65.9	224/U

SC—silver plated copper, C—bare copper, PE—polyethylene, NCV—non-contaminating vinyl, V—polyvinylchloride, TC—tinned copper, CW—copperweld Ohms Code: △ Through to 55 □ 56 Through 80 ▲ 81 Through 100 ● 101 Through 200

Above 3000 MHz, a special type of "metal insulator" can be used in rigid coaxial lines. Here the support for the center conductor is a quarter-wavelength short-circuited stub. Figure 2-19(a) and (b) show the physical and diagrammatic representation of this construction. The device operates as we saw on page 000 because the far end of a short-circuited quarter-wave line presents an open circuit. This lossless, high-impedance structure provides the support points along the center conductor. Induction and radiation losses occur mostly in open-wire lines, since the fields around their conductors may couple to adjacent absorbing–conducting objects. The open ends of coaxial lines also offer an outlet for radiating fields.

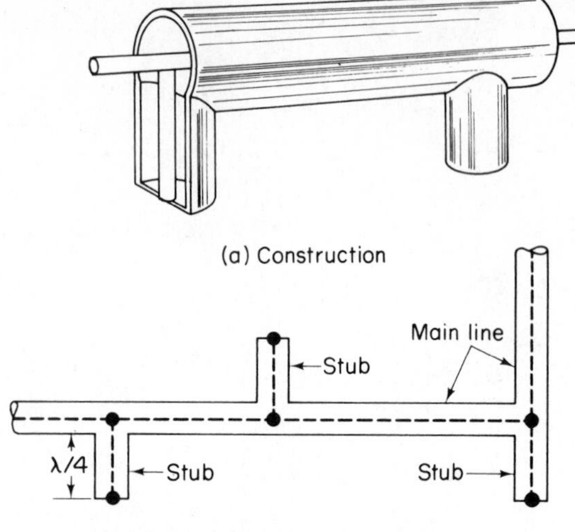

Fig. 2-19 METAL-INSULATOR QUARTER WAVE LINE SUPPORT

Wave Motion and Propagation 3

INTRODUCTION

Microwave systems deliver extremely high frequency currents, voltages, and power output, which in most cases are usually derived from stimulating or self-oscillating power inputs. This output power is passed through waveguides by traveling fields and not by electron drift as in most electronic circuits.

The fields are considered as traveling electric waves (mostly sine-wave variations) or wave motion along the inside of solid hollow conducting-surfaced tubes or pipes. The wave propagation or motion, as outlined in Chapter 1, consists of the impinged voltage or current input causing a continually moving, two-component, electromagnetic field to travel down the waveguide.

Here we should identify the three fields constituting microwave energy, that is, the main electromagnetic field and its two components. These latter are:

(a) The electric or electrostatic field due to electrical potential differences between sides or parts of a waveguide structure. These moving electrical differences produce stress in the dielectric within the waveguide and represent moving, dynamic power.

(b) The magnetic field existing at right angles to the electric field and representing the results of current flow paths along the inside surface of the waveguide.

An understanding of the nature of the fields and their action within a waveguide plus their types and variations is the primary aim of this chapter. Their construction accessories appear in Chapters 4 and 5 and important and special components are developed and described in Chapter 6.

WAVEGUIDES VERSUS TRANSMISSION LINES

MODES

It should be pointed out that transmission by wave travel in rf lines is allied to waveguide transmission in all respects up to the defining and description of the specific wave composition and the mechanics of wave travel, which in turn is defined by the lines of force making up the field configurations. These configurations are known as modes and indeed exist in coaxial lines; their use and manipulation in the latter is limited as we shall see below, because physical dimensions are mechanically too small for erecting internal accessory devices such as coupling loops, attenuators, isolators, etc.

LOSSES

A waveguide (without a center conductor) with large surface area has inherently less losses (for equal frequencies) than a coaxial line. This is due to (1) low waveguide I^2R copper losses present in both conductors in an rf line, (2) low waveguide leakage and dielectric losses (since a coaxial line must have lossy insulators supporting the inner conductor), and (3) negligible radiation since all fields are contained within the structure itself. Open-wire and some other configurations of two-wire cables have radiation losses even at relatively low radio frequencies.

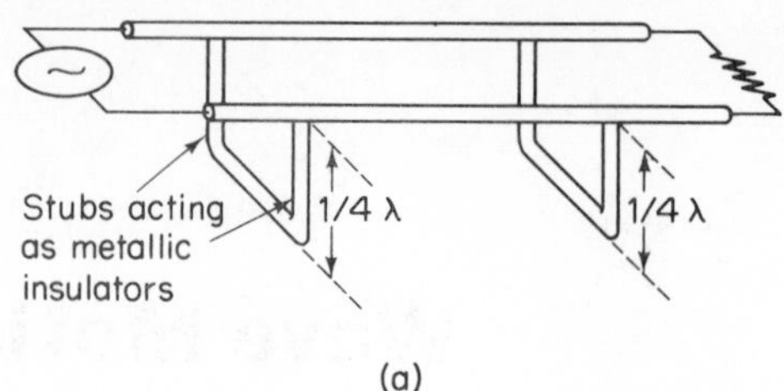

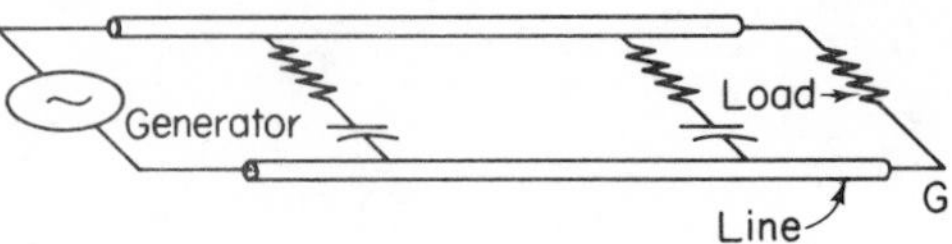

(b) Equivalent circuit at communication frequencies

Fig. 3-1 STUB SUPPORTING WAVEGUIDE CONSTRUCTION

CONSTRUCTION

Without a center conductor, waveguides are inherently simpler to construct than coaxial lines and possess greater strength, since their structure is self-supporting. However, these advantages disappear at lower frequencies because a waveguide must have a cross-sectional dimension of a half-wavelength to properly contain the electromagnetic fields. For instance, a waveguide at one MHz would have to be about 700 ft wide, and at lower radar frequencies, say at 200 MHz, the structure would be about 4 ft wide. But at 1 kMHz the advantage begins to appear, for at this frequency a guide need be only 1 in. wide. Thus we see that waveguides are bulky, cumbersome, and impracticable below about 5 MHz. Note then that microwave transmission by coaxial lines is practical extending from this point down to 1000 kHz.

Other factors bearing upon waveguide usage are the special plumbing necessary to produce couplings, bends, and twists. This structure often requires exact machine fitting, bringing the ultimate constructional cost of waveguide plumbing above that of similar rf power lines.

WAVEGUIDE AS A TRANSMISSION LINE

Simplified waveguide action is analogous to a two-wire line supported by quarter-wavelength stubs, as described in Chapter 2. We saw that transmission-line operation was possible in this case because the stubs themselves acted as metallic insulators and did not impede or affect wave propagation along the line (Fig. 3-1). This, as we remember, was because the stubs and their connecting cross members acted as insulators even though they were metal. Here, they acted as a short-circuited quarter-wave line which presented infinite impedance (at the correct frequency) to the open ends, which could therefore support the transmission-line conductors.

To develop this idea into a waveguide structure we take four steps:

1. Add a second set of stubs at the support points but on the top side of the line. Being shorted and of a quarter-wavelength, which is compatibly high impedance, they have no effect upon the line's electrical characteristics.
2. Flatten and widen the stubs and the line as in Fig. 3-2(a). No electrical harm can come from this move provided that the line spacing and the stub and conductor cross-sectional areas are maintained as in Fig. 3-1.
3. Flatten and widen the stubs, which now become little boxes spaced along the line, still without introducing any losses.
4. Finally, make the boxes fill the entire space between the stub-center location and connect their edges together by welding or soldering.

This solid wall of insulators gives a rigid rectangular waveguide structure which conducts wavefronts along the side-center-located line areas, which become rectangular cross-section two-bar conductors just as if the bars were independent transmission lines.

Now the signal-conducting-bar areas are frequency sensitive, so at higher frequencies (than those pictured), where the quarter-wavelengths are shorter and the middle conducting bar becomes wider [Fig. 3-3(a) up to a practical point], the stubs become too short and present a capacitance; at this high-frequency limit the box–bar–stub analogy does not hold.

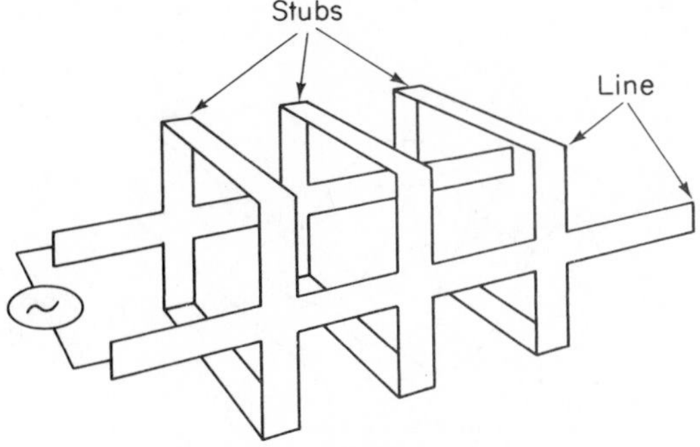

(a) Added topside stubs

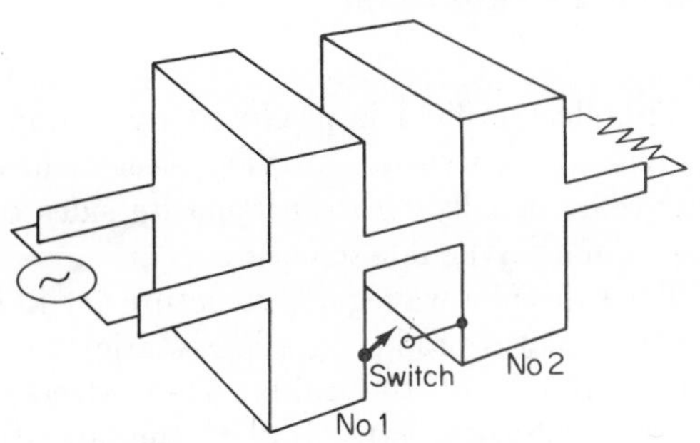

(b) Widened stubs

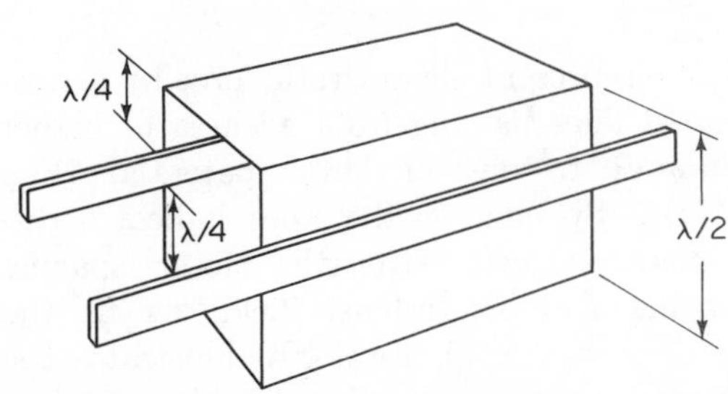

(c) Final solid stub construction

Fig. 3-2 STUB-WAVEGUIDE CONSTRUCTIONAL DEVELOPMENT

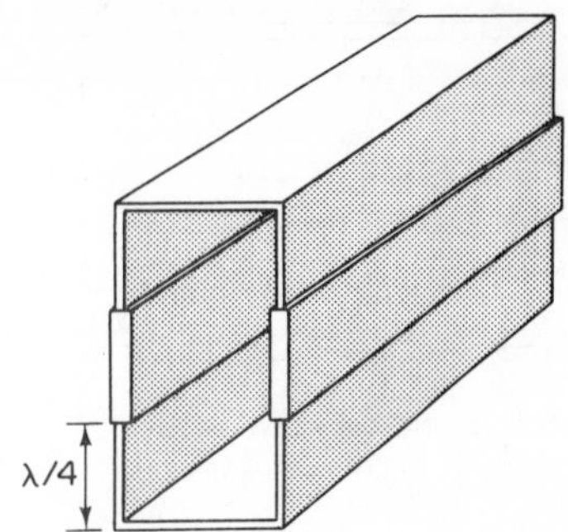

(a) Above minimum frequency

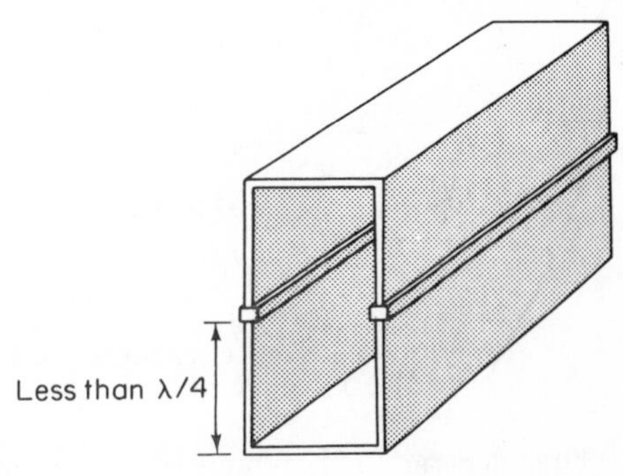

(b) Below minimum frequency

Fig. 3-3 SHORTENED STUB-WAVEGUIDE SECTIONS

At lower frequencies the sections become longer and the conducting bar becomes narrower; thus when the longer sections become less than quarter-wave [Fig. 3-3(b)], they become inductive and dissipate energy through circulatory currents. This low-frequency limit of power transfer is known as the *cutoff frequency*.

WAVEGUIDE FIELDS

The moving wave motion of transmitted power in a waveguide is generated by input voltages and currents which compositely exist as a traveling electromagnetic field. As noted above, this wave consists of two simultaneously drifting fields within the waveguide structure. They are:

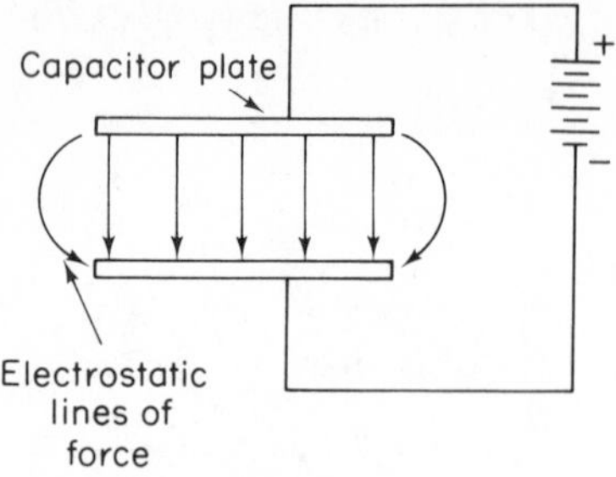

(a) Static field between plates

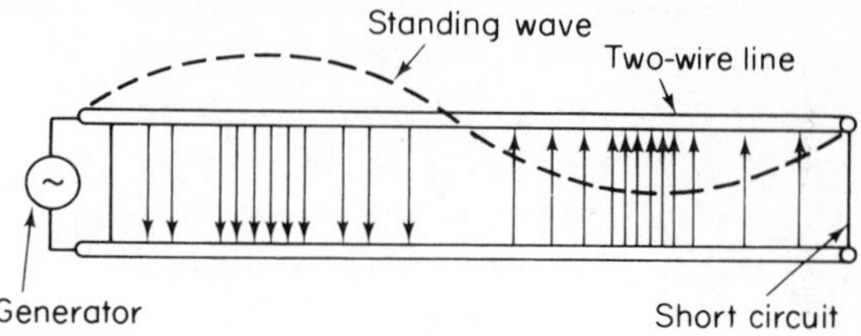

(b) Sinewave field distribution with AC signal on full wave section

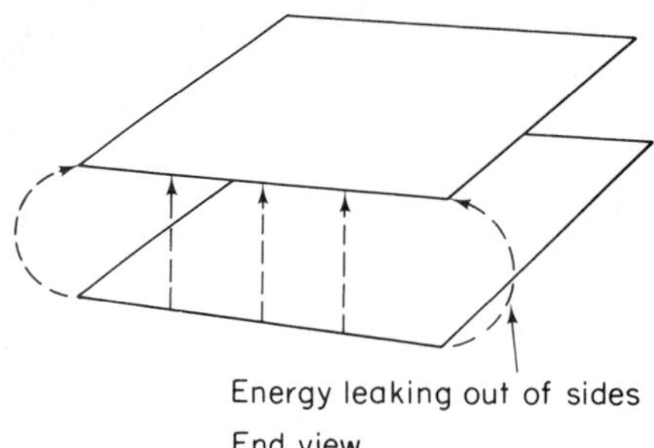

End view

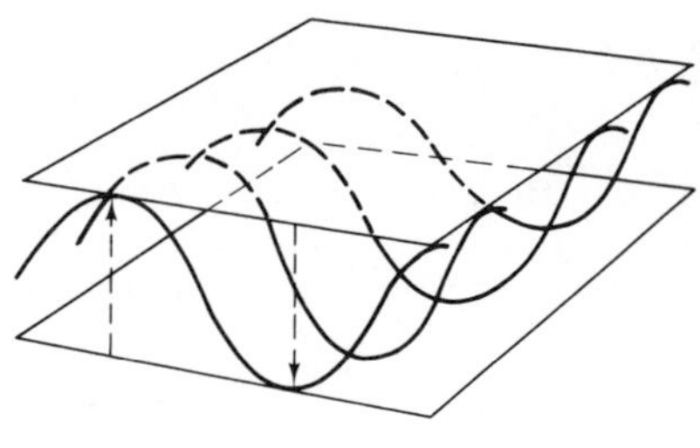

Side view

(c) Isometric projection of field distribution

Fig. 3-4 CAPACITOR PLATE ELECTRIC FIELD LINE REPRESENTATION

1. The electric field, which is a voltage-related and voltage-sensitive condition.
2. The magnetic field, which is a current-related and current-sensitive condition.

Before detailing the configuration and action of each field we must realize that

1. The fields exist simultaneously and compatibly within and upon the waveguide structure.
2. They both travel along a waveguide just as current and voltage impulses travel along a transmission line.
3. Their magnitude and variations exhibit the same maximum and minimum voltage and current variations existing in transmission-line practice, particularly the phenomenon of standing waves. These variations present stationary, space-located, and amplitude-detected voltages in the instruments used to measure them.
4. Their variations are always at 90° physical-space disposition and 90° electrical-phase disposition.

The electric field is produced by voltage differences. These are varying electrostatic potentials, usually between opposite sides of the box of a waveguide structure.

The sides of a waveguide structure act like capacitor plates producing electrostatic stress in the interior of the guide; these stresses represent physical straining of the air dielectric and mean that power is present; together with the electromagnetic waves they cause the energy to progress along the waveguide.

We represent electrostatic force by arrowheaded lines passing from an area of higher potential to one of lower potential [Fig. 3-4(a)]. Spacing of the lines indicates differences in field strength, closer spacing showing a more intense field, as in the standing wave. Figure 3-4(b) illustrates the electrostatic intensity along a full-wave line section. Note also the force-line polarity reversal. Figure 3-4(c) is an isometric representation of the field distribution, including the reversal and sine-wave distribution between parallel plates.

For convenience we call this the electric field or E field. Illustrating the E field still further, Fig. 3-5(a) shows its density distribution at half-wave frames on a full-wave line, cross-sectionally photographed, so to speak, along the two-wire box-simulated line of Fig. 3-2. Note that the intensity-line distribution completely fills the box to maximum intensity at alternate half-wave points, decreasing in intensity to the alternate minimum points, where no cross lines exist at all.

Now if the lines were visible and we could look at the E lines through the side, the top, and the end of the box, they would appear as in Fig. 3-6(a), (b), and (c); were the box completely transparent and isometrically viewed, the force lines would appear as in Fig. 3-6(d).

Note also the direction change of the force arrows in alternate halves of the line structure, and remember that these are the field-strength conditions existing at a certain instant when a

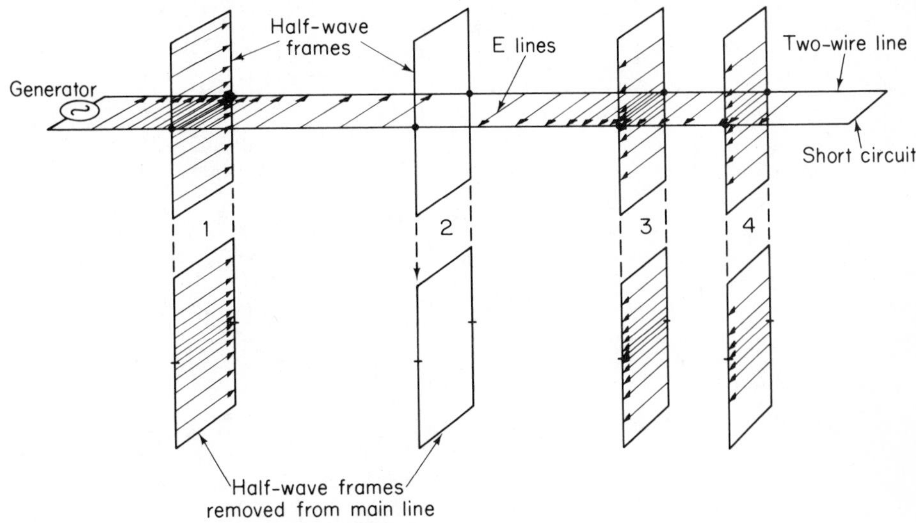

Fig. 3-5 DENSITY DISTRIBUTION OF *E* FIELD

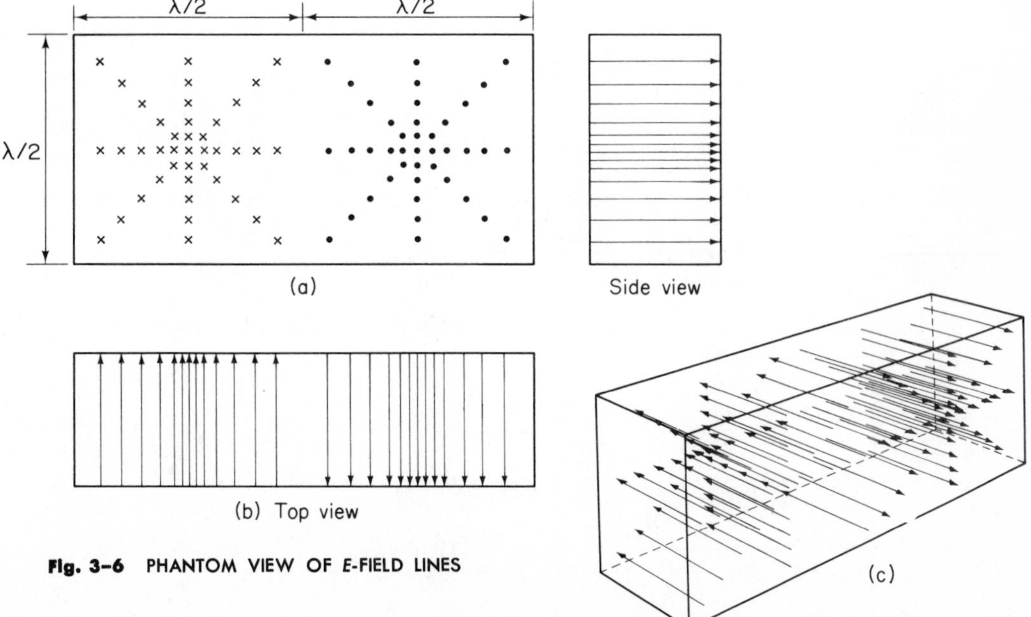

Fig. 3-6 PHANTOM VIEW OF *E*-FIELD LINES

standing wave is at its maximum on the line. At other times the voltage and E field vary from zero to the peak value, reversing every alternation of the applied voltage.

MAGNETIC FIELD. The magnetic field is formed by currents in a waveguide's metallic-coated side section. These currents generate flux lines or a field of force, a field called the H field.

As in a solenoid, flux lines generated by the single turns of a coil winding cancel each other between turns, so the sum total of the flux lines passes at right angles to the plane of the current-carrying conductors.

Now, simulating a solenoid winding by our breakdown of a waveguide into separated

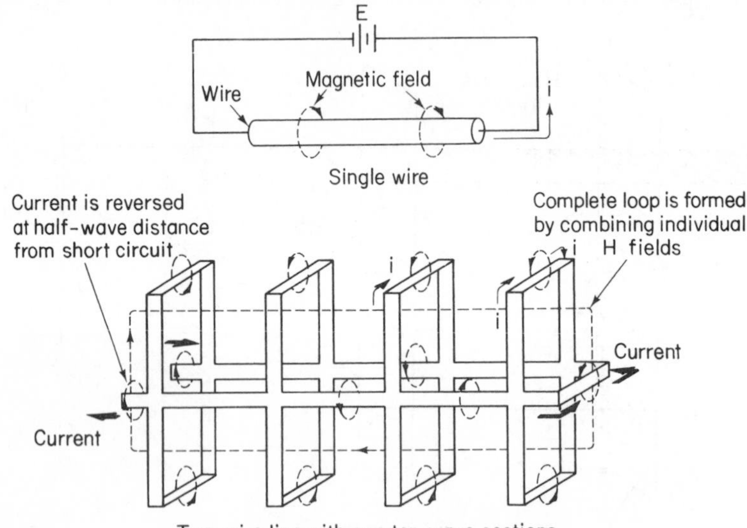

(a) In quarter-wave sections

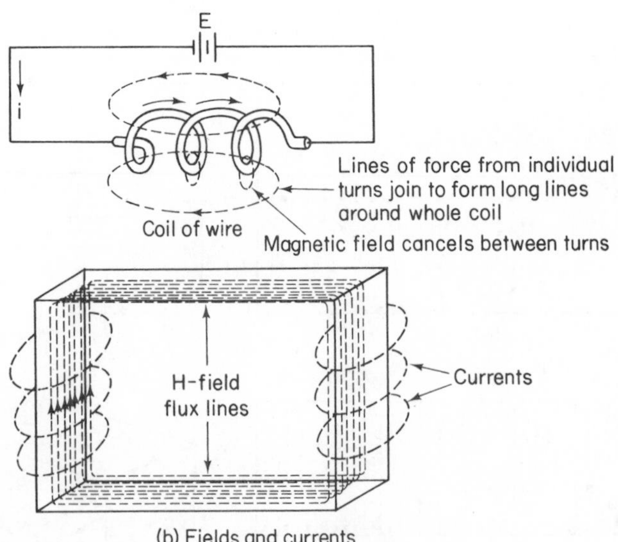

(b) Fields and currents

Fig. 3-7 *H*-FIELD LINE PICTURIZATION

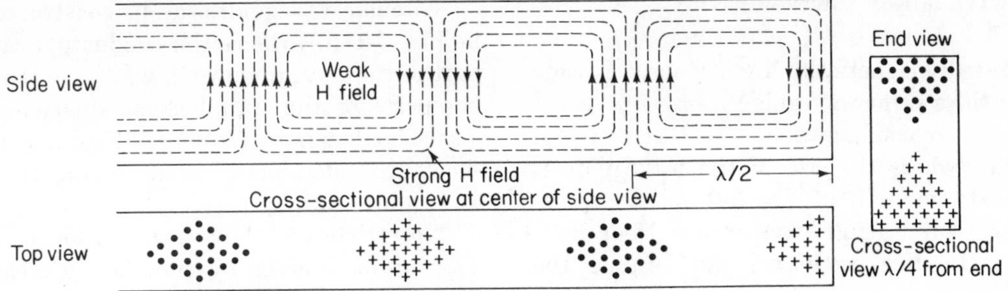

Fig. 3-8 PHANTOM VIEW OF *H*-FIELD LINE PICTURIZATION

quarter-wave supporting sections [Fig. 3-7 (a) and (b)], we see that currents in these supporting sections cause the flux loops filling the interior of each quarter-wave box of the waveguide.

Note that unlike *E*-field force lines, the *H*-field lines form complete loops, whereas the electrostatic force lines terminate at the source potential planes.

Figure 3-8 is a pseudo-visual representation by means of side, top, and end views of the magnetic *H*-field contours in a waveguide which is three half-wavelengths long; note that the field is strongest at the edges of the guide where the current is strongest and reluctance of the magnetic path is least (owing to a shorter loop). This again is an instantaneous picture of field conditions, seen at one peak of a half-cycle of ac input–all of which reverses at half-wave distances and within the alternate peaks of applied alternating current.

Composite field. Simultaneously picturing both *E* and *H* fields becomes graphically complicated [Fig. 3-9(a)]; individually, the force lines and arrows would appear as in Fig. 3-9(b), (c), and (d), giving us what is called a *dominant mode of operation*, since it is the simplest condition existing under fundamental sine-wave variations in force lines.

It should be again emphasized that the intensity variations of *E* and *H* fields are always mutually perpendicular, even in the dynamic alternating state, and that *E*-field lines terminate abruptly at the sides of a waveguide providing their electrostatic potential difference; *H* lines form loops that are

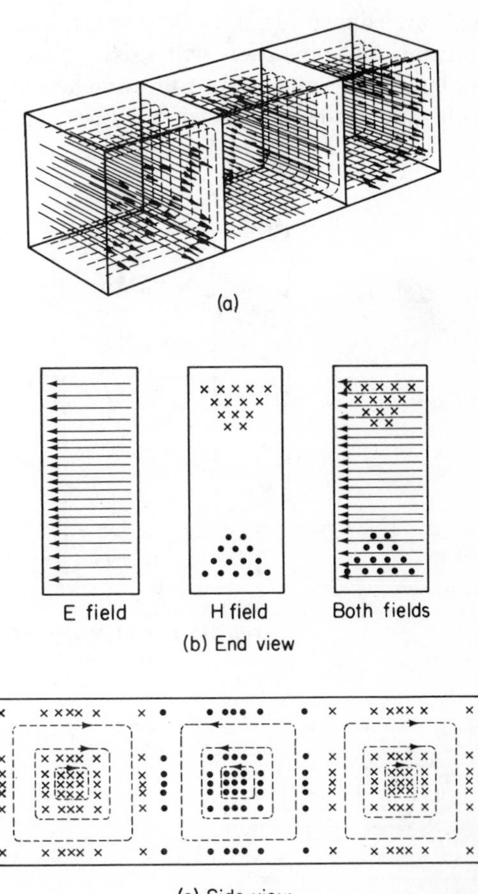

Fig. 3-9 COMPOSITE *E*- AND *H*-FIELD CONFIGURATION

generated magnetically by currents flowing in the sides of the waveguide.

Another configuration is pictured in Fig. 3-10(a), where the size of the waveguide is doubled over that of Fig. 3-9 and the cross section will be full wave rather than half-wave. In this mode we can assume the conductor plane to be spaced one quarter-wavelength from the top. Since the remaining support stub section is three-fourths λ, it is still high impedance (as is the λ/4 stub), and the conventional operation will exist. The field configuration will show a full wave across the wide dimension, as shown in Fig. 3-10(b).

Circular waveguides can be constructed as in Fig. 3-11(a), where the conductors can be part of the waveguide wall, while the remaining parts of the wall become quarter-wave insulating support sections. Figure 3-11(c) shows the dominant mode for a circular waveguide section.

Note that the above explanation of field distribution is given to show how waveguide construction is related to transmission-line operation, particularly with respect to the electric- and magnetic-field distribution in their standing-wave condition.

In the following section we discuss in greater detail how the fields interact, travel, and propagate along the waveguide structure.

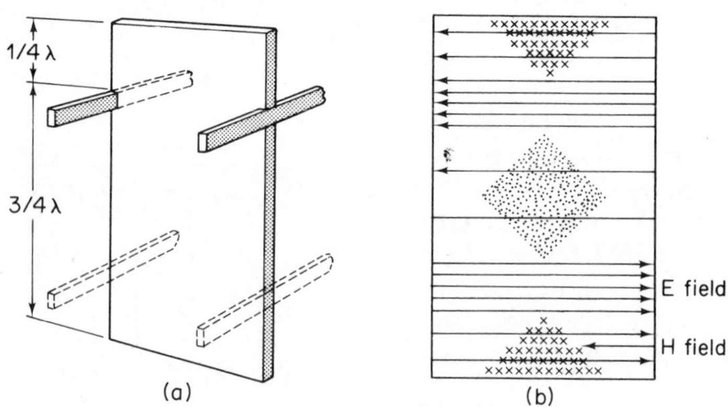

Fig. 3-10 FULL WAVE FIELD CONFIGURATION IN WAVEGUIDE

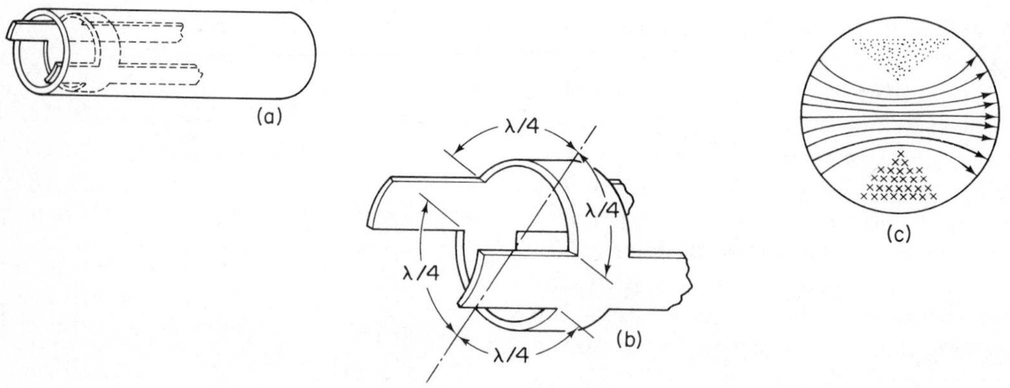

Fig. 3-11 FIELD LINES IN CIRCULAR WAVEGUIDE

WAVE MOTION IN A WAVEGUIDE

GENERAL

Wave motion is guided power transmission, not unlike radio-wave propagation in free space; in both cases an electromagnetic wave consisting of an electric field combined with a magnetic field is used. In a radio transmitter the fields are radiated into free space; in a waveguide they are merely impinged upon and guided by a hollow metal-coated tube.

In fact, it is generally conceded that ac power transmission is one concerning wave propagation in a dielectric. Wires or waveguides do not transmit electric power; they simply guide the waves which carry power through the surrounding air or other insulating medium.

WAVE GENERATION

We can best picture the action of wave motion by comparing the fields generated by a simple dipole antenna to those existing in an elementary waveguide. The two situations are almost identical, and indeed the most efficient way to generate wave motion in a waveguide is to apply signal power to a small dipole-like probe inserted in the wall of a waveguide.

Figure 3-12 illustrates the field lines generated around a dipole antenna supplied with high-frequency power. Here the E and H fields exist in exactly correct position for transmitting their energy in a forward direction. Each field exists perpendicular to the other, and, when caused to increase or decrease, their changes are 90° out of phase with each other as each field intensity moves forward and outward along the direction of propagation. We see this in Figs. 3-9 and 3-10 at the standing-wave points where the magnetic (H) lines are bunched closer or are of a maximum intensity around phantom-pictured partition locations; the electrostatic (E) lines are bunched around the midpoint between the partitions.

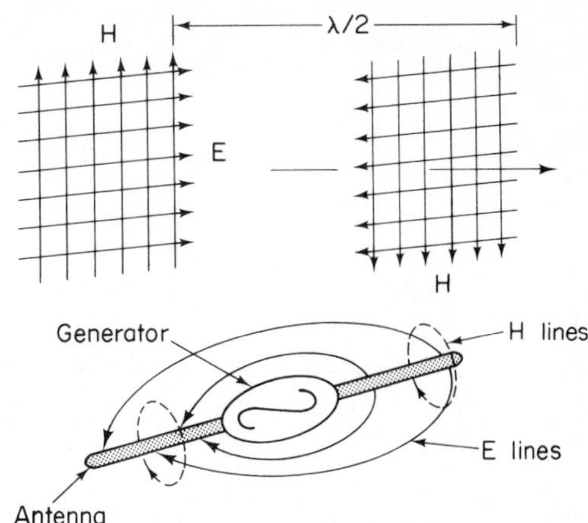

Fig. 3-12 FIELD LINE AROUND A DIPOLE

DEFINITION

Two points must be emphasized in the following:

1. The fields themselves (that is, their lines of force) are at right angles to the direction of travel. They are transverse or crosswise to propagation. Thus we call them *transverse electromagnetic*, or TEM, waves.

Figure 3-13(a) isometrically shows the physical relationship of both E and H fields as they propagate energy (at right angles) down a waveguide. Figure 3-13(b) is a plane graphic representation with dashes indicating vectors coming out of the plane of the paper and the crosses as vectors going down into it. Note that the fields add vectorially at each point in space, so if they happen to have the same polarity, they reinforce and otherwise can cancel each other.

In mode-operational descriptions in which the E-field lines are all crosswise, we have a TE or *transverse electric mode*, and in magnetic crosswise situations, a TM or *transverse magnetic mode*.

2. Combination of the energy in the E and H fields (that is, power transfer along the line) travels in a helix or corkscrew-like configuration known as a *Poynting vector*. This power combination of the traveling

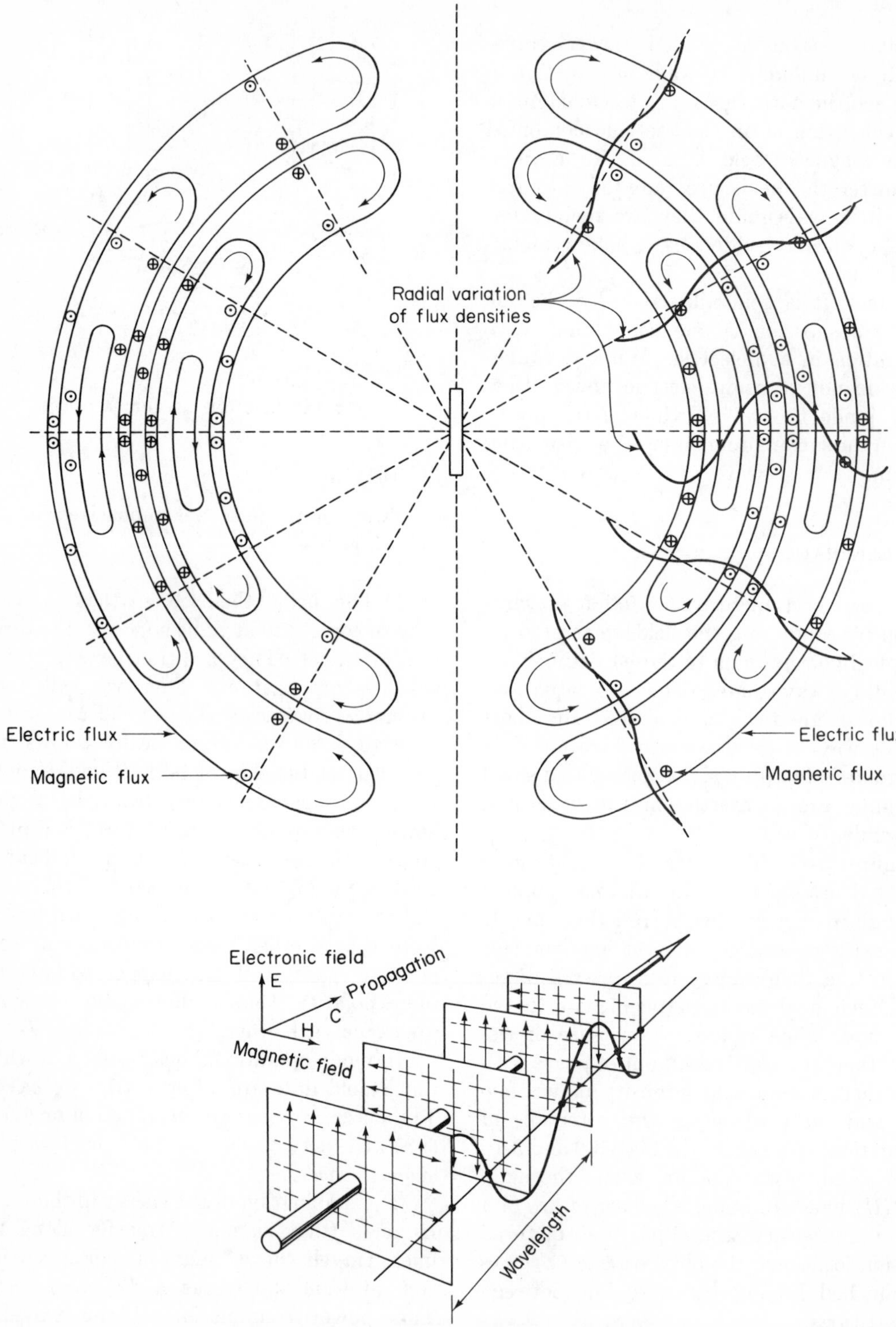

Radial variation
of flux densities
Electric flux
Magnetic flux
Electric flux
Magnetic flux
Electronic field
E
Propagation
C
H
Magnetic field
Wavelength

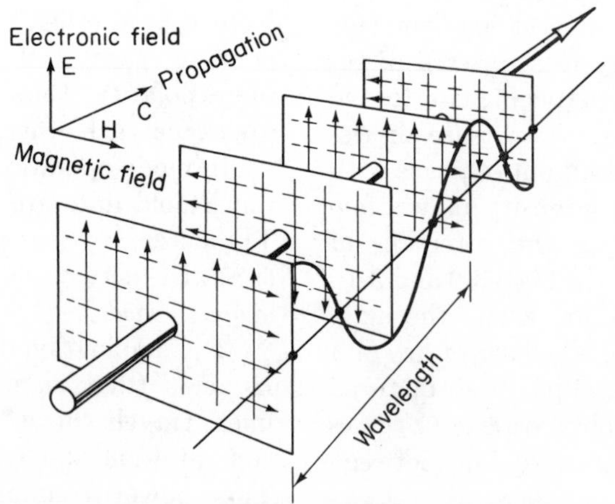

Fig. 3-13 *E*- AND *H*-FIELD PROPAGATION

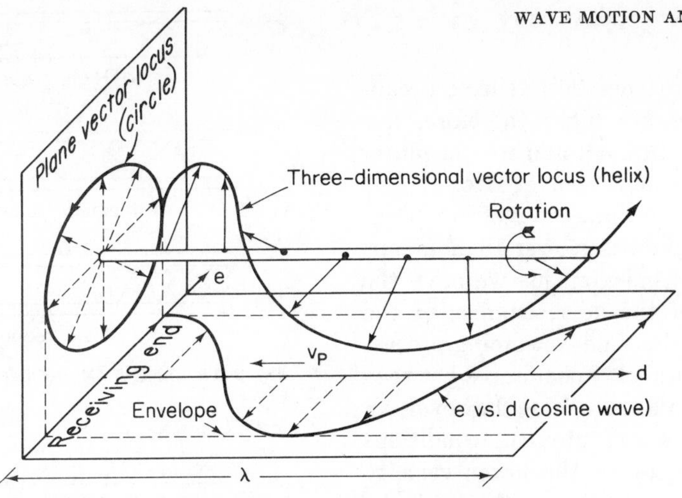

Fig. 3-14 THE POYNTING VECTOR

fields is pictured in Fig. 3-14. Its magnitude is an energy product of both electric and magnetic fields.

WAVE PROPAGATION

Both field components of an electromagnetic wave travel down a waveguide by means of reflections which occur against the side of the waveguide. Examining Fig. 3-12, we see that the main horizontally radiated energy, directly applied down to the center line of the waveguide, goes nowhere. In fact, it soon becomes rapidly attenuated and does not enter our power-transfer situation.

Both radioangularly generated fields proceed outward toward the sides of the waveguide, where they are reflected back and forth with negligible loss as they proceed forward. Figure 3-15 illustrates this passage of power, which we again emphasize has maximum and minimum E and H components, and which under mismatch or other end conditions behaves like the standing voltage waves along a transmission line. In other words, as a result of reflections the waves become concentrated at certain modal points; they also suffer some cancellation due to side or angular reflections.

Note also that the angle at which a wavefront crosses a waveguide is a function of the wavelength and the cross-sectional dimension of the waveguide. We see this in Fig. 3-16 and note that as the frequency increases, the angle of incidence becomes less and the signal travels further before it reaches the opposite wall. Also, when frequency becomes lower, the angle of reflection becomes greater,

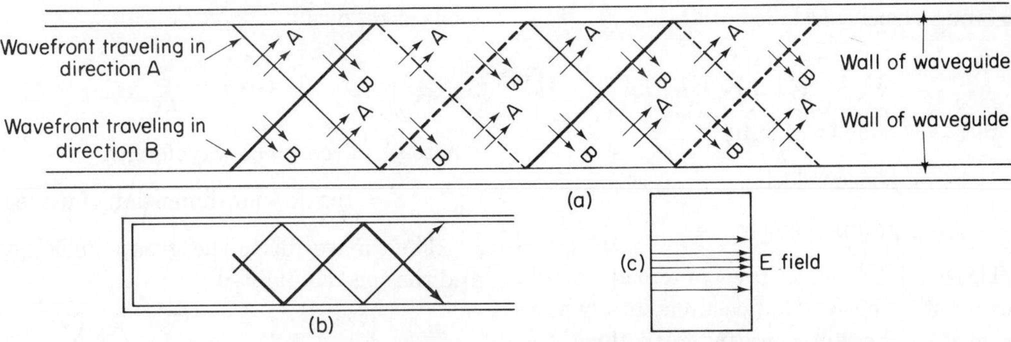

Fig. 3-15 POWER TRAVEL WITHIN A WAVEGUIDE

and when it approaches 90°, reflective cancellation or losses are high; therefore, less signal energy gets through and we encounter the cutoff frequency of the waveguide.

As in transmission lines, wavefront propagation in waveguides is slower than in air. In a two-wire transmission line we have the retarding effect of the dc resistance, of the shunt losses, plus the time-consuming process of charging the parallel capacitances between the conductors. In a waveguide, however, an entirely different set of physical conditions exists, chiefly because of the lower velocity resultant to the longer zigzag path which the wavefront must take, even though direct reflections proceed at the speed of light.

However, wavefronts assume at different instants different angular positions as they travel—as in Fig. 3-17, a group of successive wavefronts has a definite axial velocity, equivalent to one wavefront traveling the distance G. This is known as the *group velocity* since, as a result of the diagonal movement (direction of the arrow), the wavefront has actually moved down the guide only the distance G, which yields an overall lower velocity for a group of successive wavefronts.

So if a probe were inserted in the wall to detect and measure two wavefront positions, they would be a distance T apart—which would be greater than the distances L or G and seemingly would yield a wavefront velocity greater than the speed of light. This we know is impossible, so further study shows us that this travel is a phase velocity and has a mathematical relationship with the velocity of light and group velocity as follows:

$$V_c = \sqrt{V_p \times V_g} \qquad (1)$$

where V_c = velocity of light

V_p = phase velocity

V_g = group velocity

These relationships have practical importance in standing-wave measurements, where we must make appropriate corrections for two wavelengths; for it is the phase velocity

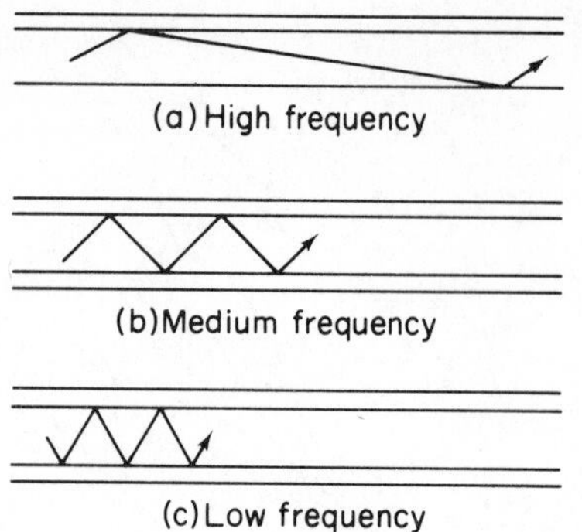

Fig. 3-16 ANGLE OF INCIDENCE IN WAVETRAVEL

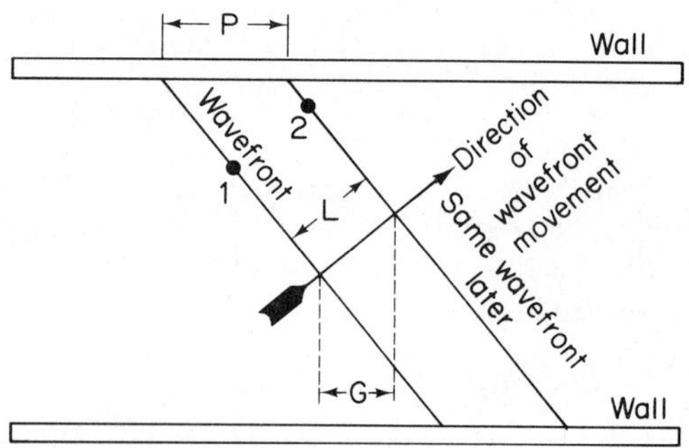

Fig. 3-17 POSITIONAL CONDITIONS IN WAVEFRONT TRAVEL

that determines the distance between maximum and minimum points. This shows up in modulation sine waves, where the envelope waves move forward at group velocity while individual rf cycles move at phase velocity.

Thus waveguide-measurement calculations must involve trigonometric measurements between waveguide dimensions and true wavelength. Thus, as shown in Fig. 3-18, the angle that the wavefront makes with the wall, θ, is related to wavelength and guide dimensions as follows:

$$\cos\theta = \frac{\lambda}{2\beta}$$

where λ = free space wavelength

β = inside wide dimension of waveguide

Then if we relate the group velocity, V_g, and the speed of light, V_c

$$\frac{V_q}{V_c} = \sin\theta = \sqrt{1 - \left(\frac{\lambda}{2\beta}\right)^2} \qquad (2)$$

and since it is possible to measure the wavelength within the guide λ_g, the wavelength in space

$$\frac{\lambda_g}{\lambda_s} = \frac{1}{\sin\theta} = \frac{1}{\sqrt{1 - (\lambda/2\beta)^2}} \tag{3}$$

$$\lambda_s = \frac{2\beta\lambda_g}{\sqrt{\lambda_g^2 + 4\beta^2}} \tag{4}$$

Another way of studying waveguide construction and operation develops when we consider the action of combining E and H fields, when they consist of incident and reflective waves, and how this combining forms resultant field areas. We illustrate this in Fig. 3-19(a) and (b), where we apply the successive E- and H-field vector configurations at half-wave intervals for incident and reflective wave. Figure 3-19(c) shows their superimposition and Fig. 3-19(d) the resultant field configuration.

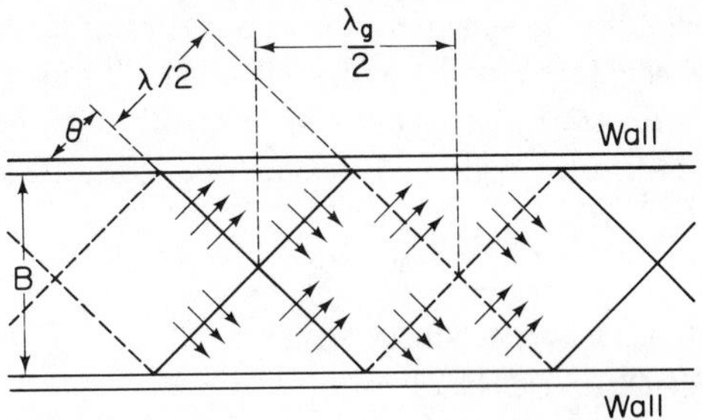

Fig. 3-18 TRIGONOMETRICAL RELATIONSHIPS ON WAVETRAVEL

Now if we place four metal plates directly coincident with the planes of zero-field intensity, we shall have built a waveguide in

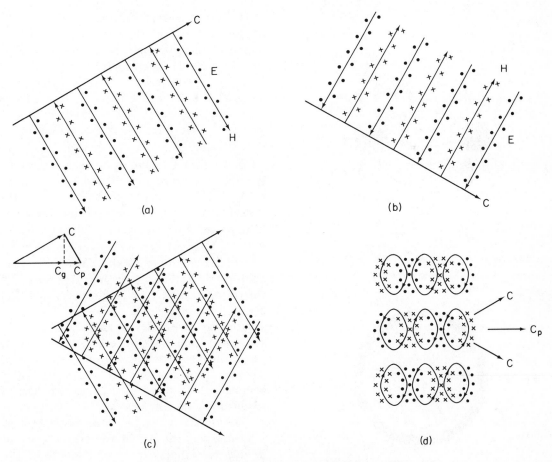

Fig. 3-19 COMPOSITE FIELD-VECTOR RELATIONSHIPS

which the interacting power-carrying fields between them will not be affected. Thus we have created a structure compatible with the incident and reflected waves traveling from an E–H source.

MODES IN WAVEGUIDE FIELD CONFIGURATIONS

Each of the possible E- and H-field configurations in a waveguide is called a *mode*; thus every structure at its particular frequency of operation can be labeled by its mode letters plus its appropriate subscripts —$TE_{1,1}$, $TM_{0,1}$, etc. After the modes have been determined, they may be examined to find which one is most useful for a given application or to find a suitable means of exciting it.

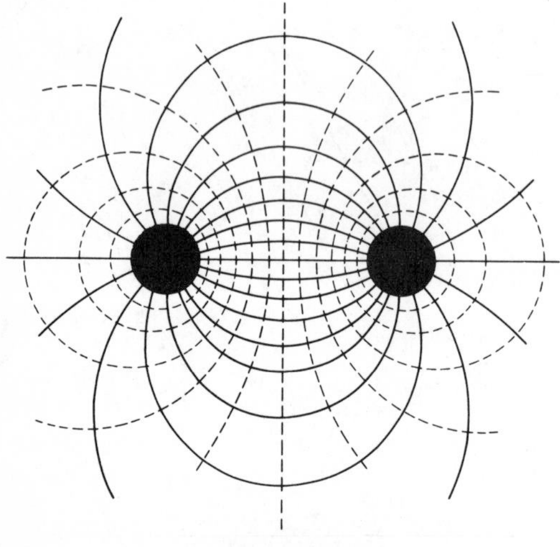

(a) Two-wire lines

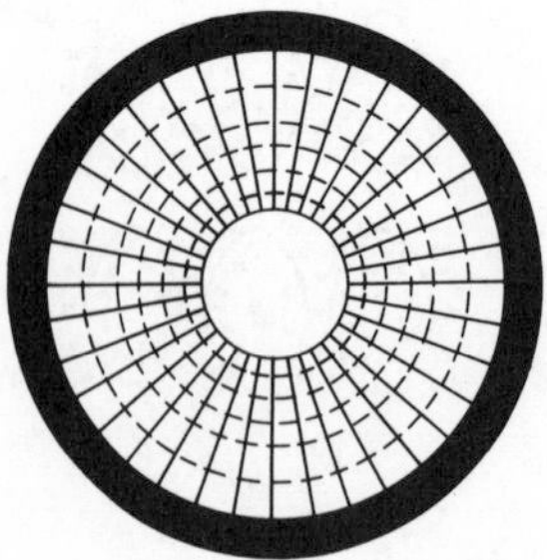

(b) Coaxial lines

Fig. 3-20 FIELDS IN TWO-WIRE AND COAXIAL LINES

The normal configuration of the electromagnetic field within a waveguide is called the *dominant mode of operation.* (See Fig. 3-10 for a rectangular waveguide. For a circular waveguide see Fig. 3-11.)

It is interesting to compare the fields in a circular waveguide to those existing in two-wire and coaxial transmission lines (Fig. 3-20). Here, as noted in Chapter 2, the wavefront mechanism of the E–H field is not used, power transmission being explained in terms of voltage and current waveforms. Any waveguide-field configuration can be classified as either a transverse electric or a transverse magnetic mode.

To review the physical facts, we have observed in examples previously described that:

1. In a transverse electric mode, all parts of the electric field are perpendicular to the length of the guide and no E line is parallel to the direction of propagation.
2. In a transverse magnetic mode, the plane of the H field is perpendicular to the length of the waveguide and no H line is parallel to the direction of propagation.

Note that a wavefront radiated by an antenna in free space or the signal in a coaxial cable is a TEM mode, since both fields are perpendicular to the direction of propagation. This mode, of course, cannot exist in a waveguide.

In addition to the TE and TM labeling, the field configuration is further described by subscripts related to the dimensions of the sides of the waveguide. Thus, in Fig. 3-11, illustrating the $TE_{0,1}$ mode, the first subscript says that there is no half-wave pattern across the b (shortest) dimension of the guide. The second subscript says that there is *one* complete half-wave pattern across the a (wide) dimension of the guide.

For circular waveguides the first subscript indicates the number of full waves existing *around the circumference* of the guide; the second subscript indicates the number of half-wave patterns that exist *across the diam-*

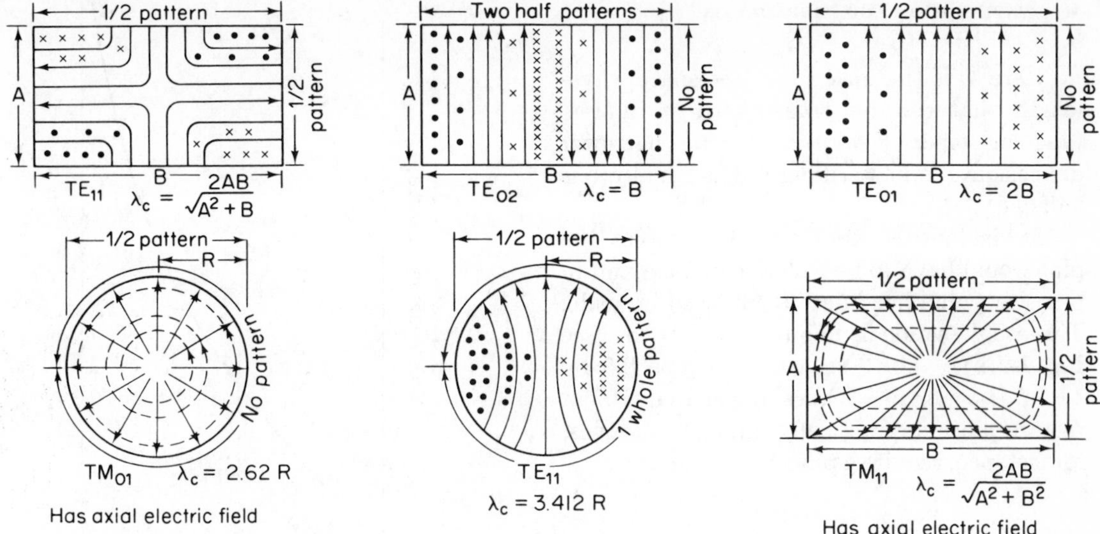

Fig. 3-21 MODES IN RECTANGULAR AND CIRCULAR WAVEGUIDES

eter [Fig. 3-12(b)]. The most common mode in circular waveguides is the $TE_{1,1}$ mode.

Figure 3-21 illustrates a number of modes in rectangular and circular structures.

NOTES ON WAVEGUIDE OPERATION

GENERAL CHARACTERISTICS

In listing rectangular waveguides and their characteristics the basic quantities are

1. Outside diameter and wall thickness.
2. Rectangular size dimension, *a* and *b* (Fig. 3-22).
3. Preferred mode of operation.
4. Cutoff wavelength.

In circular waveguides, the inside diameter, the center conductor size, the modes, and the cutoff wavelength are usually listed.

Specific design data, tables, and operating characteristics on commercial waveguides are discussed more fully in Chapter 4.

CUTOFF WAVELENGTH

The reference on page 3 to the frequency characteristics of waveguides is more conveniently translated for the design engineer in terms of wavelengths. As explained previously, a hollow-pipe wavelength is like a high-pass filter in that it will transmit only frequencies higher than a certain value known as the cutoff frequency, the value of which is related to the physical dimensions. The larger the pipe, the lower its cutoff frequency; placing these facts in terms of wavelengths we have

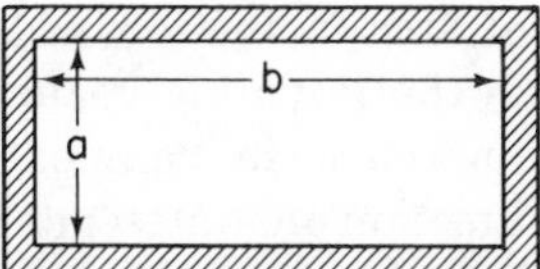

Accessory dimensional data

Wall thickness, t
Outside, (a ≠ 2t) x (b + 2t)
Preferred operational mode
Cutoff wavelength

Fig. 3-22 BASIC RECTANGULAR WAVEGUIDE DIMENSIONS

$$\lambda_c = \frac{c}{f_c} \tag{5}$$

where λ_c = cutoff wavelength

c = velocity of wave propagation in free space

For a rectangular pipe operating in the dominant mode this wavelength dimension relationship is

$$\lambda_c = 2b$$

Here λ_c is the maximum free-space wavelength that can be transmitted in a rectangular pipe of width b. The narrower dimension a is not critical but is usually one half of b.

For instance, to transmit 10-cm waves, the pipe would have to be wider than 5 cm and in practice would be about 70 cm or $2\frac{3}{4}$ in. wide. The usual section is about $1\frac{1}{4}$ by $2\frac{3}{4}$ in. [see Fig. 1-2(b)]. For 3-cm waves, a pipe 0.5 by 1.0 in. is common. Waves longer than 10 cm require pipes impractically large; 1-m waves, for instance, require a pipe $2\frac{1}{2}$ ft wide.

ENERGY TRANSFER

Energy transfer in a waveguide is also related to its cross-sectional area. Power flow per unit area in the forward direction is derived from the Poynting-vector conception (page 41), which is its component in the x (forward) direction in watts per square unit area. This is expressed as

$$P = \frac{Z_w H_0^2}{2} axb \tag{6}$$

where Z_w = wave impedance

H_0 = maximum magnetic field laid down by the source

The wave impedance is the ratio of the electric- to the magnetic-field strengths in the transverse direction. For air-filled waveguides the wave impedance is equal to 120 multiplied by a factor depending upon the wavelength of the source and the size of the waveguide.

GUIDE WAVELENGTH

Now the a and b dimensions are related to the true waveguide length λ_g; it in turn is different from the free-space wavelength λ_a. For a rectangular guide operating in the $TE_{0,1}$ mode,

$$\lambda_g = \frac{\lambda_a}{\sqrt{1 - (\lambda_a/2b)^2}} \tag{7}$$

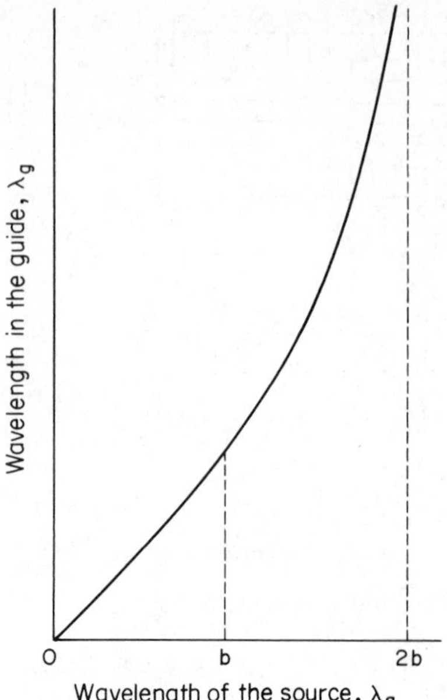

Fig. 3-23 EFFECTIVE WAVEGUIDE WAVELENGTH

For example, if a 10-cm magnetron is coupled to a waveguide operating in the $TE_{0,1}$ mode having a 7-cm cross-sectional dimension, the wavelength in the guide will be

$$\lambda_g = \frac{10}{\sqrt{1 - 10/14^2}} = 14.3 \text{ cm}$$

We can plot this relationship of the source and the true waveguide wavelength in Fig. 3-23.

Note that when λ_a is much smaller than $2b$, $\lambda_g = \lambda_a$. As λ_a approaches $2b$, λ_g increases rapidly. Above $2b$, λ_a produces an imaginary quantity indicating that propagation ceases. This is compatible with the cutoff wavelength in the $TE_{0,1}$ mode of a rectangular pipe, because $\lambda_c = 2b$. Thus, equation (7) becomes

$$\lambda_g = \frac{\lambda_a}{\sqrt{1 - \lambda_a/\lambda_c^2}} \tag{8}$$

which is valid for every mode in any guide of any cross section provided the value of λ_c corresponds to the mode and cross section being used. This is seen in Table 3-1 for a number of rectangular and circular guides.

TABLE 3-1 CUTOFF WAVELENGTHS OF RECTANGULAR AND CIRCULAR WAVEGUIDES

Rectangular		*Circular*	
Mode	λ_c	*Mode*	λ_c
$TE_{0,1}$	2b	$TE_{1,1}$	1.71 d
$TE_{1,1}$ or $TM_{1,1}$	$\frac{2}{\sqrt{(1/a)^2 + (1/b)^2}}$	$TM_{0,1}$	1.31 d
$TE_{0,2}$	b	$TE_{0,1}$	0.82 d
$TE_{1,0}$	2a	$TE_{2,1}$	1.03 d
$TE_{m,n}$ or $TM_{m,n}$	$\frac{2}{\sqrt{(m/a)^2 + (n/b)^2}}$	$TM_{1,1}$	0.82 d

The symbol d represents the diameter of the circular pipe.

WAVEGUIDE IMPEDANCE

To be more specific, the waveguide impedance referred to on page 000 can be expressed in terms of wavelengths, since it depends upon the ratio of the complex transverse electric-field intensity to its corresponding magnetic field intensity at any given transverse plane along the guide.

If we correctly terminate a waveguide so that all the transmitted power is absorbed,

$$Z_w = \begin{cases} 120\,\pi \dfrac{\lambda_g}{\lambda_c} & \text{for all TE modes} \\ 120\,\pi \dfrac{\lambda_a}{\lambda_g} & \text{for all TM modes} \end{cases}$$

Thus Z_w is the waveguide analog of the characteristic impedance of a conventional transmission line.

When a waveguide is not terminated correctly, a portion of the energy in the field incident on the termination will be reflected and standing waves will be set up in the guide. It is then that the ratio of the transverse electric and magnetic field is no longer equal to the waveguide impedance but varies along the guide. The impedance of an angular distance β_s from the termination can be shown to be

$$Z_s = \frac{Z_0 + j\,Z_w \times \tan \beta gs}{Z_w + j\,Z_0 \times \tan gs}$$

where Z_0 is the ratio of the complex electric and magnetic field at the load.

Note that this expression is exactly the same as for the ordinary transmission line, and, in fact, if the impedances that appear in the expression are interpreted as the ratio of electric- and magnetic-field intensities, instead of as voltage and current ratios, the equation has the same meaning. This coincidence makes it possible to use an ordinary transmission line as the equivalent circuit for a waveguide. To do this the mode conditions must be observed in all cases:

1. The wavelength employed in the equivalent transmission line must equal λ_g rather than λ_a.

2. Each mode being studied must be represented by a separate transmission line. Thus a waveguide may have several equivalent circuits, depending upon the modes at which it operates.

DESIGN FACTORS AND OPERATION

Engineering a waveguide obviously requires a combined knowledge of the theoretical and practical aspects of circuitry operation, construction, and manufacture. Included in this are melding of the source, load, and accessory characteristics, plus a wide knowledge of measurement techniques. These will be covered as follows: waveguide construction, Chapter 14; accessories, Chapter 5; microwave sources, Chapter 7; circuitry, Chapters 6, 7, and 9; measurement, Chapters 10 and 11; equipment and applications, Chapters 12 and 13.

Waveguide Characteristics and Construction

4

GENERAL

Waveguides, rectangular or circular, are categorized electrically and physically by a number of characteristics. Ridged and flexible waveguides represent special application and construction described in this section.

Generalized structural data can be categorized as follows:

1. Structural dimensions: OD, *a*, *b*, wall thickness.
2. Operating frequency range.
3. Mode of operation.
4. Natural wavelength λ_a.
5. Attenuation (dB/m).

For circular guides we list the following:

1. Structural dimensions: ID, diameter, wall thickness.
2. Operating frequency range.
3. Modes of operation: TE, TM.
4. Wavelengths: cutoff, natural.
5. Attenuation.

Table 4-1 lists these characteristics in a number of commonly used rectangular commercial waveguides. Figure 4-1 gives common data on circular guides. Table 4-1 illustrates the attenuation factors for rectangular guides in the $TE_{0,1}$ mode. Figure 4-1 illustrates a number of typical commercial waveguides. Greater detail upon their size relationships and how they are used in operational setups appears in Chapter 12.

RECTANGULAR WAVEGUIDES AND CONSTRUCTION

GENERAL

The rectangular waveguide simply used to carry signal or power energy is based upon the width, *a*, and height, *b*, dimensions. Construction, however, is not that simple because a guide rarely consists of a simple straight section: it usually contains bends, twists, flanges, and junctions, each of which has many variations, depending upon the application and the frequency being transmitted.

In descriptively treating these components it should be remembered that they are not signal-processing devices; they are hardware pieces that merely pass the signal while being physically compatible with the character of the main waveguide. All joining of straight and specific nonlinear assemblies is achieved by bolted, abutted flanges incorporated on both ends of each section.

General waveguide characteristics for government equipment are covered under MIL-W-85C. For commercial and design data, EIA, JAN, and International-type listings are given in Table 4-1.

LINEAR STRUCTURES

FLANGES—JOINTS

The most important item in waveguide construction is the connecting flange. We see

TABLE 4-1 RECTANGULAR WAVEGUIDE CHARACTERISTICS

EIA WG Designation WR( )	Recommended Operating Range for TE_{10} Mode		Cut-off for TE_{10} Mode		Range in $\frac{2\lambda}{\lambda c}$	Range in $\frac{\lambda g}{\lambda}$	Theoretical cw power rating lowest to highest frequency megowatts	Theoretical attenuation lowest to highest frequency (dB/100 ft.)	Material Alley	JAN WG Designation RG( )/U	JAN FLANGE DESIGNATION		EIA WG Designation WR( )	DIMENSIONS (inches)				Wall Thickenss Normal
	Frequency (kmc/sec)	Wavelength (cm)	Frequency (kmc/sec)	Wavelength (cm)							Choke UG( )/U	Cover UG( )/U		Inside	Tel	Outside	Tel	
2300	0.32-0.49	93.68-61.18	0.256	116.84	1.60-1.05	1.68-1.17	153.0-212.0	0.051-0.031	Alum.				2300	23.000-11.500	±0.020	23.250-11.750	±0.020	0.125
2100	0.35-0.53	85.65-56.56	0.281	106.68	1.62-1.06	1.68-1.18	120.0-173.0	0.054-0.034	Alum.				2100	21.000-10.500	±0.020	21.250-10.750	±0.020	0.125
1800	0.41-0.625	73.11-47.96	0.328	91.44	1.60-1.05	1.67-1.18	93.4-131.9	0.056-0.038	Alum.	201			1800	18.000-9.000	±0.020	18.250-9.250	±0.020	0.125
1500	0.49-0.75	61.18-39.97	0.393	76.20	1.61-1.05	1.62-1.17	67.6-93.3	0.069-0.050	Alum.	202			1500	15.000-7.500	±0.015	15.250-7.750	±0.015	0.125
1150	0.64-0.96	46.84-31.23	0.513	58.42	1.60-1.07	1.82-1.18	35.0-53.8	0.128-0.075	Alum.	203			1150	11.500-5.750	±0.015	11.750-6.000	±0.015	0.125
975	0.75-1.12	39.95-26.76	0.605	49.53	1.61-1.08	1.70-1.19	27.0-38.5	0.137-0.095	Alum.	204			975	9.750-4.875	±0.010	10.000-5.125	±0.010	0.125
770	0.96-1.45	31.23-20.67	0.766	39.12	1.60-1.06	1.66-1.18	17.2-24.1	0.201-0.136	Alum.	205			770	7.700-3.850	±0.005	7.950-4.100	±0.005	0.125
650	1.12-1.70	26.76-17.63	0.908	33.02	1.62-1.07	1.70-1.18	11.9-17.2	0.317-0.212 0.269-0.178	Brass Alum.	69 103		417A 418A	650	6.500-3.250	±0.005	6.660-3.410	±0.005	0.080
510	1.45-2.20	20.67-13.62	1.157	25.91	1.60-1.05	1.67-1.18	7.5-10.7						510	5.100-2.550	±0.005	5.260-2.710	±0.005	0.080
430	1.70-2.60	17.63-11.53	1.372	21.84	1.61-1.06	1.70-1.18	5.2-7.5	0.588-0.385 0.501-0.330	Brass Alum.	104 105		435A 437A	430	4.300-2.150	±0.005	4.460-2.310	±0.005	0.080
340	2.20-3.30	13.63-9.08	1.736	17.27	1.58-1.05	1.78-1.22	3.1-4.5	0.877-0.572 0.751-0.492	Brass Alum.	112 113		553 554	340	3.400-1.700	±0.005	3.560-1.860	±0.005	0.080
284	2.60-3.95	11.53-7.59	2.078	14.43	1.60-1.05	1.67-1.17	2.2-3.2	1.102-0.752 0.940-0.641	Brass Alum.	48 75	54A 585	53 584	284	2.840-1.340	±0.005	3.000-1.500	±0.005	0.080
229	3.30-4.90	9.08-6.12	2.577	11.63	1.56-1.05	1.62-1.17	1.6-2.2						229	2.290-1.145	±0.005	2.418-1.273	±0.005	0.064
187	3.95-5.85	7.59-5.12	3.152	9.510	1.60-1.08	1.67-1.19	1.4-2.0	2.08-1.44 1.77-1.12	Brass Alum.	49 95	148B 406A	149A 407	187	1.872-0.872	±0.005	2.000-1.000	±0.005	0.064
159	4.90-7.05	6.12-4.25	3.711	8.078	1.51-1.05	1.52-1.19	0.79-1.0						159	1.590-0.795	±0.004	1.718-0.923	±0.004	0.064
137	5.85-8.20	5.12-3.66	4.301	6.970	1.47-1.05	1.48-1.17	0.56-0.71	2.87-2.30 2.45-1.94	Brass Alum.	50 106	343A 440A	344 441	137	1.372-0.622	±0.004	1.500-0.750	±0.004	0.064
112	7.05-10.00	4.25-2.99	5.259	5.700	1.49-1.05	1.51-1.17	0.35-0.46	4.12-3.21 3.50-2.74	Brass Alum.	51 68	52A 137A	51 138	112	1.122-0.497	±0.004	1.250-0.625	±0.004	0.064
90	8.20-12.40	3.66-2.42	6.557	4.572	1.60-1.06	1.68-1.18	0.20-0.29	6.45-4.48 5.49-3.83	Brass Alum.	52 67	40A 136A	39 135	90	0.900-0.400	±0.003	1.000-0.500	±0.003	0.050
75	10.00-15.00	2.99-2.00	7.868	3.810	1.57-1.05	1.64-1.17	0.17-0.23						75	0.750-0.375	±0.003	0.850-0.475	±0.004	0.050
62	12.4-18.00	2.42-1.66	9.486	3.160	1.53-1.05	1.55-1.18	0.12-0.16	9.51-8.31 – – 6.14-5.36	Brass Alum. Silver	91 – 107	541 – –	419 – –	62	0.622-0.311	±0.0025	0.702-0.391	±0.003	0.040
51	15.00-22.00	2.00-1.36	11.574	2.590	1.54-1.05	1.58-1.18	0.080-0.107						51	0.510-0.255	±0.0025	0.590-0.335	±0.003	0.040
42	18.00-26.50	1.66-1.13	14.047	2.134	1.56-1.06	1.60-1.18	0.043-0.058	20.7-14.8 17.6-12.6 13.3-9.5	Brass Alum. Silver	53 121 66	596 598 –	595 597 –	42	0.420-0.170	±0.0020	0.500-0.250	±0.003	0.040
34	22.00-33.00	1.36-0.91	17.328	1.730	1.57-1.05	1.62-1.18	0.034-0.048						34	0.340-0.170	±0.0020	0.420-0.250	±0.003	0.040
28	26.50-40.00	1.13-0.75	21.081	1.422	1.59-1.05	1.65-1.17	0.022-0.031	– – – – 21.9-15.0	Brass Alum. Silver	– – 96	600 – –	599 – –	28	0.280-0.140	±0.0015	0.360-0.220	±0.002	0.040
22	33.00-50.00	0.91-0.60	26.342	1.138	1.60-1.05	1.67-1.17	0.014-0.020	– – 31.0-20.9	Brass Silver	– 97		383 –	22	0.224-0.112	±0.0010	0.304-0.192	±0.002	0.040
19	40.00-60.00	0.75-0.50	31.357	0.956	1.57-1.05	1.63-1.16	0.011-0.015						19	0.188-0.094	±0.0010	0.268-0.174	±0.002	0.040
15	50.00-75.00	0.60-0.40	39.863	0.752	1.60-1.06	1.67-1.17	0.0063-0.0090	– – 52.9-39.1	Brass Silver	– 98		385 –	15	0.148-0.074	±0.0010	0.228-0.154	±0.002	0.040
12	60.00-90.00	0.50-0.33	48.350	0.620	1.61-1.06	1.68-1.18	0.0042-0.0060	– – 93.3-52.2	Brass Silver	– 99		387 –	12	0.122-0.061	±0.0005	0.202-0.141	±0.002	0.040
10	75.00-110.00	0.40-0.27	59.010	0.508	1.57-1.06	1.61-1.18	0.0030-0.0041						10	0.100-0.050	±0.0005	0.180-0.130	±0.002	0.040
8	90.00-140.00	0.333-0.214	73.840	0.406	1.64-1.05	1.75-1.17	0.0018-0.0026	152-99	Silver	138	–	–	8	0.080-0.040	±0.0003	0.156 DIA	±0.001	–
7	110.00-170.00	0.272-0.176	90.840	0.330	1.64-1.06	1.77-1.18	0.0012-0.0017	163-137	Silver	136	–	–	7	0.065-0.0325	±0.00025	0.156 DIA	±0.001	–
5	140.00-220.00	0.214-0.136	115.750	0.259	1 65-1.05	1.78-1.17	0.00071-0.00107	308-193	Silver	135	–	–	5	0.051-0.0255	±0.00025	0.156 DIA	±0.001	–
4	170.00-260.00	0.176-0.115	137.520	0.218	1.61-1.05	1.69-1.17	0.00052-0.00075	384-254	Silver	137	–	–	4	0.043-0.0215	±0.00020	0.156 DIA	±0.001	–
3	220.00-325.00	0.136-0.092	173.280	0.173	1.57-1.06	1.62-1.18	0.00035-0.00047	512-348	Silver	139	–	–	3	0.034-0.0170	±0.00020	0.156 DIA	±0.001	–

Courtesy of Microwave Development Laboratories, Inc., Natick, Mass.

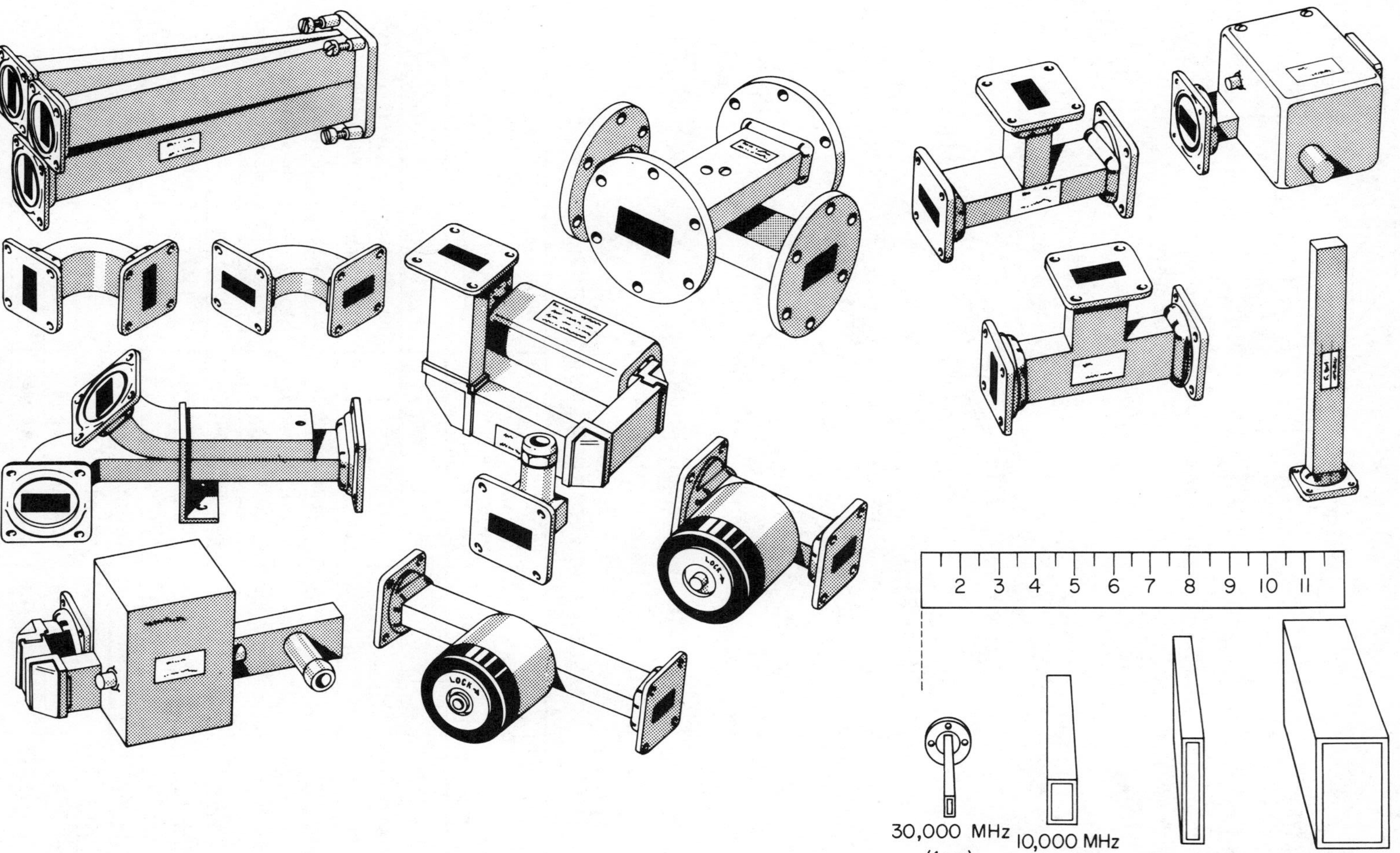

Fig. 4-1 COMMERCIAL WAVEGUIDES AND SIZE COMPARISONS

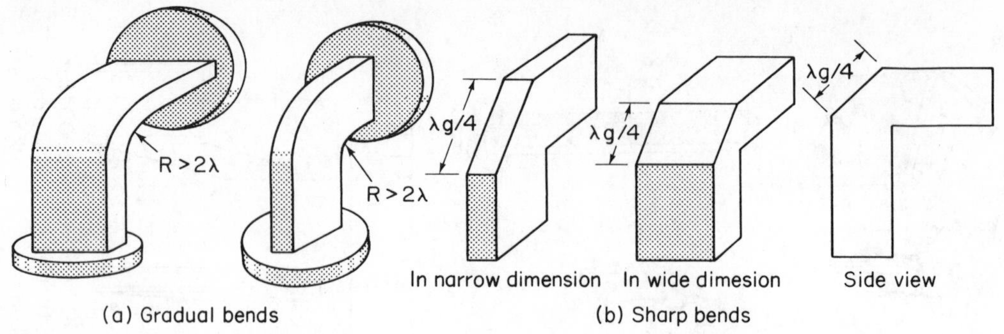

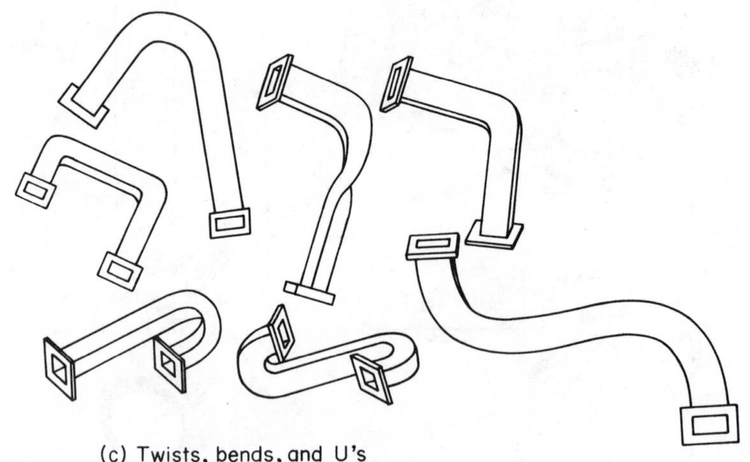

(c) Twists, bends, and U's

Fig. 4-2 FLANGE JOINTS

TABLE 4-2 STANDARD FLANGE DIMENSIONS

AN NO.	*Flange choke AN NO.*	*Use with waveguide AN NO.*	*RETMA-NO.*	*Dimensions of waveguide O.D.*	*Frequency range (kHz)*
UG-387/U		RG-99/U	WR 12	0.202 x 0.141	60.0-90-0
UG-385/U		RG-98/U	WR 15	0.228 x 0.154	50.0-75.0
UG-383/U		RG-97/U	WR 22	0.304 x 0.129	33.0-50.0
UG-599/U	UG-600/U	RG-96/U	WR 28	0.360 x 0.220	26.5-40.0
UG-595/U	UG-596/U	RG-53/U	WR 42	0.500 x 0.250	18.0-26.5
	UG-117/U	RG-53/U	WR 42	0.500 x 0.250	18.0-26.5
UG-419/U	UG-541/U	RG-91/U	WR 62	0.702 x 0.391	12.4-18.0
UG-39/U	UG-40A/U	RG-52/U	WR 90	$1 \times \frac{1}{2}$	8.2-12.4
UG-135/U	UG-136A/U	RG-67/U		$1 \times \frac{1}{2}$	8.2-12.4
UG-51/U	UG-52A/U	RG-51/U	WR 112	$1\frac{1}{4} \times \frac{5}{8}$	7.05-10.0
UG-344/U	UG-343A/U	RG-50/U	WR 137	$1\frac{1}{2} \times \frac{3}{4}$	5.85-8.20
UG-149A/U	UG-148B/U	RG-49/U	WR 187	2×1	3.95-5.85
UG-53/U	UG-54A/U	RG-48/U	WR 284	$3 \times 1\frac{1}{2}$	2.60-3.95
UG-584/U	UG-585/U	RG-75/U		$3 \times 1\frac{1}{2}$	2.60-3.95

this in Fig. 4-2, where sectionalized bends and twists are physically joined together.

Table 4-2 lists standardized flanges for use in conjunction with common waveguide dimensions.

Flange design and construction are quite precise, since both sides must be the same size and fit tightly together. Otherwise, irregularities at the joints will set up standing waves and cause loss of power; welding and machining to a few thousandths of an inch to prevent this are not uncommon. We see this precision in the dimensions specified on the rectangular flange on Fig. 4-3 and in the metal gasket of Fig. 4-3(b).

CHOKE JOINT

When it is necessary to disassemble joints for maintenance and repair, it is common to use a choke joint, shown in Fig. 4-4. In effect, the arrangement is a double flange with the two sections of waveguide series connected together by the zero impedance presented by a built-in half-wave line section. This is done on the flange at the left by slotting it a quarter-wavelength deep at the distance from the joint where the guide walls are joined. The right-hand flange is conventionally constructed.

The choke action simulates the use of rf chokes in lead-offs from power lines; in a waveguide the choke joint keeps the electromagnetic fields contained because of the high impedance offered by quarter-wavelength sections on each side of where the flanges are physically joined. Actually, this choke action reduces the critical nature of dimensions and allows separation of the abutment joint, say enough to insert a rubber gasket for pressurization. Losses are also reduced by the choke flange, and a well-designed joint may have less than 0.03-dB loss, while an unsoldered, machined, permanent joint may have losses over 0.05 dB. Actual dimensions and design data are shown in the commercial choke joints illustrated in Fig. 4-5.

The choke flange has a direct usage in the power transfer across the joint of most rotating radar antennas. Here we must separate the sections mechanically and join them electrically, so a circular waveguide with a $TM_{0,1}$ mode is used (Fig. 4-4).

However, since most radar systems use rectangular waveguides, rotation must be attained by insertion of a circular choke joint in the guide system (Fig. 4-6). Here we note that the rectangular guide is operated in the $TE_{0,1}$ mode (E lines only are shown). The E lines couple the energy from the rectangular guide into the circular guide and excite it in the $TM_{0,1}$ mode. This mode provides the required axial symmetry for rotating joints. At the top of the joint, the E lines couple the energy back into the rectangular guide that leads to the antenna where the guide is operating in the $TE_{0,1}$ mode.

Note that any circular guide that will carry the $TM_{0,1}$ mode is subject to being excited in the $TE_{1,1}$ mode, an undesirable mode for rotation. We see this in Fig. 3-19, since it would destroy the standing-wave ratio of certain angular positions. For this reason, mode-suppression rings (Fig. 4-6) are provided in the circular guide to suppress the $TE_{1,1}$ mode. In addition, a Teflon matching ring is used at each end of the joints to improve the impedance match between the circular and the rectangular waveguide sections.

JUNCTIONS—STUBS

T junctions connect one waveguide to the side of another and may be constructed by two methods:

1. An H-type junction going into the narrow dimension of the guide.
2. An E-type junction going into the broad side (Fig. 4-7).

These junctions, for the $TE_{0,1}$ mode, are of course intended to preserve the fields respectively associated with each dimension. In the H-type arrangement this is equivalent to a parallel circuit feeding signal to a wire line [Fig. 4-7(c)], by producing a short circuit at the junction (using the half-wave dimension) when introducing the incoming line signal at the far end of the connecting stub; it

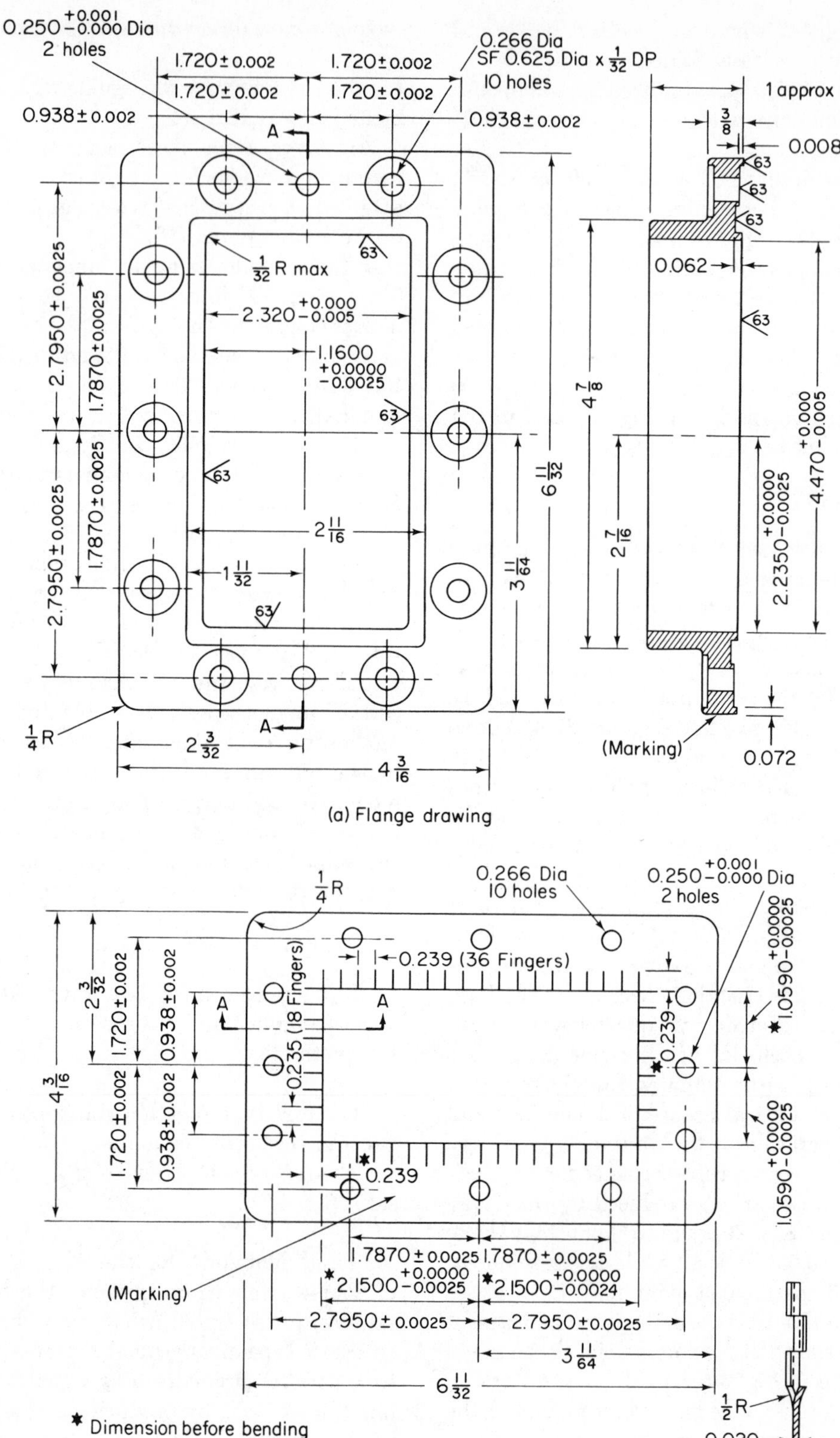

Fig. 4-3 RECTANGULAR FLANGE AND GASKET DIMENSIONS

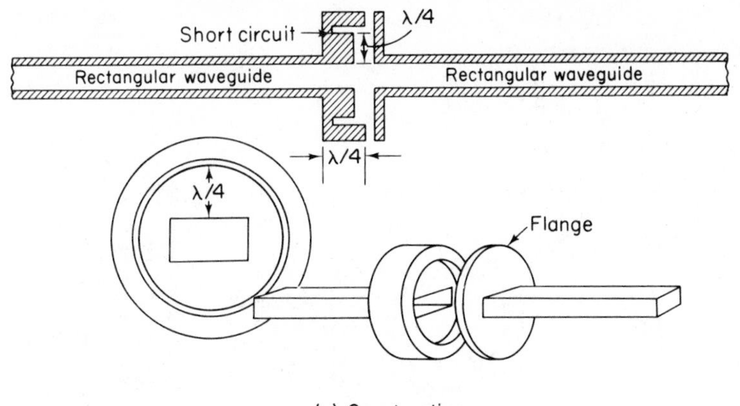

(a) Construction

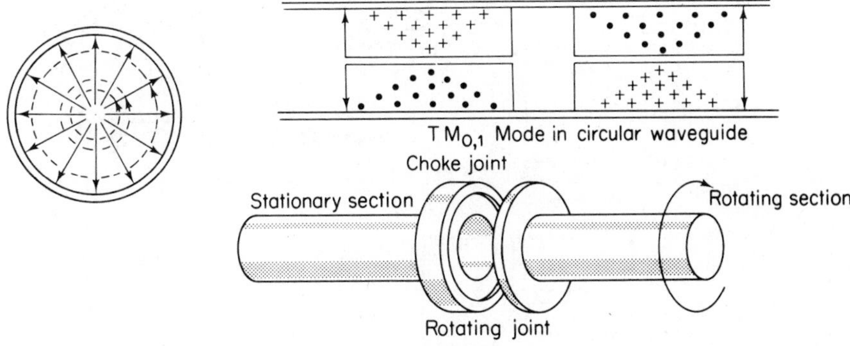

(b) Rotating choke joint and $TM_{0,1}$ mode

Fig. 4-4 CHOKE JOINT USE IN ROTATING STRUCTURE

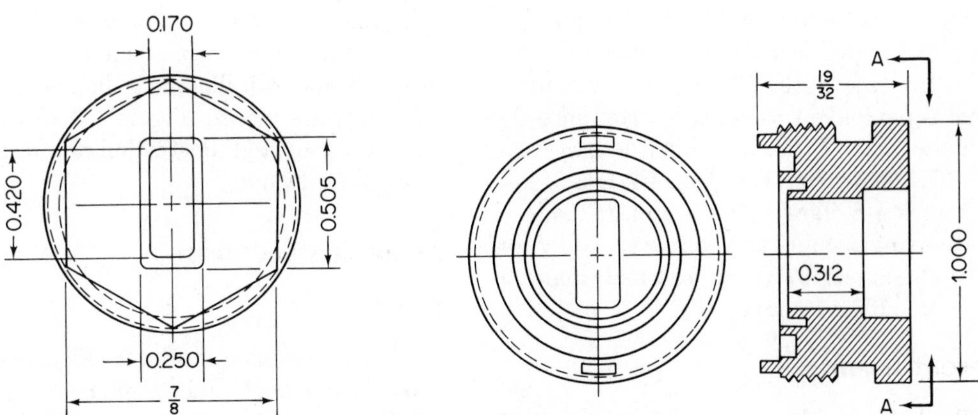

Fig. 4-5 CHOKE JOINT FLANGE DIMENSIONS

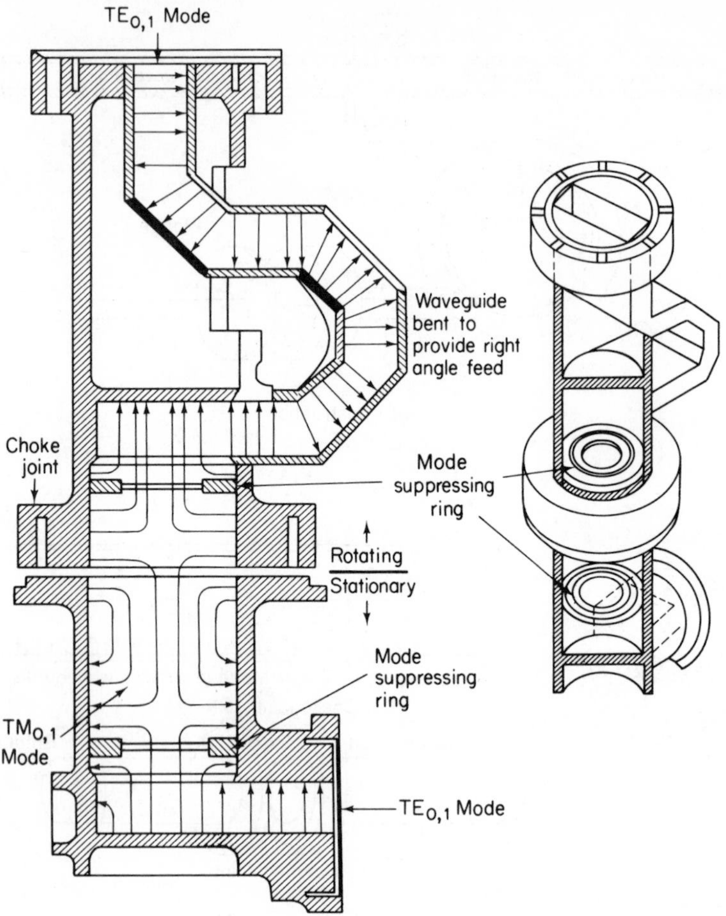

Fig. 4-6 ROUND-RECTANGULAR CHOKE JOINT TRANSITION

thus presents a short circuit or offers a means to inject a signal into the main line.

In the case of the E type, the introduced signal is equivalent to being inserted in series with the main E fields through a half-wave short-circuit connection [Fig. 4-7(f)].

Figure 4-8 illustrates E- and H-mode Y junctions plus three- and four-way configurations, all subject to circuit and stub manipulation as outlined above.

SUPPORT HARDWARE

To hold, guide, and otherwise support waveguide systems, a number of specialized mechanical devices must be used. Inspection of the generalized measurement setup shown in Fig. 4-9(a) will illustrate the need for various hardware pieces. Figure 4-9(b) is a commercial listing and illustration of a line of this type of hardware.

ANGULAR STRUCTURES

BENDS

The assemblies shown in Fig. 4-2(a) are typical E-plane and H-plane right-angle bends. Figure 4-2(b) shows a number of U, S, and angular bends.

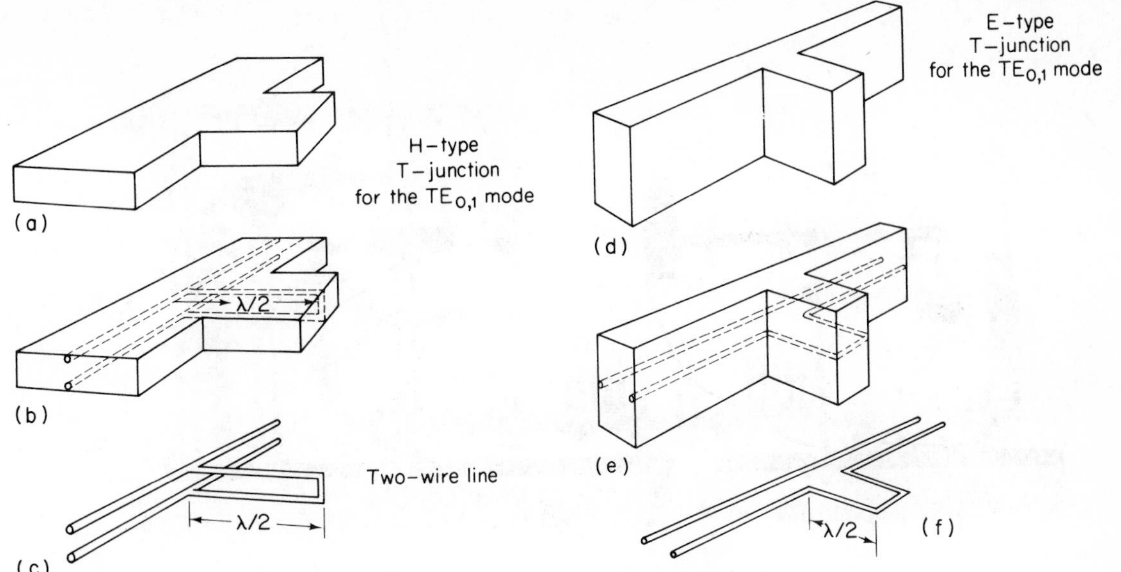

Fig. 4-7 *T* **JUNCTION CONSTRUCTION**

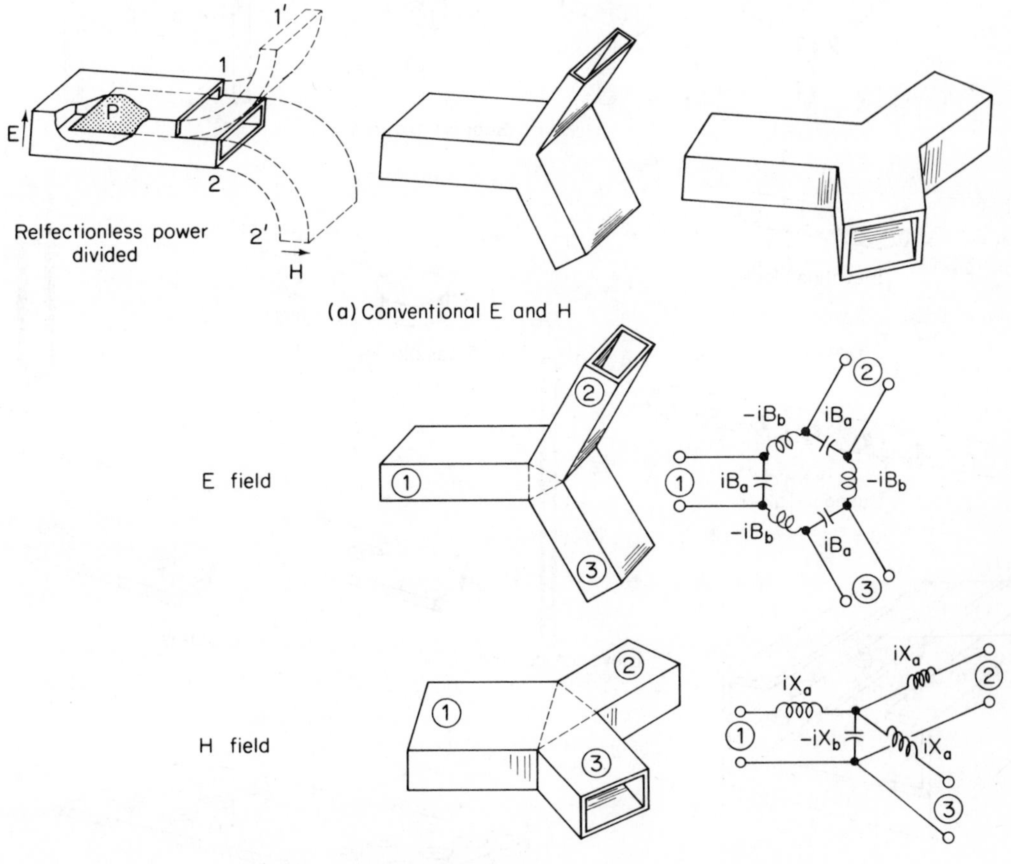

1′
1
P
E
2
2′
H
Relfectionless power
divided

(a) Conventional E and H

E field

H field

(b) Equivalent circuit

Fig. 4-8 TYPICAL *Y* **JUNCTION**

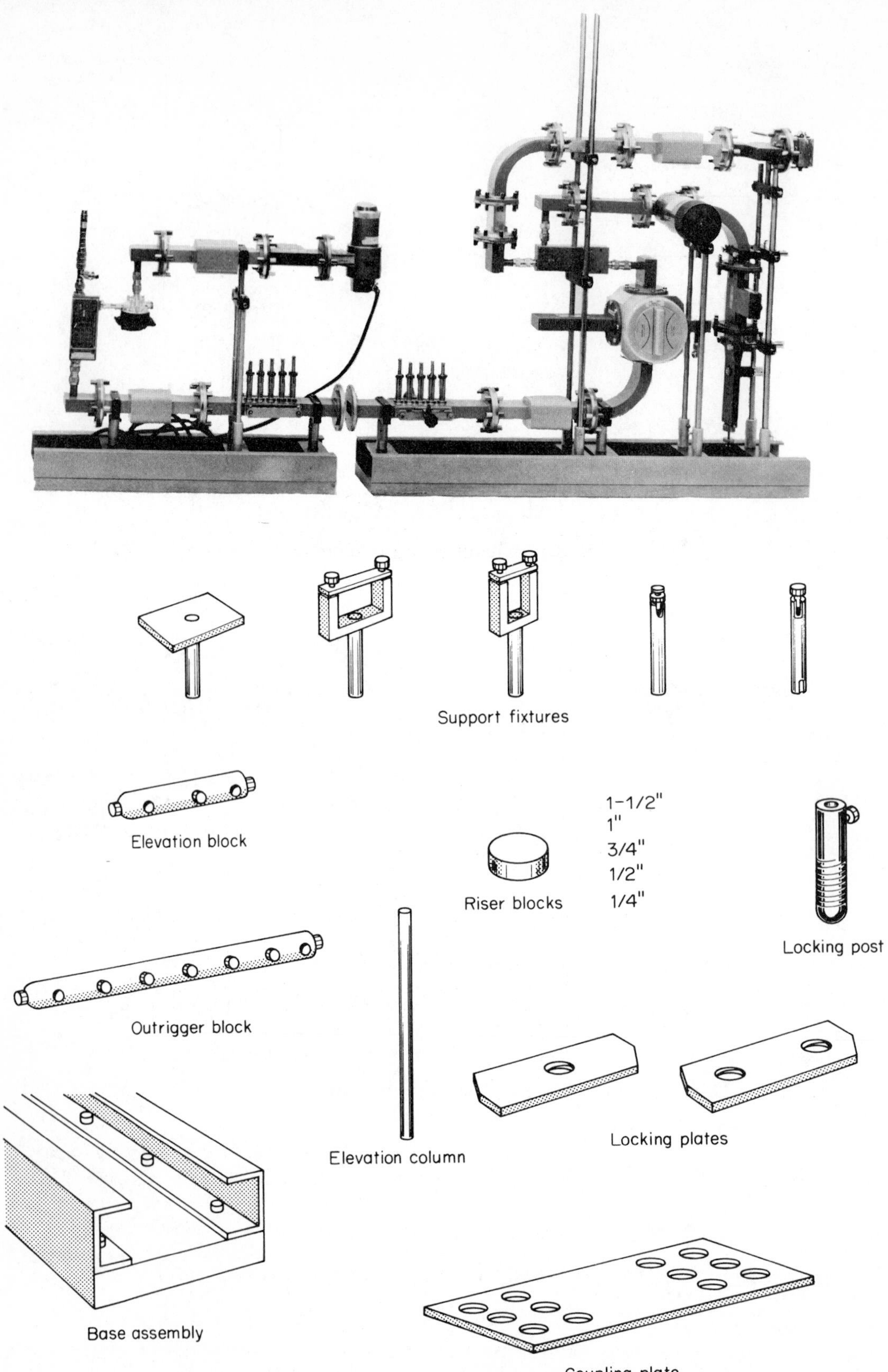

Fig. 4-9 STRUCTURAL SETUP ARRANGEMENT (Courtesy of Orbitronics, Inc.)

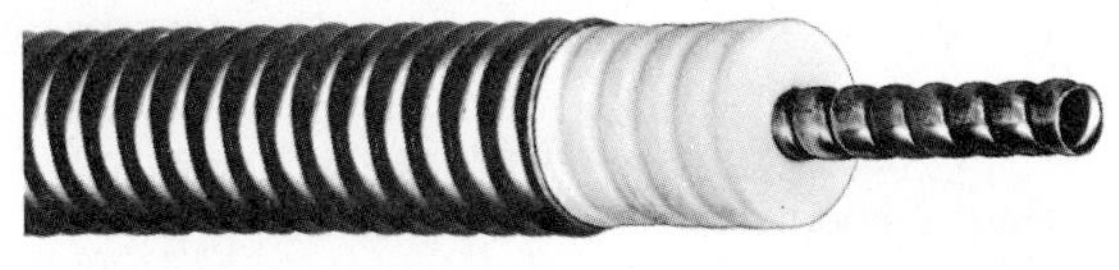

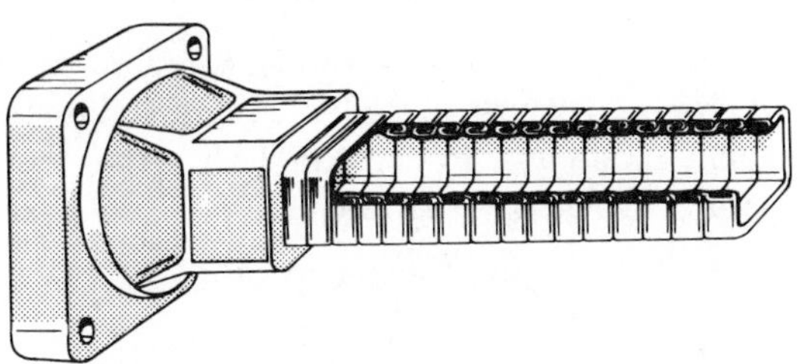

Fig. 4-10 FLEXIBLE WAVEGUIDE (Courtesy of Andrew Corp.)

TWISTS—FLEXIBLE UNITS

Figure 4-6 illustrates a simple flanged section of twisted waveguide used to rotate the magnetic field so that it is in the proper direction for matching. The mechanical twisting is graduated slowly enough so that reflections are at a minimum.

To permit using any special bend which an installation might require, flexible waveguides are manufactured like the one illustrated in Fig. 4-10. This unit is made of a spiral, wound of edge-interlocked ribbon brass, which maintains its internal waveguide dimension in any angular position. Interior-surface skin effect is reduced by chromium plating; the exterior is covered with rubber for protection against air and water.

RIDGED RECTANGULAR WAVEGUIDES

The ridged waveguide is a special design aimed at improving the bandwidth of a conventional rectangular structure; generally such guides are limited to coverage of about 40

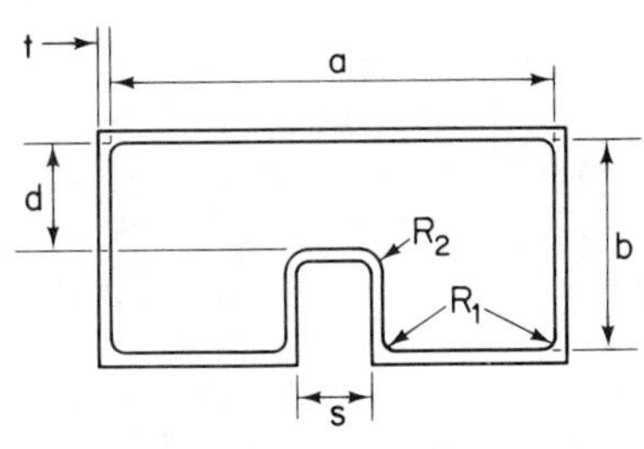

Single ridged

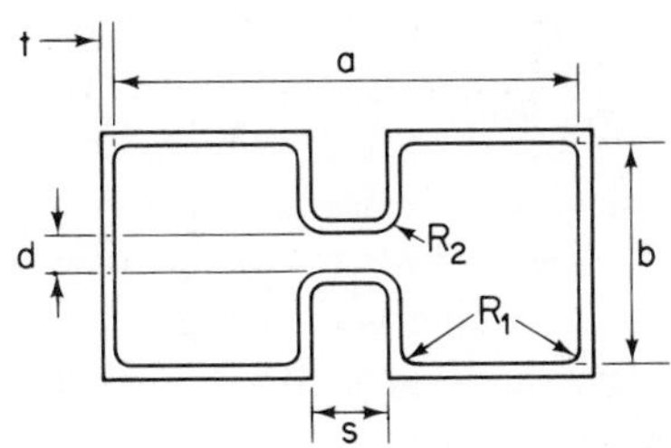

Double ridged

(a) Cross-sectional details

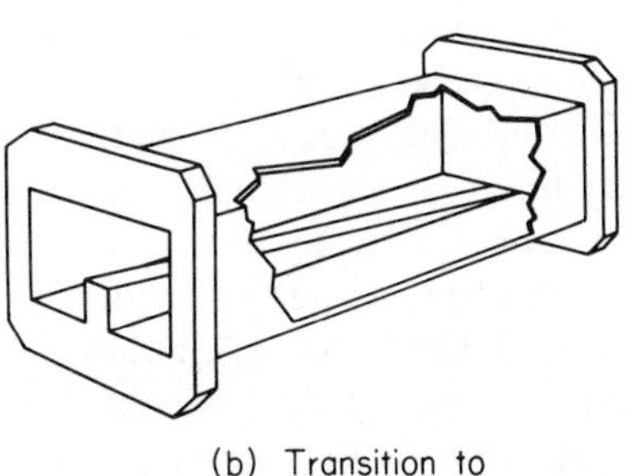

(b) Transition to rectangular

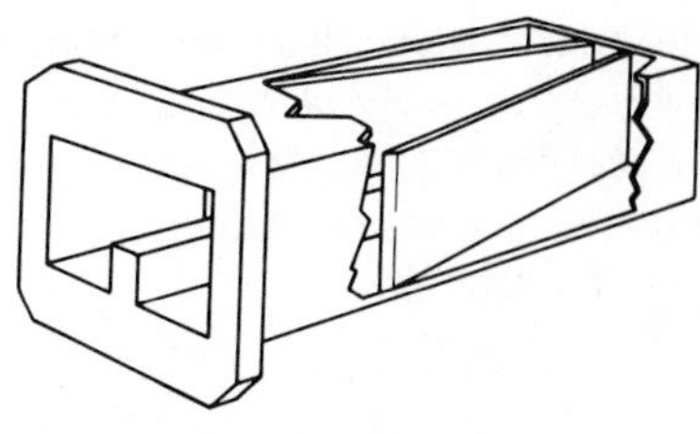

(c) Termination method

Fig. 4-11 RIDGED WAVEGUIDE CONSTRUCTION

per cent of their mean operating frequency.

Other characteristics of this variation are:

1. $TE_{1,0}$-mode cutoff frequency is lowered.
2. $TE_{2,0}$-mode cutoff frequency remains unchanged.
3. Attenuation per unit length increases with bandwidth.
4. Power-handling capacity is reduced.

Figure 4-11(a) shows typical commercial units, Fig. 4-11(b) the transition from conventional rectangular to ridged construction, and Fig. 4-11(c) ridge-guide termination.

WAVEGUIDE DESIGN

BASIC FACTS

The size of waveguide chosen for a given hardware design depends upon the operating frequency, the mode, and the amount of attenuation that can be tolerated. At a specific mode the guide must be large enough so that the cutoff wavelength is greater than the wavelength of the source. If, however, the guide has a mode so large that the cutoff wavelengths of higher modes are greater than λ_a, the higher mode may also be propagated. This is not usually advisable, mainly because it is usually impossible to efficiently extract power at the receiving end of a guide system unless the termination is fitted for one particular mode.

In radar sets the $TE_{0,1}$ mode is the most commonly used; here the best propagation is obtained if the b dimension exceeds $\lambda_g/2$. However, to exclude the $TE_{0,2}$ mode, b must be less than λ_z, so the a dimension may be as small as desired without barring the $TE_{0,1}$ mode. However, to exclude the $TE_{1,0}$ mode a should be less than $h_z/2$. Thus, the cross-sectional dimensions of a rectangular guide should lie in the ranges

$$0 < a < \lambda_{a/2} \tag{1}$$

$$\lambda_{a/2} < b < \lambda_a \tag{2}$$

If dimensions a and b lie between the specified limits, no mode except the $TE_{0,1}$ can be propagated, because the cutoff wavelengths of all higher modes are shorter than the operating wavelength. We note this by inspection of the $TE_{0,2}$ and $TE_{1,0}$ modes in Fig. 4-7. Here we see that if a and b are chosen as large as the inequalities shown in equations (1) and (2) permit, the cutoff wavelengths of the $TE_{1,1}$ mode are about $0.9h_g$ and, for still higher modes, the cutoff wavelengths are even smaller.

Typically, in practice, it has been found that a good choice yields

$$b = 0.7\lambda_a$$

$$a = 0.5b \approx 0.35\lambda_a$$

Similar conditions hold in circular waveguides, for here, in order to propagate the $TE_{1,1}$ mode and to exclude $TM_{0,1}$, the diameter must satisfy the inequality

$$0.586\lambda_a < d < 0.765\lambda_a$$

From the above it would seem that it is impossible to propagate a higher mode in any waveguide without also propagating all the modes with large cutoff wavelengths. This would be true if the polarization of the source were completely random and equally stimulating to all modes. But most sources are linearly polarized and contribute only very small energy to modes unlike their own in polarization; thus by properly orienting the source, the higher modes may be set up and propagated, while the lower modes, whose fields are polarized differently from the source, will be eliminated.

MECHANICAL

After establishing the a and b relationship and the best mode conditions, the designer can then consider constructional factors leading up to a specific waveguide design. The conventional approach is to start with the standardized types available for commercial and military usage. Table 4-1 lists these against IEC (International Electrotechnical Commission) and American JAN and EIA

TABLE 4-3 DIMENSION-MODE-CUTOFF FREQUENCY IN CIRCULAR WAVEGUIDES

IEC: R-140 Brit: WG-18 Amer: RG-91/U WR-62
Outside dimensions: 0.702 x 0.391 ± 0.002 inches
17.83 x 9.93 ± 0.05 millimetres
Inside dimensions: 0.6220 x 0.3110 ± 0.016 inches
15.799 x 7.899 ± 0.031 millimetres
a/b = 2.000
Cutoff wavelength = 2a = 3.160 centimetres
Cutoff frequency = 9487.705 Mc
Frequency tabulated: 11.00 to 19.20 GHz

f (GHz)	λ (cm)	λ_g (cm)	$1/\lambda_g$ (1/cm)	λ_g/λ	λ_g/λ	λ_g (in.)	$1/\lambda_g$ (1/in.)
11.000	2.7254	5.3849	0.18571	1.97582	0.50612	2.12003	0.47169
11.100	2.7008	5.2026	0.19221	1.92629	0.51913	2.04827	0.48822
11.200	2.6767	5.0362	0.19856	1.88149	0.53149	1.98277	0.50435
11.300	2.6530	4.8836	0.20477	1.84075	0.54326	1.92267	0.52011
11.400	2.6298	4.7428	0.21085	1.80352	0.55447	1.86725	0.53555
11.500	2.6069	4.6125	0.21680	1.76934	0.56518	1.81593	0.55068
11.600	2.5844	4.4913	0.22265	1.73784	0.57543	1.76823	0.56554
11.700	2.5623	4.3783	0.22840	1.70871	0.58524	1.72374	0.58014
11.800	2.5406	4.2725	0.23405	1.68169	0.59464	1.68210	0.59450
11.900	2.5193	4.1733	0.23962	1.65654	0.60367	1.64302	0.60864
12.000	2.4983	4.0799	0.24511	1.63307	0.61234	1.60624	0.62257
12.100	2.4776	3.9917	0.25052	1.61112	0.62069	1.57155	0.63631
12.200	2.4573	3.9084	0.25586	1.59053	0.62872	1.53876	0.64988
12.300	2.4373	3.8295	0.26113	1.57119	0.63646	1.50768	0.66327
12.400	2.4177	3.7546	0.26634	1.55298	0.64392	1.47819	0.67650
12.500	2.3983	3.6834	0.27149	1.53580	0.65113	1.45014	0.68959
12.600	2.3793	3.6155	0.27659	1.51956	0.65808	1.42343	0.70253
12.700	2.3606	3.5508	0.28163	1.50420	0.66481	1.39794	0.71534
12.800	2.3421	3.4889	0.28662	1.48963	0.67131	1.37359	0.72802
12.900	2.3240	3.4297	0.29157	1.47581	0.67760	1.35029	0.74058
13.000	2.3061	3.3730	0.29647	1.46266	0.68368	1.32797	0.75303
13.100	2.2885	3.3187	0.30133	1.45015	0.68958	1.30656	0.76537
13.200	2.2712	3.2664	0.30614	1.43823	0.69530	1.28600	0.77760
13.300	2.2541	3.2162	0.31092	1.42685	0.70084	1.26624	0.78974
13.400	2.2373	3.1679	0.31566	1.41599	0.70622	1.24722	0.80178
13.500	2.2207	3.1214	0.32037	1.40560	0.71144	1.22890	0.81374
13.600	2.2044	3.0765	0.32504	1.39566	0.71651	1.21124	0.82560
13.700	2.1883	3.0332	0.32968	1.38614	0.72143	1.19419	0.83739
13.800	2.1724	2.9914	0.33429	1.37701	0.72621	1.17773	0.84909
13.900	2.1568	2.9510	0.33887	1.36825	0.73086	1.16182	0.86072
14.000	2.1414	2.9119	0.34342	1.35983	0.73538	1.14642	0.87228
14.100	2.1262	2.8741	0.34794	1.35175	0.73978	1.13152	0.88377
14.200	2.1112	2.8374	0.35244	1.34397	0.74407	1.11709	0.89519
14.300	2.0965	2.8019	0.35691	1.33648	0.74823	1.10310	0.90654
14.400	2.0819	2.7674	0.36135	1.32927	0.75229	1.08952	0.91783

(a) 3 cm.

IEC: R-40 Brit: WG-11A Amer: WR-229
Outside dimensions: 2.418 x 1.273 ± 0.005 inches
61.42 x 32.33 ± 0.12 millimetres
Inside dimensions: 2.2900 x 1.1450 ± 0.0046 inches
58.17 x 29.083 ± 0.12 millimetres
a/b = 2.000
Cutoff wavelength = 2a = 11.634 centimetres
Cutoff frequency = 2576.865 Mc
Frequency tabulated: 3.000 to 5.100 GHz

f (GHz)	λ (cm)	λ_g (cm)	$1/\lambda_g$ (1/cm)	λ_g/λ	λ/λ_g	λ_g (in.)	$1/\lambda_g$ (1/in.)
3.000	9.9931	19.5196	0.05123	1.95331	0.51195	7.68487	0.13013
3.050	9.8293	18.3768	0.05442	1.86960	0.53487	7.23496	0.13822
3.100	9.6707	17.3991	0.05747	1.79915	0.55582	6.85003	0.14598
3.150	9.5172	16.5499	0.06042	1.73894	0.57506	6.51569	0.15348
3.200	9.3685	15.8030	0.06328	1.68682	0.59283	6.22164	0.16073
3.250	9.2244	15.1392	0.06605	1.64121	0.60930	5.96031	0.16778
3.300	9.0846	14.5440	0.06876	1.60095	0.62463	5.72600	0.17464
3.350	8.9490	14.0063	0.07140	1.56512	0.63893	5.51430	0.18135
3.400	8.8174	13.5173	0.07398	1.53302	0.65231	5.32176	0.18791
3.450	8.6896	13.0699	0.07651	1.50408	0.66486	5.14563	0.19434
3.500	8.5655	12.6585	0.07900	1.47785	0.67666	4.98368	0.20065
3.550	8.4449	12.2786	0.08144	1.45397	0.68777	4.83409	0.20686
3.600	8.3276	11.9262	0.08385	1.43213	0.69826	4.69535	0.21298
3.650	8.2135	11.5981	0.08622	1.41208	0.70817	4.56620	0.21900
3.700	8.1025	11.2918	0.08856	1.39361	0.71756	4.44557	0.22494
3.750	7.9945	11.0047	0.09087	1.37654	0.72646	4.33257	0.23081
3.800	7.8893	10.7351	0.09315	1.36072	0.73491	4.22641	0.23661
3.850	7.7868	10.4811	0.09541	1.34601	0.74294	4.12643	0.24234
3.900	7.6870	10.2414	0.09764	1.33230	0.75058	4.03205	0.24801
3.950	7.5897	10.0146	0.09985	1.31950	0.75786	3.94276	0.25363
4.000	7.4948	9.7996	0.10204	1.30752	0.76481	3.85812	0.25919
4.050	7.4023	9.5955	0.10422	1.29628	0.77144	3.77774	0.26471
4.100	7.3120	9.4012	0.10637	1.28572	0.77777	3.70127	0.27018
4.150	7.2239	9.2162	0.10851	1.27579	0.78383	3.62841	0.27560
4.200	7.1379	9.0396	0.11062	1.26641	0.78963	3.55888	0.28499
4.250	7.0539	8.8708	0.11273	1.25756	0.79519	3.49243	0.28633
4.300	6.9719	8.7093	0.11482	1.24920	0.80052	3.42885	0.29164
4.350	6.8918	8.5546	0.11690	1.24127	0.80563	3.36794	0.29692
4.400	6.8135	8.4061	0.11896	1.23375	0.81053	3.30951	0.30216
4.450	6.7369	8.2636	0.12101	1.22662	0.81525	3.25340	0.30737
4.500	6.6621	8.1266	0.12305	1.21984	0.81978	3.19946	0.31255
4.550	6.5888	7.9948	0.12508	1.21338	0.82414	3.14756	0.31771
4.600	6.5172	7.8678	0.12710	1.20723	0.82834	3.09757	0.32283

(b) 10 cm.

type numbers. This listing gives the frequency range plus the a, b, and inner dimensions.

For more specific data, tables that include more specific dimensions, frequencies, and wavelength information are compiled for each type. We illustrate this in Table 4-3 for typical 3 and 10 radar guides (with an a/b ratio of 2:1), where details extend into the waveguide machining tolerances on both outside and inside dimensions. Also provided for specific frequencies are optical and waveguide wavelengths plus their ratios and inverse numbers in both inches and centimeters.

ATTENUATION

Primarily, attenuation in a waveguide is caused by conductor losses plus dielectric losses. Conductor losses are caused mainly by resistivity of material within the skin depth. Skin depth is defined as the depth at which the current density has fallen to $1/e$ (or 0.3679) of its surface value. Its value is given by

$$\delta = \frac{1}{2\pi}\sqrt{\frac{\lambda_c \rho}{30\mu}} \tag{1}$$

where δ = skin depth

λ_c = free-space wavelength

ρ = resistivity of conductor

μ = permeability of conductor

The attenuation factor can be defined as the relation between attenuation (in dB/ft) of a waveguide of given dimensions to an electromagnetic wave of a given frequency and known modal pattern. This factor can be given for $TE_{1,0}$ dominant-mode transmission in an air-dielectric, rectangular, copper waveguide as follows:

$$\alpha_c = \frac{0.01107}{x^{3/2}} \frac{\frac{1}{2}\frac{x}{y}\left(\frac{f}{f_{c0}}\right)^{3/2} + \left(\frac{f}{f_{c0}}\right)^{-1/2}}{\sqrt{\left(\frac{f}{f_{c0}}\right)^2 - 1}} \text{ dB/ft} \tag{2}$$

where x = long inner dimension of guide (in.)

y = short inner dimension of guide (in.)

f = operating frequency

f_{c0} = cutoff frequency

The 0.01107 constant is valid for copper. For any other material, merely multiply the constant by the ratio of the resistivity of the new material to the resistivity of copper:

$$\text{Constant} = (\rho_m/\rho_c) \times 0.01107 \tag{3}$$

where ρ_m = resistivity of guide material

ρ_c = resistivity of copper

$= 1.72 \times 10^{-6}$ Ω/cm

Dielectric losses are caused by imperfections in the dielectric material that fills the waveguides. In the case of an air dielectric, the dielectric losses are negligible, but for a waveguide filled with a dielectric other than air, they can be substantial. The dielectric losses are due to the complex nature of the dielectric constant of an imperfect dielectric.

Table 4-4 presents a generalized summary of waveguide attenuation in terms of physical dimensions and wall material.

POWER-CARRYING CAPACITY OF WAVEGUIDES

The power capacity of waveguides is limited only by the breakdown field strength of the dielectric filling the waveguide. For a rectangular waveguide the power at the breakdown field strength is given by

$$P_{bd} = E^2_{\max}(6.63 \times 10^{-4}) \cdot x \cdot y \cdot \frac{\lambda_c}{\lambda_g} \quad \text{watts} \tag{4}$$

where $E_{\max}$ = breakdown field strength (V/cm)

x and y = inner dimensions of waveguide (cm)

λ_c = free-space wavelength

λ_g = waveguide wavelength

TABLE 4-4 WAVEGUIDE ATTENUATION CHARACTERISTICS

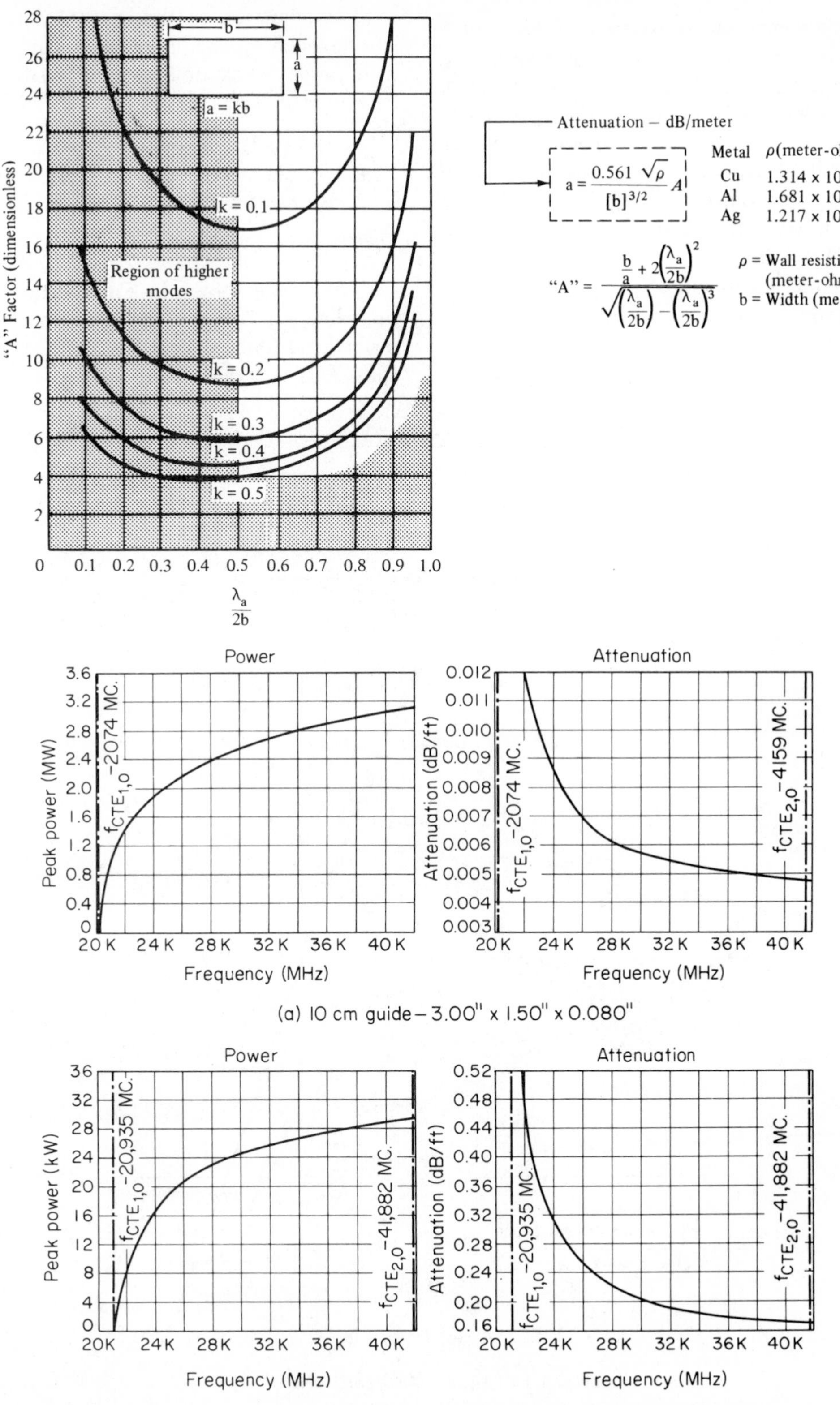

Fig. 4-12 PEAK POWER HANDLING CAPABILITY IN RECTANGULAR WAVEGUIDE

The power-carrying capacity of a waveguide can be increased substantially by pressurizing the waveguide and filling it with certain gases.

Thus we also see that the larger inside diameter of a waveguide increases the breakdown voltage capability; for instance, a 10-cm (S-band) low-power waveguide might be $\frac{1}{2}$ by 2 in. while a high-power structure would be 1.5 by 3 in.

Plotting an analog of equation (4) shows how peak power capabilities vary with frequency in a 1.5- by 3-in. waveguide (Fig. 4-12).

WAVEGUIDE EXCITATION

The structure at the end of a waveguide is influenced by the injection of signal power. This process may be accomplished by one of three methods:

1. By electric fields.
2. By magnetic fields.
3. By electromagnetic fields.

ELECTROSTATIC FIELDS

The items shown in Fig. 4-13(a) are set up when an rf signal is applied to a small probe or antenna inserted in a waveguide. As the fields alternate and set up concurrent

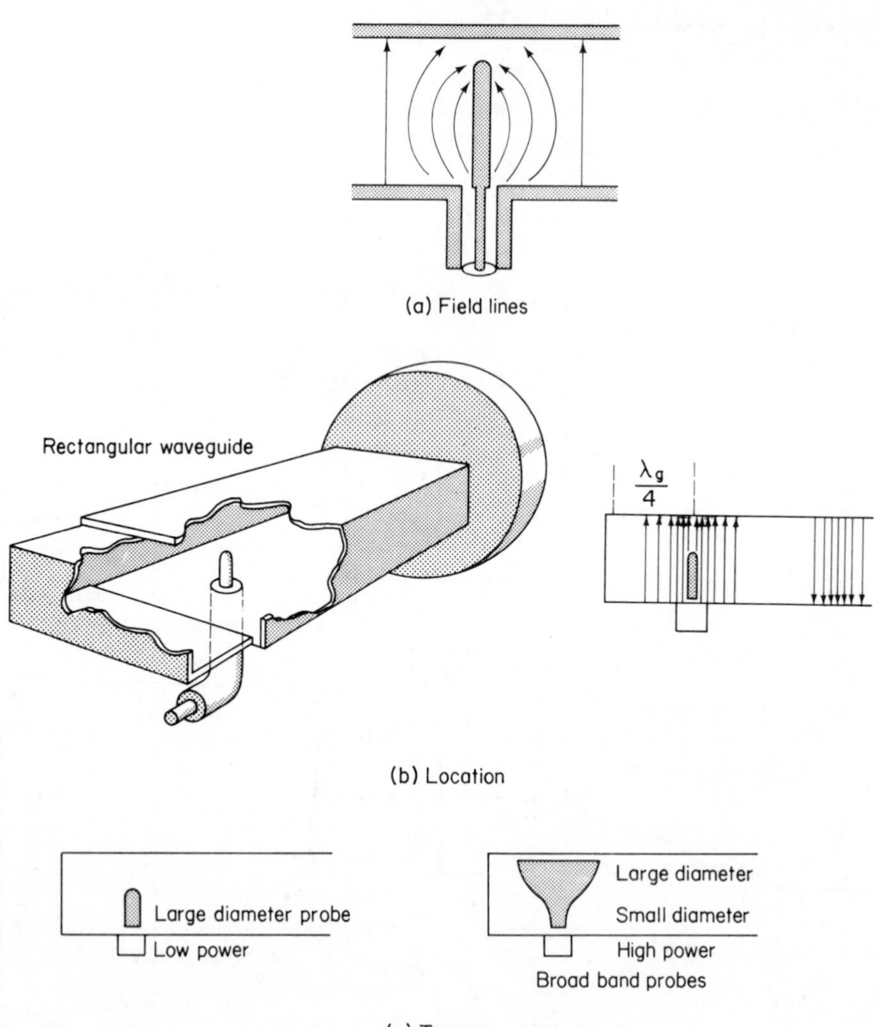

Fig. 4-13 PROBES AND FIELD LINES

compatible magnetic fields, quanta of energy become detached with each alternation and travel, wavelike, forward through the waveguide. For best coupling through the waveguide-transmission mode, a probe would be located in the middle of the wide side and a quarter-wavelength from the shorted end of the guide [Fig. 4-13(b)]. Excitation could also be achieved with the probe at the three-quarter-wavelength point.

A wide variation of adjustments and manipulations is used in probe injection; we may, for instance, reduce its length, retract it by means of a movable plunger, shield it, or change its physical contour (using conical or doorknob shapes) for broad-band or high-power operation [Fig. 4-13(c)].

In the extraction of power, probe techniques generally follow the above procedures in reverse.

MAGNETIC EXCITATION

This move creates H fields by placing a current-carrying loop inside the end of a waveguide. The magnetic field built up around the loop expands if the frequency of the stimulating current is correct and, in this manner, transfers energy to the waveguide. This energy in the right mode, when optimally coupled to the magnetic-field pattern along the guide walls, simultaneously creates E fields (also at the correct mode) and completes the wave-propagation mechanism previously described.

Figure 4-14(a) illustrates how a physical loop, receiving power from a coaxial current-

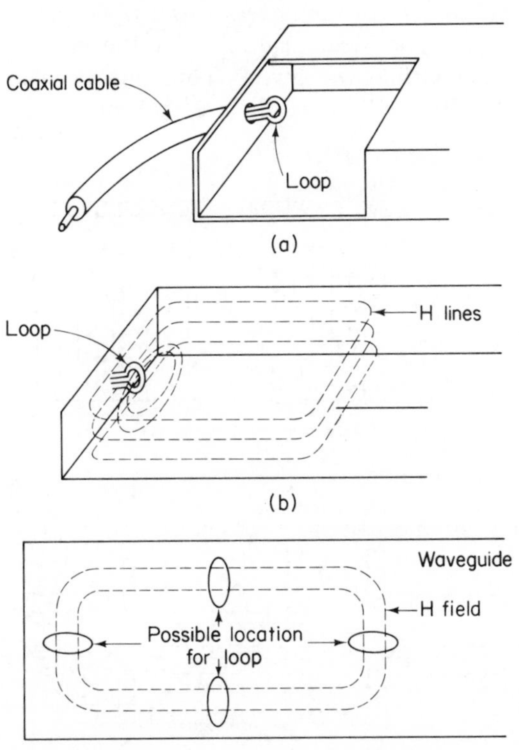

Fig. 4-14 FIELD LINES WITH LOOP

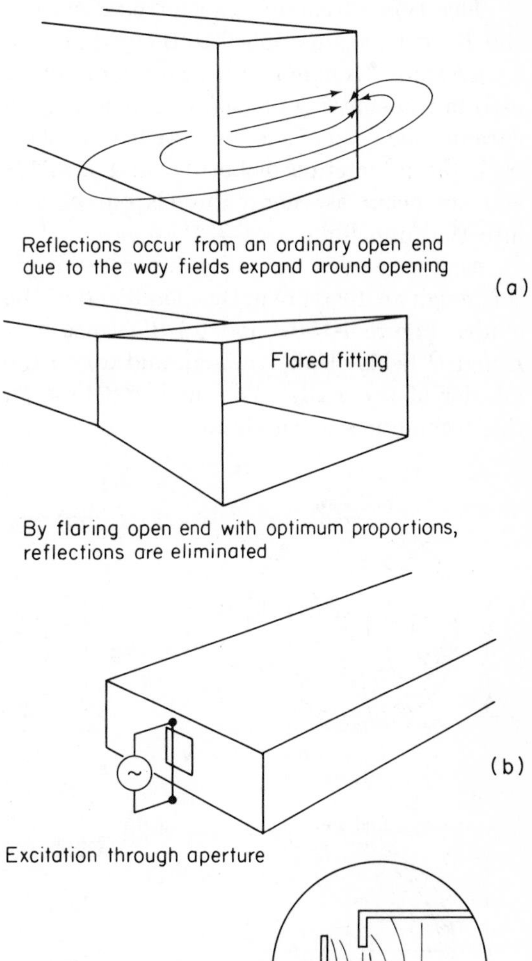

Fig. 4-15 WAVEGUIDE APERTURE

carrying cable, creates H lines that coincide with those existing around the end plate of the guide [Fig. 4-14(b)]. Alternate positions of loop location appear in Fig. 4-14(c).

To control the amount of coupling, the loop size, position, or orientation can be varied to fit individual design requirements and conditions. Conversely, energy can be extracted by a loop positioned at the far end of an excited waveguide.

ELECTROMAGNETIC COUPLING

This type of energy transfer uses both E and H fields, again applied to the end area of a waveguide. A common method, more often used in transmitting or radiating power, is to flare the open end of a guide so that reflections from the interaction of both E and H fields will not occur as energy travels out of and into the funnel-like area.

Another way of excitation (or transmittal) is through an aperture in the closed end of the guide. Figure 4-15(b) and (c) illustrate how E and H fields "leak" through and excite the interior of the guide. For further details on this technique see page 152.

CIRCULAR WAVEGUIDES

GENERAL

Circular-waveguide usage is minimal, except in special cases where mechanical and mode symmetry is important (as in the rotating joint of Fig. 4-4). The circular construction presents electrical problems because of its symmetry, as discussed below.

Referring back to Chapter 3, we see that the nodal classification system calls for the first subscript to denote the number of full-wavelength patterns around the circumference of the guide, while the second subscript denotes the number of half-wavelength patterns across the diameter of the guide.

These dominant modes, shown in the $TE_{1,1}$ and $TM_{0,1}$ configurations of Fig. 4-16, are most widely used. We can also calculate the cutoff wavelength with direct reference to the inside diameter of the guide. See Table 4-5.

Now, operationally, circular waveguides are less stable because their mode patterns are not as easily restricted as in rectangular waveguides. In the latter, E and H fields are inherently at right angles, as are the sides of its compatible waveguide. Thus they become naturally confined in their proper place. In

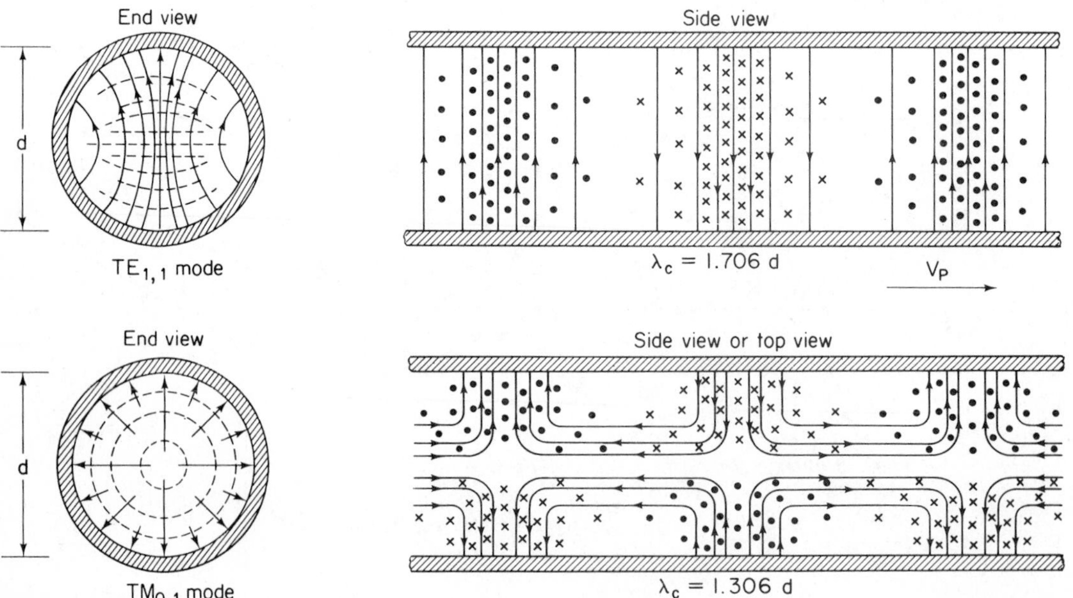

Fig. 4-16 CIRCULAR WAVEGUIDE MODES

TABLE 4-5 CIRCULAR WAVEGUIDE CHARACTERISTICS

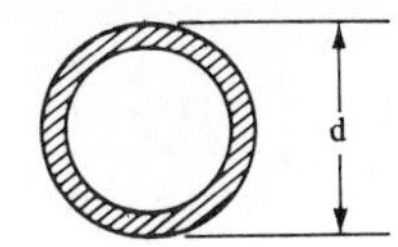

Size (I.D.) (in.)	d(cm)	λ(cm)	$TE_{1,1}$ Mode			$TM_{0,1}$ Mode			$TE_{2,1}$ Mode	
			λ_g(cm)	λ_c(cm)	Atten. (dB/m)	λ_g(cm)	λ_c(cm)	Atten.* (dB/m)	λ_g(cm)	λ_c(cm)
$\frac{15}{16}$	2.38	3.20	5.17	...	0.18	...	...	...	...	...
		3.30	5.54	4.06	...	...	3.11	...	...	2.45
		3.40	6.15	...	...	...	...	...	...	...
0.930 (A.T.)	2.36	3.20	5.27	...	0.12	...	...	...	...	...
		3.30	5.75	4.03	...	...	3.08	...	...	2.43
		3.40	6.36	...	...	...	...	...	...	...
$1\frac{3}{16}$	3.02	3.20	4.08	...	0.08	5.49	...	0.135	...	...
		3.30	4.30	5.15	...	6.04	3.94	...	...	3.11
		3.40	4.52	...	...	6.73	...	...	...	...
$2\frac{1}{2}$	6.35	9.1	16.8	...	...	...	...	...	...	...
		9.8	23.0	10.83	0.12	...	8.29	...	...	6.54
		10.7	70.0	...	...	...	...	...	...	...
$2\frac{3}{4}$	6.99	9.1	14.1	...	...	11.80	...	...	...	...
		9.8	17.2	11.92	0.047	...	9.12	...	...	7.19
		10.7	24.2	...	...	...	...	...	...	...
3	7.62	9.1	12.8	...	...	22.4	...	...	...	...
		9.8	14.9	13.00	0.030	56.4	9.95	...	...	7.85
		10.7	18.8	...	...	...	...	...	...	...
4	10.16	9.1	10.7	...	...	12.5	...	...	18.5	...
		9.8	11.9	17.33	0.012	14.5	13.27	0.02	28.0	10.46
		10.7	13.6	...	...	18.1	...	...	...	...

*Attenuation calculated for brass waveguides (except for the case of the 0.930 inch aluminium guide).

CIRCULAR WAVEGUIDE CUTOFF FREQUENCIES

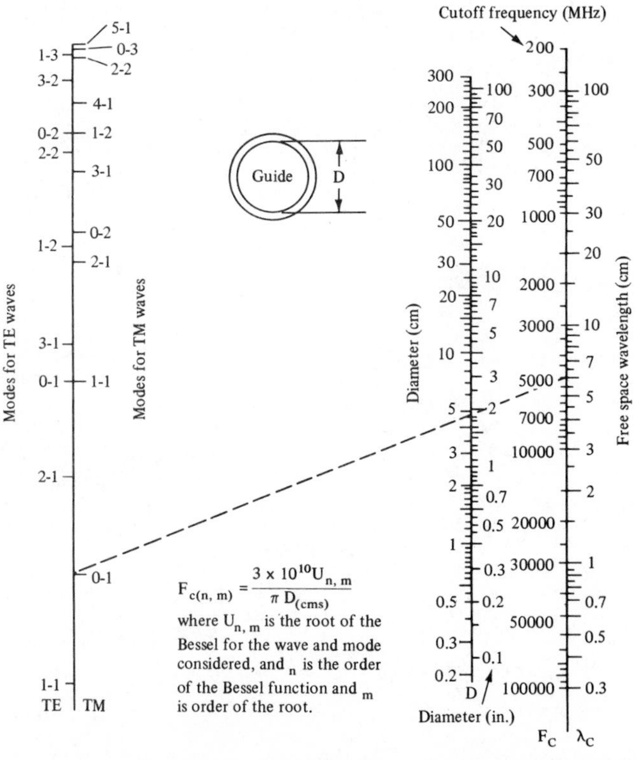

$$F_{c(n,m)} = \frac{3 \times 10^{10} U_{n,m}}{\pi D_{(cms)}}$$

where $U_{n,m}$ is the root of the Bessel for the wave and mode considered, and $_n$ is the order of the Bessel function and $_m$ is order of the root.

TABLE 4-6 2- and 10-CM WAVEGUIDE DATA

Size (ID)	*d(cm)*	λ_c *(cm)* $TE_{1,1}$ *mode*	λ_c *(cm)* $TM_{0,1}$ *mode*	λ_c *(cm)* $TE_{2,1}$ *mode*
$\frac{15}{16}$	2.38	4.06	3.11	
$1\frac{3}{16}$	3.02	5.15	3.94	3.11
$2\frac{1}{2}$	6.35	10.83	8.29	
$2\frac{3}{4}$	6.99	11.92	9.12	7.19
3	7.62	13.00	9.95	7.85
4	10.16	17.33	13.27	10.46

a circular guide, since there are no corners, the confinement is not as specific, and excitation into spurious modes is more apt to occur.

Later we shall also observe that coupling and operational devices (the magic tee, hybrids, etc.) are most easily fashioned when the waveguide walls are perpendicular planes.

Table 4-6 gives representative circular-waveguide dimensions for three commonly used modes.

Microwave Signal Components 5

INTRODUCTION—COUPLING DEVICES

In all microwave systems the transmitted, received, processed, and measured signal travels through accessory components which are attached to, integral with, or in some way necessary to the system operation. We have touched upon these and their analysis in Chapter 4, with description covering joints, and flanges plus the various types of straightaway junctions which we regarded more as structural than functional signal processing devices.

Functionally, these waveguide structures are usually coupling devices, since by using them signals are added, subtracted, canceled, or mixed in the course of measurement or transmission. In this chapter we have deliberately omitted resonant devices (windows and tuned posts) and attenuators even though they have coupling properties.

We saw in Fig. 4-1 a collection of four-, five-, and six-way structures which provided coupling and signal processing of special application to particular equipments. A simplified listing of these waveguide-junction-allied coupling structures reveals two main categories:

1. Magic tees or hybrids.
2. Directional and couplers.

In addition to these, we may process junction signals by using ferrite materials built into waveguide structures. These are generically included under isolators and circulators described at the end of the chapter.

HYBRID JUNCTIONS

These devices derive their name from usage in combining *E*-plane and *H*-plane tee structures. As noted later, they use coupling devices utilizing the circuit premises associated with equivalent shunt and series injection of signal or power into the collinear sections of a waveguide transmission path.

MAGIC TEE

When both waveguide *E* and *H* signal fields meet in a tee junction, such as shown in Fig. 5-1, the assembly is called a hybrid, and the one to be described is known as a magic tee. Since the *E* and *H* fields are always at right angles, we can do this physically and still keep the fields independent by adding just one more arm or port located directly at the junction, making the assembly a four-port device. Note that adding the fourth port

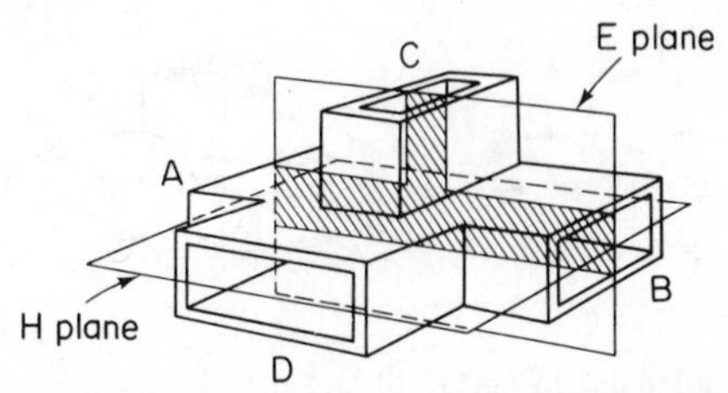

Fig. 5-1 BASIC MAGIC *T* HYBRID CONSTRUCTION

makes the device a three-dimensional device, because we have positioned another half-wave stub-like arm on the top side of the flat, H-plane junction. In effect it superimposes the E and H energy field, but since they are at right angles, they do not interact. In this respect the arrangement is analogous to a balanced bridge circuit, used in telephone communication.

Next we show wave and field relationships of the E and H planes in Fig. 5-2 by splitting the assembly and drawing a phantom view of them broken apart.

Here, going back to the explanation of waveguide junctions, we see that the stub equivalent of the input coupling to the H arm is a shunt-connected injection of signal into the H field; the stub-equivalent coupling of input to the E arm is a series-connected injection of signal [see Fig. 5-3(a)] for these equivalencies in the H and E junctions).

So, shunt-wise, energy fed into the H arm (port D) divides equally, appearing at both collinear ports A and B with none appearing at port C. And, series-wise, energy coupled into port C divides between the adjacent collinear ports A and B only [Fig. 5-3(b)]. Thus energy into D cannot excite the dominant mode of C because its E component lies on C's vertical plane, and no transfer can take place if physically right-angle linkages are not made [see Fig. 5-3(c)]. Phase relationships for inputs to ports C and D are precisely the same as when inputs are fed to simple series or shunt tees.

Briefly, magic tees are used as power dividers in balanced bridge circuits and for phase-related bridge measurements. In a balanced-mixer circuit, a matched detector is placed in each of the two collinear arms; local oscillator power fed into the shunt arm then becomes completely decoupled from the signal power fed into the series arm. Figure 5-4 illustrates a number of commercial magic tee assemblies.

In a phased-bridge application, to avoid reflection errors, the input signal is fed to the shunt arm, enabling reflections from the test piece to be matched out by introducing an equal reflection at the phase-shifter input. Setup adjustments arrange for elimination of spurious recirculating reflection power by arranging for line reflections to differ by

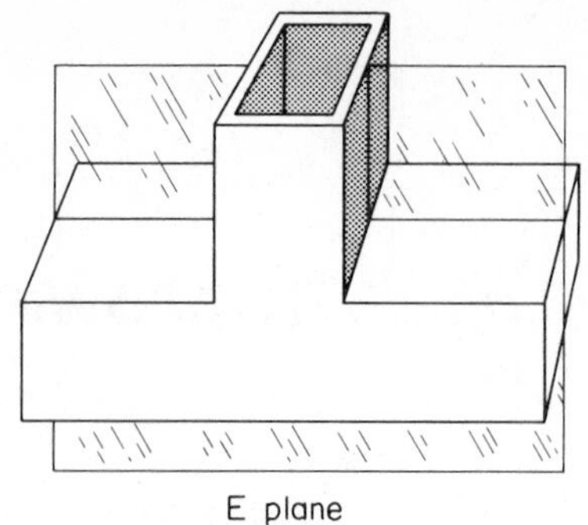

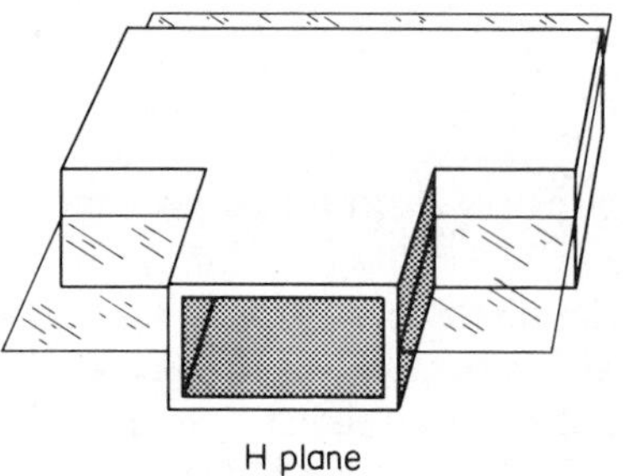

Fig. 5-3 ENERGY DIVISION IN MAGIC *T*

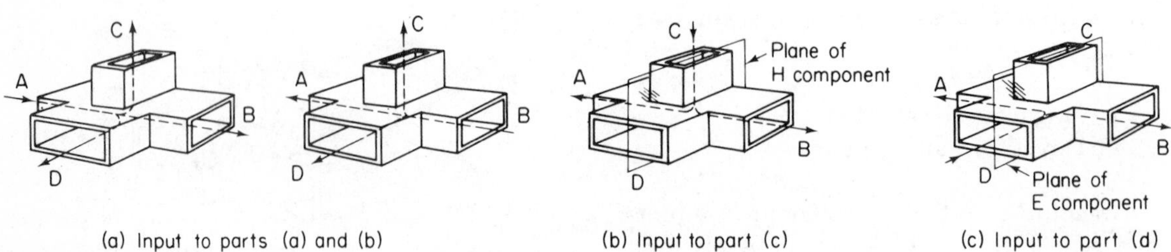

Fig. 5-2 PHANTOM VIEW SHOWING SPLIT OF MAGIC *T* FIELDS

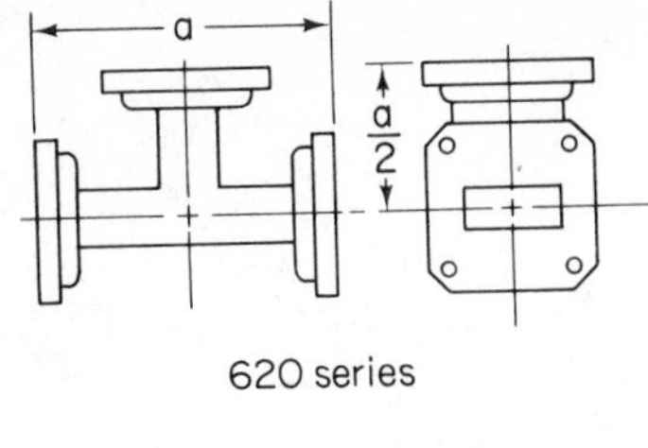

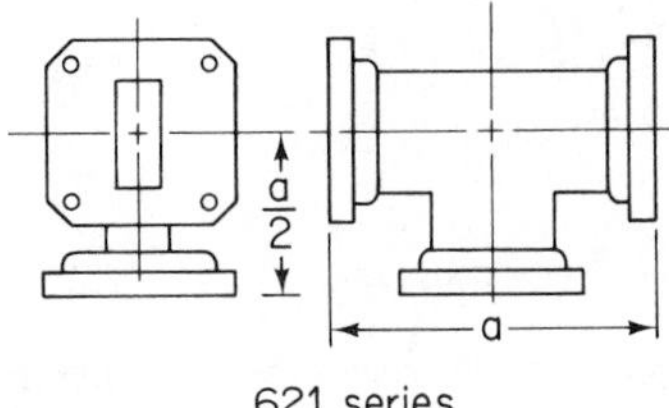

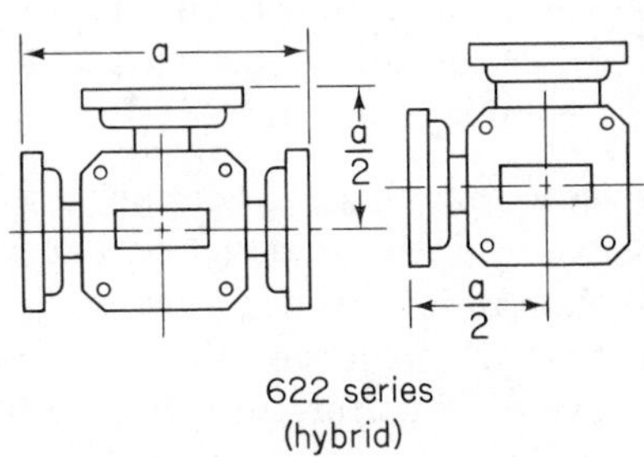

Fig. 5-4 COMMERCIAL MAGIC *T* ASSEMBLIES

$\lambda_a/4$ so that all reflected energy will appear at the load and near the input (see Chapter 11).

RAT-RACE HYBRID JUNCTION

This device allows special coupling in waveguide structures, capitalizing upon the maneuver of folding a waveguide tee back upon itself. Doing this produces a ring-like structure; so mechanically we can place the structure flat-wise using H-plane operation with shunt connections, or we can place it upright using E-plane operation with series connections.

Now, electrically, in conducting the folding operation and correctly locating the ports, the overall circular length must be compatible with the operating wavelength. Thus we see that the reentrant construction allows a fourth port to be added to the usual three-arm tee.

Figure 5-5 is a skeleton view of the typical rat race; two conditions must be built into the structure.

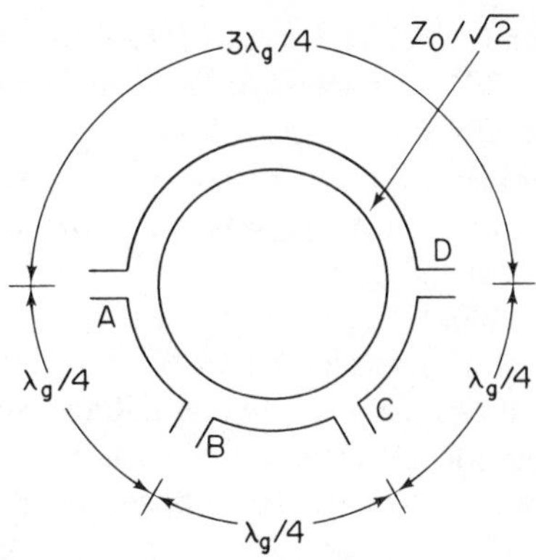

Fig. 5-5 BASIC RAT RACE CONSTRUCTION

1. The mean circumference must be $1.5\lambda_g$.
2. The four ports must be $\lambda_g/4$ apart.

Now, feeding signal to port A splits the energy at the junction, half of it passing clockwise and half counterclockwise. At ports B and D the outputs appear simultaneously (in phase) but cancel at port C, delivering no output. Thus energy divides equally between B and D, port C becoming isolated; also, if we apply input to port C, it divides equally between B and D, leaving A isolated.

In a series structure with a given port impedance Z_0, the impedance of the whole ring should be $0.707Z_0$; for a shunt-folded structure ring, impedance should be $1.41Z_0$.

The rat race can be used to combine two signals as well as to divide one signal in half; unequal signals will give additive output at two ports—a different signal at the isolation port.

WAVEGUIDE COUPLERS

GENERAL

Couplers are flanged, built-in waveguide assemblies that tap off signal power or energy for measurement or monitoring microwave performance. They may measure transmitter power, VSWR, feed signal to a receiver, detect reflected power, or perform a number of other specialized operations.

Couplers may be directional, measuring only forward-traveling power, or bidirectional, measuring both forward and reflected power. The most common bidirectional coupler, for instance, consists of a length of main waveguide with two auxiliary sections of waveguide mounted on each of its opposite walls, one section to tap off incident power and the other to tap off or measure reflected power. Figure 5-6 illustrates the placement of the auxiliary piece of waveguide in a simple directional assembly.

Three other features enter coupler construction:

1. There must be coupling slots for passage of the energy from some other auxiliary source or equivalent device. These must pierce both the main and auxiliary walls so that the holes are common to both main and auxiliary sections.

2. An internal probe used to detect power for indication or for measurement; this usually extends through the side of the auxiliary waveguide and terminates forming an output terminal for externally located coaxial connectors.

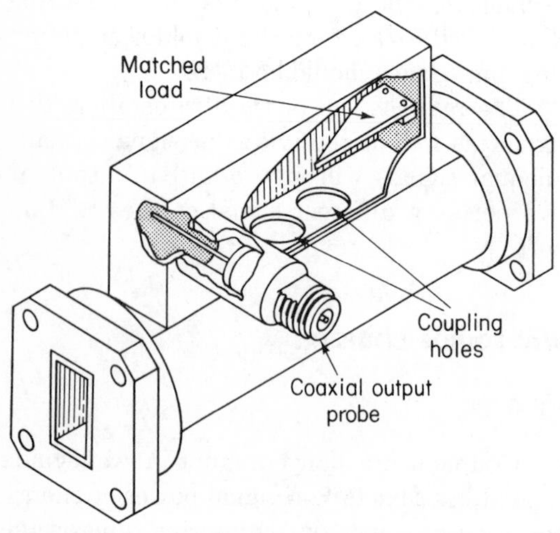

Fig. 5-6 SIMPLE COUPLER ASSEMBLY

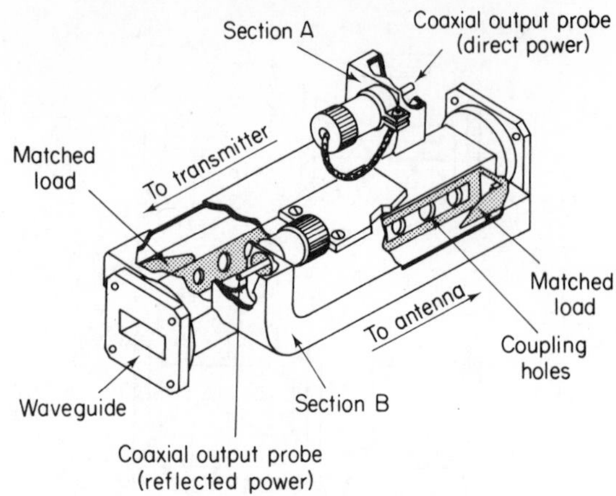

Fig. 5-7 BI-DIRECTIONAL COUPLER

3. An internal load within the auxiliary sections which absorbs reflected power from one of the coupling lead-ins.

The various methods of coupling generally describe the coupler construction and usage. A brief description of the various types is given below, although the principle of operation is covered only for the two-hole or slot type.

COUPLER TYPES

1. Two-hole directional units are simple side-by-side, broad-wall H-field structures described below.

2. Bidirectional units have four holes using double side-by-side wall structures with coupling holes or slots on each side (Fig. 5-7).

3. Top-wall units are narrow-wall, side-by-side, E-field couplers.

4. Single-hole crossed-guide units are common broad-wall units with single slots or holes (Fig. 5-8).

5. Branching-guide couplers use a common wall instead of coupling holes (Fig. 5-9).

6. Strip-line couplers are parallel, ground-plane, metallic strips running internally within the guide structure.

7. Short-slot couplers (Fig. 5-10).

8. Bifurcated couplers (Fig. 5-11).

COUPLER OPERATION

Figure 5-12(a) illustrates the overall and internal structure of a bidirectional coupler

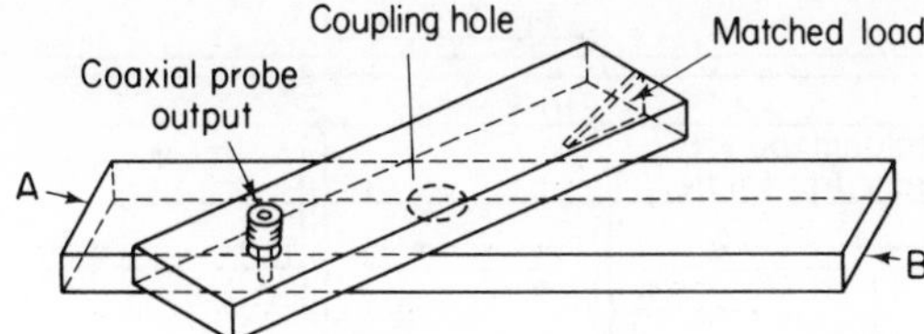

Fig. 5-8 SINGLE HOLE CROSSED FIELD COUPLER

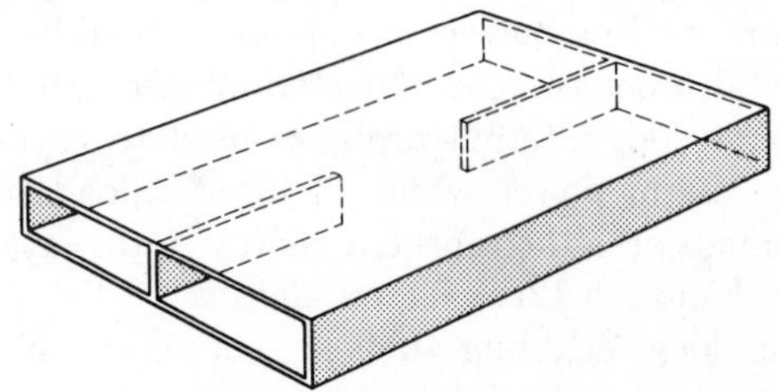

Fig. 5-10 SHORT SLOT COUPLER

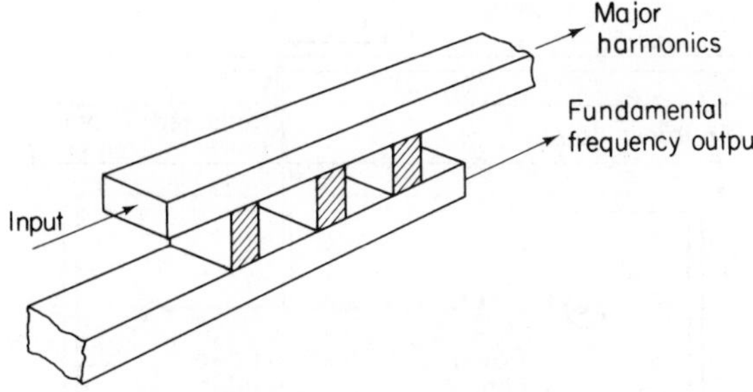

Fig. 5-9 BRANCHING GUIDE COUPLER

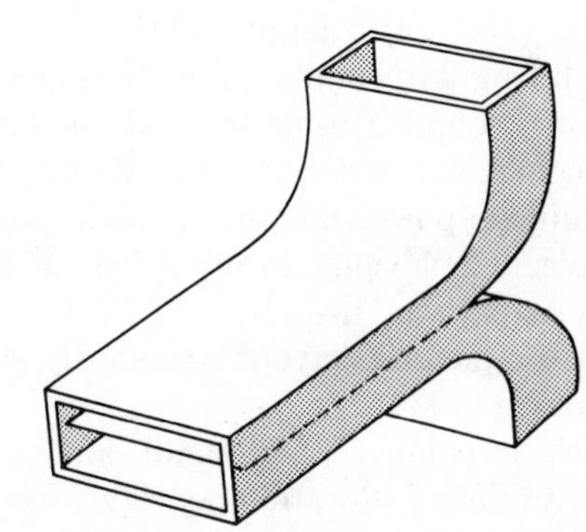

Fig. 5-11 BIFURCATED COUPLER

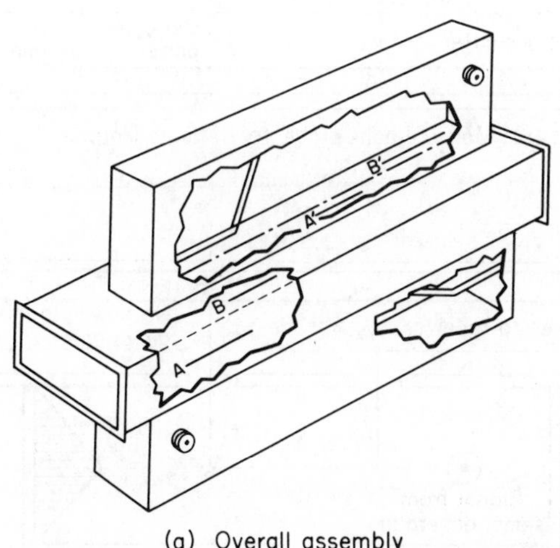

(a) Overall assembly

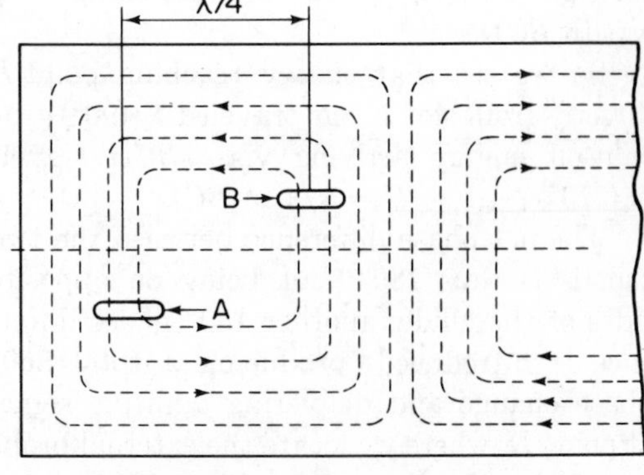

(b) Coupling slot arrangement

Fig. 5-12 BASIC BI-DIRECTIONAL COUPLER

with the auxiliary sections mounted on the opposite, flat, voltage walls of a main section. The lower coupler diverts, by means of the coupling slots, a portion of the incident, forward-moving power; the upper section, by its slots, processes the reflected power. In the former, an internal probe is used to extract the useful power while an internal-load septum absorbs the reflected energy in the latter.

Figure 5-12(b) shows details of the slot locations enabling *H*-field coupling. Note that the coupling slots *A* and *B* are a quarter-wavelength apart and are on the opposite sides of the guide center line—a design feature which, as we note by the direction of flux arrows, samples outer-phase or 180° phase different components of the magnetic field.

Figure 5-13(a) diagrammatically shows the path length of coupling flux lines in a unidirectional coupling using point A_0 as a reference, coincident with slot *A*. Energy-path lines indicate power flow in opposite sides of the waveguide, keeping in mind that distance *AD* is equal to distance $BC(\lambda/4)$, even though the pictorial layout necessarily shows them longer.

Now, dynamically, a small amount of energy traveling toward the right, couples through slot *A* and then splits. A portion passes across the wall and then a quarter-wavelength to point *D* (see solid-line energy-coupling paths). One quarter-wavelength farther on, the rest of the energy couples through slot *B* to points *C* and *D*, completing a path *BCD*.

So we see that energy reaching point *D* directly from slot *A* has traveled $\lambda/(90°)$; the split-off energy arriving via *ABCD* travels $\frac{3}{4}\lambda(270°)$—$(A - B = \lambda/4 + BCD = \frac{1}{2}\lambda)$.

The net phase difference between the two signals is thus 180°; but being on opposite sides of the guide, another 180° phase difference is introduced, producing a total 360° phase change and delivering additive signal at point *D*, where we locate the external probe by which output power is extracted.

Now, energy from slot *A* can also split (at point A_0) and travel by way of path *ADC* (dashed lines), while whole *B*-coupled energy

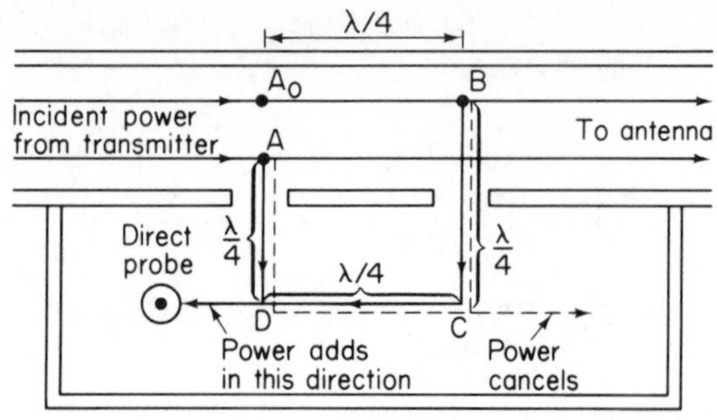

(a) Basic signal paths

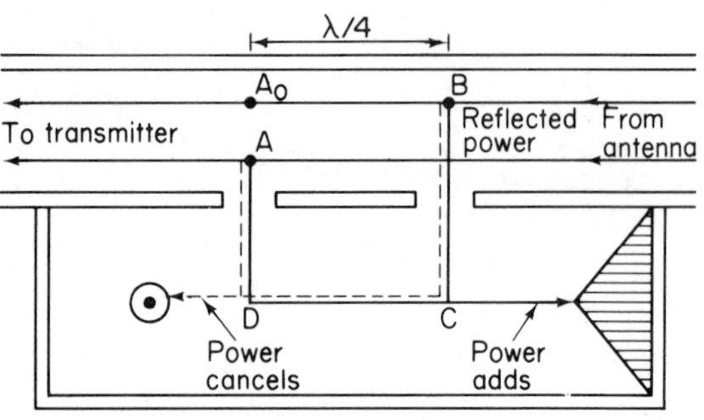

(b) Reflected signal

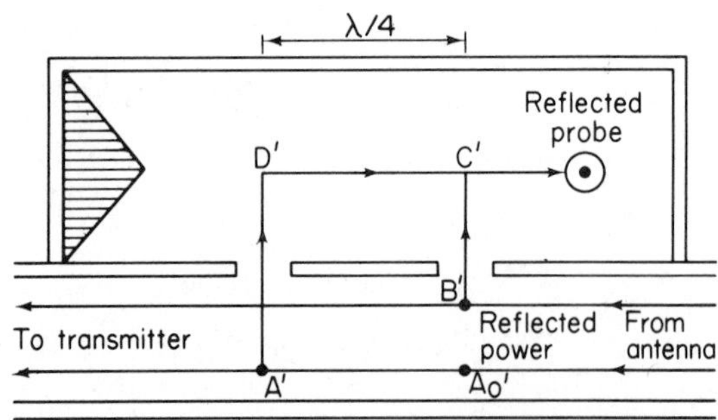

(c) Measurement probe for reflected power

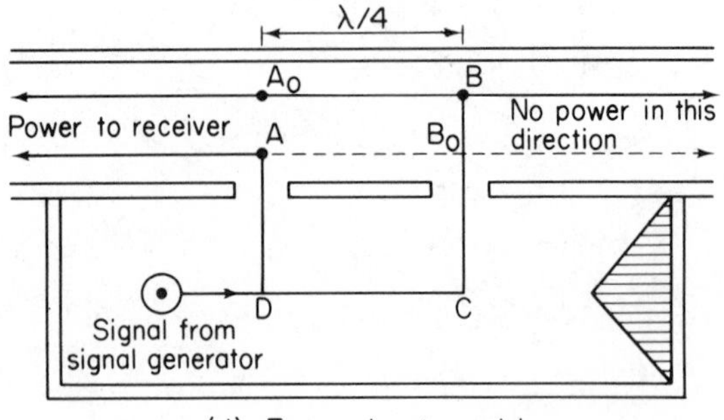

(d) Energy input provision

Fig. 5-13 DETAILS OF COUPLER OPERATION

can split at C, traveling the path A_0DC ($\lambda/2$). These paths are of equal length, but since the energy coupled through slot D is 180° out of phase with that from A, the fields cancel and the main external stimulation to the signal-extraction probe is a true measure of transmitted energy.

Reflected power [Fig. 5-13(b)] at point C goes through a similar addition of coupled components and is absorbed if a dissipating load is placed at C. Similarly, reflected power at the probe is self-canceling. We would arrange such a coupler (on the opposite wall of the main waveguide) as shown in Fig. 5-13(c). Here a probe is installed which measures the reflected power.

Thus, we see that a coupler has directional capabilities and can be so constructed that it measures power transmitted in only one direction. Two auxiliary waveguide sections, turned in opposite directions of course, can be used to measure both incident and reflected energy. In this case we would physically arrange the assembly as shown in Fig. 5-10(a).

Energy may also be fed into a waveguide by means of the coupler probe. Figure 5-11(d) illustrates the mechanism using a signal-generator source applied to the lower waveguide; paths similar to those illustrated in Fig. 5-13(a) deliver small amounts of power leftward, say toward a receiver.

DESIGN—CONSTRUCTION

In building and measuring a coupler, the two main factors are, first, the amount of coupling (that is, how much outgoing power is tapped off) and second, how much unwanted power appears across the reflection-dissipating load located within the waveguide structure. This might be called the fourth port of the coupler, which normally appears as a three-port device.

Coupling is dictated by the size and location of the holes in the waveguide walls; it is commonly expressed in decibel ratio of the input power to the output-coupled power; thus an output-coupled power of $\frac{1}{100}$ would designate a 20-dB coupler, while as much as 50 per cent output coupling would be a 3-dB coupler.

Now, directivity is much more difficult to design into a coupler, since even over a narrow band, perfect structures are difficult to fabricate. This factor is a measure of the decibel ratio of coupling power to unwanted power generated by reflections. The unwanted power thus represents a small portion of the input power, and in a 20-dB coupler with a 20-dB directivity, it would represent 1/10,000 of the input power (40 dB).

An interesting variation is the Bethe or single-hole crossguide coupler pictured in Fig. 5-8. Here waves in the auxiliary guide are generated through a single hole which includes signals produced by both the electric and magnetic fields. Because of the phase relationships involved in the coupling process, the signals generated by the two types of coupling cancel in the forward direction but reinforce in the reverse direction. Thus power entering at point A is coupled to the coaxial output, while power entering at point B is absorbed in the dummy load. If the two waveguides were paralleled, the magnetic component would be coupled to a greater degree than the electrostatic, and the directivity would be poor. But by placing the auxiliary guide at the proper angle, the amplitude of the magnetically excited wave is made equal to that of the electrostatically excited wave (without changing the latter) and good directivity is obtained. In this design the angle used depends upon the frequency of operation.

Coaxial couplers are useful in high-power radar and communication equipment, where spurious transmitter harmonics are extremely difficult to control in rectangular guides because of their tendency to be sensitive to excitation of several spurious modes. Slot coupling can yield excitation at the first higher $TE_{1,1}$ mode in a rectangular guide when coupled from the TEM mode of a coaxial line. The slot relationship is seen in Fig. 5-14(a); the commercial design of simple directional and dual directional couplers is illustrated in Fig. 5-14(b).

Strip-line couplers using relatively simple configurations are commercially available in

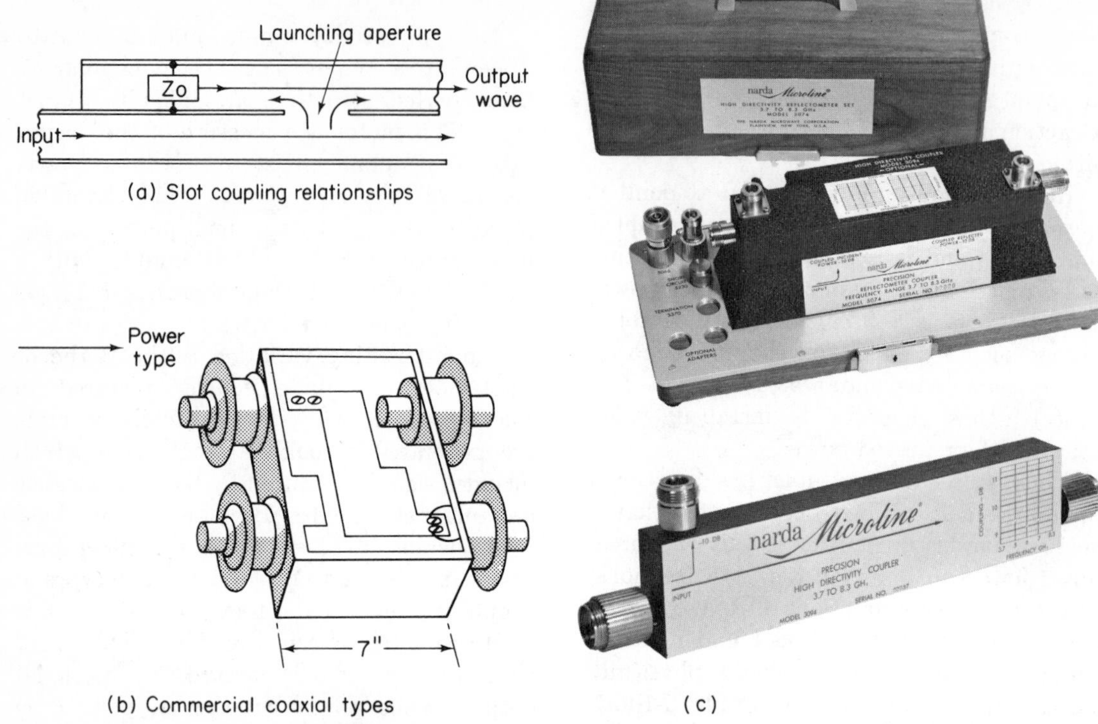

(a) Slot coupling relationships

(b) Commercial coaxial types

(c)

Fig. 5-14 COAXIAL AND STRIPLINE COUPLERS
(Courtesy of Narda-Microline.)

solid-state microwave switching and amplifier assembly [Fig. 5-14(c)].

OTHER COUPLING DEVICES

Excluding resonant cavities and attenuators, these can be broadly categorized under circulators, isolators, modulators, and duplexers. The first two are operationally similar to physically stimulated and constructed couplers (that is, they distribute signals among multiport arrangements). The latter two are properly considered under switching and in transmitter circuitry. However, they depend upon the action of ferrite materials (mostly cores and wafers) described in the following brief introduction.

FERRITE CHARACTERISTICS

Unlike most other magnetic as well as some nonmagnetic substances, ferrites have unusual magnetic properties that promote their adaptation and use in microwave equipment. The first difference from magnetic metal-like iron and nickel is that ferrites are oxide-based compounds containing iron, zinc, manganese, cobalt, aluminum, or nickel which are formed by firing (at 2000°F or more) powdered oxides of the materials and pressing them into specific shapes. This specialized processing gives the materials added characteristics making them possess ceramic insulator properties, a quality unlike iron at low frequencies but which makes the materials usable at microwaves. This is due to the high internal molecular resistance, which acts to reduce eddy-current losses.

Ferrites owe their characteristics to atomic structure, capitalizing upon a fundamental property of atoms whereby both electrons and protons spin upon their own axes. As in the solar system, the electron, like the earth, rotates upon its own axis while it travels in orbit around the sun. Now, as an electron spins, it creates a magnetic field along its own spin axis, creating an associated current flowing in a loop centered at the electron axis. Thus materials with atoms having more spinning electrons than others create substances with magnetic properties, and when placed in a static magnetic field, all the electrons align themselves and the material becomes magnetized. Any spinning body acts like a gyroscope and has a property called *precession*. It is produced by applying another sideways magnetic force to the electron; it will cause the body to move in another off-center local orbital path at right angles to the sideways field so that it "wobbles" around its central axis. This wobble is another suborbital path, which usually has a resonant frequency called the *natural precession frequency* and in most ferrites exists in the presence of a dc field at between 3 and 9 GHz.

This application of the combination of a steady polarizing dc field and an rf field has directional properties upon wave motion when applied to ferrites introduced into a waveguide. We use them in several devices to reduce reflected power, for modulation, and in switching devices.

An isolator allows energy to flow in one direction but absorbs energy traveling in the opposite direction. This is accomplished using the rotational effects produced by combining a permanent dc magnetic field with an ac field.

We depict this in Fig. 5-15(a) with the apparent rotation of a magnetic wave (without ferrite effects) traveling from right to left. We see this by inspecting the magnetic field at point X_1, which is off the center line of the guide. At T_1 the magnetic field is pointed upward; with the wave point T_2 moved to position X_2 it is pointed toward the right. When the T_3 wave point reaches X_3 the field is downward, and when point T_4 on the wave reaches X_4, the field is pointed leftward.

Thus, as the wave passes X_1 the magnetic field appears to rotate in a clockwise direction; this will happen to any point off-center in the waveguide. Similarly, a wave passing from left to right will make the magnetic field appear to rotate counterclockwise.

Next, let us place a section of ferrite in the waveguide at X_1 while surrounding the waveguide with a permanent magnet. If the ferrite's natural precession resonance is at the same frequency as the signal energy passing through the waveguide, and if they are stimulated by a wave passing from left to right, the electrons will precess or produce rotating magnetic fields in a counterclockwise direction. We saw, however, that the field rotation of a magnetic field under the influence of a normal waveguide signal produces a clockwise field rotation.

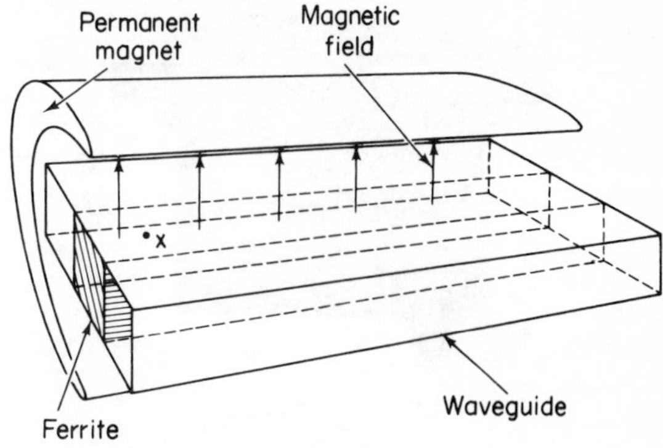

(a) Basic construction

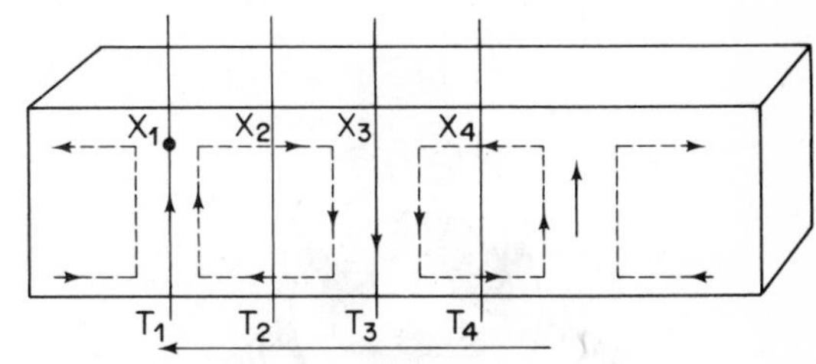

(b) Rotation of magnetic field

Fig. 5-15 ISOLATOR ACTION

Naturally, the two fields will cancel, and signals will pass more easily from right to left than in the opposite direction. The normal loss of incident wave is usually less than 1 dB; the isolation or reduction of "wrong-way" power varies in commercial units from 10 to 30 dB.

Construction-wise, isolator magnets and ferrites may be rectangular, coaxial, or strip line. Figure 5-16(a), (b), and (c) show representative models of each type. Most of those shown include a permanent magnet within the

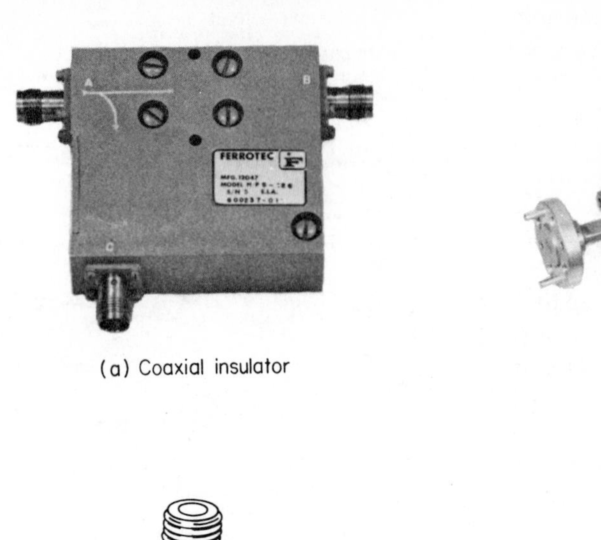

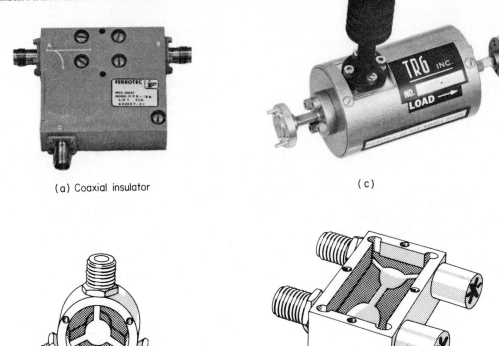

(a) Coaxial insulator

(c)

(b) Printed isolators

Fig. 5-16 ISOLATOR ASSEMBLIES

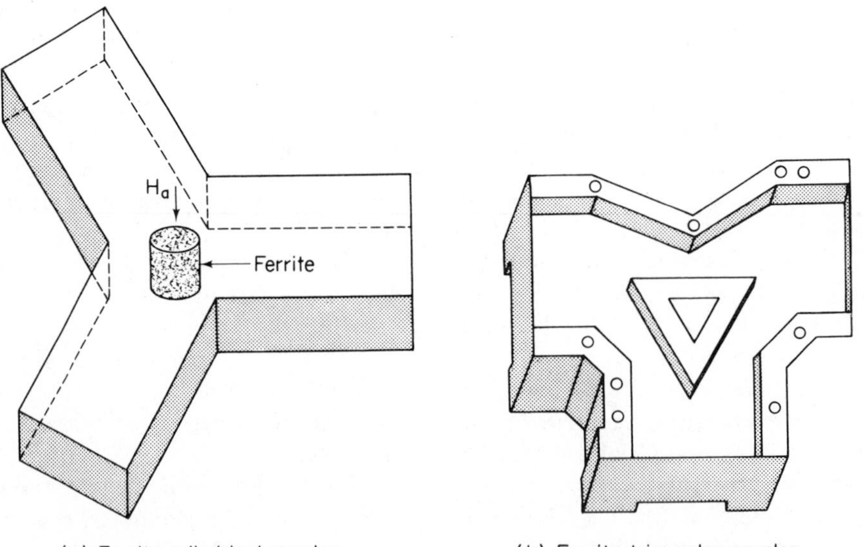

(a) Ferrite cylindrical coupler

(b) Ferrite triangular coupler

Fig. 5-17 CIRCULATOR CONSTRUCTION (Courtesy of Ferrotec Inc. and TRG Inc.)

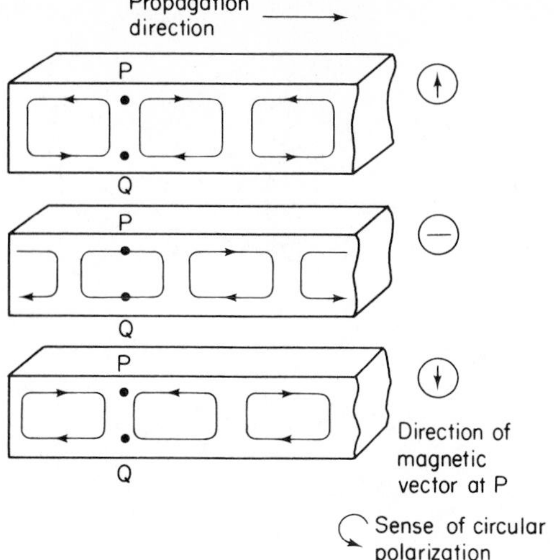

Fig. 5-18 SKEWED FIELD IN CIRCULATOR

guide and handle less than 10 W of signal power. Typical frequency ranges are from 1 to 8 GHz, with peak power-handling capacities in some units running as high as 5 to 10 kW.

CIRCULATOR

This unit is a round ferrite-based waveguide junction. Input–output terminals make the circulator a three- or four-port device in which energy introduced at one port circulates only to a second port, while energy introduced at the second port circulates only to the third. In a circular, three-port assembly, energy to port three is fed back to port one.

A circulator functions because the rf magnetic fields are deliberately skewed or bent by the introduction of internal ferrite slugs at or near the junction. Figure 5-17(a) is a simplified illustration of the ferrite placement at the Y junction between three waveguide ports. A permanent magnetic bias is usually employed to establish exact magnetic conditions for best operation. Figure 5-17(b) illustrates a ferrite triangle used in a commercial structure.

The skewing of the *H* field on the top and bottom halves of a waveguide starts with the rf magnetic field described in Fig. 5-18 where (similar to Fig. 5-15) a left-to-right moving *H* field produces opposite polarization to the magnetic vectors of points *P* and *Q*. Placing a ferrite slug laterally but adjoining these fields adds to the polarization along oné edge of the guide (say for the *P* area) and reduces it along the *Q*-field area.

These two effects increase the polarization and velocity of propagation of the wave on one side and decrease it on the other, thus skewing the wave right at the Y junction. Adjustable biasing field and correct ferrite design make the energy "bend" directionally into the next adjoining port.

This relation of the plane of wave polarization (commonly known as Faraday rotation) operates a circulator above the precession spin resonant frequency of the electron. When rf energy enters the ferrite material, the magnetic motion of the electron precesses as usual, but, being at a different frequency, produces a resultant rf magnetic vector which is adjusted to specifically meet the angular requirements of funneling energy to the next port.

Some circulator applications are shown in Fig. 5-19. We may have in Fig. 5-19(a) a circulator simultaneously connecting a transmitter and a receiver to a single antenna. A variation of this is seen in Fig. 5-19(b), where transmitted energy circulates as shown by the solid arrows. Transmitter 1 feeds the antenna while received energy is directed to receiver 1. If the magnetic field is reversed, energy will circulate as shown by the dashed arrows. Here the output of transmitter 2 is reflected from the filter of receiver 1 and fed to the antenna. The received signal is reflected from the filter for transmitter 1 and fed to receiver 2.

A circulator with one port terminated becomes an isolator of exceptionally low loss. Two circulators, with one terminated by a matched load, can be used in tunnel-diode or parametric amplifiers.

Strip-line circulators offer somewhat simpler construction; Fig. 5-20(a) and (b) show how ferrite discs are used in coaxial line junctions. Figure 5-20(c) illustrates typical field

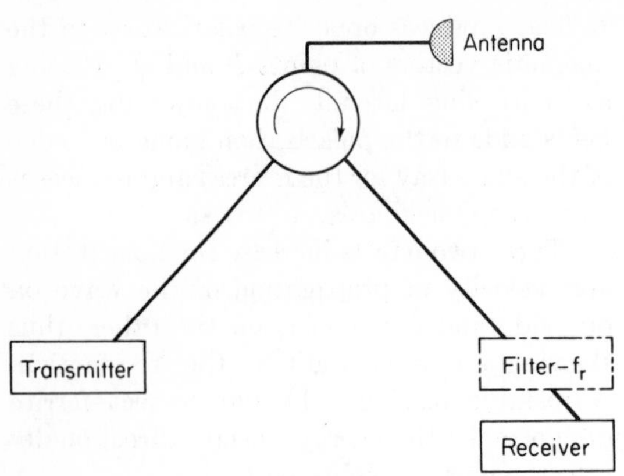

(a) Circulator using single antenna

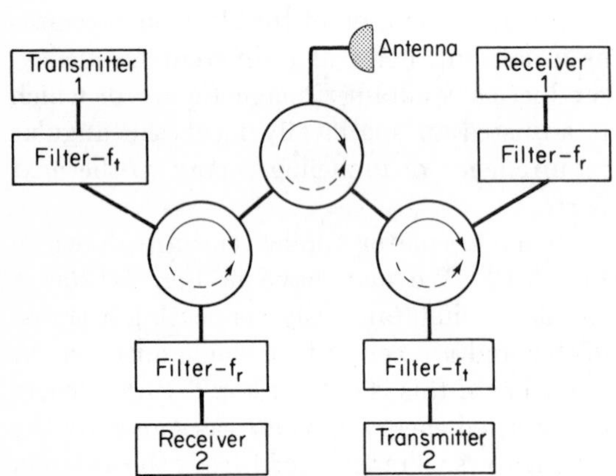

(b) Standby transmitter–receiver using common antenna

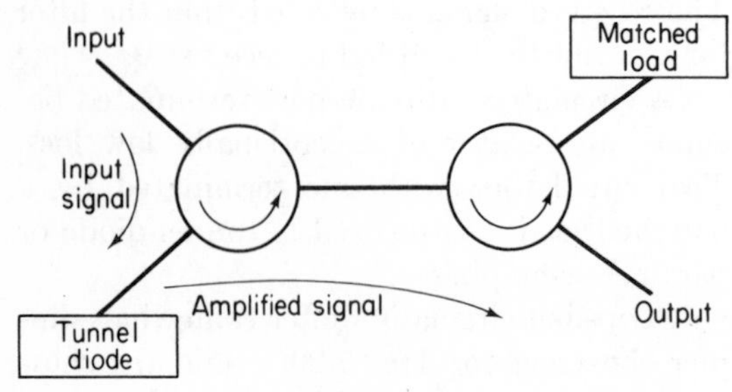

(c) Amplifier insulation from output reflections

Fig. 5-19 CIRCULATOR APPLICATIONS

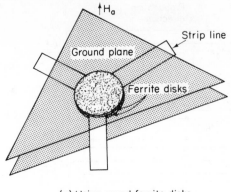

(a) Using round ferrite disks

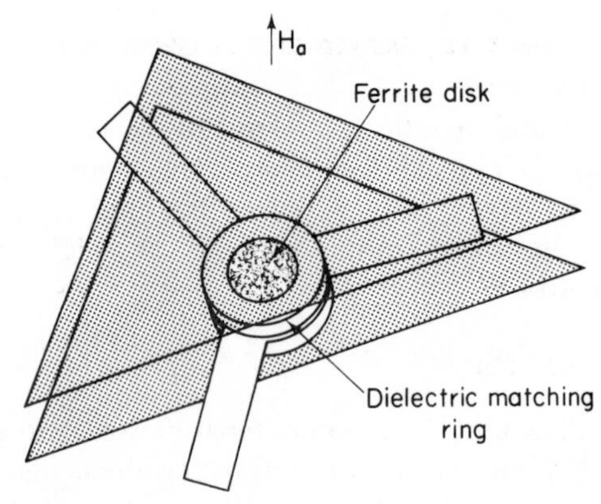

(b) Using dielectric matching ring

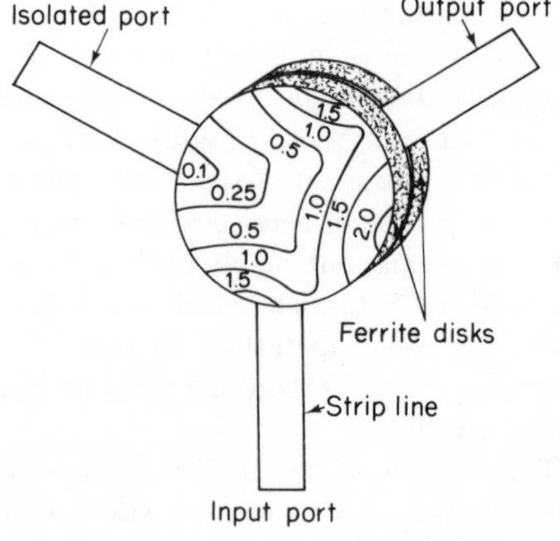

(c) Typical field control

Fig. 5-20 STRIP-LINE CIRCULATOR DESIGN

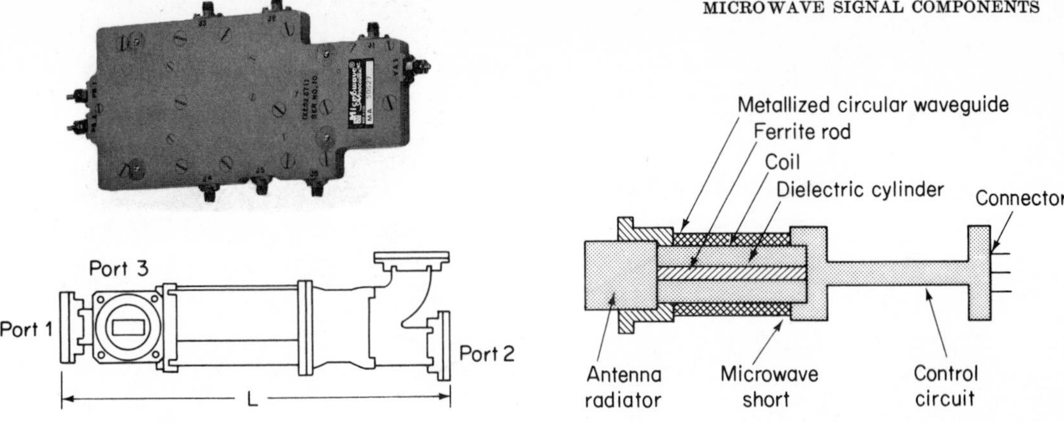

Fig. 5-21 COMMERCIAL CIRCULATOR ASSEMBLIES (Courtesy of Microwave Associates.)

Fig. 5-22 PHASE SHIFTER CONSTRUCTION

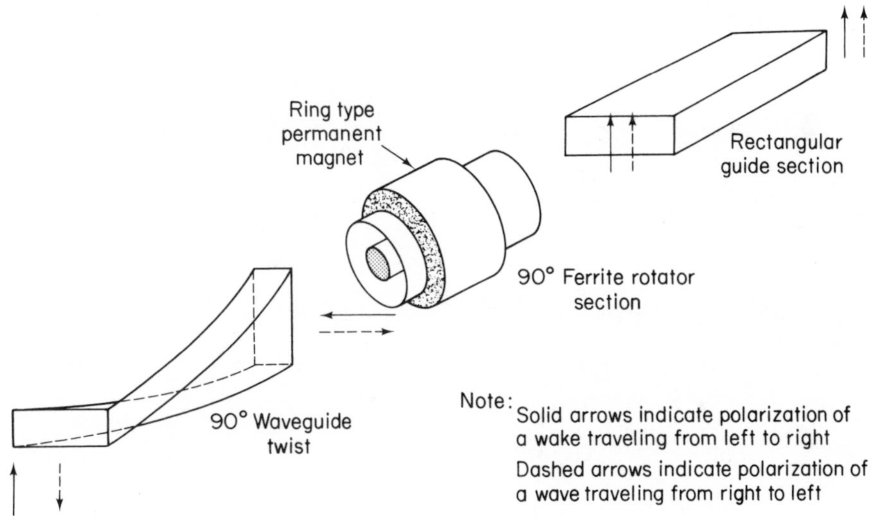

Fig. 5-23 EXPLODED VIEW OF GYRATOR

contours. Figure 5-21 illustrates commercial circulators.

SPECIAL FERRITE DEVICES

Two devices utilize Faraday rotation for specialized applications.

1. Phase shifter. This is a miniature device used in large multiple assemblies to individually feed multielement phased antenna units. It uses a circularly symmetrical waveguide loaded with ferrite material so that an applied axial magnetic field causes the microwave permeability to vary and to produce the desired phase shift. Figure 5-22 illustrates a commercial X-band phase shifter.

2. The gyrator is a two-port device in which the phase shift through the device is one direction and differs from the phase shift in the other direction by 180°. This device consists of a 90° ferrite rotator section. To keep the input and output polarization the same, a 90° waveguide twist is usually used in conjunction with the rotator section. Figure 5-23 shows an exploded view of a gyrator where the signal enters the waveguide twist and is rotated counterclockwise 90°. The ring magnet provides a lengthwise field and the ferrite rotates the wave clockwise 90°.

6 Microwave Cavities and Filters

CAVITIES

GENERAL

Cavities are solid, completely enclosed chambers which can be considered to be resonant circuits of waveguide technology; at microwaves they probably enjoy a wider usage than in the low-frequency regime because, besides operating fundamentally at their resonance, they can act as simple devices for transmission, coupling, filtering, and mixing.

Indeed, a cavity can be considered to be an integral section of a waveguide; this analogy is quite vivid when we consider that a cavity is merely a chamber constructed by crosswise slicing a waveguide at the minimum voltage points and closing the ends by soldering two plates across the open ends. Figures 6-1(a) and (b) show the field current and voltage distribution in a cavity constructed in this manner, plus similar conditions in cylindrical and square cavities. Figure 6-2 pictures four other ways of looking at the resonant circuit–cavity analogy as follows:

1. Figure 6-2(a) shows a broad, flat section of conductor bent into a semiclosed U shape. When the conductor's distributed capacitance between legs are at resonance with its inductance, we have a pseudo cavity.
2. Figure 6-2(b) constructs a cavity from an assembly of half-turn loops in parallel.
3. Figure 6-2(c) shows quarter-wave line sections assembled in parallel to produce structural resonance.

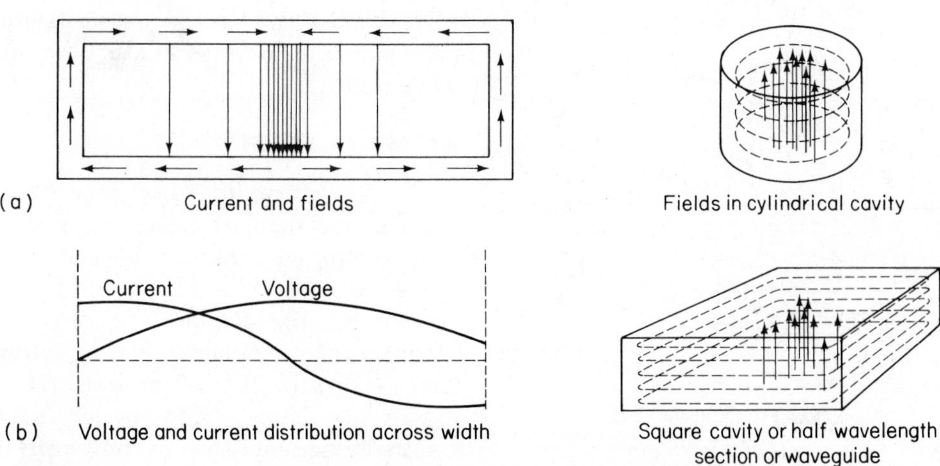

Fig. 6-1 CAVITY FIELD DISTRIBUTION

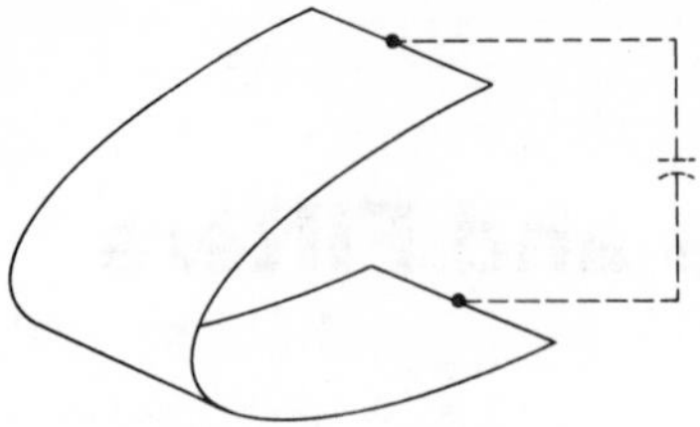

(a) Bent U-shaped flat container

(b) Parallel half-turn loops

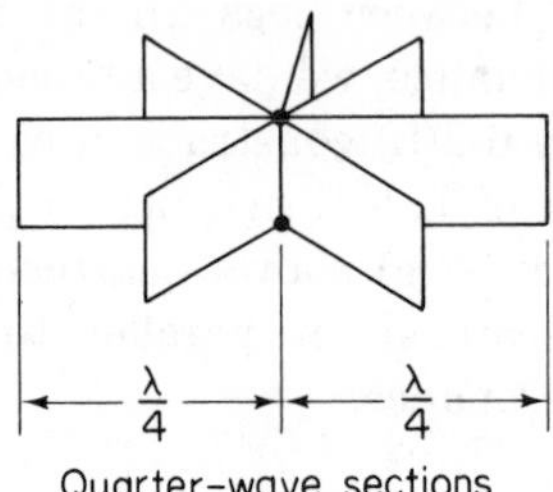

(c) Paralled quarter-wave sections

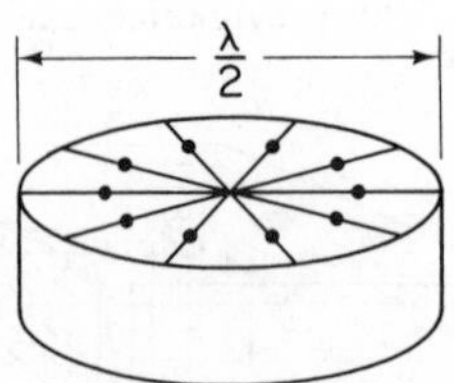

(d) Covered metal container

Fig. 6-2 CAVITY-RESONANT CIRCUIT ANALOGY

4. Figure 6-2(d) connects all-distant, equipotential points in Fig. 6-2(c) and our structure arrives at a solid cover to the parallel, quarter-wave sections; we have thus constructed an equivalent, closed metal container cavity.

MODES

The modes in a cavity are identified by the same numbering system used in waveguide, except that a third subscript is used to indicate the number of patterns of the transverse field along the axis of the cavity (perpendicular to the transverse fields). For example, the cylindrical cavity shown in Fig. 6-2(e) is a form of circular waveguide where the axis is the center of the cylinder and the transverse field is the magnetic field, marked TM. Around and parallel to the circumference there is a constant H magnetic field.

Thus, since there is no circumferential field pattern, the first subscript is zero; but with one complete field pattern across the diameter, the second subscript is 1; and since there is also a continuous magnetic H field through the center (without pattern), the third subscript is zero. A complete labeling reveals a $TM_{0,1,0}$ mode.

In the half-wave closed-end section of rectangular waveguide described in Fig. 6-1(b), the basic mode is the dominant one, that is, $TE_{0,1}$. Added to this is the plane of the E field on which there is a full half-wave, so the complete mode description is $TE_{0,1,1}$.

Table 6-1 lists the three basic cavities and gives the formula for calculating their natural wavelength.

TYPES AND ADJUSTMENT

Generally, cavity types may be categorized as rectangular, cylindrical, and coaxial; the basic forms of rectangular and cylindrical types are described in Figs. 6-1 and 6-2. When we shorten the cylinder, we can construct chambers of the doughnut and cylindrical-ring types shown in Fig. 6-3(a) and (b).

Tuning the cavity to specific frequencies is usually accomplished by physically changing the cavity size or by introducing metallic or ferrite plugs within the chamber.

TABLE 6-1 CAVITY TYPE AND WAVELENGTH CALCULATION

Shape	*Mode*	*Resonant wavelength*	*Applications*
Cube	$TE_{m, n, p}$ or $TM_{m, n, p}$	$\lambda_k = \frac{2b}{\sqrt{p^2 + m^2 + n^2}}$	Echo box
Cylinder	$TE_{0, 1, 1}$	$\lambda_k = \frac{2}{\sqrt{1/\ell^2 + 1.49/r^2}}$	Echo box wavemeter
Coaxial cylinder	TEM_p	$\lambda_k = 2\frac{\ell}{p}$	Wavemeter
Sphere with re-entrant cones	TEM_1	$\lambda_k = 4r$	T-R box

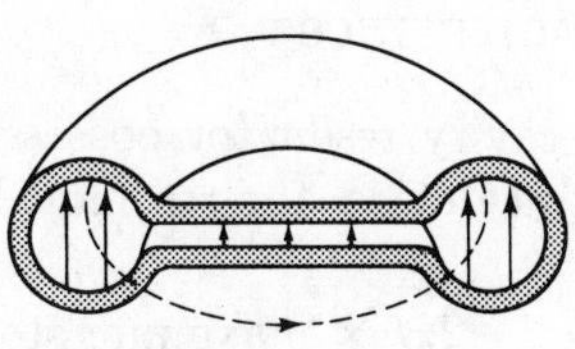

(a) Doughtnut-shaped

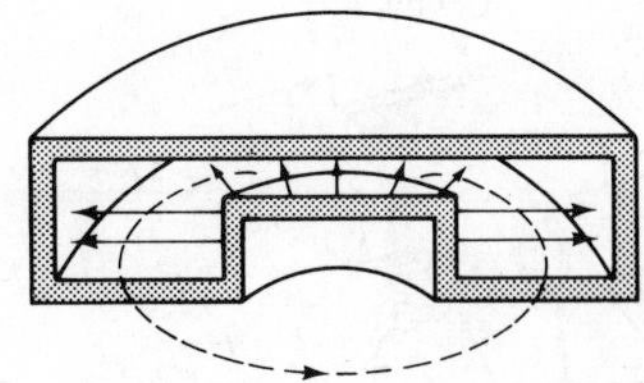

(b) Cylindrical ring (Q–26,000)

Fig. 6-3 DOUGHNUT AND RING CAVITY STRUCTURES

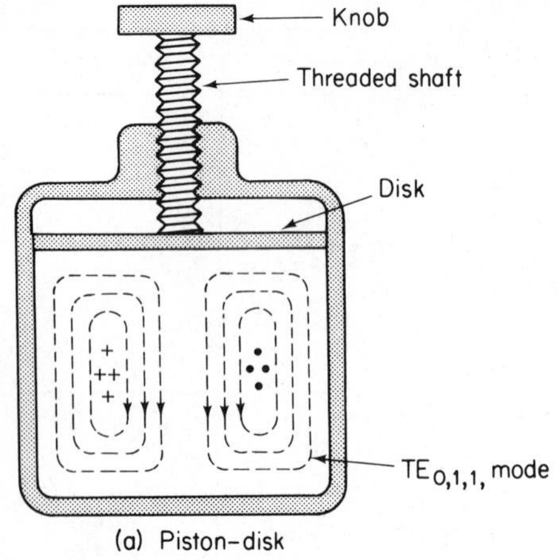

(a) Piston-disk

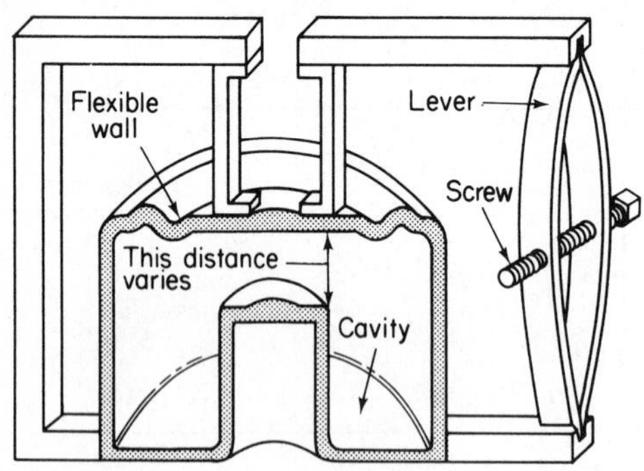

(b) Flexible wall

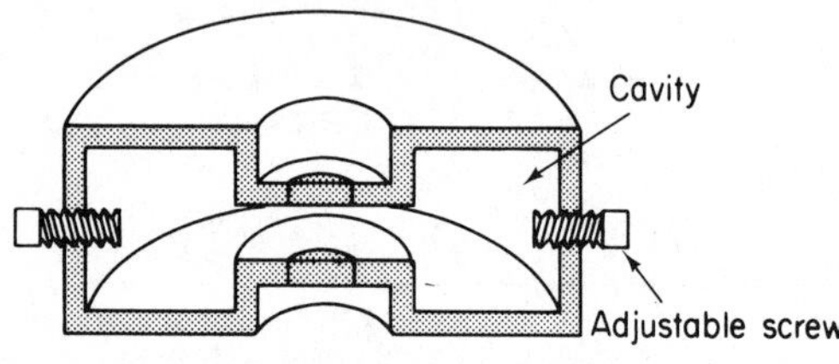

(c) Movable wall plugs

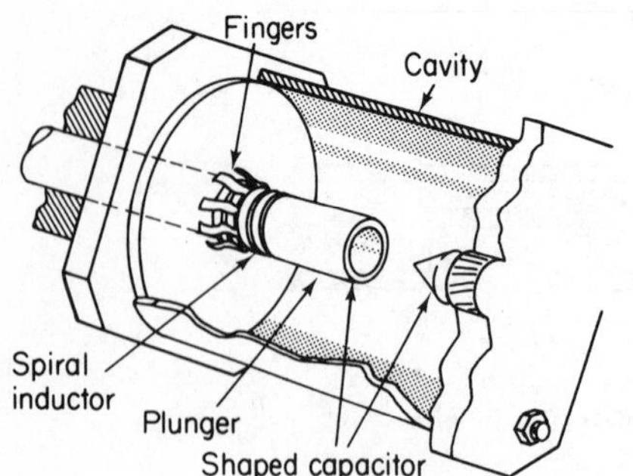

(d) Inductor-capacitor mechanism

Fig. 6-4 CAVITY TUNING SYSTEMS

Figure 6-4(a) illustrates how a cylindrical cavity may be shortened by the movement of an internal adjustable disk or piston. Using a cavity with a flexible wall, we can alter resonant frequency by leverage mechanisms as shown in Fig. 6-4(b). Movable plugs inserted through the wall of a cylindrical-ring cavity produce change in resonant frequency [Fig. 6-4(c) and 6-4(d)]. An ingenious tuning system [Fig. 6-4(g)] extends the range of a tunable cavity to 20 or 30 to 1. Here we use a combination of a spiral inductance and a capacitive probe mounted on a movable plunger. At higher frequencies the inductor acts as a shorting plunger (as if it were withdrawn) and as a capacitor plate as it is extended into the cavity.

COUPLING

Exciting or removal of energy from a cavity for use in operating circuits is usually accomplished by either a probe or loop introduced into the appropriate field of a cavity wall. Note that the method and positioning of coupling to a cavity determines whether it is to act at series resonance to absorb power or at parallel resonance to reinforce power.

Figure 6-5(a) shows placement of a probe in the E field of a cylindrical cavity having the $TM_{0,1,0}$ mode. Figure 6-5(b) shows how a magnetic loop is placed to extract H field energy from the same cylinder. Aperture coupling from waveguide-to-cavity-to-waveguide is shown in Fig. 6-5(c). Here the holes may be irises, holes, or slots which act to intercept current flow in the common walls between guide and cavity.

CHARACTERISTICS

A cavity resonator possesses, as does any resonant device, the resonant loss factor Q or

$$Q = \frac{2\pi f \times \text{maximum stored energy}}{\text{power loss}}$$

This factor is measurable, as we shall see in Chapter 11 and varies between 1000 and 3000 in commonly used cavities.

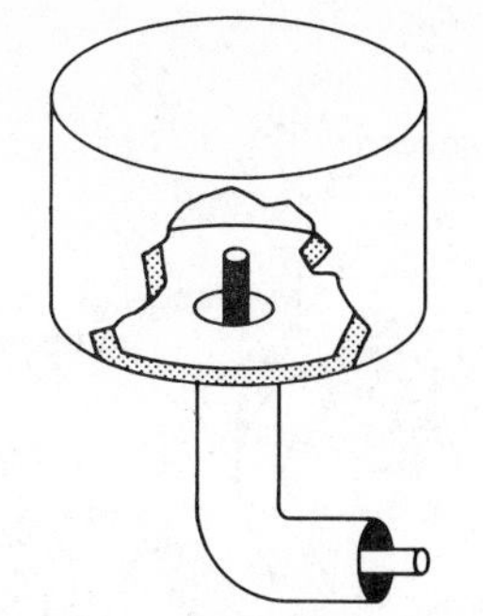

(a) Probe coupling $TM_{0,1,0}$ mode

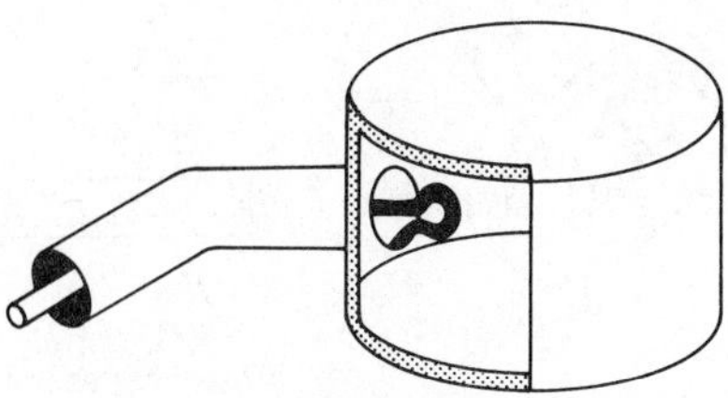

(b) Loop coupling $TM_{0,1,0}$ mode

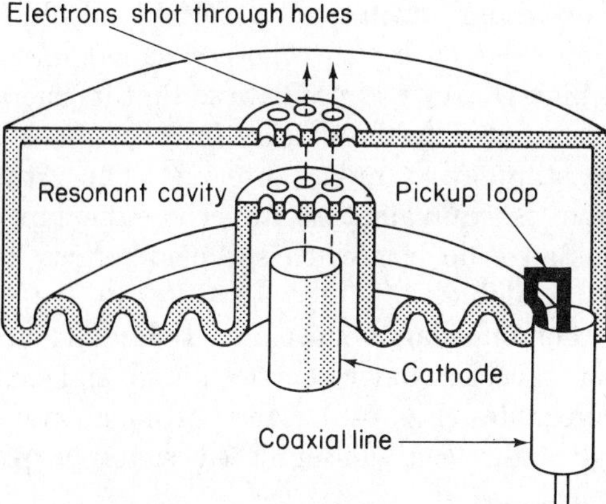

(c) Injection and output through a flexible ring

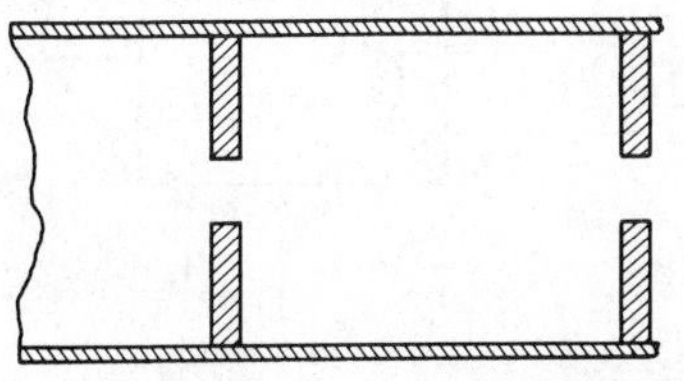

(d) Aperture coupling

Fig. 6-5 CAVITY COUPLING DEVICES

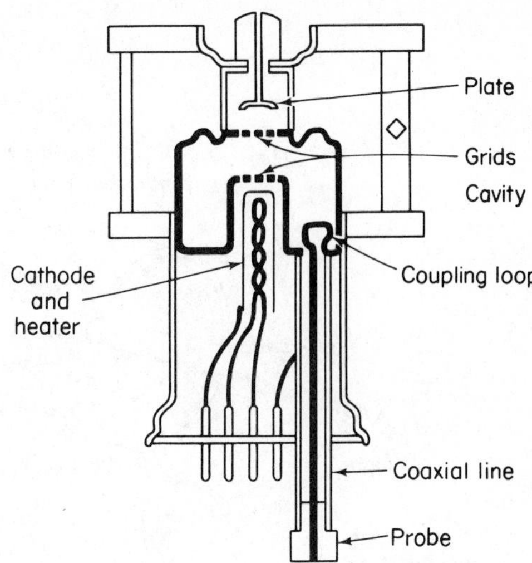

(a) Klystron cavity and coupling loop

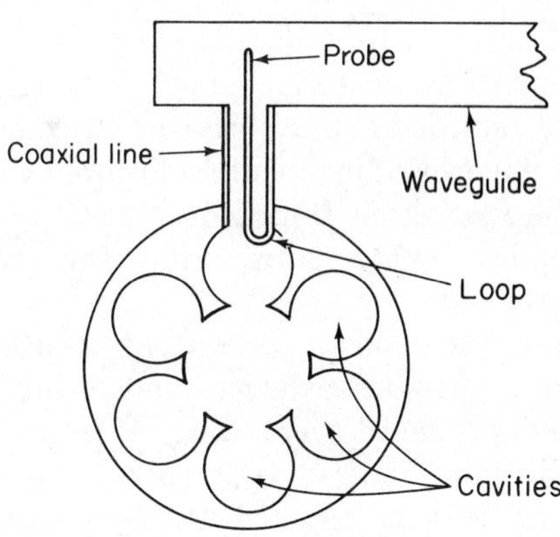

(b) Magnetron cavity and coupling loop

Fig. 6-6 KLYSTRON AND MAGNETRON CAVITIES

USES OF CAVITIES

Functionally, microwave cavities involving tuned circuits are divided into (1) oscillator, (2) amplifier, (3) filter circuits, and (4) hybrid arrangements. These align various cavity types into equipment and measurement uses.

Klystron and magnetron oscillators provide good examples of working cavities. (Also see Chapter 7.) Figure 6-6(a) shows a typical klystron assembly with the electron-

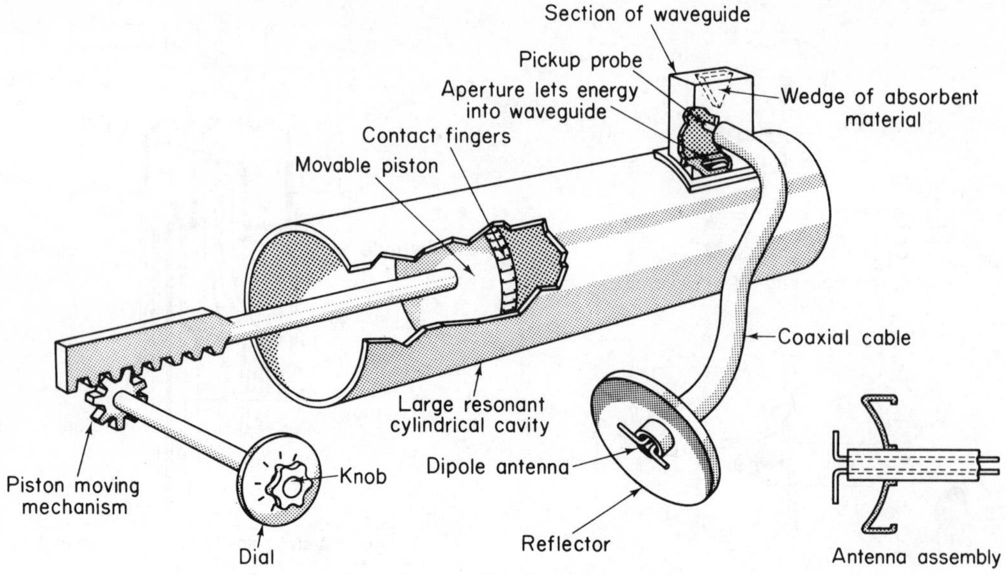

Fig. 6-7 CAVITY RINGING CIRCUIT

tube structure centering about a doughnut-shaped built-in resonant cavity, all complete with frequency adjustment and coupling loop.

The magnetron [Fig. 6-6(b)] uses a series of coupled cavities surrounding the central tube structure.

In a wave meter (see Chapter 10) for frequency-measuring devices and techniques, the cavity is tuned over a range of frequencies by adjustable rods; output energy from the coupling loop is rectified and stimulates a milliammeter.

Cavity range circuits are used to supply external energy after being stimulated by a sharp, high-power voltage pulse. This sudden input (applied periodically) causes the cavity circuit to continue to oscillate or "ring" for a period of a dozen or more oscillations (say 10 to 30 μsec). These adjustable frequency oscillations are radiated through an attached antenna and are used for comparison with oscillations generated in other resonant circuits. Figure 6-7 shows the complete ringing-circuit assembly, commonly called an *echo box*.

In radar testing and measurement the echo box serves as a target simulator when the main radar pulse is used to stimulate it. Main radar output at a suitable distance, when received by an echo-box signal, causes its high-Q cavity to oscillate so that its energy is reradiated (after a delay interval) back to the stimulating radar antenna. The return signal is equivalent to an echo—the box is thus an echo box when specified targets are not available.

For impedance matching, the cavity acts as a transition device between large and small waveguides (Fig. 6-8). Here, although standing-wave-reflection-stimulated signal is pre-

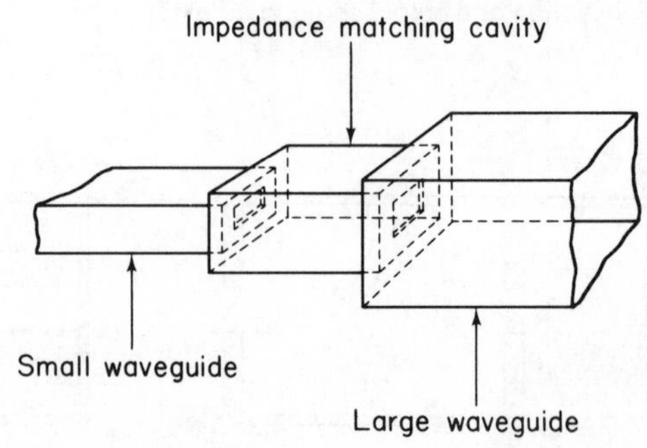

Fig. 6-8 CAVITY BASED FILTERS

TABLE 6-2 CAVITY Q CALCULATIONS

Shape	*Mode*	Q	*Maximum Q*	*Conditions for maximum Q*
Cube	$TE_{m,n,p}$ $TM_{m,n,p}$	$Q_k = \frac{\lambda_k}{\Delta} \frac{\sqrt{p^2+m^2+n^2}}{8}$ provided $p, m, n > 0$		
Cylinder	$TE_{0,1,1}$	$Q_k = \frac{\lambda_k}{\Delta} \frac{\sqrt{r^2/\ell^2+1.49}}{2} \times \frac{r^2/\ell^2+1.46}{r^3/\ell^3+2.92}$	$Q_k = 0.67\frac{\lambda_k}{\Delta}$	$\frac{r}{\ell} = 0.5$
Coaxial Cylinder	TEM_p	$Q_k = \frac{\lambda_k}{\Delta} \frac{p}{4+\frac{\ell}{b}\frac{1+b/a}{\ell n\, b/a}}$	$Q_k = \frac{\lambda_k}{\Delta} \frac{p}{4+3.6\frac{\ell}{b}}$	$\frac{b}{a} = 3.6$
Sphere with re-entrant cones	TEM_1	$Q_k = \frac{\lambda_k}{\Delta} \frac{1}{4+3.3\frac{\csc\theta_0}{\ell n \cot\theta_0/2}}$	$Q_k = 0.11\frac{\lambda_k}{\Delta}$	$\theta_0 = 33.5°$

sent in the intermediate waveguide section, none exists in the terminating sections, so we have created, in effect, transformer action stimulated by the inserted guide section.

RESONANT CAVITIES

Resonant cavities are commonly used for single elements in filter circuits, as will be discussed in the following section. Major cavity characteristics, in order to simplify calculation of their performance when inserting into filter assemblies, are seen in Table 6-2.

MICROWAVE FILTERS

GENERAL

The usual microwave filter is a signal frequency acceptance or rejection device inserted in a waveguide system to allow transmission or to reject transmission over a specified frequency range. The range of frequency is called a *band* and we have, roughly speaking, three pass bands and three stop bands when we consider all frequency signals from zero to infinity.

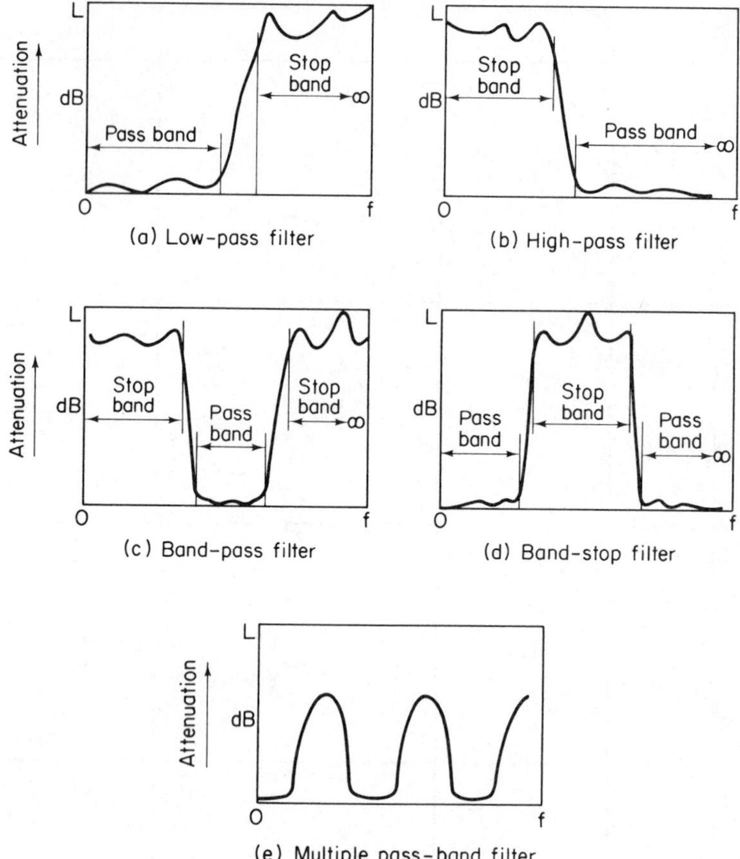

Fig. 6-9 BASIC FILTER CHARACTERISTICS

This is illustrated in Fig. 6-9; plotting attenuation on a vertical scale we have a *low-pass filter*, which transmits all signals from direct current up to a specified point where it meets a stop band, which attenuates all frequencies from there on up to infinity.

A high-pass filter attenuates all frequencies up to a specified point and passes all frequency signals above it.

Now, if the filter is selective over a given band of frequencies, it offers either pass-band or stop-band qualities [Fig. 6-9(c) and (d)].

Note that these filter characteristic curves, unlike low-frequency performance curves, have attenuation irregularities in both stop- and pass-band areas; this is chiefly due to the high Q of the filter element and also to the fact that microwave circuits employ multi-element sections, all ideally requiring simultaneous function, which is a condition very difficult to obtain. This is particularly true with high Q circuits where, in addition to being unable to control loss factors, a particular circuit element may be impossible to isolate for measurement purposes.

DESIGN—CONSTRUCTION

Briefly, microwave filters may be either waveguide or stripline in nature; discussion of the former is more applicable since design specifics are better correlated with the rectangular waveguide background covered up to this point.

The usual filter structure is made by direct assembly of a number of resonant waveguide

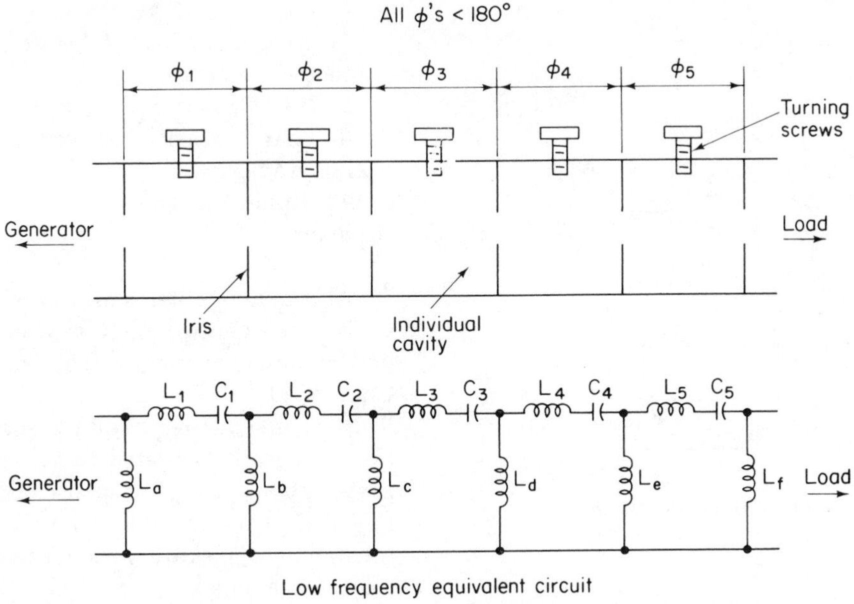

Fig. 6-10 FILTER DESIGN ARRANGEMENTS

cavities, each cavity being coupled to an adjacent cavity by an aperture or an iris. A typical arrangement correlated with the equivalent low-frequency circuit is shown in Fig. 6-10. The series resonant circuits are represented by the cavities while the irises constitute the shunt inductance.

Approaching physical quantities within the normal dimensions of most waveguides, we require knowledge, measurement, and tailoring of five items:

1. Resonant wavelength of the individual cavities.
2. Loaded Q of the individual cavities.
3. Equivalent susceptance of the irises.
4. Mutual coupling between adjacent cavities.
5. Input and output impedances.

The resonant frequency of a rectangular waveguide (Fig. 6-10) is given by

$$\lambda_g = \frac{4}{\sqrt{(l/a)^2 + (m/b)^2 + (n/z_0)^2}}$$

where l = number of half-wave variations of field along the x axis

m = number of half-wave variations of field along the y axis

n = number of half-wave variations of field along the mutually right-angle axis

a = waveguide width (E field)

b = waveguide height (H field)

z_0 = waveguide impedance

In design, the exact resonant frequency is modified by practical considerations; for instance, final cavities are cut somewhat shorter than the calculated length because the tuning screw tends to lower the frequency. Also, end sections likewise differ slightly in frequency from internal sections because of capacitive loading.

The loaded Q of a cavity depends upon the skin depth, the resonant wavelength, and the cavity dimensions. Inasmuch as it also depends upon other less controllable factors of

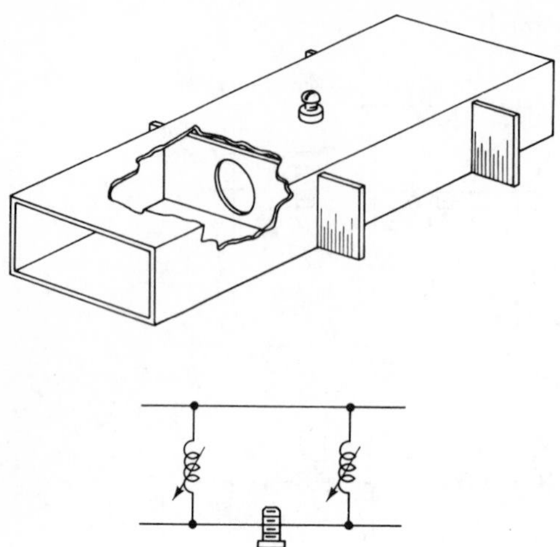

Fig. 6-11 ADJUSTABLE IRIS DESIGN

construction, such as surface plating, smoothness, joint configuration, etc., design prediction upon loaded Q cannot be too firmly established.

Iris design is also a highly unpredictable process depending upon the relation of hole size and shape to waveguide dimensions and guide wavelength. To compensate for variations, most irises have some form of adjustability to arrive at optimum coupling. Figure 6-11 shows one method of installing an adjustable-iris system.

As noted above, input and output sections must be tailored to compensate for termination and transition to the waveguide impedance. Figure 6-12 shows overall construction of a number of commercial microwave filters.

Measured overall characteristics, besides determining the frequency-bandwidth characteristics, are concerned with:

1. Steepness of the sides of the response characteristic curve.
2. Insertion loss.
3. VSWR when assembled with its compatible waveguide.
4. Ripple response or Tchebycheff behavior.

All the above items are more or less self-explanatory and will be treated in more detail under measurement procedures covered in Chapter 11.

Item 4, however, requires some explanation because it is related to the variations in pass- and stop-band response. Referring to Fig. 6-13(a) and (b), the variations in response resembling ripples are called Tchebycheff variations. Designers aim to make these ripples all equal by loading or decoupling the cavities; in doing this some variations are reduced at the expense of steepness of the sides of the main curve or, in effect, by decreasing the attenuation.

Fig. 6-12 COMMERCIAL FILTER ASSEMBLY CHARACTERISTICS (Courtesy of Microlab/FXR.)

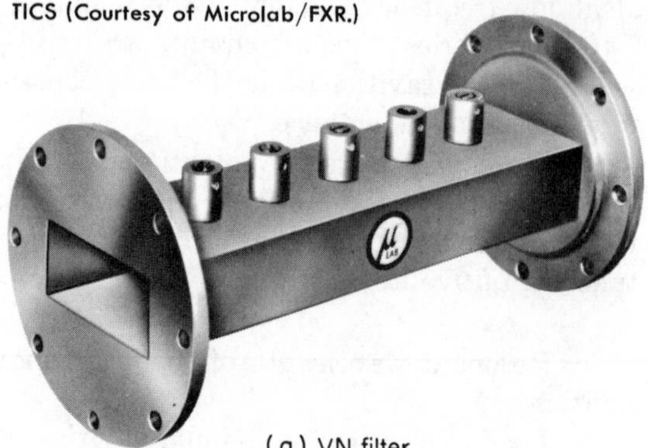

(a) VN filter

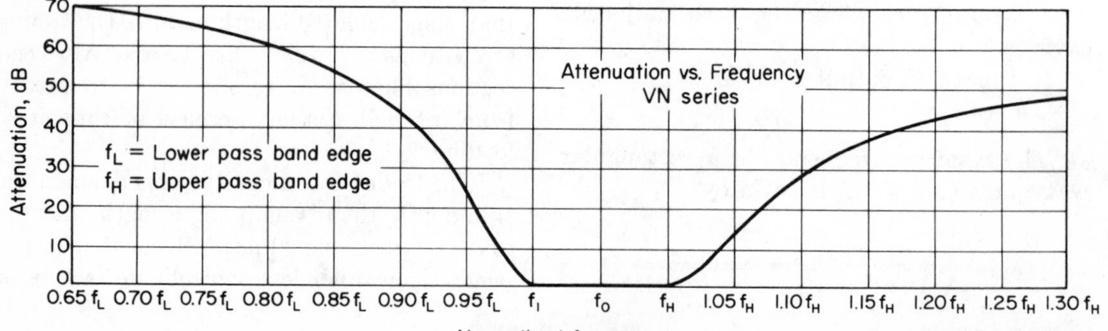

(b) Characteristics

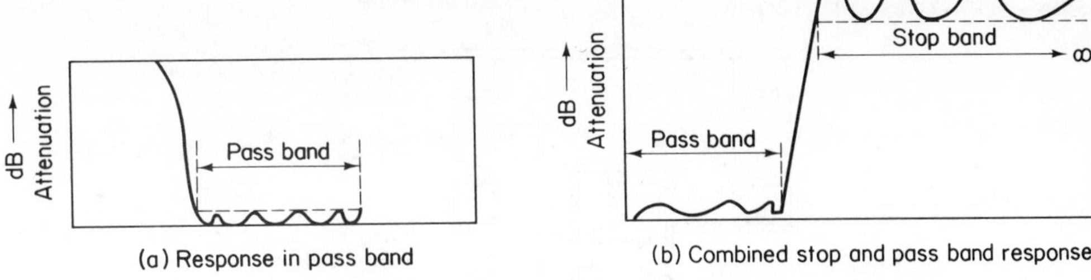

(a) Response in pass band

(b) Combined stop and pass band response

Fig. 6-13 TSCHEBYSCHEFF RESPONSE CURVES

STRIP-LINE FILTERS

These units stem directly from strip-line waveguide construction and are merely half-wavelength resonators directly coupled through series capacitors. Figure 6-14(a) and (b) show how a cross section of basic strip line plus an equivalent circuit of the series and parallel capacitances exists at a gap. Figure 6-14(c) and (d) show the physical dimensions of a working model plus its band-pass characteristic.

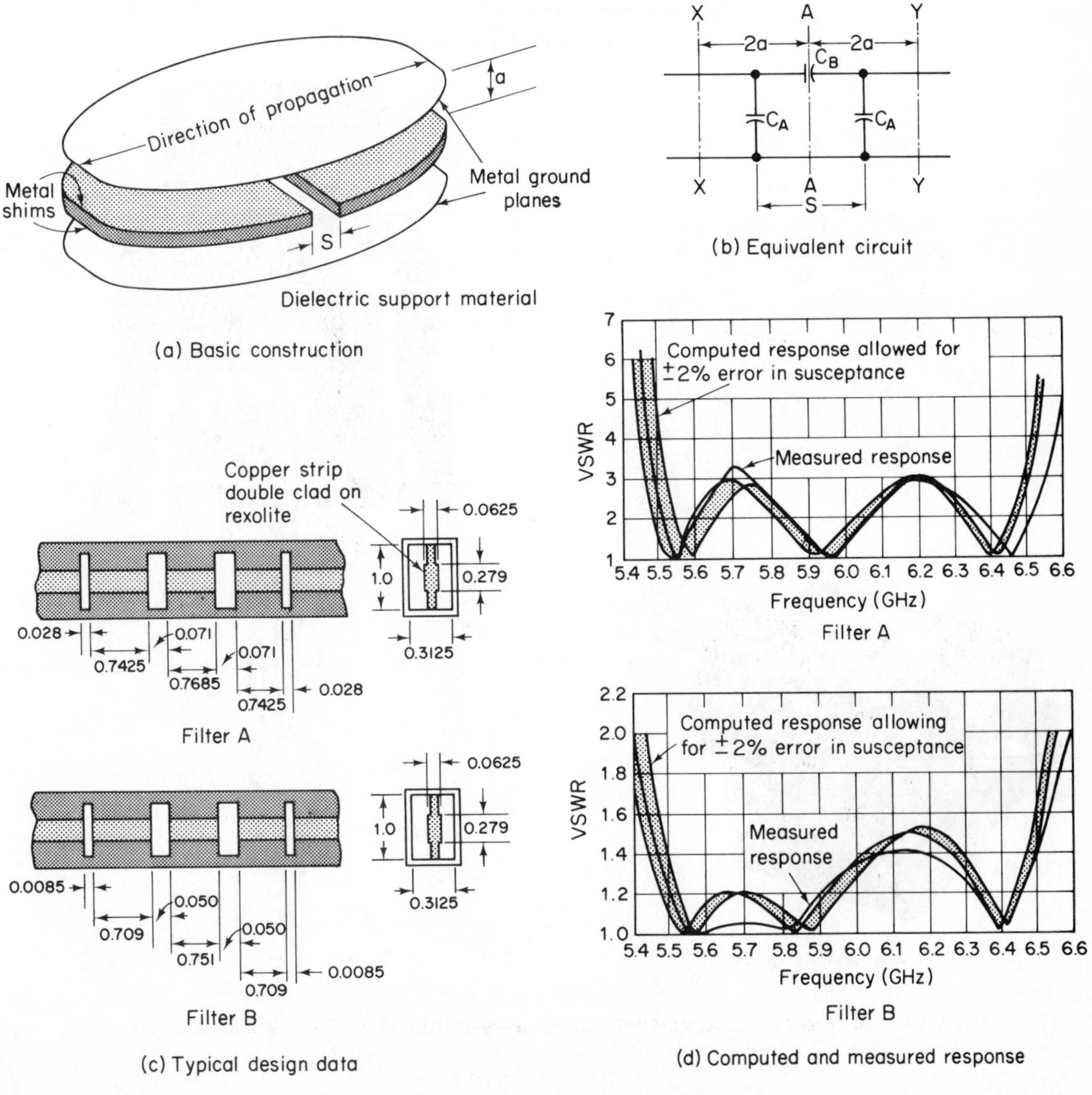

(a) Basic construction

(b) Equivalent circuit

(c) Typical design data

(d) Computed and measured response

Fig. 6-14 STRIP-LINE FILTER DESIGN

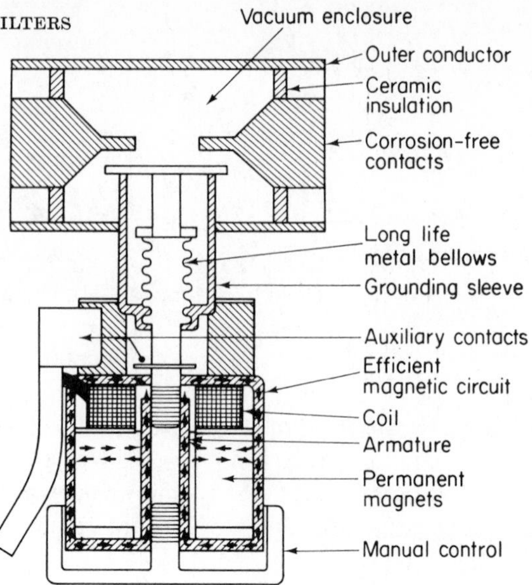

(a) Relay and plunger type

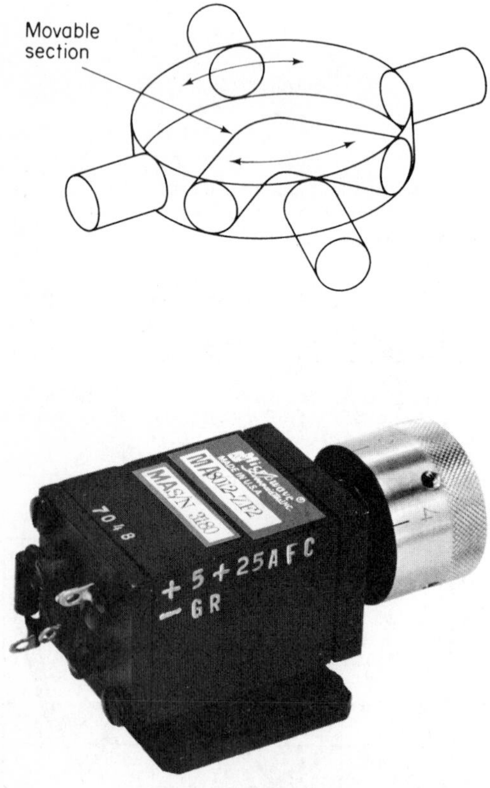

Movable section

+5 +25 A F C
— G R

(b) Rotating line section

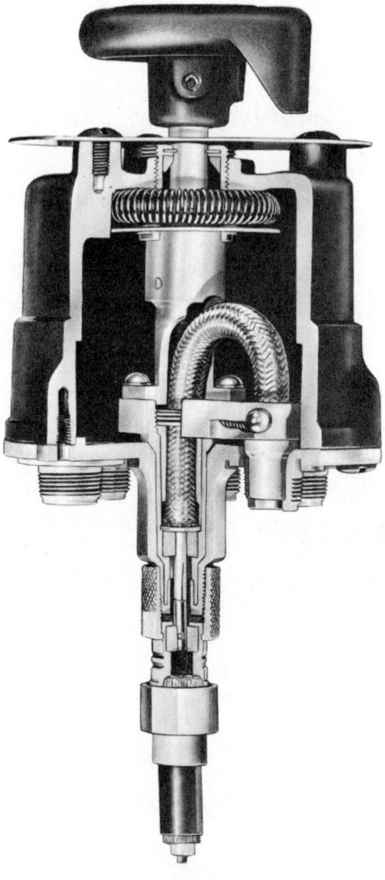

(c)

Fig. 6-15 MECHANICAL COAXIAL SWITCHES (Courtesy of Microwave Associates and Microlab/FXR.)

MICROWAVE SWITCHES

Microwave systems employ four basic methods of diverting, channeling, short-circuiting, or by-passing a signal or transmitted power:

1. By mechanical switches.
2. By diode or other solid-state devices.
3. Through ferrite-controlled signal diversion.
4. By gas tubes.

MECHANICAL SWITCHES

Mechanical switches may be generalized under coaxial and waveguide-absorption types. Coaxial switches may use either plunger or rotary mechanisms, while waveguides use rotary, vane, and shutter devices to bridge or transfer power.

Figure 6-15(a) shows a relay-operated plunger type of coaxial switch, operating the shorting element in a vacuum. Figure 6-15(b) uses a rotating drum or rotor containing a short section of coaxial line which is spring loaded at its center for contact between two adjacent input lines. Most mechanical switches are adapted only for relatively limited use in the lower microwave frequencies.

Waveguide-absorption switches insert obstacles within the structure to absorb or divert energy flow. We may use a shutter [Fig. 6-16(a)], a post [Fig. 6-16(b)], or a fin [Fig. 6-16(c)] to accomplish this.

Figure 6-17(a) shows the shutter type commonly used to block off detector crystals or in applications where excessive reflections or deterioration of VSWR are of no consequence.

Figure 6-17(b) illustrates a switch utilizing a sliding shuttle or fin which diverts input power from the input port to one or another output ports.

SOLID-STATE SWITCHES

A microwave switching diode internally placed directly across the correct field contour within a waveguide can act, when properly biased, as a switch. Since diode structures

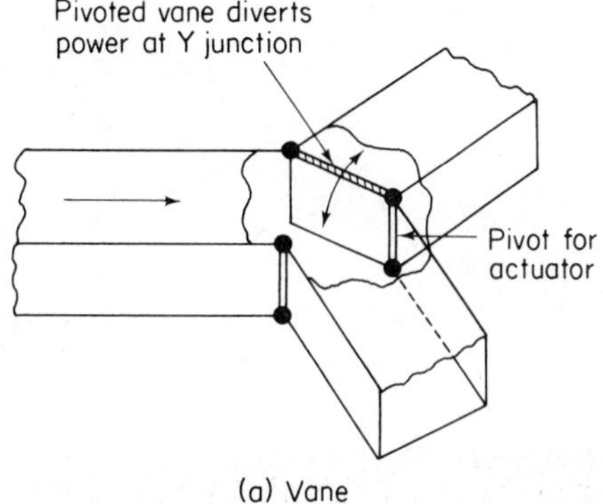

(a) Vane

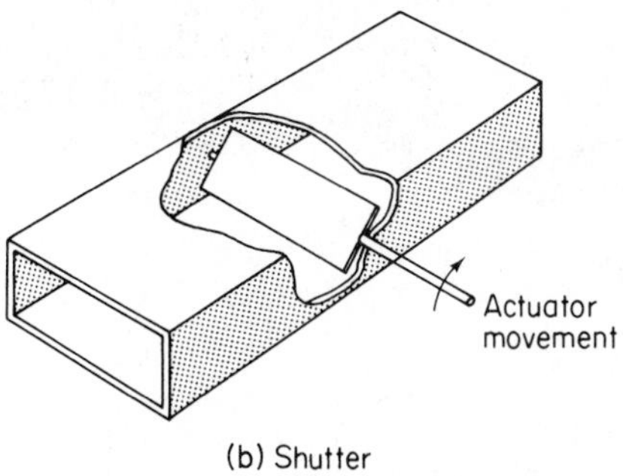

(b) Shutter

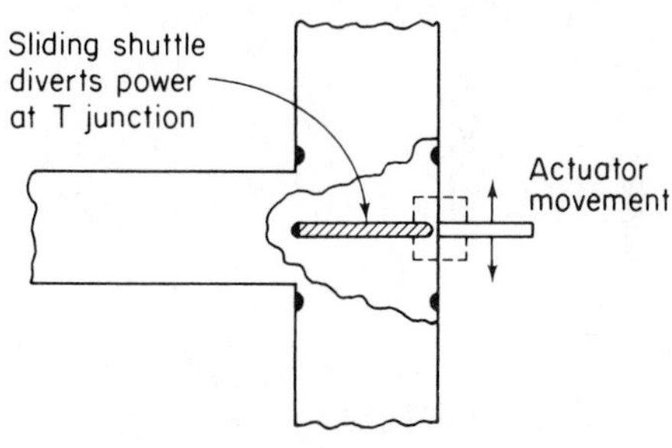

(c) Sliding shuttle

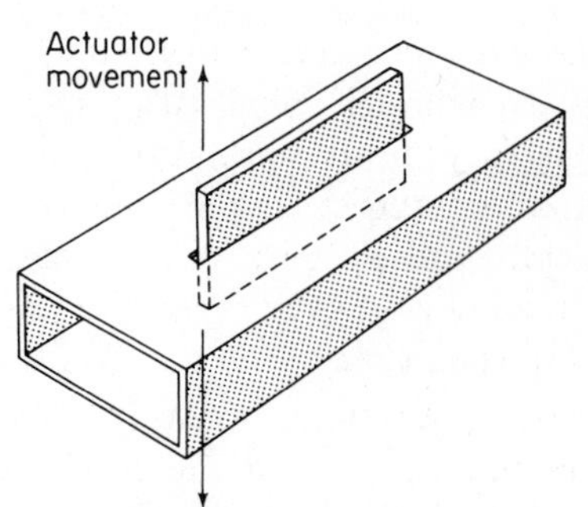

(d) Inserted fin

Fig. 6-16 WAVEGUIDE SWITCH SYSTEMS

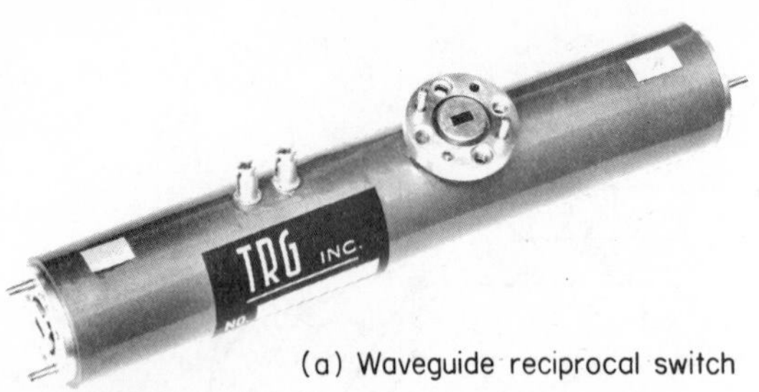

(a) Waveguide reciprocal switch

(b) Waveguide ferrite switch

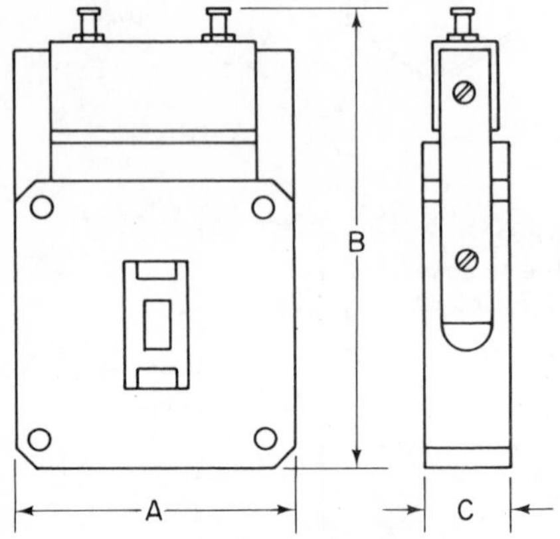

(c) Waveguide shutter-type switch

Fig. 6-17 COMMERCIAL WAVEGUIDE SWITCHES

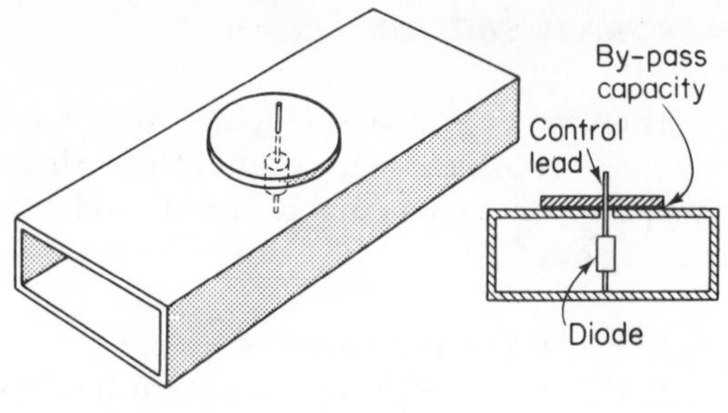

(a) Layout

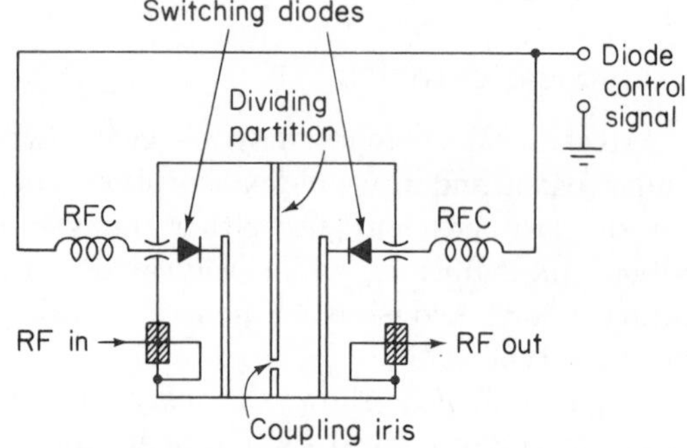

(b) Switching circuit

Fig. 6-18 DIODE SWITCHING CIRCUITS
(Courtesy of TRG Inc. and Ferrotec Inc.)

are minute elements having low package and electrode capacitance plus negligible self-inductance, they can be specially constructed to operate at gigahertz frequencies; they are thus critically and electrically ideal elements to switch microwave power directly under and at operating conditions.

Electrically, provision must be made to isolate the bias-carrying diode lead with a bypass capacitor or through use of an rf choke. Figure 6-18(a) shows a shunt arrangement of a diode and its bypass capacitor; Fig. 6-18(b) shows two switching diodes mounted on the center conductor of two iris-coupled coaxial cavities. The diodes are controlled by bias signals isolated by rf chokes as they pass through the cavity wall. Table 6-3 shows representative circuit constants of a number of switching and varactor diodes. Figure 6-19(a), (b), are representative commercial switching units.

FERRITE SWITCHES

The action of a permanent magnet upon a signal-carrying waveguide containing ferrite slabs, fins, or vanes was discussed in Chapter 5 when describing the action of isolators and circulators. In these, of course, the permanent magnetic field is so applied that the losses are at a minimum.

If the field is reversed, the losses become high, and if properly absorbed, we have established a short-circuiting switch. Figure 6-20

TABLE 6-3 DIODE SWITCHING CIRCUIT CONSTANTS

Parameter	*Symbol*	*Si Junction MA450 Type*	*Ge Silver Bonded Varactor Diodes*			*Units*
			GSB1A	*GSB1B*	*GSB2*	
Lead inductance	L_s	2	3	3	3	nh
Package capacitance	C_p	0.4	0.1	0.1	0.1	pf
Zero bias junction Capacitance	C_o	0.6–2.0	0.3	0.3	0.3	pf
Cutoff frequency	f_e	90-120	60 (min)	100 (min)	40 (min)	kMc/sec
Breakdown voltage	V_B	6	6	6	11	volts

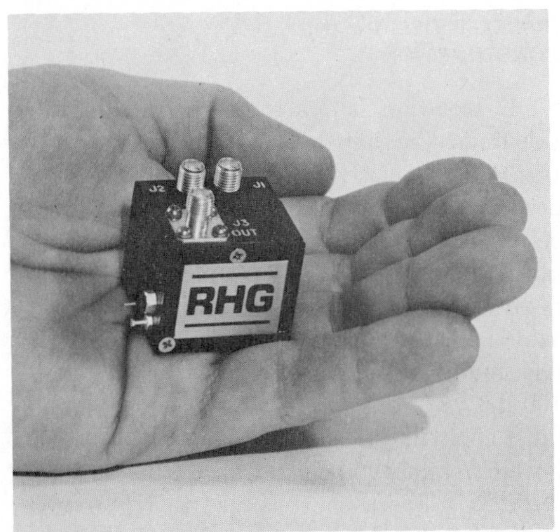

(a) Diode switch

(b) Coaxial diode switches

(c) Diode test signal switches

Fig. 6-19 COMMERCIAL SOLID-STATE SWITCHER CIRCUITS (Courtesy of Scientific Atlanta, Inc., RHG Electronics Lab., and Narda-Microline.)

illustrates the construction of an absorption-type ferrite switch. Here we have two ferrite slabs externally located in a standard waveguide. A thin resistance card is inserted between the ferrite slabs perpendicular to the input rf field. In the normal condition, without a permanent magnetizing field, rf energy is transmitted without loss through the guide. Under magnetization the rf energy stimulating the ferrite is coupled into the resistance card, resulting in power attenuation. This trans-material effect is known as *tensor permeability*.

GAS SWITCHES

A waveguide section filled with ionizable gas and arranged with close-spaced electrodes

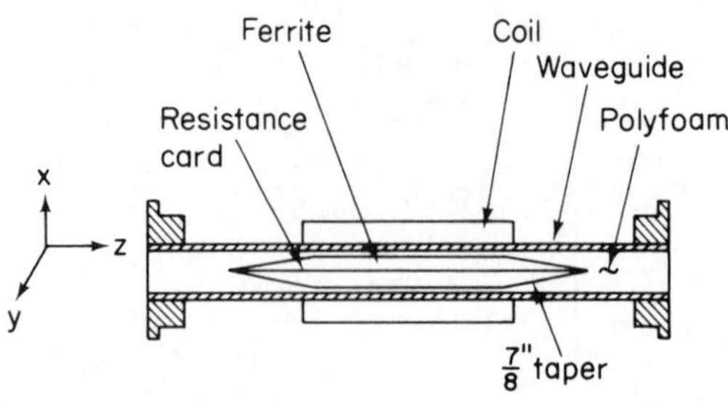

Fig. 6-20 ABSORPTION TYPE FERRITE SWITCH FOR WAVEGUIDE

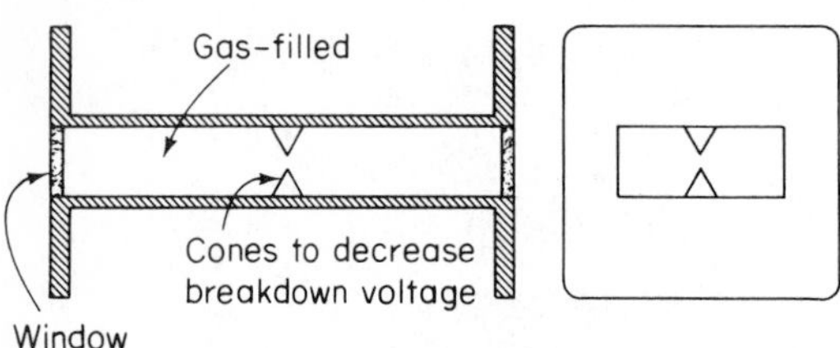

Fig. 6-21 GAS SWITCHING TUBE

can act as a short-circuiting switch; when the signal voltage or applied external voltage ionizes, the gas breakdown occurs.

Such a unit, with the gas sealed within it by glass end windows, will pass microwave energy without losses. The device is power selective since internally projecting electrode points are so spaced that the desired peak power produces breakdown (Fig. 6-21).

DUPLEXERS AND DIPLEXERS

These devices are essentially switch-oriented units; their functional operation follows the basics covered in the previous paragraph on gas switches.

A duplexer is a device that specifically and physically switches an antenna between transmitter and receiver by means of a gas tube or ferrite.

A diplexer switches an antenna between transmitter and receiver by means of selective filters. We can see this in the generalized layout of Fig. 6-22, where we have pictured the overall Y or three-post character of a transmit–receive system using a single antenna.

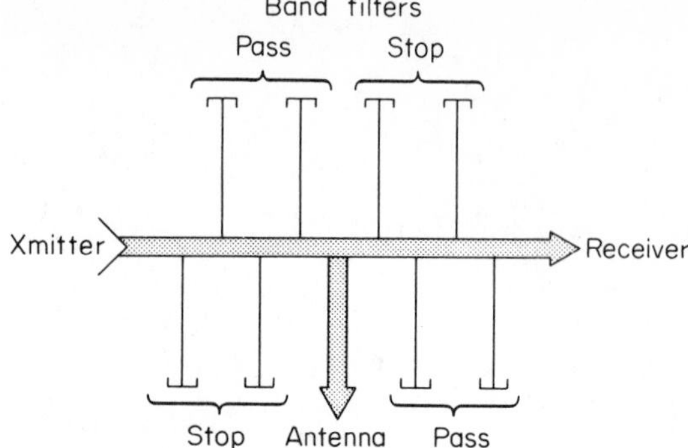

Fig. 6-22 DIPLEXER SWITCHING LAYOUT

DIPLEXER DESIGN AND CONSTRUCTION

1. Isolation is the basic requirement for this device in that its structure must prevent transmitter energy from reaching the associated receiver and also must protect the receiver from massive transmitter-output signals. This must be accomplished over the bandwidth requirements of the receiver (the same as the transmitter passband) and usually requires rejection factors of from 40 to 100 dB. In other words, the receive terminal must look like a short circuit to the transmitter frequency and the transmitter terminal must look like a short circuit to the receive frequency. Such short circuits usually have a 20-dB front-to-back isolation.

2. VSWR in the pass band under environmental changes must be below 1.25:1 so that the transmitter load appears constant under all conditions and prevents intermodulation when communication frequencies are used.

3. Insertion loss should be kept low (under 0.4 dB). In a high-power transmitter this wasted power factor can become expensive, besides bringing up heat-dissipation problems.

4. Bandwidth and tunability. In communication systems, wide, sharply defined pass and stop bands must be compatible with easy, practical tunability. In other words, the stubs or cavities must not be too complicated or too sensitive for installation or maintenance.

5. Harmonic rejection. The diplexer must provide some attenuation against harmonics of the transmitter's fundamental frequency, although most communication transmitters

use harmonic filters across their output terminal.

6. Power capabilities. In transmitters of the 100-kW and over range, physical size must be capable of sufficient thermal dissipation without becoming ungainly.

GENERAL DESIGN

Table 6-4 summarizes diplexer type and construction. We can conclude from the above tabulation that we use:

1. Coaxial or waveguide stubs for lower UHF.
2. Stop- or pass-band tuned cavities for upper UHF.
3. Hybrid, folded T, or circulator units for short microwave.

FOLDED-T SWITCHING CONFIGURATION (FIG. 6-23)

This system uses the fact that folded T hybrid-waveguide sections split input signals equally and also act as directional couplers. If we apply transmitter signals through one arm of a folded T, the output splits equally and in phase between the two colinear arms of the T; when passing signals through two lines L_1 and L_2, one line's wavelength being different from the other's folded T antenna terminal, we deliver full transmitted power, since the signals will reinforce and equal the original input power.

The receiver terminals at the X leg of number 1 folded T are shielded from transmitter energy by the directivity of the hybrid number 1.

Now, to receive, antenna signals are split and are phase shifted, going backward through lines L_1 and L_2 so that they arrive in phase for full reception at the receiver terminal.

DUPLEXER DESIGN AND CONSTRUCTION

Early duplexers, particularly in radar, used both TR and ATR circuits (called boxes) in single antenna operation of a transmitter–receiver complex. The TR box (including the gas tube) acted to prevent the transmitter power from reaching the receiver, while the purpose of the ATR box and circuit (including a TR tube) was to disconnect the transmitter during the receiving period. The TR and ATR tubes are gas-filled switching tubes, TR denoting *transmit–receive* and ATR denoting *anti-transmit–receive.*

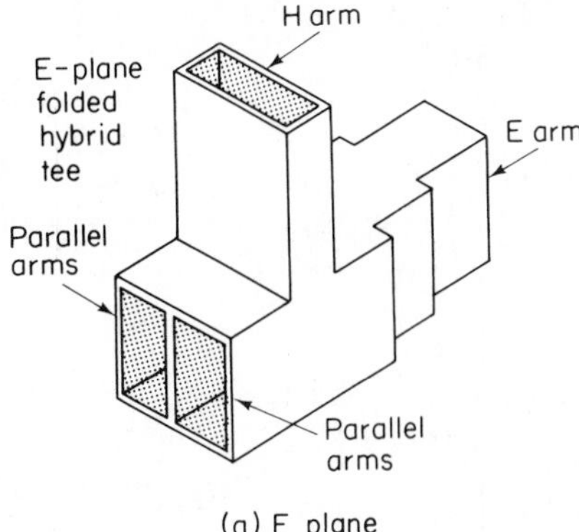

(a) E plane

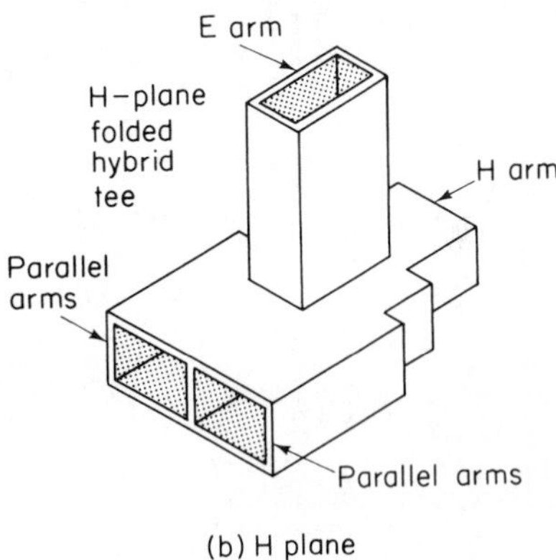

(b) H plane

Fig. 6-23 FOLDED *T* WAVEGUIDE SWITCHING ASSEMBLY

Figure 6-24(a) simulates the waveguide or coaxial line as a two-wire circuit with all the duplexer components; here the receiver transmission line is in parallel with the transmitter line and shows a spark gap (as a simplified gas tube) placed at the receiver terminals at $\lambda/4$ from the transmission-line junction. This is the structure of the TR box.

One quarter-wavelength from the receiver-line junction, going toward the transmitter, an additional quarter-wavelength line is placed in parallel with the transmitter; this line, also terminated in a gas-tube spark gap, is the ATR box.

TABLE 6-4 DIPLEXER TYPES

	Type description	Sketch	Type of line	Tuneability	Environmental	Size	Power	Pass band VSWR	Pass band loss	Stop band attenuation	Harmonic rejection
BAND STOP	Stubbs on a "T"	Xmitter, Ant	Coax or Waveguide	Very good field tuneable over full band Any freq. separation	Rugged Not sensitive to environment	Large tuning shorts must be accessible	Will stand very high average and peak power. Needs no cooling	Depending on No. of stubs typical–4 stubs 1.2 over 0.05% band	Good Determined entirely by total length of guide or Coax used	Any isolation level can be obtained. Depends on No. of stubs. Units can be cascaded. Band width limited. Typical–4 stubs 85 dB at 0.05% 150 dB at f_0	Good
	Hybrid and stubs	Ant, Xmitter, Rec	Coax or Waveguide	Very good field tuneable. Limited by bandwidth of hybrid	Rugged Not sensitive to environment	Coax is used at low freq. Stubs may be 20 ft long depending on freq. separator	Can be used up to the full rating of the hybrid	Depending on No. of stubs and hybrids typical–2 stubs 1.2 over 0.05% band	Good Depends on No. of stubs and hybrid	Any isolation level can be obtained depending on No. of stubs addition isolation obtained from hybrid ≈ 30 dB–waveguide ≈ 40 dB–Coax	Good
BAND PASS	Band pass filter on "T"	Xmitter, Rec, Ant	Waveguide Coax at very low freq.	Waveguide not field tuneable. Filters have to be interchanged. / Coax units can be field tuned	Sensitive to temperature	More compact	At high average powers radiating fins or water cooling must be used	Very good 1.2 up to 20% band Band width establishes No. of cavities	Medium Depends on No. of cavities and quantity of transmission line	No band width problem. Level set by No. of cavities	Medium
	Band pass filter and hybrid	Ant, Xmitter, Rec	Coax or Waveguide	Waveguide not tuneable. Filter and hybrids have to be interchanged. / Coax units can be field tuned	Sensitive to temperature	Can be made very compact	Filters are limiting factor. Not hybrid	Very good Limited by band width of hybrid	Medium Depends on No. of cavities and hybrid	No band width problem. Addition isolation obtained by hybrid	Medium
FOLDED "T"	Folded "T"	Ant, L_1, L_2, Rec, Xmitter	Waveguide	Field tuneable by replacing line length	Rugged Not sensitive to environment Nothing to deteriorate with time	Size in between band pass and stubs type	Highest power rating Will handle full power of folded "T"	Depends on length of L1 and L2 Typical 1.2 over 0.03%	Very good Depends only on amount of transmission line involved	Medium Maximum isolation obtainable is isolation of hybrid	Very good
CIRCULATOR	Circulator and band pass filter	Xmitter, Rec, Ant	Waveguide with Coax Preselector	Broad band has bandwidth of circulator. Must use Coax preselector type	Limited by behaviour of ferrite	Compact	Low to medium power At high average power circulator performance deteriorates	Depends on circulator and preselector filters	Medium Loss in circulator increases with higher average powers	Medium Isolation of circulator may be as low as 10 – 20 dB Relatively little protection from rec. noise generated by transmitter	Medium

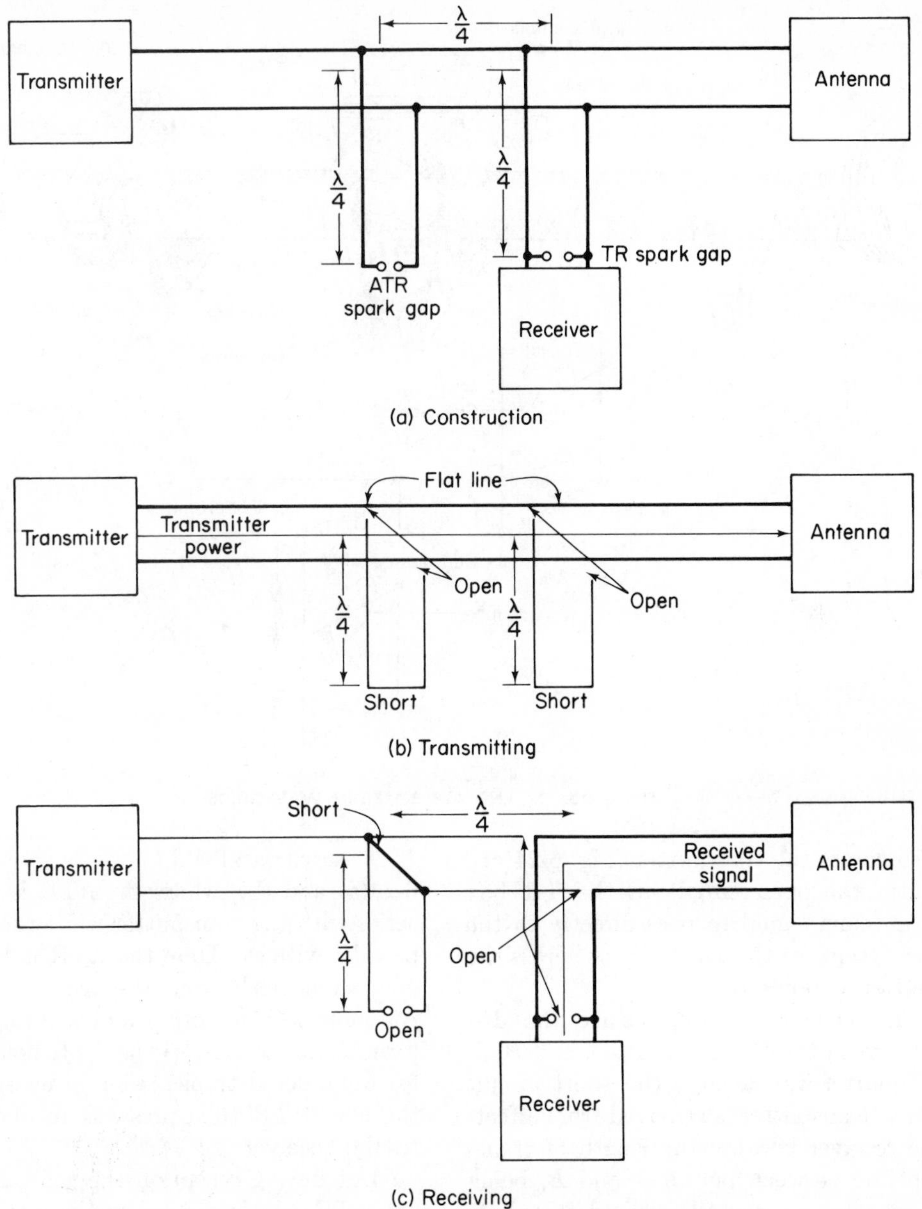

Fig. 6-24 TR AND ATR ACTION IN DIPLEXER SYSTEMS

During transmission, both spark gaps fire [Fig. 6-24(b)] and cause the TR and ATR circuits to act as shorted quarter-wave transmission lines, thus reflecting an open-circuit, maximum impedance structure at both junctions with the main transmission line and allowing full transmitted power to pass to the antenna.

The receiver terminals are at this time short circuited and no transmitter power reaches the receiver circuits. During reception, the amplitude of the received power is not enough to fire either gap, so the ATR circuit is now an unterminated quarter-wave transmission line, which, however, reflects a short circuit back across the transmission line

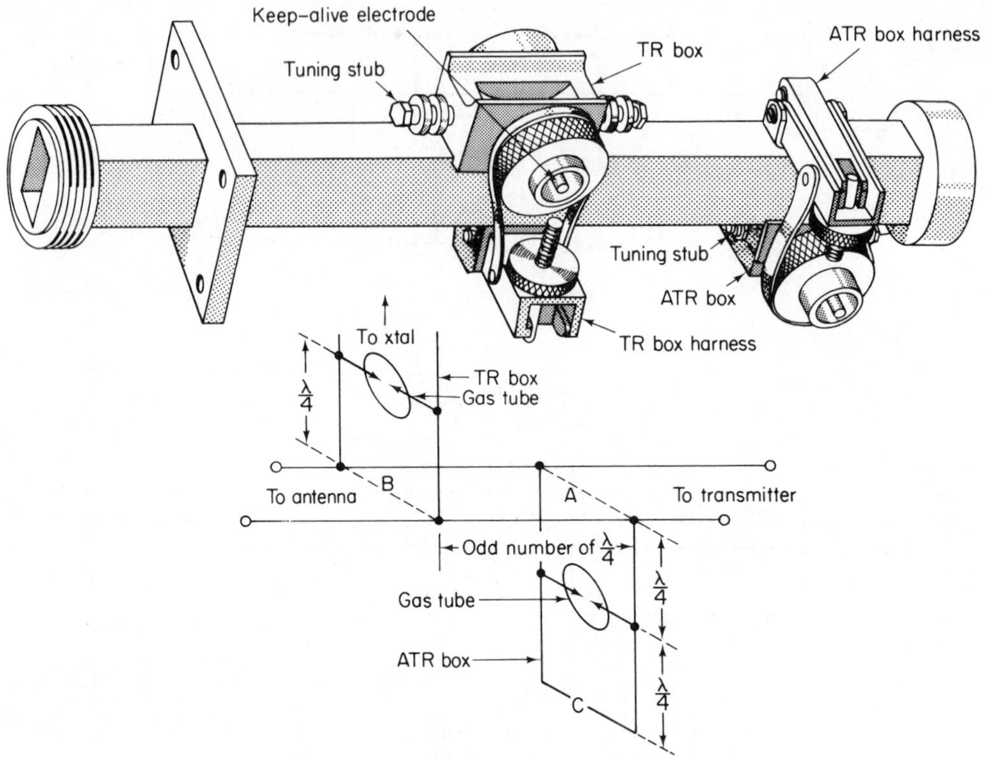

Fig. 6-25 TR AND ATR BOXES IN WAVEGUIDE

and also across the transmitter [Fig. 6-24(c)]. Here also the open circuit in the TR box allows antenna signal to pass directly to the receiver except for the short circuit across the transmitter at point B.

Simultaneously, however, since the distance between the boxes (points A and B) is also a quarter-wavelength, the short circuit across the transmitter is removed from affecting the receiver because the length of transmission line between points A and B, being also a quarter-wavelength, causes the short circuit to make this section look like an open circuit. The receiver thus automatically isolates itself and opens up its own shorting bar for full reception.

Figure 6-25(a) illustrates a typical assembly with the main waveguide supporting the two boxes, their adjustable tuning stubs, and the TR-tube sockets. Note that switching will occur when the distance between points A and B is a distance equal to any odd number of quarter-wavelengths.

A series parallel duplexer is shown in Fig. 6-26(a) and (b), where the ATR is placed in series with the transmitter while the TR is in parallel with it. Here the ATR is located at one wavelength from the receiver junction. At transmission both tubes are ionized and form short circuits, so the ATR box is sealed for unhindered transmission of energy. Also, the closed TR tube presents an open circuit to the receiver.

But during reception the gas tube within the ATR box is not ionized, and the device reflects a high impedance at A; and, since the distance from A to B is a full wavelength, it also presents a high impedance at B. Receiver energy, of course, passes directly from the antenna to its terminals.

FERRITE DUPLEXERS. Ferrite duplexers are used in low- and moderate-power radar systems, following the principle of operation described in a circulator (see Chapter 5). The mechanism makes use of ferrite-actuated Fara-

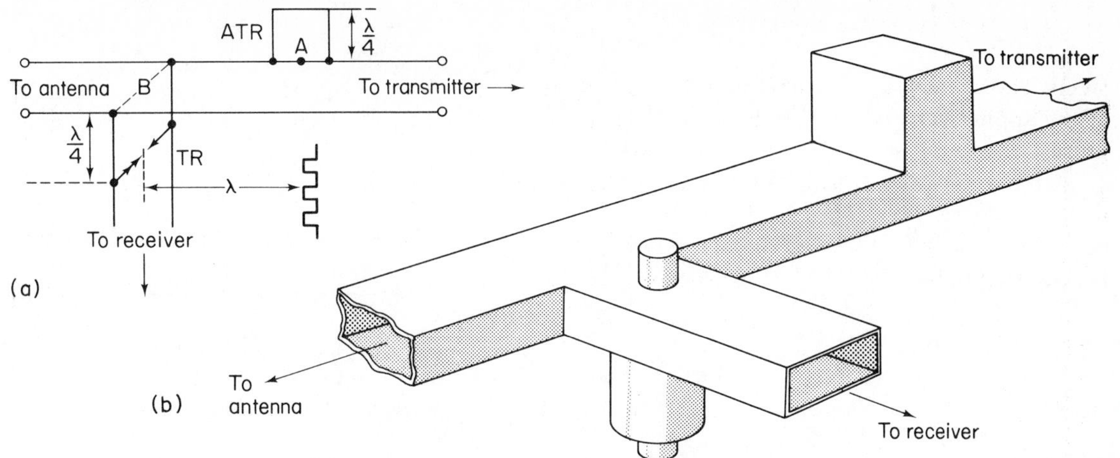

Fig. 6-26 WAVEGUIDE DUPLEXER

day rotation and uses the folded T and slotted hybrid arrangement shown in Fig. 6-27. Basically, the folded T section splits the transmitter power into two equal parts but uses these signals in two different ways—the first for transmission, and the second, which arrives at the receiver 180° out of phase, for eventually canceling the first.

For transmission, both outputs from the legs of the folded T signal go through two ferrite-loaded waveguide sections, one of which shifts leg 1 energy 90°, while its companion, receiving energy from leg 2 of the folded T, passes energy in original phase. Traveling from here through a short-slot hybrid, leg 1 signal passes unchanged in phase while leg 2 signal receives a 90° shift backward, which reconstructs the original signal to full strength for transmission, having produced no

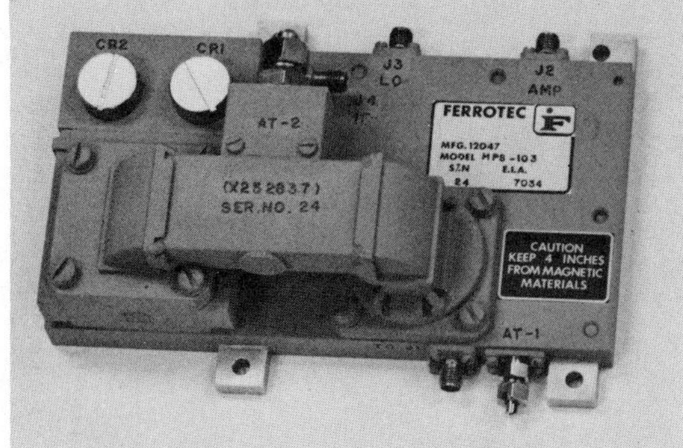

Fig. 6-27 FERRITE DUPLEXER
(Courtesy of Ferrotec Inc.)

signal at the receiver, as we shall see below.

Antenna reception is similar in that signal reaching the short slot hybrid is split two ways, signal 1 passing through the ferrite section and entering the folded T in the original phase. Signal 2 in going through the short slot hybrid is shifted 90° plus another 90° in the same direction in the other ferrite section, so it, too, reaches the antenna terminals in full reinforcement.

Receiver signals during transmission, however, are subject to two phase reversals in opposite directions and consequently cancel, so no transmitter power reaches the transmitter terminals.

PIN diodes can successfully be used in duplexer design by incorporating an array within the input-output combining section of a folded magic-tee so that the phase shift referred to above can be accomplished when the diodes are biased and in the switched condition. Figure 6-27 illustrates a "C" band Phase-Shifter-Duplexer which must be programmed.

TR tubes are usually filled with a mixture of argon gas and water vapor and require a negative voltage continuously applied to an auxiliary electrode (Fig. 6-28). This is called the *keep-alive potential.* This negative voltage maintains a steady glow discharge, providing a continuous supply of ions and free electrons so that when transmitter energy is applied to the electrodes, they fire or break down more quickly. This breakdown time must be relatively short, usually better than 0.01 μsec after application of the radar-transmitted pulse.

After removal of the radar pulse, the device requires about 5 μsec to deionize in order to be ready for received signal or another transmitter pulse.

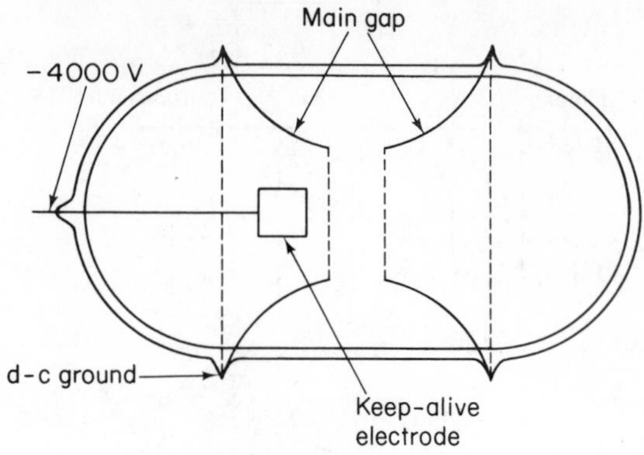

Fig. 6-28 TR TUBE MECHANISM

7 Microwave Electron Tubes

THE FIELD

Microwave power devices (excepting heavy current switches) are predominately electron tubes using or operating upon a cathode-generated stream of electrons. These tubes may be grouped into three broad categories:

1. Resonant-rotational field tubes—that is, magnetrons (sometimes called cross-field devices).
2. Linear beam field-operated tubes—klystrons and traveling-wave tubes.
3. Power-grid tubes, mostly planar triodes.
4. Gaseous and solid-state switching tubes.

Categories 1 and 2 operate upon an electron beam by means of various magnetic and beam-focusing systems; they generate microwave power by several subsystems described below (see Table 7-1). Categories 3 and 4 follow more conventional electron-tube practices.

MAGNETRON

This power-diode oscillator was one of the early microwave power generators used in radar, countermeasures, communication, telemetry, industrial heating (including cooking), and diathermy. Inasmuch as the cathode-beam generating section is built inte-

TABLE 7-1 MICROWAVE TUBE CATEGORIES AND APPLICATIONS

	Categories	*Application*
Resonant-rotational field	Magnetrons VTM'S CFA amplitrons	Power generators – heating Swept signal sources
Linear beam field	Klystron reflex– TWT BWO	Power sources Signal amplifiers – wide band high gain Measurement source Swept signal
Power-grid planar	Ceramic triodes Pencil type	Transmitter output
Gaseous switch	Thyratrons	Power switching Pulse modulations

gral with the magnetic structure, its space factor and simplicity of auxiliary-control apparatus give structural advantages over klystrons and TWT's in many applications. Also its compact size, light weight, high efficiency, and high power capabilities find extensive use in mobile and airborne radars.

Physically, the magnetron, which is a diode, consists of a cylindrical cathode surrounded at a specific spacing by a cylindrical anode; the anode is segmented into resonant cavities extending out radially from the center of the combined cathode–anode structure (Fig. 7-1). The space between the cathode and anode is called the *interaction space* and is where the electrons, acted upon by the combined electric and magnetic fields, gyrate and oscillate in the process of delivering output microwave energy.

For operation under static conditions, we apply a strong dc magnetic field parallel to the cathode structure, through the familiar double-horseshoe magnets common to magnetron construction. We also apply ac voltage between anode and cathode.

Next, when an electron is emitted, if the

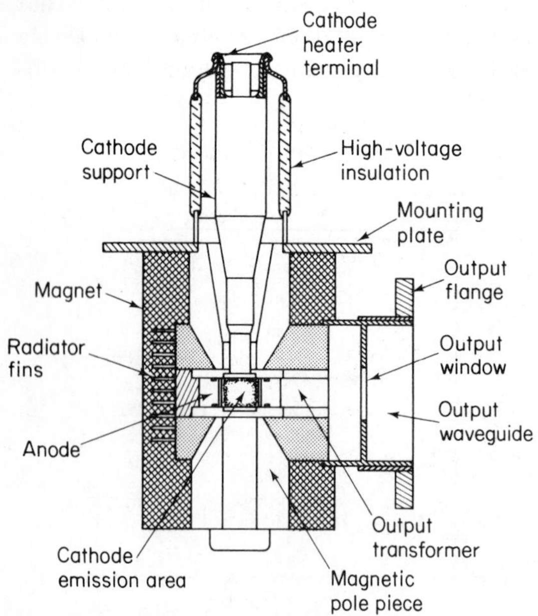

Fig. 7-1 BASIC MAGNETRON STRUCTURE

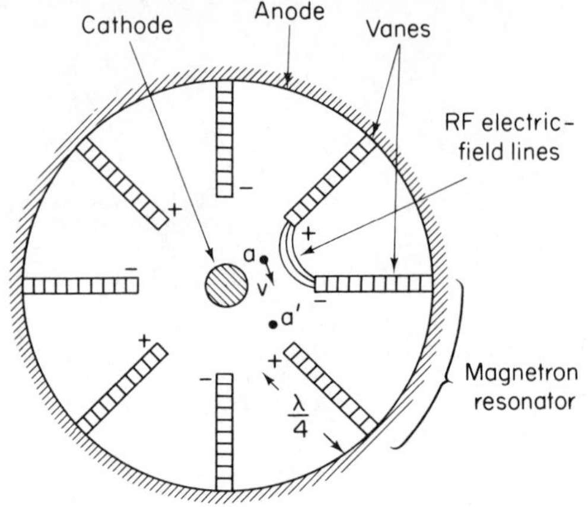

(a) Basic cavity construction

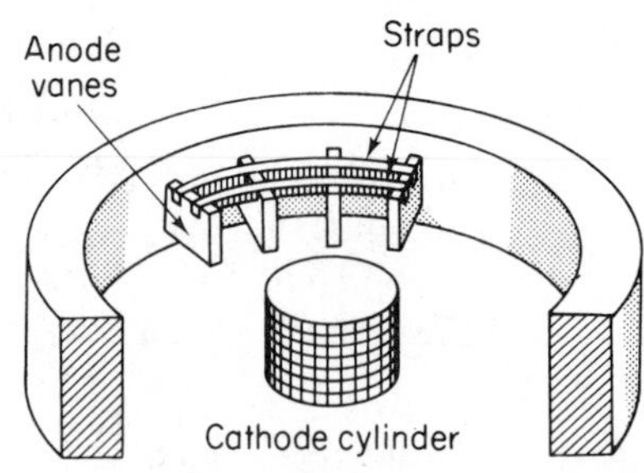

(b) Strapped anode construction

Fig. 7-2 MAGNETRON CAVITY ARRANGEMENT

surrounding, radially located cavities are not excited by rf, and if the static field and applied anode voltage are correctly proportioned, these forces (being opposite) can keep an electron suspended somewhere in the interaction space and cause it to rotate around the central axis. In this equilibrium condition, therefore, a cloud of electrons continuously circles the cathode and extends out to a certain radius in the interaction space. Beyond this radius there is no space charge, and if the anode voltage is raised, the radius of rotation increases and no electrons eventually reach the anode.

Now, spacing vane-separated resonant cavities around the inner circumference of the anode introduces longitudinal fields (cross fields) which also act upon the electron's rotational path. Figure 7-2 shows the arrangement of such cavities, which can be thought of as quarter-wavelength-long transmission lines shorted at their outermost ends and producing a maximum field at the inner vane tips. These varying fields "wobble" the path of the rotating electrons by imparting and absorbing electronic energy as they spiral outward from the cathode and pass by alternately polarized vanes. The alternations are adjustable at the selected microwave frequency and represent generated energy which is transformed and coupled through the iris output window to the transmitting waveguide. The clouds of emitted electrons are thus bunched or grouped as they travel in their spiral path, and, in effect, form alternate densely and thinly populated areas, similar to the spokes of a wheel, which rotate in synchronism with the rf field.

The device is thus an oscillating energy-conversion device which turns dc field and potential into rf output. Its basic frequency is determined by the physical dimensions of the resonant cavities (see annular anode strap rings in Fig. 7-3) with some tuning adjustment possible by anode potential variations as described below.

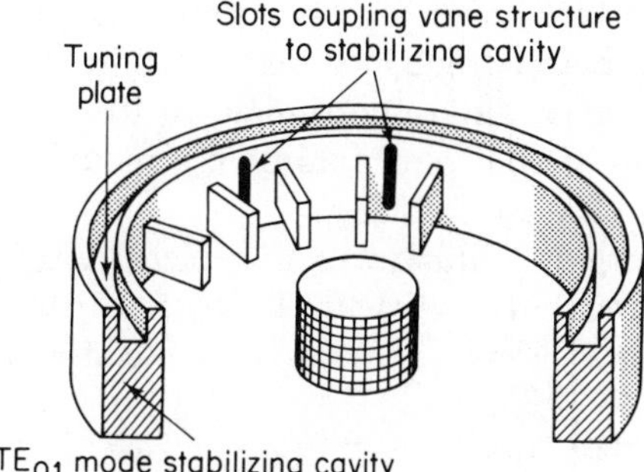

(a) Vanes and coupling slots

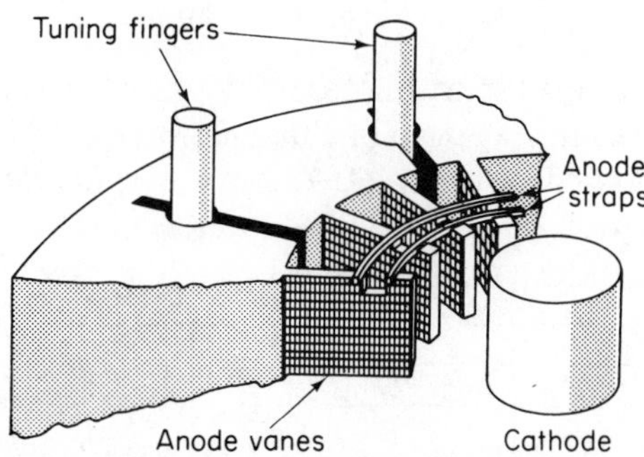

(b) Strapped vanes and tuning fingers

Fig. 7-3 COAXIAL MAGNETRON VANE AND TUNING CONSTRUCTION

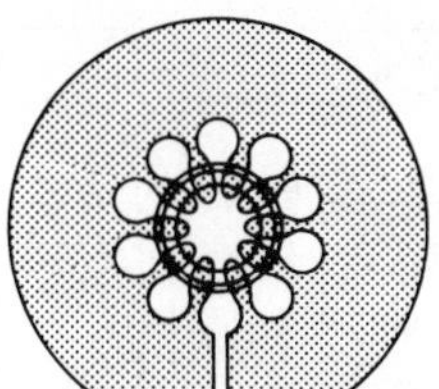
(a) Hole and slot design

(b) Rising sun construction

Fig. 7-4 HOLE SLOT AND RISING SUN CONSTRUCTION

CONSTRUCTION

The main cavity types described below fall into three basic types:

1. Vane construction (Fig. 7-2). Fine construction of this cavity type is usually accomplished by adjustable tuning fingers inserted in the base of the cavity [Fig. 7-3(a) and (b)]. This construction is also called coaxial.
2. Hole and slot cavities [Fig. 7-4(a)] use cylindrical holes instead of wedge-shaped cavities but are coupled to the interaction area by radial slots, which in turn are interconnected by annular rings.
3. Rising-sun construction [Fig. 7-4(a)] uses alternate large and small radially placed cavities (without straps) aimed at increasing circuit Q and improving the stability of specific mode operation.

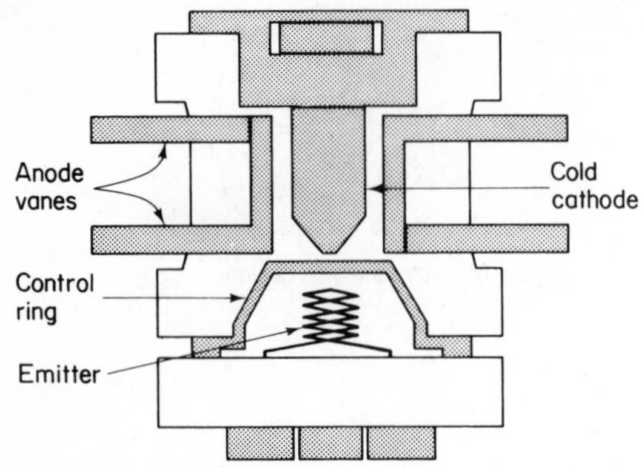

Fig. 7-5 VOLTAGE TUNABLE MAGNETRON

OPERATIONAL TYPES

Functionally, the magnetron is designed to operate under

1. Pulse conditions.
2. Continuous wave (CW).
3. Voltage tuning.
4. Special types.

PULSED MAGNETRONS. These are most extensively used in radar and are anode modulated by high voltage (0.5 to 10-μsec duration pulses) applied as trigger pulses and having duty cycles ranging from 0.005 to 0.02 sec. These units can deliver peak powers of around 1 MW in the S and C bands.

CW MAGNETRONS. At high power (1 to 30 KW) these units are used for commercial heating, cooking, and industrial processing. Low-power, voltage-tunable units are used in fast-tracking local oscillators.

VOLTAGE TUNABLE MAGNETRONS. These tubes are so constructed that variation in the anode–cathode voltage produces directly proportional changes in the generated, cavity-determined oscillations. This is accomplished by controlling the supply of electrons to the interaction area so that the emitter is isolated from back-bombardment and other influences of the anode potential.

Figure 7-5 shows how emitter electrons are funneled by action of the control ring in and up toward the interaction space surrounding the cathode. The stream of electrons thus passes straight upward along the sides of the cathode instead of being emitted from it. Their control and oscillating excursions are thus more directly and accurately controlled by the anode's electric field. This feature produces flexibility of operation and allows for flat power-frequency response, high efficiency, and the opportunity to frequency modulate the output by applying modulating signal to the control ring.

Figure 7-6 illustrates a typical array of commercial magnetrons. Table 7-2 lists excerpts from a manufacturer's specifications.

CHARACTERISTICS

Of chief electrical interest to the designer is a magnetron's performance chart, showing the effects on power output and efficiency of changes in anode voltage and current. Figure 7-7 is typical of this specification.

THE RIECKE DIAGRAM. Figure 7-8 shows dynamic operation of a magnetron as the load is varied; when plotted on a circular load-

TABLE 7-2 MAGNETRON CHARACTERISTICS

Min. Peak Power	*Frequency (MHz)*	*Operating Conditions*				*Max Duty (%)*	*Weight (lbs)*
		E_F	I_f *(amp)*	E_b *(Kv)*	I_b *(amp)*		
175	4900 → 5800	13.5	2.5	21.5	22	0.1	25-35
225	8800 → 9400	13.75	3.35	21.5	27.5	0.1	10
300	9206 → 9400	13.75	3.35	27.5	27.5	0.1	14
2	406-450	6.5	55	55	97.2	0.002	220
40-100	35,000-500	6.3	4 → 8	15-20	13-20	0.04 → 0.1	14
100 → 400 (CW)	1000-8000	5.5	17	3-4	0.25mA	CW	18

Fig. 7-6 COMMERCIAL MAGNETRON MODELS (Courtesy of Litton Industries.)

impedance chart, curves show variation in VSWR with load.

Other definitions concerning performance are

PULLING FIGURE. This is defined as the maximum frequency excursion of the unit as its VSWR varies from 1 to $1\frac{1}{2}$. It is a measure of the coupling between the magnetron and its load.

PUSHING FIGURE. This is a measure of the instantaneous frequency change produced by a given variation in anode current at a constant load. We obtain this factor by observing output on a spectrum analyzer as the magnetron pulse current is being changed.

THERMAL STABILITY. This tells how much the center frequency (Hz) shifts per degree of anode temperature change.

PULSE STABILITY. This is an index concerning the percentage of spectrum lines (on a sweep display) missing (30 per cent less than normal) with change in normal amplitude, shorter pulses, or incorrect center frequency.

BANDWIDTH. This is defined by the megacycles between the first "zeros" on a spectrum

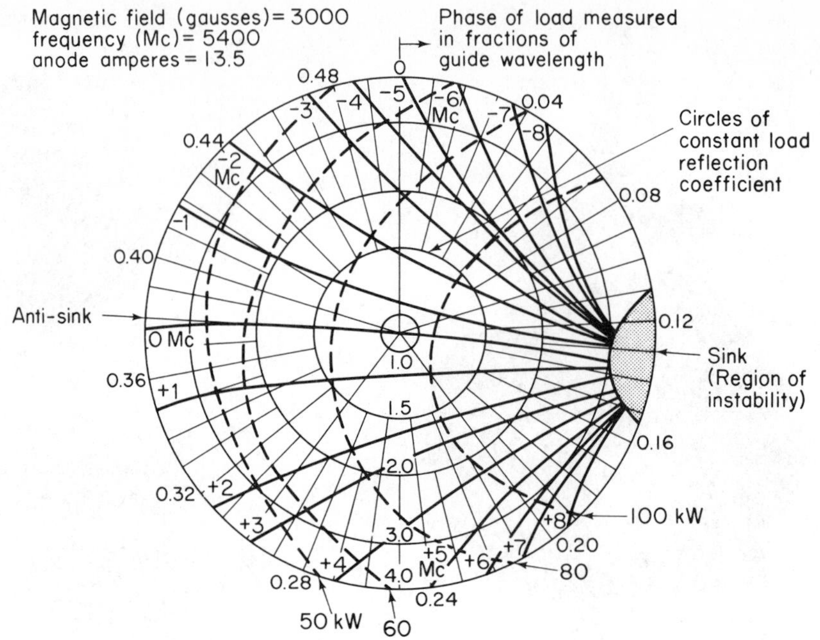

Fig. 7-7 RIECKE PERFORMANCE DIAGRAM

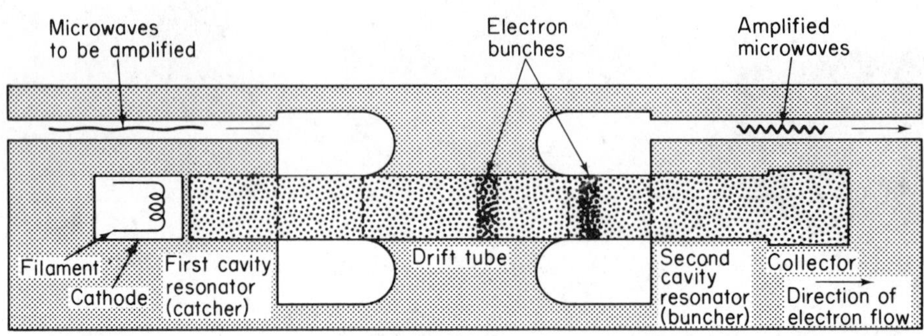

(a) Basic diagram

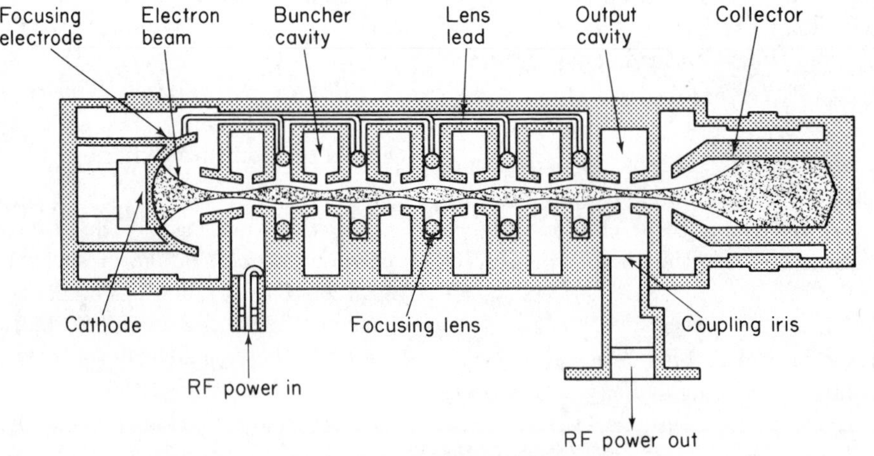

(b) Operating components

Fig. 7-8 SIMPLIFIED KLYSTRON BUNCHING MECHANISM

analyzer display, or by the difference in frequency between quarter-power points on a frequency-response curve.

NORMAL LOAD IMPEDANCE. This is determined by phase relationships with respect to a given reference point made on VSWR measurements of the output line (see Chapter 11).

OTHER CONSIDERATIONS

As inferred in the specifications of Fig. 7-6, power rating of the filament and anode circuits must be adhered to within reasonable operating tolerances. Less obvious, however, are items such as

1. Pulse shape and duration.
2. Anode power-supply regulation.
3. Cathode temperature.
4. Cathode insulation.
5. Load effects.
6. Physical factors.

Details of these are discussed in the following:

1. The performance of a magnetron is critically dependent upon the exact shape of its exciting pulse. The maximum and minimum rate of rise time on the leading edge of the pulse must often be specified for correct moding and, in the case of a thyratron, where arcing may occur. Also, the top of the firing pulse should be reasonably flat with a minimum of spikes on the front edge. Modulator design can compensate for many of these aberrations.

2. Heater and anode supply voltages must be specifically regulated because low, cold heater resistance can produce damaging initial-current surges.

3. Cathode temperature. During magnetron operation an appreciable fraction of the pulse input power is dissipated in the cathode as a result of back-bombardment from the space charge. Thus during operation the heater voltage must be automatically reduced to compensate.

4. Cathode insulation between the body of the magnetron and the cathode and heater terminals is a critical area; glass or ceramic insulating material in the stem must be kept free of dirt, salt spray, or other atmospheric contaminants—preferably wiped clean every few days.

5. Load and external accessories. Stability and even starting performance may be affected by load mismatch and "long-line" effects when the tube is mounted some distance from the antenna. Highly reflective components such as an ATR tube must be carefully matched through the output for proper operation.

6. Physical parameters. When mounting or using a magnetron we must consider stray magnetic fields, strainless mounting, good grounding, sufficient shielding, freedom from shock and vibration, plus firm and solid terminal connections.

KLYSTRON

BASICS

Klystrons are electron-beam-modulated vacuum tubes, called velocity modulated tubes or linear beam tubes. Mainly, the klystron beam has a long-enough path so that the velocity of its electrons can be alternately increased and decreased by cavity resonators surrounding the beam. This speeding up and slowing down of the electrons causes them to arrange themselves and travel in bunches. Changes in the amount of bunching represent modulation, which is applied near the start of the beam (say at the first cavity) and increases in subsequent cavities as the beam travels through them.

An expanded view of this process appears in Fig. 7-8, describing a two-cavity model. Here we consider the klystron to be an elongated vacuum tube consisting of two sections enclosed in resonator cavities. The electron beam is generated by a filament-cathode at the left-hand end of the structure as it passes through a first cavity, through the body of the tube called the *drift section*, through a second cavity, and finally to an electrode called the *catcher*.

In simplified language, the first input-signal coupled cavity produces an oscillating electric field which alternately slows down and then accelerates the electrons, bunching them, so to speak, so that they travel in "blobs" through the drift area. From here on the grouped blobs excite the second catcher cavity and at this point promote stronger or amplified oscillations which are removed in the form of output power. The remaining electrons pass onward to the collector and are returned to the cathode through the external power-supply circuit. Practically, the structure has other elements which improve performance. Figure 7-9 illustrates a working two-cavity model.

First, the indirectly heated cathode emits electrons that are focused into a sharp beam by a control grid operating at a relatively low potential. Second, the beam is accelerated by high, positive voltage applied to grid 1. Third, the beam passes through buncher grids G_2 and G_3, very closely spaced, and also connected to the edges of the buncher cavity. It is this combination of cavity and grids which creates the measure bunching action. Fourth, the input-signal voltages are superimposed on this cavity–bridge structure and provide the modulation for the bunching that has been produced. Fifth, after traveling through the drift space, bunches of electrons arrive at catcher grids G_4 and G_5, where they deliver energy to excite the catcher cavity, the edges of which are connected to these same grids. An output coupling loop in the catcher cavity extracts signal energy from the bunched electron. Finally, the leftover electrons pass to the collector.

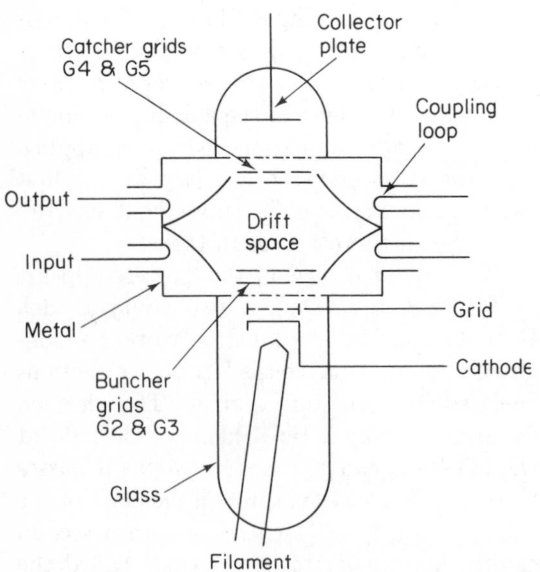

Fig. 7-9 TWO-CAVITY KLYSTRON STRUCTURE

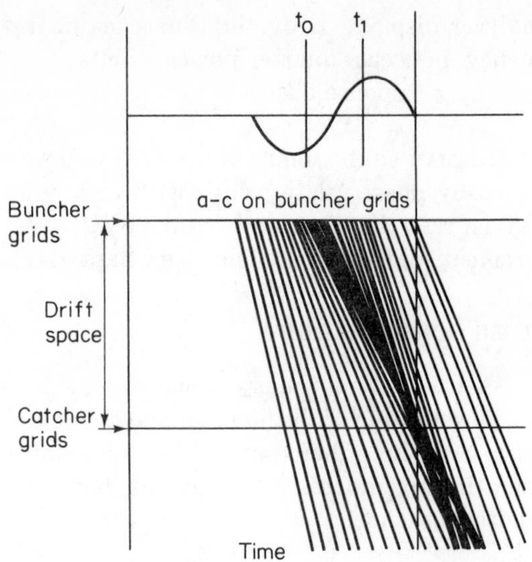

Fig. 7-10 BUNCHING IN KLYSTRON DRIFT SPACE

We can picture the bunching in Fig. 7-10, where the passage of electrons through the drift space is represented by lines starting at the top of the diagram (at the buncher grid) which pass downward, tilted toward the right, in their passage to the catcher grids. The signal-voltage waveform at the top of our drawing is proportional to the variations in the bunching of the electrons; this signal waveform represents different electron velocities, as shown by the slope of the electron bunching lines as they progress through the cathode. Increased crowding of the lines at the catcher over that existing at the buncher represent power amplification of the input signal. Note also that electrons that arrive at the buncher grids between times t_0 and t_1 also form a single bunch at the catcher grid; and since there is only one bunch of electrons in one cycle, the frequency of bunches arriving at the catcher grids is the same as the input frequency.

In design and operation the resonant cavities must be accurately tuned, the drift space must not be too long, or bunches will disappear, accelerator grid and collector voltages must be stabilized and precisely adjusted, and input and output coupling must be tailored for input and output.

Using multicavity structures, coupling from catcher to buncher cavities is arranged for optimum efficiency and stability in the design of the high-powered units described below.

REFLEX PRINCIPLE

Single-cavity klystrons for oscillators and low-power applications use the reflex principle of bunching electrons in microwave oscillating circuits. To analogize the principle we might say that if, in the above two-cavity illustration, some of the catcher energy were fed back in correct phase to the buncher, the unit would oscillate. In a reflex klystron, in effect, a negative voltage is placed upon the collector in a single cavity bunching assembly which acts to repel the leftover electrons reaching this electrode, which we now call a *repeller*.

Figure 7-11 illustrates the arrangement of electrodes, the cavity, and the voltages necessary for reflex operation. Here the electrons emitted by an indirectly heated cathode are

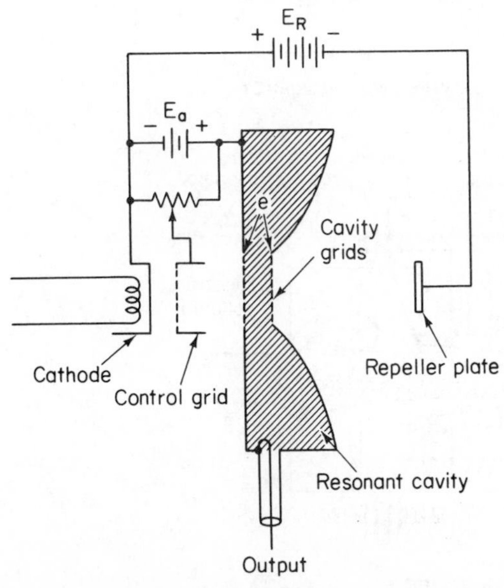

Fig. 7-11 REFLEX KLYSTRON CONSTRUCTION

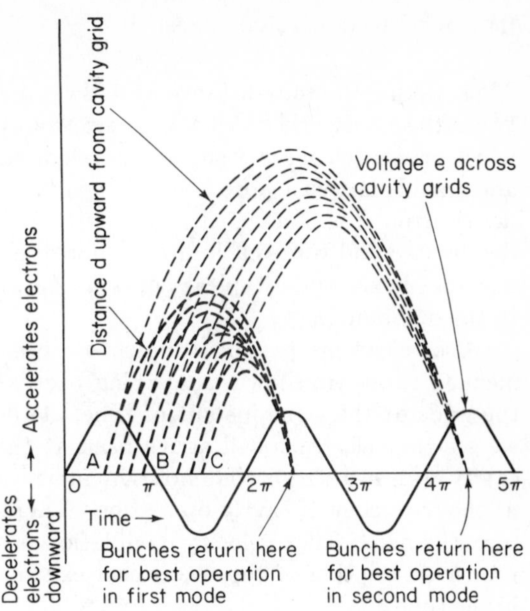

Fig. 7-12 TRANSIT TIME BUNCHING EFFECTS

accelerated toward the cavity grids by the positive voltage E_a and are controlled by some portion of E_a applied to the control grid.

Most of the electrons pass through the central and cavity grids and continue on toward the repeller plate. In this region, however, they meet opposing forces of negative repeller voltage and accelerating cavity-grid voltages so that they are stopped, reverse direction, and pass back to the grids, where they are collected by the grids themselves, the control grids, or the shell.

Now this passage of electrons through the closely spaced cavity grids, while the cavity is oscillating because of resonance, results in bunching. This is because (1) the time that is required for the electrons to pass through the short distance between the grids is small compared to the period of oscillation, (2) electrons that enter the space between the grids when *e* (the field strength) is zero encounter no field and pass on unhindered, (3) those entering when *e* of the right-hand grid is positive with respect to the left-hand grid are speeded up, and (4) those entering when *e* of the left-hand grid is positive with respect to the right-hand grid are slowed down.

Figure 7-12 shows how bunching and repulsion occur at various positions of electrons during selected times of the transit cycle.

Thus, taking the zero distance as midway between the cavity grids, we have (1) electron A, which arrives when e is positive, is accelerated, and travels farther before being turned back; (2) electron B is unaffected; (3) electron C is decelerated and turns back after a shorter excursion—hence these typical paths as shown in the diagram.

Also, electrons passing through at intermediate times are shown as arriving back at the grids at the same instant of time. Thus we see that electrons will arrive back at the cavity grids in a stream that varies in intensity at the frequency of cavity oscillation. Therefore, the name reflex-velocity modulation is a proper description of the nechanism existing in the tube.

Examination of Fig. 7-12 shows another mode of operation occurring when bunches of electrons may also return to the grids on the second positive swing of e. Thus the time in transit for the average electron R may be $\frac{3}{4}$, $1\frac{3}{4}$, $2\frac{3}{4}$ cycles, etc.—each time being the determinant for the particular mode.

Also, we can see the almost entire dependence of operation (particularly the selection of modes) upon the ratio of repeller to grid voltage, since the original electron velocity depends upon (1) E_a and the distance that the electron travels before turning back, and (2) the speed with which it returns, a factor controlled by the difference between E_a and E_r.

Figure 7-13 shows power output and frequency of oscillation as a function of repeller voltage for three modes of operation. Note

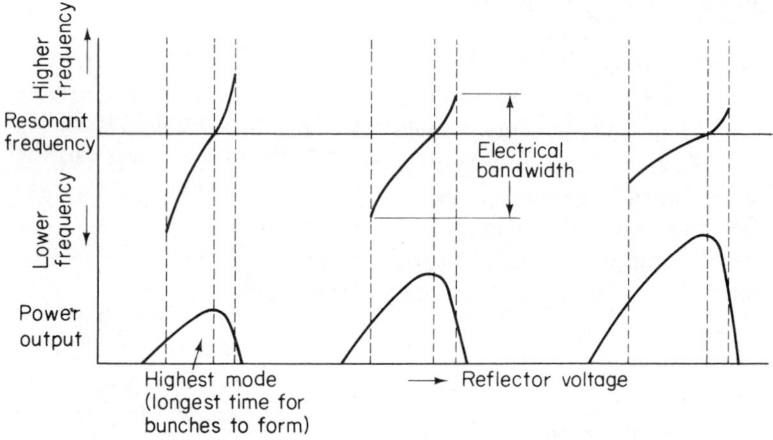

Fig. 7-13 REPELLER VOLTAGE VERSUS POWER AND FREQUENCY

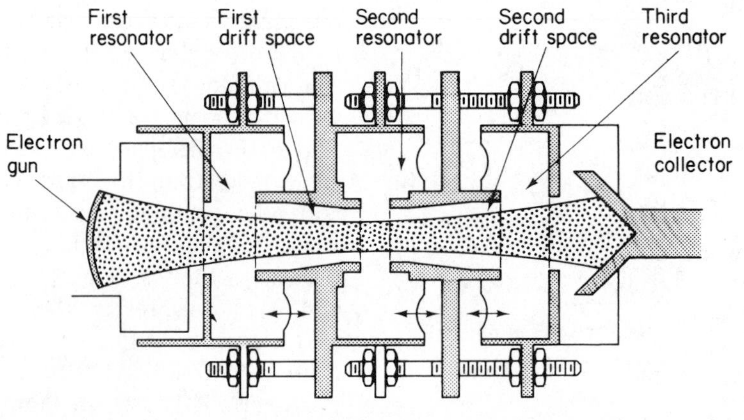

Fig. 7-14 THREE-CAVITY KLYSTRON MODELS (Courtesy of Litton Industries.)

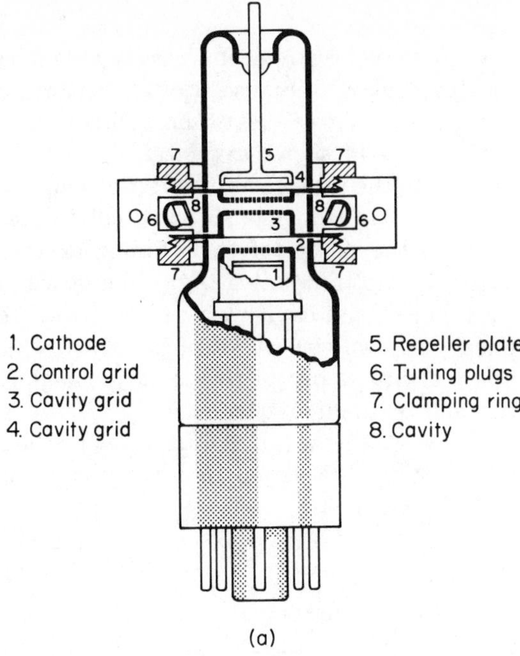

(a)

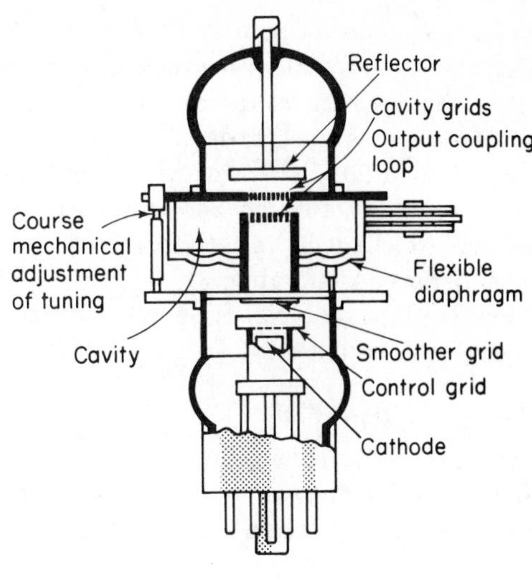

(b) Flexible diaphragm

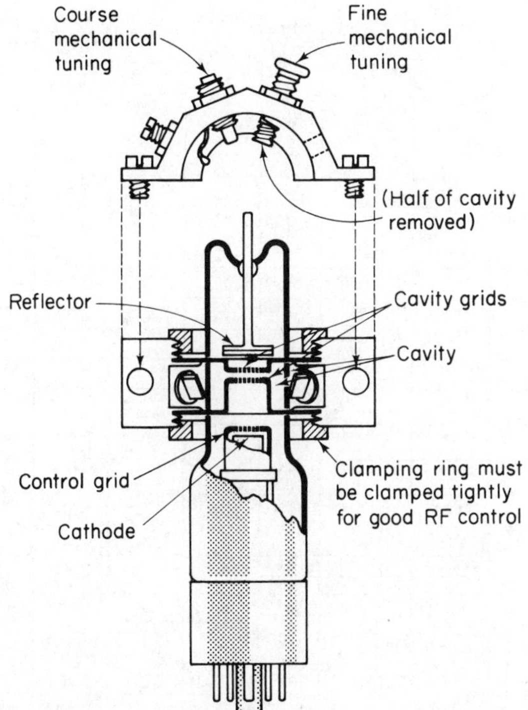

(c) Tuning plug plus cavity distortion

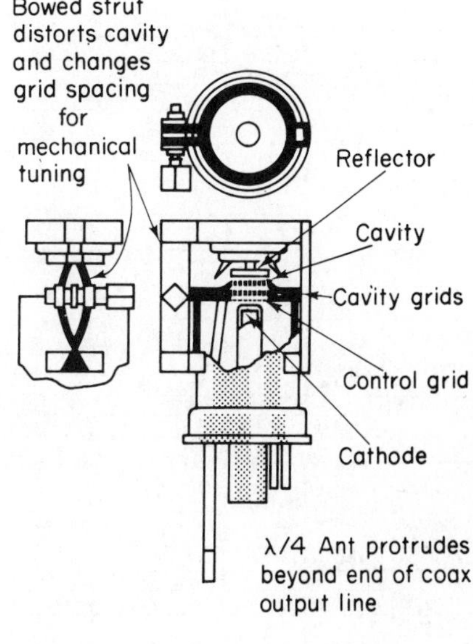

(d) Bowed strut mechanism

Fig. 7-15 KLYSTRON TUNABLE CAVITIES

that the frequency of points of maximum output is the same for all three modes, being, of course, the resonant-cavity frequency. Output decreases as the mode of operation increases, and although bandwidth at the higher modes is greater, the repeller voltage is very sensitive to changes and thus makes lower mode operation preferable.

Practical commercial-tube configurations will be discussed below.

CONSTRUCTION—TYPES

In general, klystrons may be categorized under multicavity and single-cavity types. Powerwise this automatically means that high-powered klystrons of necessity are multicavity devices, because cavity separation immediately allows greater heat dissipation.

Figure 7-14 shows a typical three-cavity unit; note the provision for water cooling.

Reflex klystron models show chief differences in the design of their tuning cavities. Figure 7-15(a) illustrates a typical unit with a tunable external cavity where the coarse frequency control consists of plugs which, when screwed into or out of the cavity, change its size and resonant frequency.

Figure 7-15(b), (c), and (d) show methods of distorting the actual physical size of the main cavity.

TABLE 7-3 KLYSTRON CHARACTERISTICS

Tube type	Maximum frequency range	*TYPICAL OPERATION*										Type of cavity	Type of tuning	Type of output coupling	Tube type
	(megacycles)	Frequency (megacycles)	Wavelength (Centimeters)	Power output average min. (milliwatts)		Reflector potential (dc volts)	Resonator potential (dc volts)	Focus or control grid potential (dc volts)	Electronic tuning e. ref./P.O. (megacycles)	Maximum temp. coef. (MC/C°)	Filament current at 6.3 V (Amperes)				
RK-6037	5120-5430	5280	5.68	30	20	−80 to −160	300	0 to −200	40 min.	....	Klystron-0.470 Tuner-0.825	Self Contained	Thermal Capacitive	Waveguide RG-49/U	RK-6037
QK-404	5925-6450	6200	4.83	120	100	−195 to −275	300	....	20 Average	−0.10 to +0.10	0.440	Self Contained	Mechanical Capacitive	Waveguide RG-50/U	QK-404
5976	6200-7425	6750	4.45	110	85	−78 to −158	300	....	32 min.	−0.10 to +0.10	0.440				5976
QK-422	6200-7425	6750	4.45	100	85	−78 to −158	300	....	32 min.	−0.10 to +0.10	0.440	Self Contained	Mechanical Capacitive	Waveguide RG-50/U	RK-422
2K26	6250-7060	6660	4.51	100	80	−70 to −115	300	....	32 min.	0 to −0.20	0.440	Self Contained	Mechanical Capacitive	Waveguide RG-50/U	2K26
RK-6390	6870-10,750	3-3/4 Reflector Mode-3/4 Cavity Mode					1250	2 to 16	6 min.	±0.025	0.635	External	Mechanical Inductive Non-Contacting Plunge Preferred	Coaxial Cable	RK-6390
		6870-9200	4.37-3.26	80	55	−190 to −420									
		4-3/4 Reflector Mode-3/4 Cavity Mode													
		9200-11,000	3.26-2.73	80	55	−115 to −360									
2K25	8500-9660	9370	3.21	32 22	20 15	−128 to −183 −75 to −120	300	....	55 Average 64 Average	0 to −0.20	0.440	Self Contained	Mechanical Capacitive	Waveguide RG-52/U	2K25
2K45	8500-9660	9660	3.11	32	20	−95 to −145	300	−35 to +15	70 Average	....	0.762	Self Contained	Thermal Capacitive	Waveguide RG-52/U	2K45
6116	8500-9660	8500-9660	3.53-3.11	32	20	−95 to −145	300	0 to −280	70 Average	....	0.50	Self Contained	Thermal Capacitive	Waveguide RG-52/U	6116
RK-6294	8500-10,000	9300	3.23	70	50	−130 to −170	300	....	48 min.	0 to −0.20	1.20	Self Contained	Mechanical Capacitive	Waveguide RG-52/U	RK-6294
RK-6295	8500-10,000	9300	3.23	70	50	−130 to −170	300	....	48 min.	0 to −0.20	1.20	Self Contained	Mechanical Capacitive	Waveguide RG-52/U	RK-6295
RK-6409	8500-10,000	9300	3.23	70	50	−130 to −170	300	....	48 min.	0 to −0.20	1.20	Self Contained	Mechanical Capacitive	Waveguide RG-52/U	RK-6409
QK-414	9660-10,250	10,000	2.99	20	15	−128 to −183	300	....	50 min.	−0.10 to +0.10	0.440	Self Contained	Mechanical Capacitive	Waveguide RG-52/U	QK-414
QK-448	12,000-13,800	2-3/4 Reflector Mode-3/4 Cavity Mode					300	....	60 min.	±0.05	0.675	Self Contained	Mechanical Capacitive	Waveguide RG-91/U	QK-448
		12,000-13,800	2.50-2.17	85	50	−100 to −275									
RK-6178	15,750-16,250	4-3/4 Reflector Mode-3/4 Cavity Mode					300	....	75 min.	−0.20 to −0.60	0.675	Self Contained	Mechanical Capacitive	Waveguide RG-91/U	RK-6178
		15,750-16,250	1.95-1.84	25	20	−100 to −200									
QK-246	15,000-16,200	4-3/4 Reflector Mode-3/4 Cavity Mode					1500	−20 to −120	25 min.	....	0.675	Self Contained	Mechanical Capacitive	Waveguide RG-91/U	QK-246
		15,800-16,200	1.90-1.85	51	20	−152 to −173									
QK-431	15,000-17,000	15,000-17,000	1.76-2.00	25	20	−60 to −210	300	....	75 min.	−0.20 to −0.60	0.675	Self Contained	Mechanical Capacitive	Waveguide RG-91/U	QK-431
QK-336	15,000-17,000	15,000-17,000	1.76-2.00	20	15	−60 to −200	300	....	40 min.	....	0.675	Self Contained	Mechanical Capacitive	Waveguide RG-91/U	QK-336
QK-306	18,000-22,000	18,000-22,000	1.66-1.36	40	10	−80 to −220	1800	−20 to −100	40 Average	−1.0	0.650	Self Contained	Mechanical Capacitive	Waveguide RG-53/U RG-66/U	QK-306
*6253	18,000-22,000	18,000-22,000	1.66−1.36	40	10	−80 to −220	1800	−20 to −100	40 Average	−1.0	0.650				6253
2K33	22,000-25,000	22,000-25,000	1.36-1.20	40	10	−80 to −220	1800	−20 to −100	40 Average	−1.0	0.650	Self Contained	Mechanical Capacitive	Waveguide RG-53/U RG-66/U	2K33
*6254	22,000-25,000	22,000-25,000	1.36-1.20	40	10	−80 to −220	1800	−20 to −100	40 Average	−1.0	0.650				6254
QK-289		27,270-30,000	1.10-1.00	20	10	−50 to −200	2250	−20 to −250	45 Average	To be Specified	0.650	Self Contained	Mechanical Capacitive	Waveguide RG-96/U	QK-289
QK-140 QK-290		29,700-33,520	1.01-89.5	20	10	−50 to −200	2250	−20 to −250	45 Average	To be Specified	0.650	Self Contained	Mechanical Capacitive	Waveguide RG-96/U	QK-140 QK-290
QK-291		33,520-36,250	0.895-0.826	18	5	−50 to −200	2250	−20 to −250	45 Average	To be Specified	0.650	Self Contained	Mechanical Capacitive	Waveguide RG-96/U	QK-291
QK-292		35,100-39,700	0.855-0.755	10	5	−50 to −200	2500	−20 to −200	45 Average	To be Specified	0.650	Self Contained	Mechanical Capacitive	Waveguide RG-96/U	QK-292
QK-226 QK-293	34,900-42,800	37,100-42,600	0.810-0.704	5	...	−50 to −200	2500	−20 to −200	To be Specified	To be Specified	0.650	Self Contained	Mechanical Capacitive	Waveguide RG-97/U	QK-226 QK-293
QK-227 QK-294	40,000-51,800	41,700-50,000	0.72-0.60	5	...	−50 to −200	3000	−20 to −200	To be Specified	To be Specified	0.650	Self Contained	Mechanical Capacitive	Waveguide RG-97/U	QK-227 QK-294
QK-295	Two Tubes Necessary To Cover From 50,000 to 60,000 Mc.		0.60-0.50	To be Specified		−50 to −200	3500	−20 to −200	To be Specified	To be Specified	0.650	Self Contained	Mechanical Capacitive	Waveguide RG-98/U	QK-295

*With UG/425/U Output Flange.

Reflex klystrons

These rugged, air cooled, tunable power sources feature high softening point alumina silicate glass sealed to molybdenum. This allows the tube to be processed at 700°C and enables an extremely high vacuum to be obtained.

Tube type	Frequency range (Gc)	Minimum tuning range (Mc)	Power output (W) (minimum)	Maximum cavity potential with respect to cathode (kVdc)	Maximum cathode current (mA)	Cooling
L-3710 (8RK13)	35 ± 2	2000	0.03	2.3	15	Air
L-3848 (8RK17)	35 ± 2	2000		2.6	25	Air
L-3642 (12RK3)	23 ± 2	500	0.10	3.5	30	Air
L-3692 (12RK4)	21 ± 3	500	0.35	3.5	30	Air

(a) Reflex klystrons and characteristics

Fig. 7-16 COMMERCIAL MODELS OF INTEGRAL AND WAVEGUIDE KLYSTRONS (Courtesy of Litton Industries.)

DESIGN—OPERATION

Briefly, the main consideration in klystron design is one of frequency stability in the reflex types and of power dissipation in the multicavity designs. A number of engineering developments over the years have improved electron-stream convergence in the drift tube, more efficient combinations of beam voltage and current, more efficient placement and use of cavities, better amplification, and application to millimeter waves.

Table 7-3 lists a number of commercial klystron models and the main characteristic of each. Figure 7-16 illustrates a composite photograph of current commercial klystrons.

TRAVELING-WAVE TUBE (TWT)

This tube comes under the category of linear beam devices. (See page 000.) It is characterized by an electric-field circuit that acts longitudinally along an electron beam and extracts energy from electron groupings to provide useful amplification and power output.

The TWT, unlike the klystron, does not use resonant fields but provides continuous focusing action along the beam by space charge, plus electrostatic and special magnetic methods. The beam-focusing principle contributes to the tremendous bandwidth characteristics of this tube.

Modern tubes are now produced having high gain, low noise, and particularly wide bandwidth and high power. These are due to:

1. Improved focusing systems (at hundreds of cw watts and peak megawatts),
2. Better metal–ceramic construction giving power from high-helix-temperature operation plus stability under extreme environmental conditions, shock, and vibration.
3. Efficient integral, internal attenuators and matching circuits giving high gain over an octabe band.
4. Improved shielding giving over 60-dB gains.
5. Improved beam formation for low noise capabilities.
6. Reduction in overall size and weight allowing for airborne and mobile applications.

Table 7-4 illustrates an interesting comparison between noise in TWT tube systems and current micro-wave configurations.

CONSTRUCTION

Construction of the basic TWT (Fig. 7-17) consists of an electron gun, a single-wire helix, an attenuator winding integral with the helix,

TABLE 7-4 COMPARATIVE TWT NOISE FIGURES

	Device	*Noise figure (dB)*	*Gain (dB)*	*Instantaneous bandwidth (% of octave)*	*Saturated power output*	*Special requirements*
TWT S	Traveling-wave tube A-1207V	4.5	25	30	3 MW	Solenoid-weight 20 lbs Permanent magnet
TWT S	Traveling-wave tube A-1207V4	1.0	20	10	1 MW	Solenoid in liquid nitrogen. 4,500 gauss field
TWT S	Traveling-wave tube A-1173	11-15	30	100	10 MW	Periodic permanent magnet
	Triode 416A	14-20	10	1-5		
	Crystal mixer	8-15	−6 to −8	0.5		
	SS parametric analyzer	1-6	15-25	1-10	−15 dBm	Requires pump r-f generator
	Maser	0.03	30	0.1	3 μW	Requires cryostat
	Tunnel diode amplifier	4-5	20	10	−20 dBm	dc supply (low impedance)

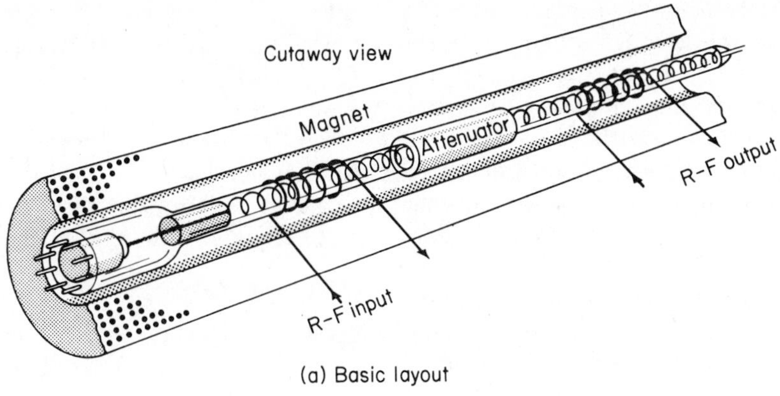

(a) Basic layout

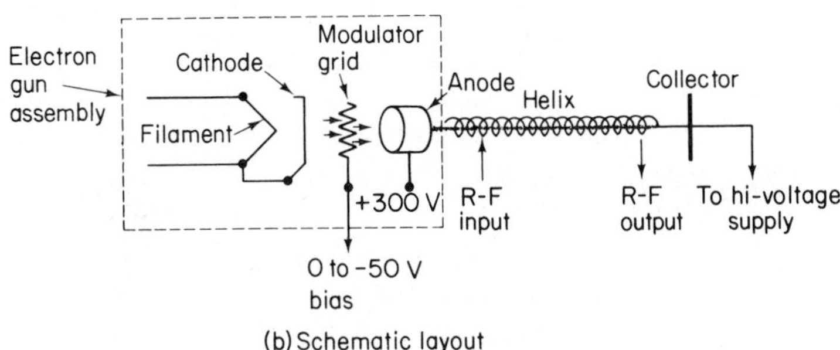

(b) Schematic layout

Fig. 7-17 BASIC TWT CONSTRUCTION

input–output coupling devices, and a collector—all enclosed in a glass envelope and the whole assembly surrounded by a focusing magnet.

The electron gun, attached to one end of the helix, produces a focused beam of electrons directed through the center of the helix. This beam, similar to the beam in a cathode-ray tube, produces electrons which are initially focused by a control grid and accelerated by the electron-gun anode. The helix is simply a wire conductor formed with a uniform spiral in order to bunch the electrons as they travel through the helix. Attenuator windings, integral with the helix, are directional couplers, or, in effect, check valves serving to prevent reflections from passing back down the helix. Rf input and output windings are placed, respectively, at the cathode and collector ends of the helix. The collector, of positive potential and connected to the far end of the helix, gives final acceleration to the beam.

OPERATION

Briefly, the main action occurs when electrons enter the helix and become "bunched," which is caused by the alternate acceleration and deceleration of the input rf wave. Because the interaction is continuous and cumulative, the amplitude of the rf signal grows as it travels down the helix.

Actually, the helix represents a common type of delay line, with a diameter and turn per unit length so selected that as the wave travels at its usual velocity along the coiled wire itself, its velocity along the axis of the helix is about one tenth the velocity of light. Now, at certain points where the wave velocity equals the beam velocity, the longitudinal fields due to currents in the helix oppose those created by electrons in the beam, causing the electrons to slow down (at other points the electrons are speeded up), producing the bunching. This means that energy is transferred from the beam to the wave traveling down the helix. Figure 7-18 shows a simplified diagram of the bunching that accompanies the traveling wave and the beam travel. Figure 7-19 shows more details of the tube's physical relationship and the buildup of beam currents.

Note that within the tube there is a lossy wire attenuator placed near the midpoint of the traveling-wave envelope; another type of attenuator is a split cylinder of graphite placed around the center of the structure (Fig. 7-17). These devices serve to isolate the input from the output by attenuating all waves but allow the electron bunches to pass through unhindered. Note in Fig. 7-19 the amplified buildup due to larger bunches getting through to the right of the attenuator.

DESIGN FACTORS

Most important in TWT design improvements has been the development of periodic permanent magnet focusing systems. These

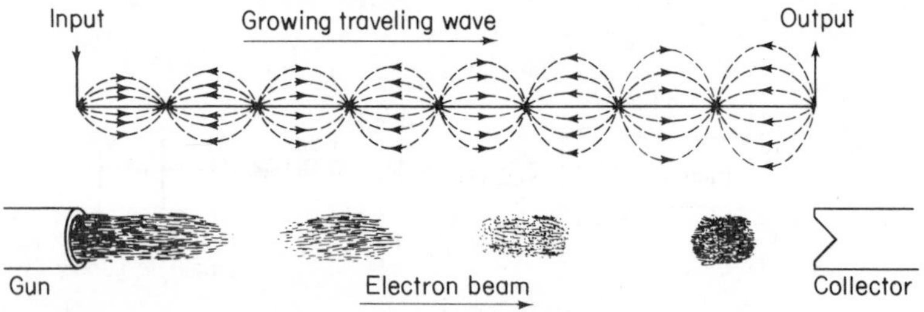

Fig. 7-18 BUNCHING IN A TWT

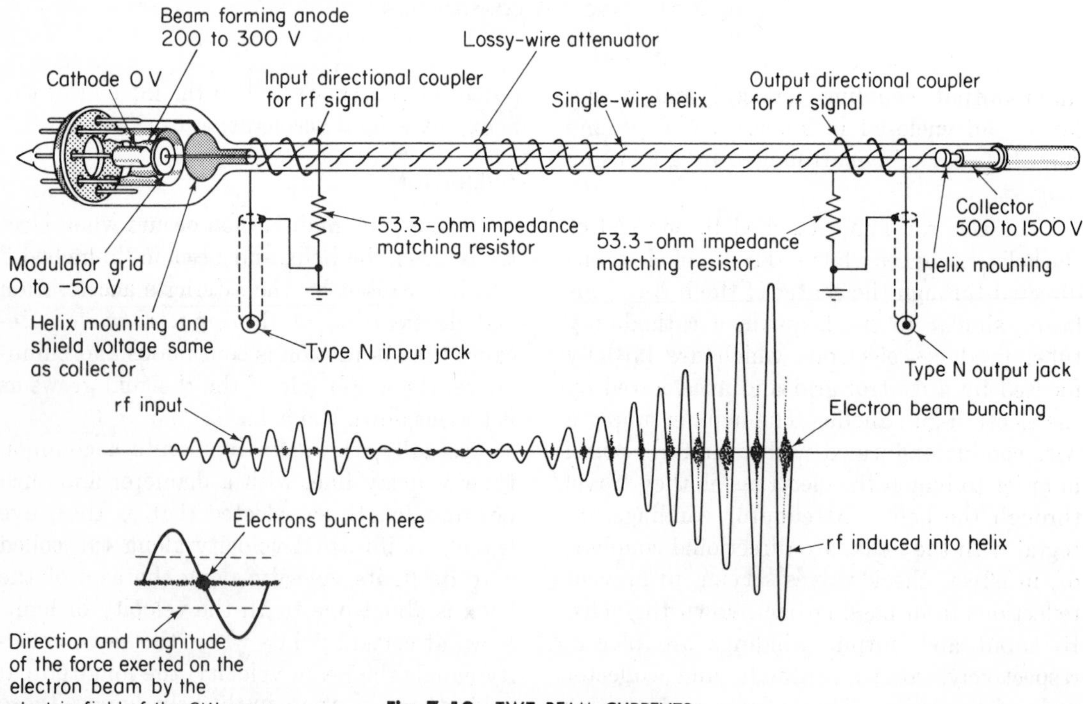

Fig. 7-19 TWT BEAM CURRENTS

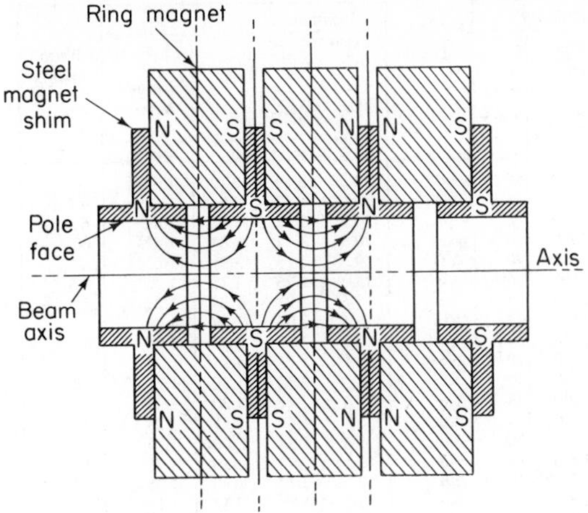

Fig. 7-20 TWT MAGNETIC FOCUSING

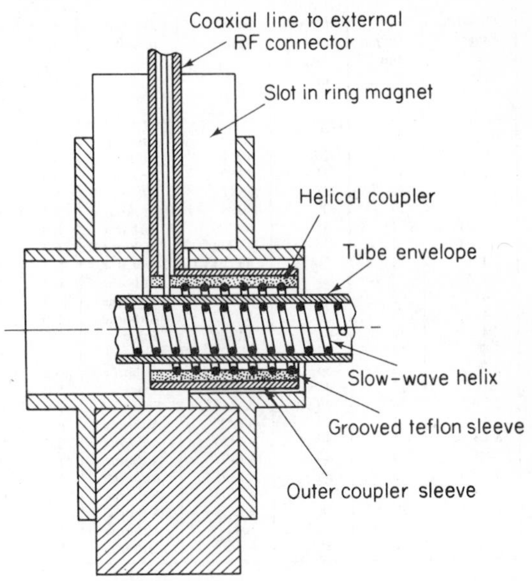

Fig. 7-21 TWT HELICAL COUPLER

arrays use thin, lightweight ring magnets plus steel magnet shims arranged in successive assemblies along the tube structure (Fig. 7-20). These magnets produce a scalloped envelope magnetic field which varies sinusoidally along the central axis. This tailored contour is particularly efficient in confining the electron beam and finally results in designs permitting the use of magnets some 5 to 10 times smaller than the older power-driven solenoids.

Other design improvement items are

1. Electron guns with parabolic or dished cathodes which generate convergent beams with diameters smaller than the cathode itself and with particularly high-density electron streams.

2. Special-control grid geometry, which allows full control of the beam prior to convergence.

3. Helical couplers for input and output signals, when incorporated integrally with the permanent periodic magnet structure, give particularly efficient operation and precise impedance matching (Fig. 7-21).

4. Parallel-flow and low-noise guns, which pass the electron beam through a series of apertures which eliminate some of the shot noise originating in the high-temperature cathodes.

5. Electrostatic focusing, in which, by using two bifilar, interleaved helices identical in diameter and turns per inch, and placing them at different potentials, focusing is achieved in the same manner as that of a system of identical converging lenses spaced at close intervals. This unit, known as the *estiatron*, eliminates focusing magnets; it is particularly adapted to low-powered applications.

MODELS—APPLICATIONS

The TWT's chief attribute is that of lightweight, low-noise amplifier performance with broad-band frequency response; in general, it is not a high-powered device, seldom being used above the low-kilowatt range because of the difficulty in coupling to the helix.

Recent models have shown design advances in low noise amplification up to 5 GHz having over 25-dB gain and noise figures under 3 dB. With further steps toward operational simplicity, these units are supplied with built-in power supplies.

A summary of typical performance characteristics of tubes now used in space programs

TABLE 7-5 TWT CHARACTERISTICS (Courtesy of RCA.)

Frequency Range[c] GHz	RF Output Min. W	Gain (Small Signal) Min. dB	RCA Type[a]	Environment[b]	Heater Voltage Typ. V	Heater Current Typ. A	Collector Voltage Typ. V	Collector Current Typ. mA	Control Grid	Dimensions[d] Clearance (Includes Conn.) Max. in L	W	H	Capsule Nominal in Dia.	Weight Approx. lb
0.75-1	20	30	A1317	Ground	6.3	0.95	1650	55	No	20	2.19	2.12	1.62	5
1-2	1	27	4021	Airborne	6.3	1.85	900	25	Yes	15.6	1.52	2.5	1.52	4.5
1-2	10	25	4053	Ground	6.3	1.75	2200	70	Yes	20.5	3.12	3.88	1.62	6.5
1.7-2.3	18	28	7642	Ground	6.3	1.75	2000	70	No	20.5	3.12	3.88	1.62	6.5
1.7-2.7	17	29	4054	Ground	6.3	1.75	2200	70	Yes	19	2.12	3.88	1.62	6.5
1.9-4.1	1	35	A1309	Airborne	6.3	1.4	1100	30	No	13	1.76	1.75	1.25	2.5
1.9-4.1	1	35	A1311	Airborne	6.3	1.3	1100	35	Yes	15.5	1.94	1.84	1.25	2.5
2-4	1	30	A1201V1	Airborne	6.3	1.3	1145	20	Yes	15.4	2.03	3.34	1.06	3
2-4	2	38	A1138V1	Airborne	6.3	1.3	1250	20	Yes	15.5	2.03	1.62	1.06	3
2-4	2	33	A1314	Airborne	6.3	1.5	1150	35	No	13	1.76	1.75	1.25	2.5
2-4	5	30	A1312[e]	Airborne	5.0	0.5	800	30	No	13	2.72	1.30	1.25	1.5
2-4	20	30	A1320[e]	Space	5.0	0.5	1000	38	No	9.5	2.72	1.30	1.25	1.6
2-6[g]	2	33	A1323	Airborne	6.3	1.35	1600	30	No	15.8	1.86	2.45	1.25	3.3
2-6[g]	3	35	A1310	Airborne	6.3	1.35	1100	30	No	15.4	1.94	2.00	1.25	3.25
2.2-3	13	35 sat.	4056[e]	Space	5.0	0.5	1000	38	No	13	2.72	1.30	1.25	1.6
2.5-4.5	5-45[h]	40 sat.	A1318[e]	Space	5.0	0.5	800-1400[h]	20-70	No	12	2.72	1.30	1.25	1.5
4.4-5	1	37	A1205	Airborne	6.3	1.35	1600	27	No	15.6	1.88	2.42	1.12	3.25
4.4-5	15	30	A1359[e]	Airborne	5.0	0.52	1050	40	No	13	1.5	1.8	0.76	7
5-10	4	40	A1397[e]	Airborne	6.3	0.70	2300	40	No	11	2.0	1.75	1.50	3
5.8-6.4	3	35	A1358	Airborne	6.3	1.35	1500	27	No	15.2	1.94	2.0	1.25	3.3
7.5-11	1	32	A1203	Airborne	6.3	0.85	2950	12	No	15.2	3.81	2.69	1.75	6
7.9-8.4	20	46	A1378	Airborne	6.3	0.7	3300	50	No	11	2.0	1.75	1.50	3
8-12	1	32	4041	Airborne	6.3	0.85	3000	12	Yes	15	2.2	3.25	1.75	6
12-18	1	35	A1399	Airborne	6.3	0.3	3000	20	No	11	1.5	1.5	1.5	3

shows that we have

Power output: 2 to 35 W
Frequency: 2 to 10 GHz
Bandwidth: 20 to 100%
Gain: 30 to 35 dB
Efficiency: 30 to 50%
Weight: 1 to 3 lb
Volume: less than 50 in.^{3}
Temperature range: 50 to 185°C
Life: 50,000 hr

Table 7-5 lists a number of commercial types and their characteristics, and Fig. 7-22 shows a commercial model.

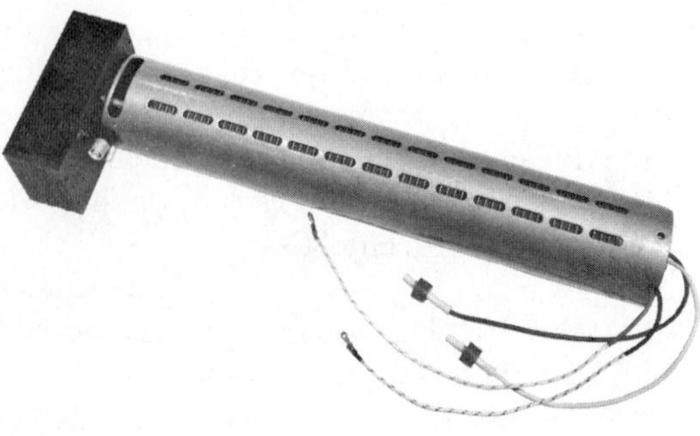

Fig. 7-22 COMMERCIAL TWT OSCILLATORS (Courtesy of RCA.)

BACKWARD-WAVE OSCILLATORS

A backward-wave oscillator (BWO) is a self-oscillating, low-power TWT delivering tunable, stable, wide-range frequency performance. The structure is similar to a TWT having a traveling, bunched, electron beam and a self-contained helix (Fig. 7-23).

With incremental feedback of wave-propagated energy, the structure is such that a forward portion of space waves in the helix, which are generated by movement of the bunched electrons, cancel when they reach the right-hand end of the helix. Also, the portion of the incrementally bunch-generated helix waves that travel backward is reinforced. This means that power is transferred to the input, having traveled from right to left, even though the beam is traveling from left to right.

Basically, the feedback is dependent upon the phase relationship of the generated and propagated waves produced in the helix; the mechanism can be analogous to a chain of identical regenerative feedback loops (Fig. 7-24). Along the top of the chain is a series of transmission-line sections (sections of the helix windings) which will support a wave

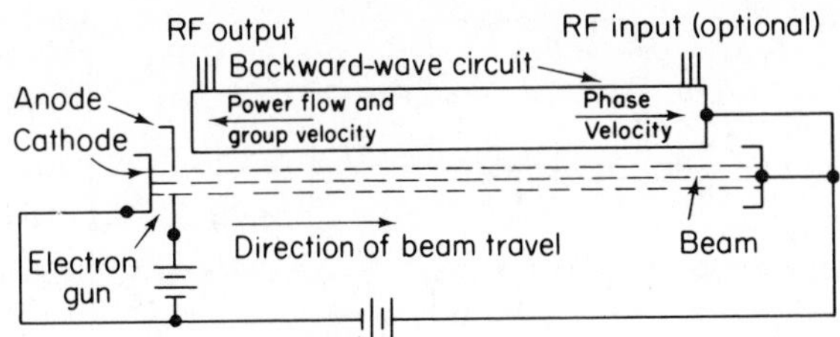

Power flows in opposite direction to electron stream

(a) Elements

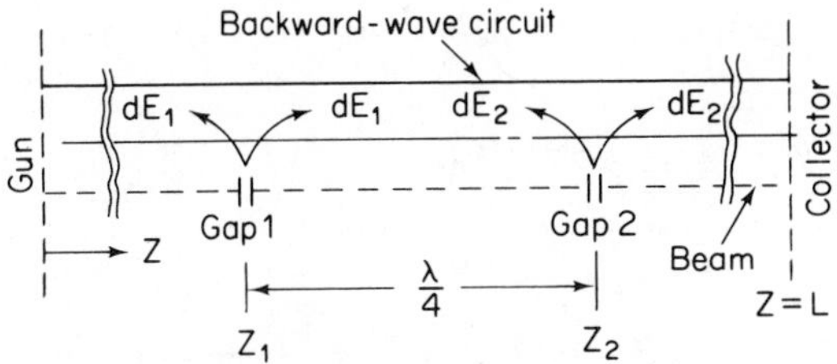

Electron stream induces an electromagnetic field in the backward-wave

(b) Beam and field travel schematic

Fig. 7-23 BWO STRUCTURE

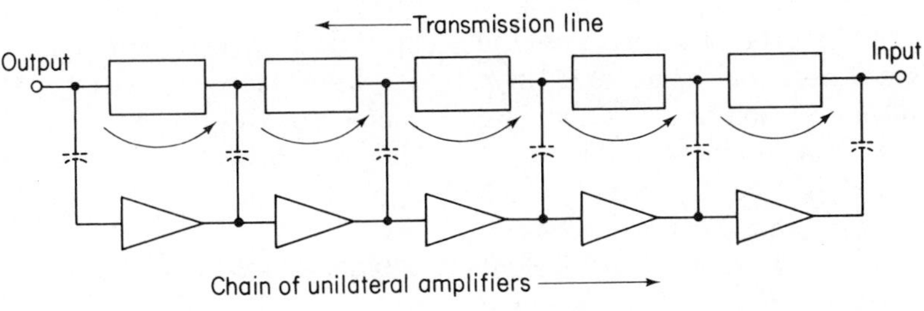

(a) Schematic loops

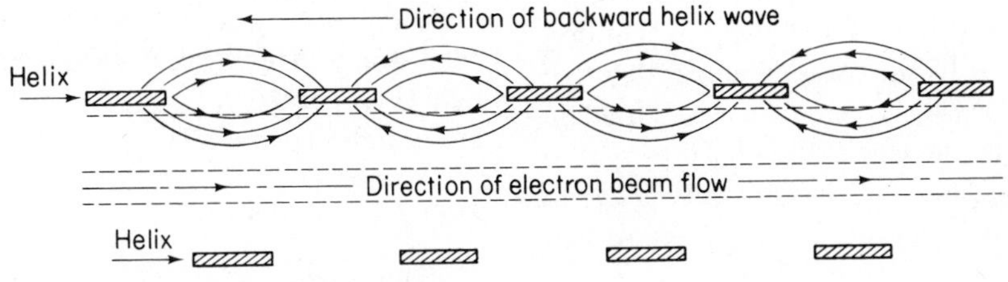

(b) Beam-induced backward wave

Fig. 7-24 BWO FEEDBACK LOOPS

TABLE 7-6 BWO DATA

Sylvania Number	*Freq. kMc*	E_{c1} *V D–C to Cath (Grid 1)*	E_{c2} *V D–C to Cath (Grid 2)*	E_w *V D–C to Cath (Helix)*	I_k *ma D–C*	E_b *V D–C to Ew (Collector)*	P_0 *mw*	*Focusing*
6699	1 to 2	Connect to cathode	+55	90 660	25	+180	70 1000	Solenoid
6496	2 to 4	Connect to cathode	+100	165 1300	25	+180	70 600	Solenoid
7096	2 to 4	Connect to cathode	+150	150 1400	40	150	100	P.M.
BW-623	4 to 8	Connect to cathode	+100	240 2400	10	+180	35 150	Solenoid
6902	18 to 26.5	0 to 600 ①	300 to 800 ①	500 2450	16	+ 90	5 to 35	P.M.
BW-1757	26.5 to 41.0	0 to 600 ①	300 to 800 ①	500 2450	12	+ 90	2 to 10	P.M.
BW-1779	62 to 77	0 to 475 ①	300 to 950 ①	1000 2600	17.5	Tied to helix ②	1 mw min 3 mw avg	P.M.

1 Circuit design values – refer to individual tube date sheet for exact values.

2 Above 40 kMc tubes do not use helix slow-wave structure. They use ridge - guide (Karp) structure. Collector connected internally to slow-wave structure

Up to 12.5 kMc tubes are glass with solenoid focusing (except 7096). Above 12.5 kMc they are metal-ceramic with P.M. focusing.

moving either to the right or left. Along the bottom of this chain is a series of unilateral amplifiers in which the signal can pass only in the left-to-right direction. Thus each loop consists of the transmission line, the coupling capacitances, and a unilateral amplifier.

In operation, positive feedback, which leads to regenerative amplification and oscillation, makes use of a wave going from right to left in the transmission line when the phase delay in a single loop is just equal to one cycle; these voltages are cumulatively additive. However, under these conditions, each amplifier's output is canceled by the line's incremental voltages being coupled back through the cross-connecting capacitances. Thus no wave voltage passes down the chain of series-connected amplifiers, although the electron beam does move forward in its conventional manner.

As in any electron oscillator, the start of operation is usually spontaneous, owing to noise, start-switching, etc., provided that feedback conditions are correct. Operating frequency depends upon phase relationships of backward-wave interaction and hence upon the beam electron velocity; this, in turn, is controlled by the cathode-to-helix voltage and is proportional to the square root of applied voltage.

DESIGN—OPERATION

BWO's use basic TWT construction and operation minus, of course, the attenuators. They employ periodic permanent magnets, hollow beams, conventional control, and accelerator and collector electrodes. Table 7-6 lists the operating characteristics of a number of commercial models.

APPLICATION

The most useful feature of a BWO is its ability to sweep or to electronically scan a complete octave of frequencies by the adjustment of a single voltage. In addition, a BWO

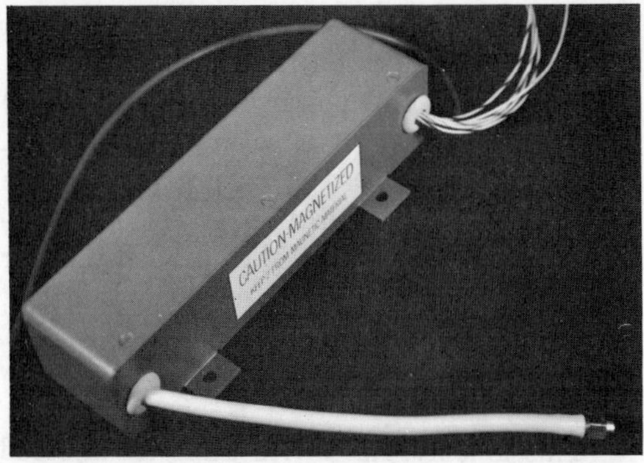

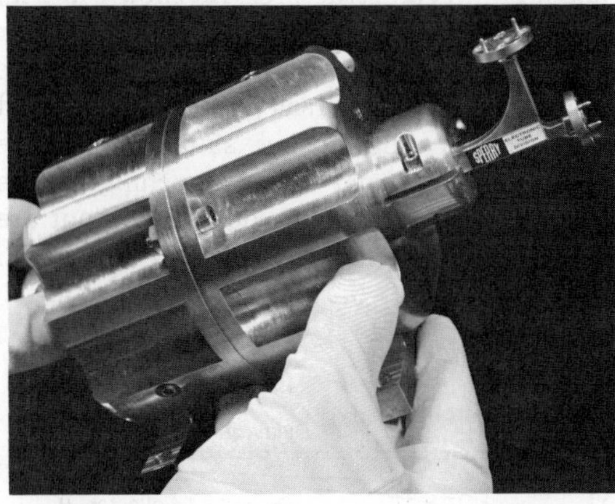

Fig. 7-25 BWO MODELS (Courtesy of Litton Industries.)

is capable of being modulated by pulse, FM, and 100 per cent AM. Circuitwise, these units are more fully described in Chapter 11. Figure 7-25 illustrates the construction in a number of current commercial models.

CROSSED-FIELD TUBES

Actually, the conventional magnetron is a crossed-field oscillator. It operates because of the composite action of an electric field combined with a magnetic field upon electron trajectories (see page 000). Its particular attributes are its high efficiency (40 to 70 per cent), simplicity and economy of construction and manufacture, relatively compact size (and low weight), low anode potential, ease of mechanical tuning, and absence of auxiliary equipment such as electromagnets, shielding, etc. Development of crossed-field magnetrons into better oscillators and into amplifier devices has expanded the magnetron regime far enough to warrant establishment of another category.

COAXIAL MAGNETRONS

This construction increases the frequency stability of a conventional magnetron by nearly 10 to 1 while maintaining high efficiency and power-handling capabilities. It is built using a stabilizing cavity coaxial with the main magnetron structure as shown in Fig. 7-26.

STABILITRON

The stabilitron is an improved coaxial magnetron incorporating an internal network within the evacuated vacuum-tube bottle; this network, known as a *platinotron*, is a physical structure similar to that used as the basis of the *amplitron*, described next. Use of it greatly improves a magnetron's pushing and pulling figure and reduces start–stop jitter.

AMPLITRON

The amplitron is a crossed-field structure delivering power gain through use of an electron-emitting cathode surrounded by a vane-type cathode–anode; the vanes act

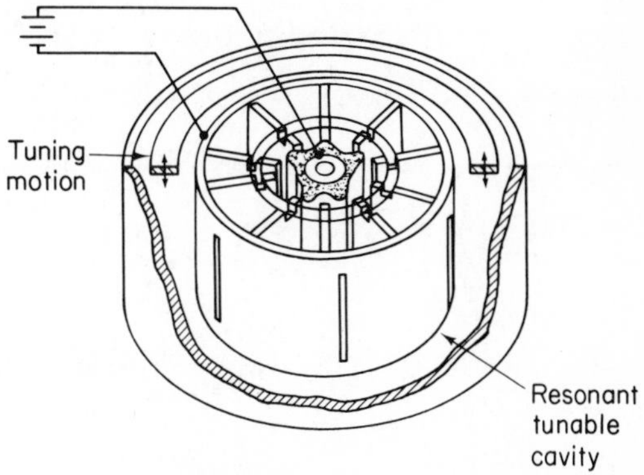

Fig. 7-26 COAXIAL MAGNETRON

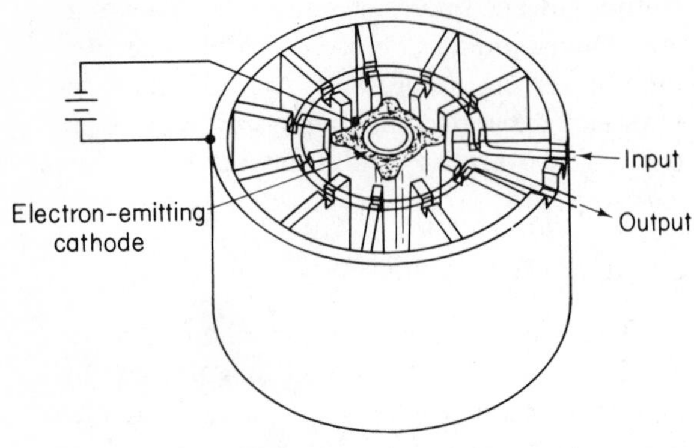

Fig. 7-27 AMPLITRON STRUCTURE

simultaneously to collect electrons and operate as a slow-wave rf circuit terminated by coupling to the input and output vanes of the radial structure (Fig. 7-27). With applied anode potential and using the conventional magnetic field, emitted electrons leave the cathode and spiral outward in concentric paths increasing speed with anode potential until the velocity of the electrons in their orbits of increasing diameter becomes synchronous with the rf wave induced at the input vane (which is normally progressing around the wave structure). At synchronism and with final collection by the anode, interaction occurs, and the electrons surrender their original energy in the form of heat to the anode and rf energy to the vanes, so built-up power appears at the output terminal.

The amplitron's power gain (10 to 20 dB) is similar to that of a saturated amplifier, delivering no increase of output power with input drive beyond certain levels of operation; its efficiency is excellent (70 to 80 per cent) and phase shift with frequency is low.

FORWARD-WAVE BEAM AMPLIFIERS

This is a hybrid device using an injected beam together with delay-line bunching in a magnetron-type structure. As shown in Fig. 7-28, a delay line, or actually an rf network positioned properly with respect to a negative electron (called the sole), receives injected beam input while being subjected to the magnetron's coaxial field. Bunching of the beam and buildup of the delay-line input are achieved in the same manner used in a TWT.

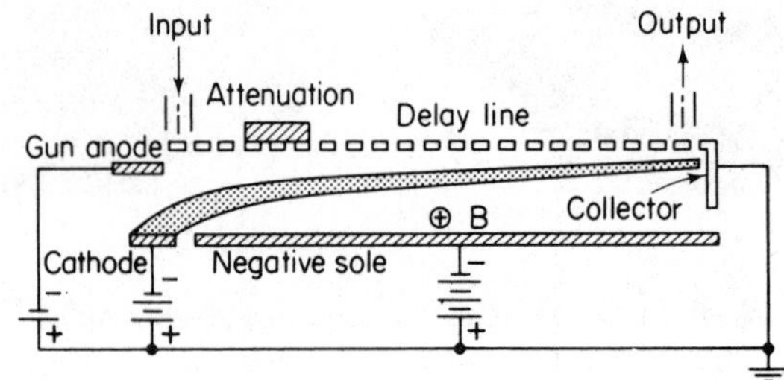

Fig. 7-28 FORWARD WAVE BEAM AMPLIFIER

Impedance—Matching—Smith Chart 8

JUNCTION TECHNOLOGY

GENERAL

In a microwave system efficient transmission of energy is a primary design requirement. This invariably calls for precise electrical matching of all components or of any element to its adjoining component or elements. This is not solid-state junction technology as in transistors, diodes, and other semiconductors and does not concern atomic surface action. It is power transfer or "connection" technology aiming to produce precise impedance matching at smooth, lossless transition when waveguides are abutted to loads, couplers, instruments, or any other element.

We might call this regime *skin-effect technology*, because effects are due to the extreme sensitivity of microwave transmission to minute surface discontinuities which cause random, unpredictable, high-frequency perturbations. These rapid changes in voltage and current along the transmission surfaces cause losses and induce unwanted reflections, both of which cause inefficient transmission. They are particularly in evidence due to the fact that unwanted or unusual currents are skin currents induced and magnified by their high-frequency nature and existing at every junction, nook, and cranny caused by physical connection.

INSERTION LOSS

For this reason, in every component or piece of waveguide the metallurgy of the surface plating, its depth, contour, and smoothness, all affect electrical performance. Inefficiency particularly shows up when these surfaces join or abut each other so that features in the construction of the flange, that is, tolerances on the dimensions of the transmission aperture, length of the adjoining legs, radius of bends or turns, etc., plus the details of its attachment to another flange, are more noticeable. This loss factor (over and above reflection losses due to impedance mismatch) is the insertion loss and is present at any juncture; it also exists as a design factor in any instrument or device which is part of a waveguide structure, such as slotted lines, wave meters, probes, etc.

IMPEDANCE MATCHING

This *"plumbing" junction technology* is intimately associated with and cannot easily be separated from the procedures and devices installed to maintain continuous and unvarying impedance across any junction. The reflections due to mismatch are an impedance phenomenon depending more upon the physical facts of electrical wavelength and the magnitude of the loads, termination, coupling, etc.

Both the junction loss factor and the mismatch reflection factors are always carefully measured when designing or constructing any microwave plumbing system. Junction-loss or insertion-loss figures are covered in flange and junction hardware listed in Table 4-1;

the entire regime of impedance matching will be discussed below.

WAVEGUIDE-IMPEDANCE DEVICES

All waveguide matching devices represent some coupling off of transmitted energy, the coupling off dependent upon the plane in which the coupling device is inserted. Briefly, matching devices may be

1. Pure resistive termination load.
2. Transformer.
3. Coupled resonant.
4. Reactive.

RESISTIVE-LOAD TERMINATION

Since in a waveguide there is no specific place to connect a resistor, we accomplish specific loading by various bulk devices. Figure 8-1 illustrates three methods of spreading the losses at the end of a line. Figure 8-1(a) shows a waveguide filled with graphited sand designed so that currents are caused to flow as the fields enter the sand; the resultant I^2R heat generated within and around the conducting particles presents a correct terminating resistance. In Fig. 8-1(b) we attempt to do the same thing by placing a resistive rod, having endwise some dimensional properties, at the center of the E field. Concentrated E-field voltage at this point causes loss currents through the resistor and reflects loading resistance across the structure. Figure 8-1(c) uses a wedge of high-resistant material secured in a plane perpendicular to the magnetic lines of force so that when they link across the wedge, voltages are generated, currents flow, and I^2R heat losses present an electrical load across the guide.

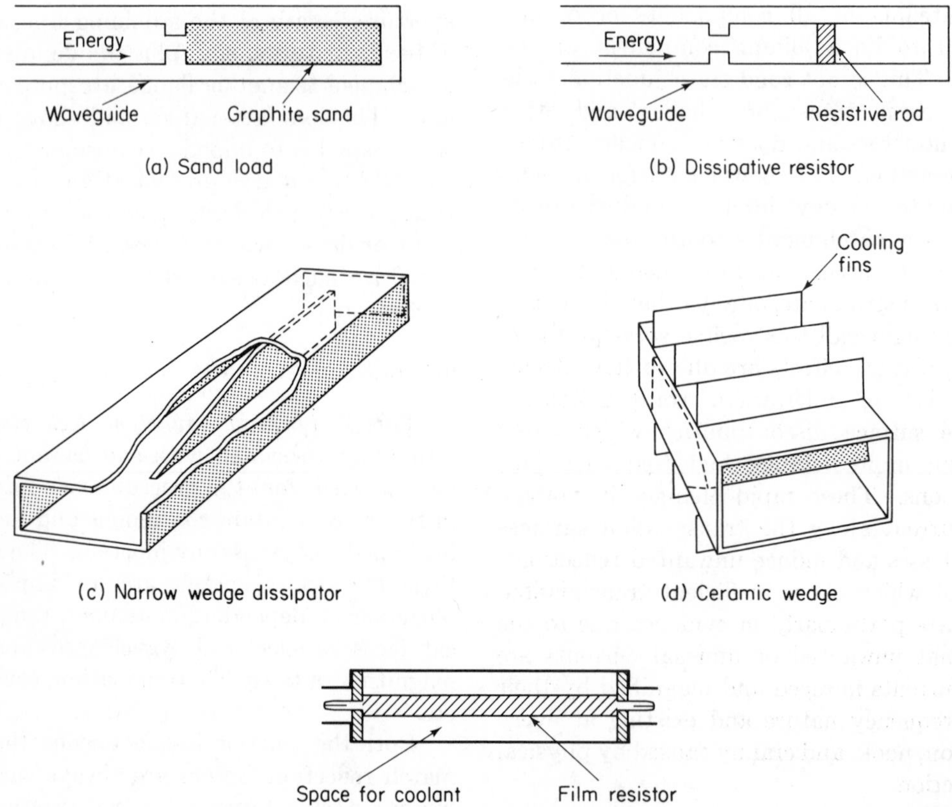

Fig. 8-1 RESISTIVE LOAD METHODS

Figure 8-2(a) shows how a load resistance may be matched to a waveguide through a quarter-wavelength section. If the junction is at a point Z_1 on the main waveguide not equal to the surge impedance Z_0, load resistance Z_R is transformed up to the value Z_1 by means of the quarter-wave section. This transformation is, of course, effective only over a restricted frequency range.

Figure 8-2(b) shows a tapered wide-frequency band line transformer where the dimensions of the guide are varied gradually, thus changing its characteristic impedance from that of the main line to that of the load.

It is fitting that we review matching and transformer characteristics of half-wave and quarter-wave sections in transmission lines. The transformer character of a transmission line is evident, because the appearance of the load as viewed from the input end of a low loss line is not a constant but varies with the length of the line. By proper choice of line length, a capacitive load, for instance, can be made from the input end to appear as though it were any one of a number of predetermined impedances. We can, by line selection, make it look like the original, larger or smaller than it, or, as in an inductance of either of two magnitudes, as a resistance of either of two values or as a wide combination of impedance components.

In mismatched half-wave lines the value of the sending-end impedance repeats itself every half-wavelength. A line that is a half-wavelength or any integral multiple half-wavelength long will therefore match any two impedances that are equal.

A quarter-wavelength of transmission line of characteristic impedance Z_c will match any two resistive impedances of which Z_c is the geometric mean. Thus a generator of known resistive internal impedance R_x may be matched to an unequal load resistance R_y [Fig. 8-3(a)] by a quarter-wavelength of interconnecting line whose characteristic impedance is

$$Z_c = \sqrt{R_x \times R_y}$$

At microwave frequencies, the generator impedance is frequently the characteristic

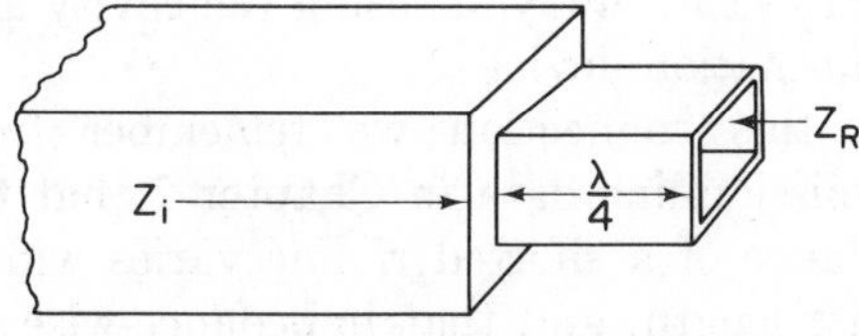

(a) Quarter-wave dissimilar sections

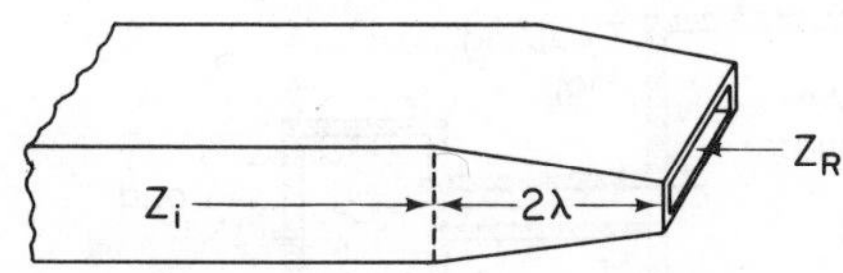

(b) Tapered section transformation

Fig. 8-2 WAVEGUIDE TRANSFORMER LOADING

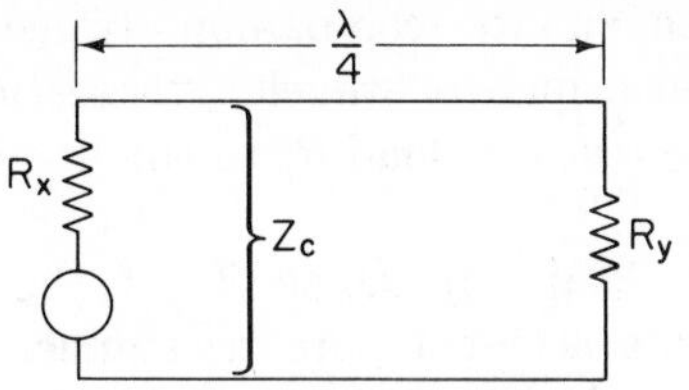

(a) Simple quarter-wave line matching

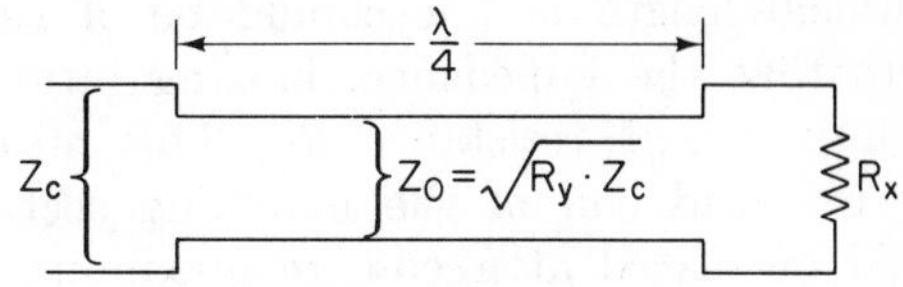

(b) Additional quarter-wave section for matching load to source

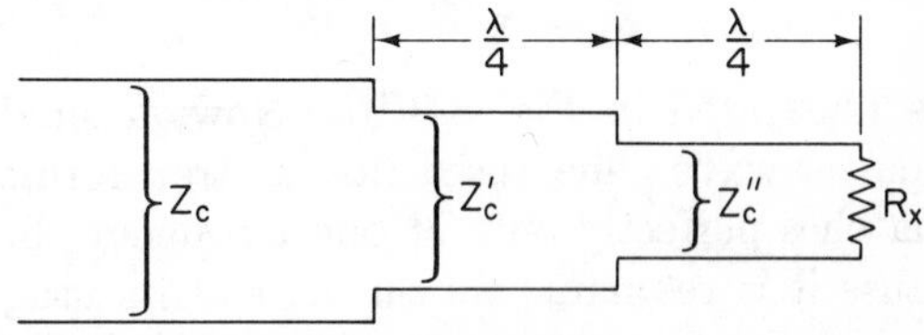

(c) Equivalent match simulating tapered sections

Fig. 8-3 BASIC QUARTER-WAVE TRANSFORMATIONS

impedance of its companion transmission line Z_c; so a quarter-wavelength section will match the resistive load R_y to the line impedance Z_c.

If the load impedance Z_c of the transmission line is not a pure resistance, it still may be matched to Z_c by a quarter-wave line plus an additional length of line between Z_1 and the quarter-wave matching section. This additional length of line should be of such length that the impedance, looking into its input, is a pure resistance R_2. This means that the load end of the matching section should be placed at a voltage maximum or minimum in the input line, at which points the input impedance looking toward the load is real. The various impedances must then satisfy the relation

$$Z_x = \sqrt{Z_c \times R_2}$$

as illustrated in Fig. 8-3(b). Now, a single quarter-wave line used as a transformer matches perfectly only at one frequency, because it is resonant; we can extend its usage and match over some increased bandwidth by placing in tandem two or more properly selected quarter-wave matching sections [Fig. 8-3(b)].

Physically, two transmission lines of different impedances can be matched by using a length of tapered transmission line whose impedance varies continuously but slowly throughout its length. Figure 8-3(c) illustrates the tapering diameter used to interconnect such sections.

COUPLED RESONANT CIRCUIT

A loaded resonant cavity can act as a transformer by using input coupling direct from probe or iris in the main waveguide and then coupling to the output load from a probe pickup located within the cavity. Figure 8-4 illustrates such a layout using the cavity as a transformation device.

In this connection we remember from transmission-line data in Chapter 2 that the impedance of a shorted rf line varies widely over its length, and that impedance-wise we can typically match a 500-Ω line to a 70-Ω line

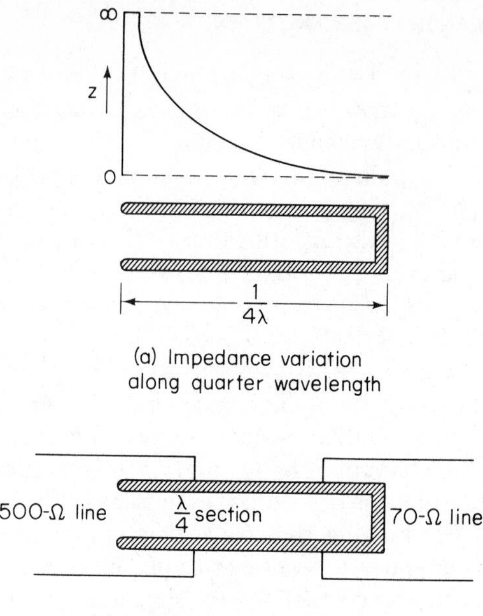

(a) Impedance variation along quarter wavelength

(b) Impedance transformation at tapped points

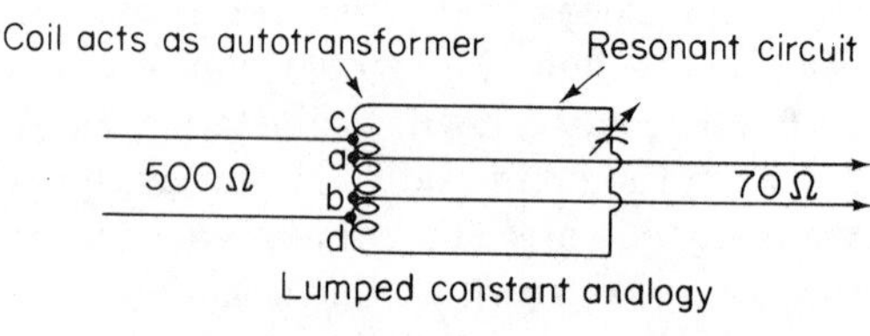

(c) Auto-transformer analogy

Fig. 8-5 TRANSMISSION LINE TRANSFORMATION

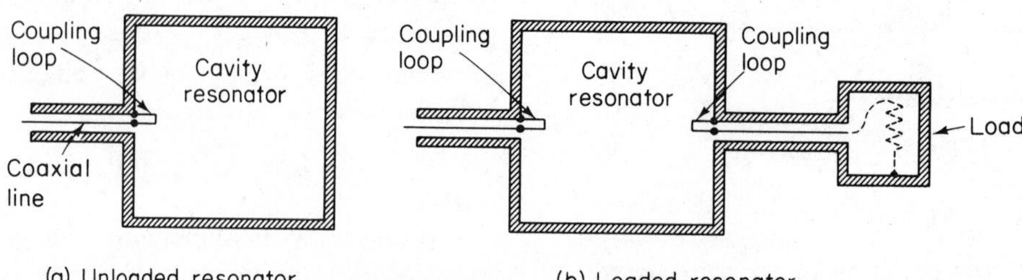

(a) Unloaded resonator

(b) Loaded resonator

Fig. 8-4 RESONANT CAVITY TRANSFORMATION

by connecting both in at the correct point on a quarter-wave section. Figure 8-5(a) shows this connection, which in terms of lumped constant parameters is the same as placing taps on a tuned auto transformer [Fig. 8-5(b)]; this whole arrangement is directly analogous to the tuned cavity mechanism.

REACTIVE DEVICES

Reactive devices can be capacitive, reactive, or tuned circuits, stubs mounted in the E or H planes, or obstacles deliberately mounted within the guide itself. Figure 8-6(a) shows series and shunt stubs as they would be mounted in a line.

The fins and apertures in Fig. 8-6(b) are, respectively, perpendicular to the H and E fields and are equivalent to shunt and capacitive inductances placed across the guide. These plates set up local oscillation in higher modes, and since the waveguide is too small to support them, the dominant mode currents

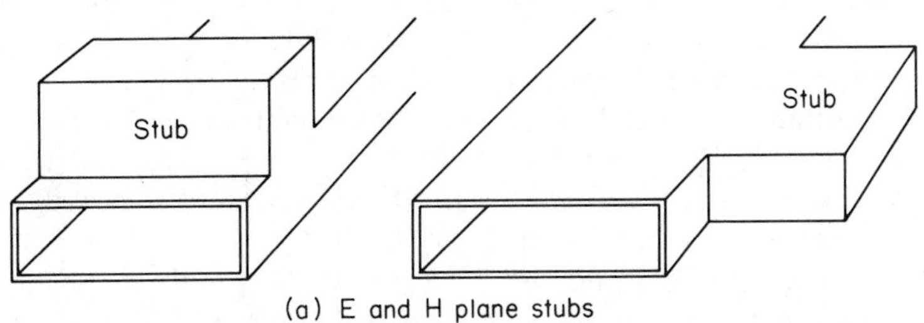

(a) E and H plane stubs

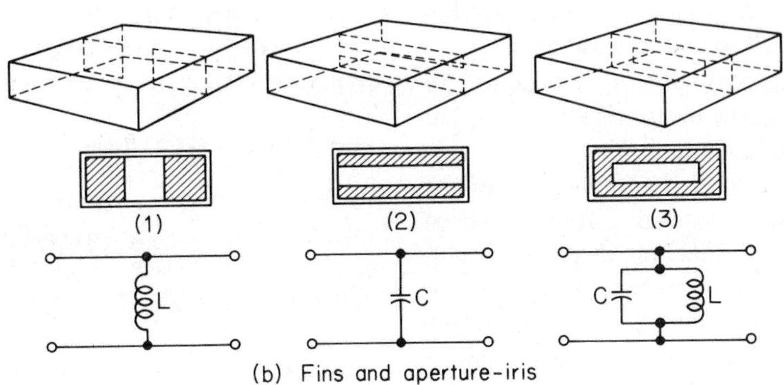

(b) Fins and aperture-iris

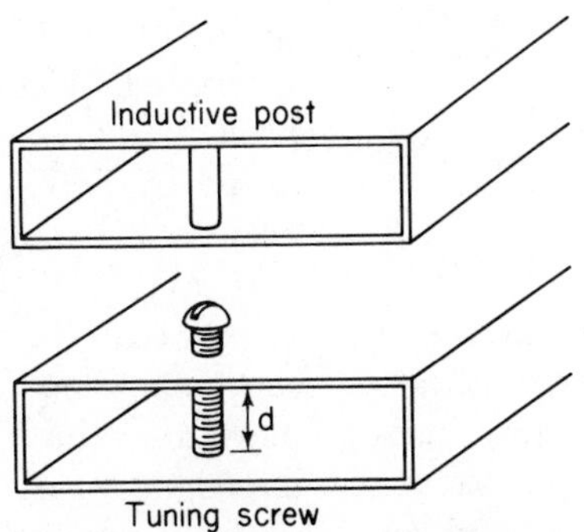

(c) Adjustable reactive element

Fig. 8-6 REACTIVE WAVEGUIDE DEVICES

at these frequencies are not propagated but remain in the vicinity of the plates. Changing the fin dimensions in each of these arrangements varies the reactance so that impedance matching with adjoining elements can be controlled.

If we combine both *E*- and *H*-field fins [Fig. 8-6(b)], we have formed an iris and, in effect, introduced a shunt resonant circuit across the waveguide—equivalent to a high resistance depending upon the size of the iris opening and its position with respect to the *H*- and *E*-field configurations.

The post or tuning screw is a very common form of adjustable reactance. Figure 8-6(c) shows the screw as a reactive post, varying with the depth of insertion *d*. When *d* is small, the screw will act as a shunt capacitance, and when *d* is increased to the point of being a quarter-wavelength, resonance occurs, with penetration beyond this point converting the post into an inductive reactance.

In simplified form, the screw actually behaves somewhat like a small antenna (with images in the top and sides of the guide) in series with the capacitance between the end of the screw and the bottom of the guide. If this capacitance were absent, the impedance of the screw might behave as an antenna whose length is equal to that of the screw and possesses an image with respect to the top of the guide. The screw would then be capacitive when its insertion distance is less than a quarter-wavelength and inductive if the insertion distance is greater than a quarter-wavelength.

WAVEGUIDE-IMPEDANCE CONSIDERATION

GENERAL

Waveguide impedance cannot be easily calculated or obtained by a single, direct measurement. It must be determined indirectly from a two-step procedure consisting of first a VSWR measurement and, second, an interpretation of it into numerical terms through transmission charts, usually the Smith charts.

This measurement technology is described in Chapters 10 and 11 along with the three other basic microwave measurements: (1) frequency, (2) attenuation, and (3) power. As we know, it is a combined amplitude and physio-electrical measurement concerning the position of maxima and minima voltage nodal points along a transmission line or waveguide. Although transmission-line impedance can be measured at low microwave frequencies by bridge methods and, for most microwaves, by a reflectometer, the VSWR techniques blend in conveniently with the facts of waveguide operation which follow.

It should be noted that although the following background material relating to impedance is slanted heavily in terms of coaxial transmission lines, it also basically applies to waveguide phenomena, even though the operating quantities become more consistently complex in nature. Thus slotted lines for VSWR measurement and Smith-chart interpretations are equally as powerful (and sometimes more convenient) in waveguide analysis as in coaxial transmission.

IMPEDANCE VARIATIONS

Sending-end impedance, Z_s, of a transmission line or of a waveguide depends upon the load impedance, (Z_r), and how far (electrically) the load is located from the source.

Z_s can thus have four ranges of value: when Z_r is equal to (1) Z_c, the characteristic impedance, (2) a short circuit, (3) an open circuit, and (4) any complex impedance.

(1) and (4). When Z_r equals Z_c, the line is perfectly matched and no reflections are generated. There would thus be no voltage peak and voltages along a 50-Ω line, terminated by a 50-Ω resistor, when the sending end is a complex impedance, Z_s equals $R + jX$, Z_c should be its conjugate, $R - jX$.

(2) and (3). If we located the open or short circuit at different distances along the line from the sending end, we would find impedance varying in a series of parabolas each spaced one half-wavelength apart [Fig. 8-7(a) and (b)].

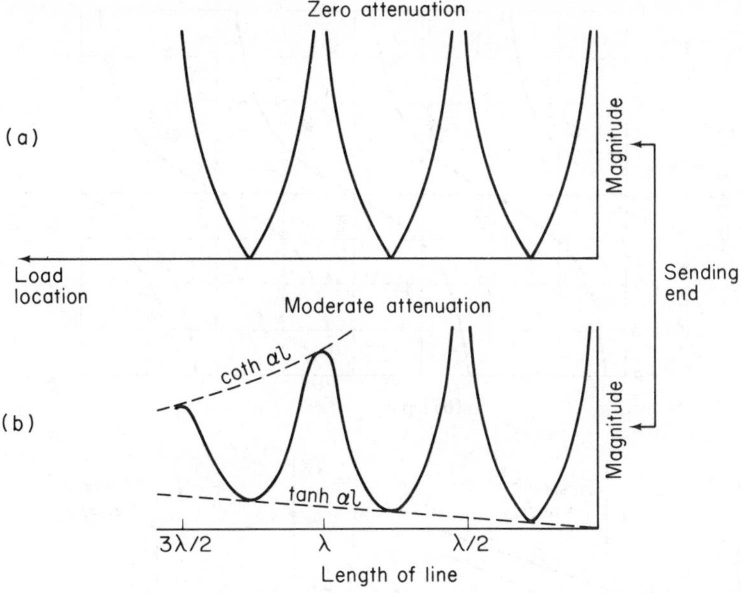

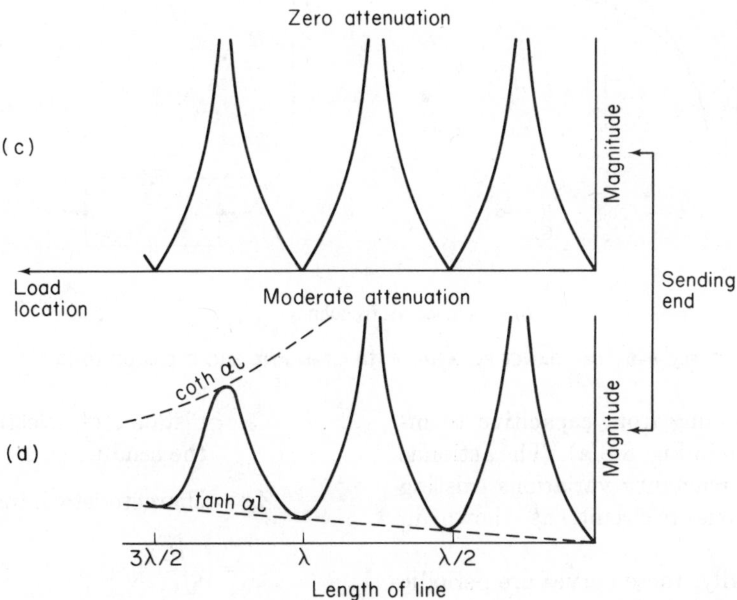

Fig. 8-7 IMPEDANCE VARIATIONS WITH OPEN AND SHORT CIRCUIT LOAD

For zero load the sending end would appear as an open circuit, decreasing to a short circuit one quarter-wavelength away; a short-circuit load produces a sending-end short circuit which varies up to infinity one quarter-wavelength away. These variations taper off in a line that has losses [Fig. 8-7(c) and (b)].

Actually, the algebraic sign of the sending impedance changes periodically every quarter-

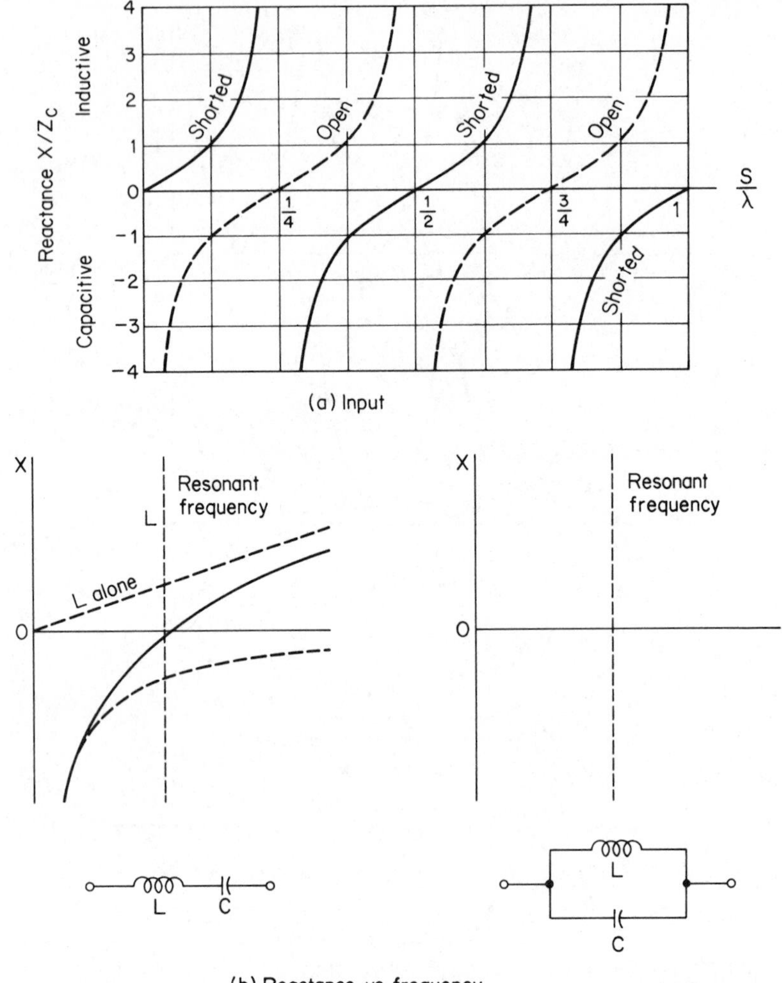

Fig. 8-8 IMPEDANCE POLARITY WITH OPEN AND SHORT CIRCUIT LINES

wavelength, changing from capacitive to inductive, as shown in Fig. 8-8(a). The action is identical to the reactance variations existing in a tuned resonant circuit, as shown in Fig. 8-8(b).

Mathematically, these curves are periodic plots of the tangent function defined by the equation

$$Z_s = \frac{Z_R + jZ_C \tan \beta S}{Z_C + jZ_R \tan \beta S} Z_C$$

where β = imaginary or phase constant part, of which is the propagation constant

s = distance of selected points from the sending end

Z_C = characteristic impedance

$$= \sqrt{\frac{L}{C}}$$

Of, if we normalize the sending-end impedance by dividing it by the characteristic impedance, the equation for transmission will be

$$\frac{Z_S}{Z_C} = \frac{Z_R/Z_C + \tanh \gamma \times S}{1 + Z_R/Z_C \tanh \gamma \times S}$$

and for zero losses,

$$\frac{Z_s}{Z_C} = \frac{Z_R/Z_C + j \tan \beta S}{1 + j(Z_R/Z_C) \tan \rho S}$$

where γ = propagation constant

$= a + j\beta$

a = the real component of γ

As we shall see later, the practical way of dealing with impedance quantities is graphically by means of transmission charts. Plotting these equations by charts gives quick, visual indications of resistive and reactive quantities without calculating individual points from the trigonometric or hyperbolic expressions in the above equations.

These steps are necessary when using firm, numerical values existing in practice and when inserting them in transmission charts.

Summarizing, we can see that the impedance of a lossless line can be a variable (unless terminated in its characteristic impedance) depending, first, on how far (electrically) the point we measure is from the sending end and, second, upon what is the magnitude and complex impedance of the terminating load.

We also note (in a lossless line) that the variation is cyclical over half-wavelength sections of the line, and that once we have determined a value, it repeats itself at periodic points every half-wave from the point we have measured. When losses are present, its magnitude decreases (slightly at microwaves) as we proceed downward toward the load point.

Now, measurement of a particular point impedance is possible just because of the mismatch existing at the selected points. It is the reflections caused by a load (which is never perfectly matched) that we can electrically measure in amplitude and locate physically and electrically along a line, and this enables us to determine from transmission charts any desired impedance.

Actually, we detect the reflections because they generate physically stationary or "standing wave" having maximum and minimum voltage points at specific distances along a half-wavelength of slotted line or waveguide which has been inserted for measurement purposes in the main system being measured (see Chapters 10 and 11).

Impedance-wise, we are interested in three procedures: a single measurement, one calculation, and the single resultant chart manipulation. In detail, (1) we must measure the voltage standing-wave ratio (VSWR) and its electrical position on the half-wave slotted section, (2) we must calculate the reflection coefficient, and (3) we must interpret these three on a Smith transmission chart for a final impedance characteristic (see Fig. 8-9).

These procedures come first in any microwave design, test, or manipulation. Second, we compare VSWR ratios along a line for attenuation and other minor factors.

VSWR

This quantity represents the ratio between two measured voltages, $E_{\max}$ and $E_{\min}$, determined as a voltage detector is moved along a slotted transmission line or waveguide. It is always greater than unity and commonly it is expressed as

$$\rho = \frac{E_{\max}}{E_{\min}}$$

In final stages of achieving perfect match it may practically be as low as 1.01–1.05; in preliminary measurements it may go as high as 10–20:1.

REFLECTION COEFFICIENT (Γ)

This quantity is specifically defined as the numerical ratio between the reflected voltage and the incident voltage. It is unity (or may be considered not to exist) when a line or waveguide is terminated in a load impedance equal to its characteristic impedance, or when $Z_R = Z_C$. If a line is short circuited ($Z_R = 0$), $\Gamma = -1$, the voltage wave is totally reflected and in phase operation at the receiving end, and it cancels the incident wave at this point. Another way of saying the same thing is to observe that the receiving-end voltage, being the sum of the incident and

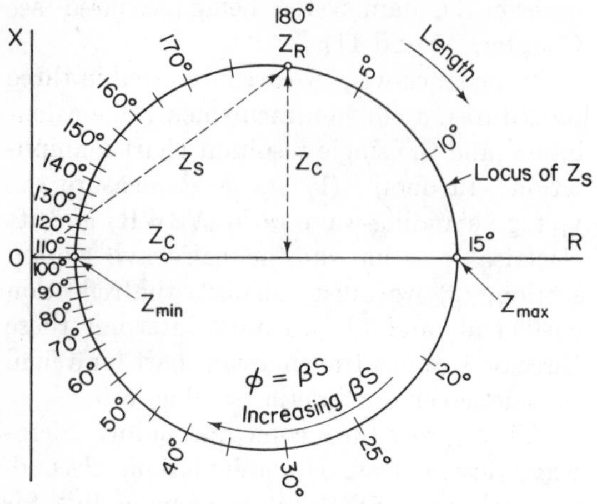

Fig. 8-9 INPUT IMPEDANCE VECTOR LOCUS

reflected waves, is zero, as it must be to exist as a short circuit.

Also, if the line is open at the receiving end ($Z_R = \infty$), then $\Gamma = 1$ and the reflected wave is in phase with the incident wave, so we have voltage doubling, since the two waves reinforce each other. However, current is reflected from the open end with phase reversal and cancels, so the net receiving-end current, being the vector sum of two equal and opposite components, is zero, as we would expect with an open circuit.

Now, if the load impedance is pure reactance ($Z_R = jX$), there is also a complete reflection ($\Gamma = 1$), a conclusion we reach from the fact that the average power consumed by a reactance is zero, so all power must be reflected—the phase of the reflection (ψ) being anywhere in the entire 360° range depending upon the sign and magnitude of the reactance.

Practically, we use the numerical value of Γ which is related to the VSWR. This expression is

$$\Gamma = \frac{\rho - 1}{\rho + 1}$$

where ρ = standing-wave ratio.

Further determination and use of Γ will be discussed in Chapters 10 and 11.

IMPEDANCE CHARTS

RECTILINEAR COORDINATES

The graphical description of the two-dimensional impedance of a transmission line (that is, of the combined sum of $R + jX$ components along its length) comes down to a circle diagram, whether the picture be painted in rectilinear or Cartesian coordinates or in modified coordinates used in the Smith chart.

In X–Y coordinates the graphical plot of Z_S locus is naturally a circle (Fig. 8-10) because it is made up of increasing and decreasing line resistance and waveguide losses plotted along the X axis, and varying plus or minus jY components plotted vertically above and below from the end of the R component. These two variations with line length (in angular terms $= \beta s$) give a circular locus of Z_S over 180° travel which is one half-line wavelength.

Note that maximum and minimum input impedances are respectively resistive and equal to $Z_C \times \rho$ and Z_C/ρ, where ρ equals the standing-wave ratio. Hence the points on the circle also represent the measured voltages across Z_{max} and Z_{min} and lie on the horizontal axis at distances $Z_C \times \rho$ and Z_C/ρ, respectively, to the right and left of the origin where these points are resistive-impedance circle points farthest from and closest to the origin. Thus each impedance-locus circle is also a circle of constant wave ratio.

To make a chart universally applicable to transmission lines or waveguides having any characteristic impedance and terminated in any load impedance, we must normalize the actual operating resistance and reactance in ohms to some small integral ratio, say between 1 and 5, by relating it to the characteristic impedance of the line. These normalized chart units, Z, R, and jX, are converted by dividing each operating actual R and jX by Z_C; thus $Z = R + jX$ equals $Z/Z_C = R/Z_C + jX/Z_C$. Conversely, final quantities are converted back to real operating ohms by multiplying by Z_C.

Now for a practical working chart and also for convenience, the basic circular plot

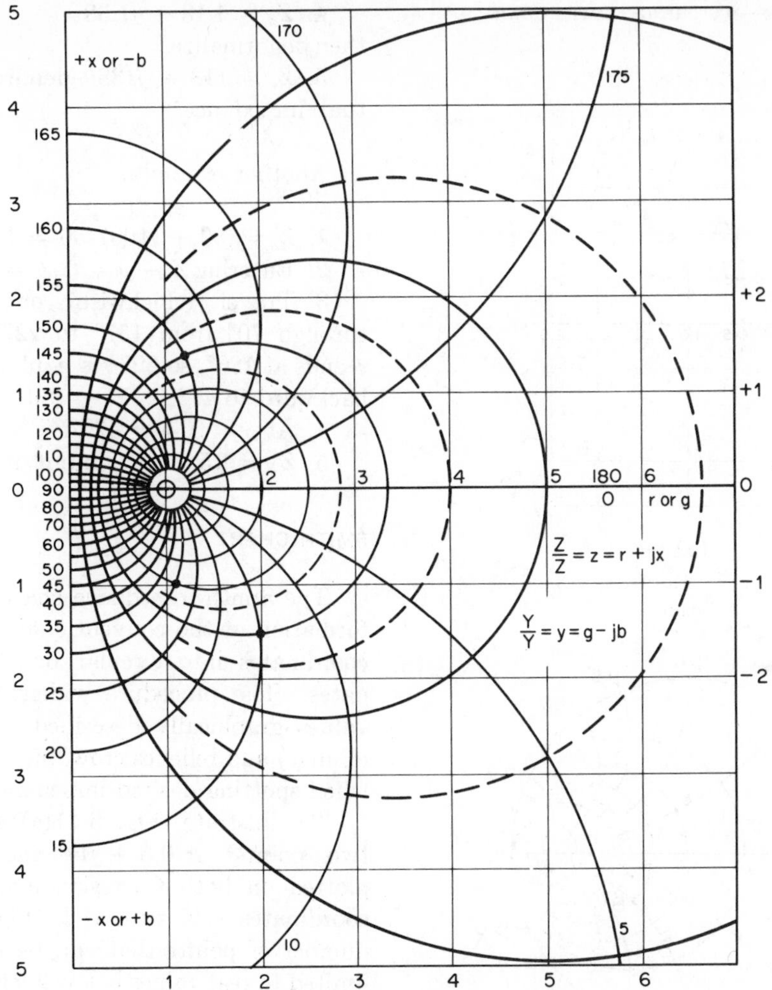

Fig. 8-10 IMPEDANCE CIRCLE DIAGRAM

of Fig. 8-10 is rotated so that the angular distances βs are referenced to the point of maximum impedance, because this is where we measure the standing-wave voltage. Thus the working chart consists of a series of eccentric circles drawn for constant impedance and standing-wave ratios.

Also, a second family of "angle" circles are drawn orthogonally to the constant ρ circles, so we can determine input impedances at remote angular points where they cut the standing-wave impedance circles (Fig. 8-10).

Note that the locus for $\rho = 1$ is the single point $1 + j0$, equivalent to a circle of 0 radius with center at $1 + j0$. This is consistent with the fact that the input of a line terminated in its characteristic impedance is equal to the characteristic impedance regardless of the length of the line.

The circle diagram may be used for admittance determination where we convert all impedances to admittances and express them in mhos of normalized chart units equal to the individual mho admittances divided by the characteristic admittance in mhos. The point of maximum admittance is a point of minimum voltage and vice versa, and as the length of the line is increased, its input admittance point moves clockwise around a constant ρ circle as in the impedance diagram.

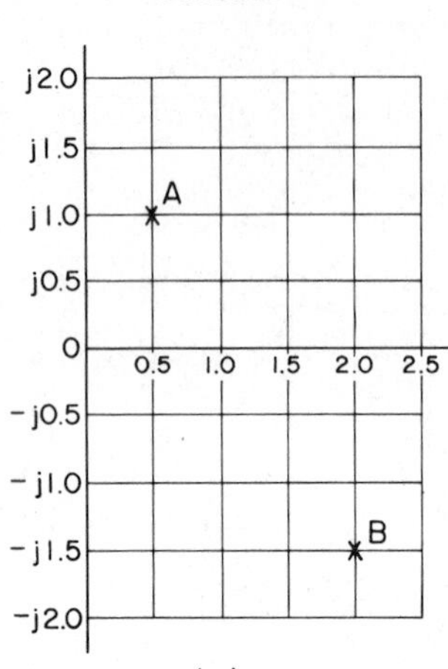

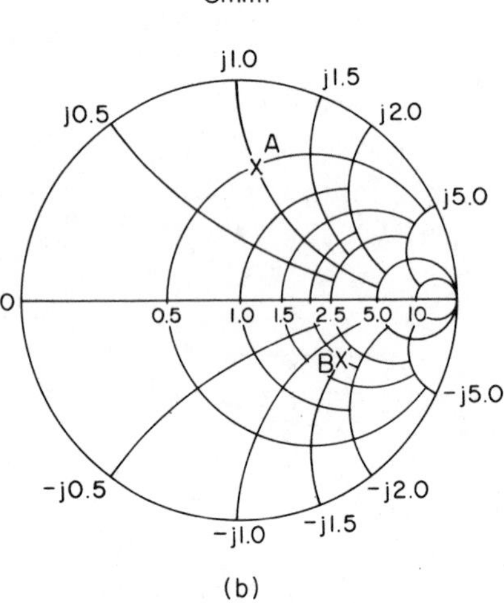

Fig. 8-11 CARTESIAN AND SMITH CHART COMPARISONS

A typical calculation using Fig. 8-10 calls for finding the input impedance of a lossless transmission line with characteristic impedance of 100 Ω, a load impedance of 200 + $j150$ Ω, and 140 electrical degrees in length.

We shall give an example. Given $Z_c = 100$, $Z_R = 200 - j150$, $\beta s = 140°$. The steps in the solution are as follows:

1. $Z_R = (200 - j150)/100 = Z - j1.5$ (normalized impedance).

2. Plot Z_r—arriving at $\theta = \beta s = 15°$ on $\rho = 3.33$ circle—see ⊙ point on Fig. 8-8.

3. Travel 140° or the $\rho = 3.33$ circle to (point Δ on 155° circle)—

4. $Z_s = 1.18 + j1.38$ then denormalize.

5. $Z_s = 118 + j138$—denormalized (actual) impedance.

Another example:

1. $Z = (50 + j100)/50 = 1 + j2.00$.

2. Entering Z_R—$\rho = 6, \phi = 157°$.

3. Travel (clockwise) on $\rho = 6$ circle through 70° from 157° to 227°. Note that ϕ ends at 180°, so 227° is equivalent to going backward to 47° (227° − 180° = 47°).

4. At $\rho = 6$, $\phi = 47°$, $Z_S = 0.30 - j0.90$.

5. $Z_S = 15 - j45$ (input impedance).

SMITH CHART

The Smith impedance chart uses a transformation of the conventional Cartesian XY coordinates into circular or "bent" coordinates. The procedure widens the range of values graphically described (usually up to infinity) and relieves crowding of angle circles when spotting desired impedance points.

To illustrate, Fig. 8-11(a) and (b) show two points: $A(0.5 + j1)$ and $B(2 + j1.5)$ plotted on both Cartesian and Smith-chart coordinates. Note that in Fig. 8-11(a) the number of points that can be represented is limited to real values below 2.5 and imaginary distances of less than 2. The Smith chart, Fig. 8-11(b), shows values from zero to well over 5, and on commercial $8\frac{1}{2}$- by 11-in. charts the device extends both coordinates to around 50.

Another example of its convenience and range is the plot of reactance versus frequency for a series resonant circuit, as shown in Fig. 8-12 for XY versus Smith coordinates. Note the range of reactances extending to infinity.

In the enlarged charts for practical usage (Fig. 8-15) the point 1,0, which corresponds to the characteristic impedance of the line, is now located at the center of the chart. Concentric circles centered at this point become paths of constant load impedance and constant standing-wave ratio. These correspond to the eccentric circles clustered around the point $1 + j0$ in Fig. 8-10.

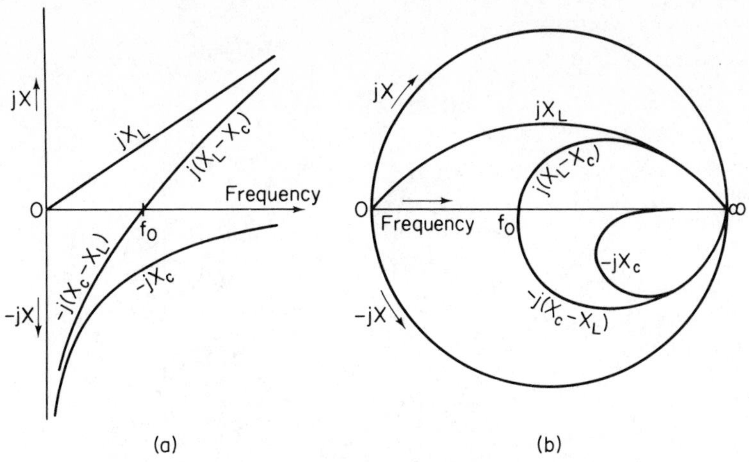

Fig. 8-12 REACTANCE PLOT VERSUS FREQUENCY

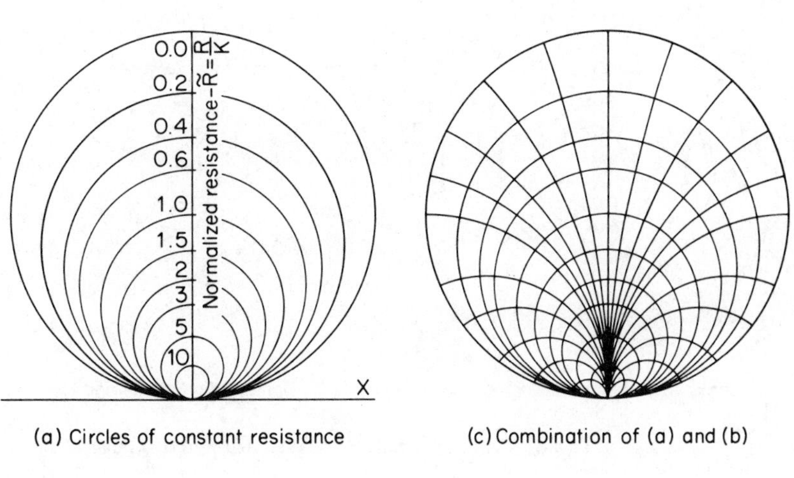

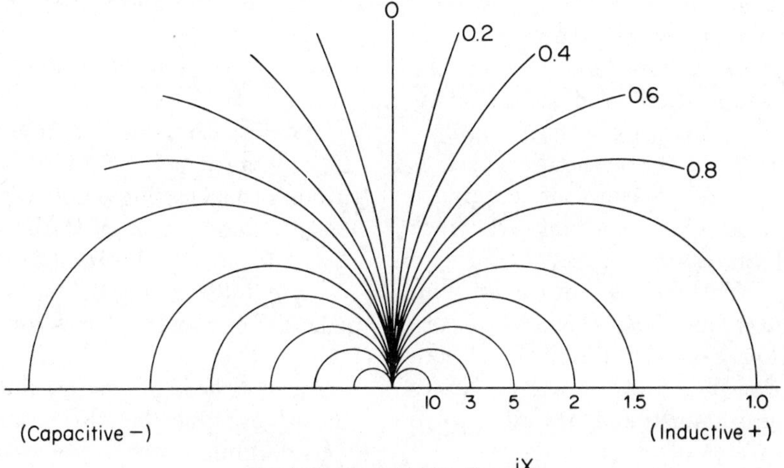

Fig. 8-13 BREAKDOWN OF SMITH CHART COORDINATES

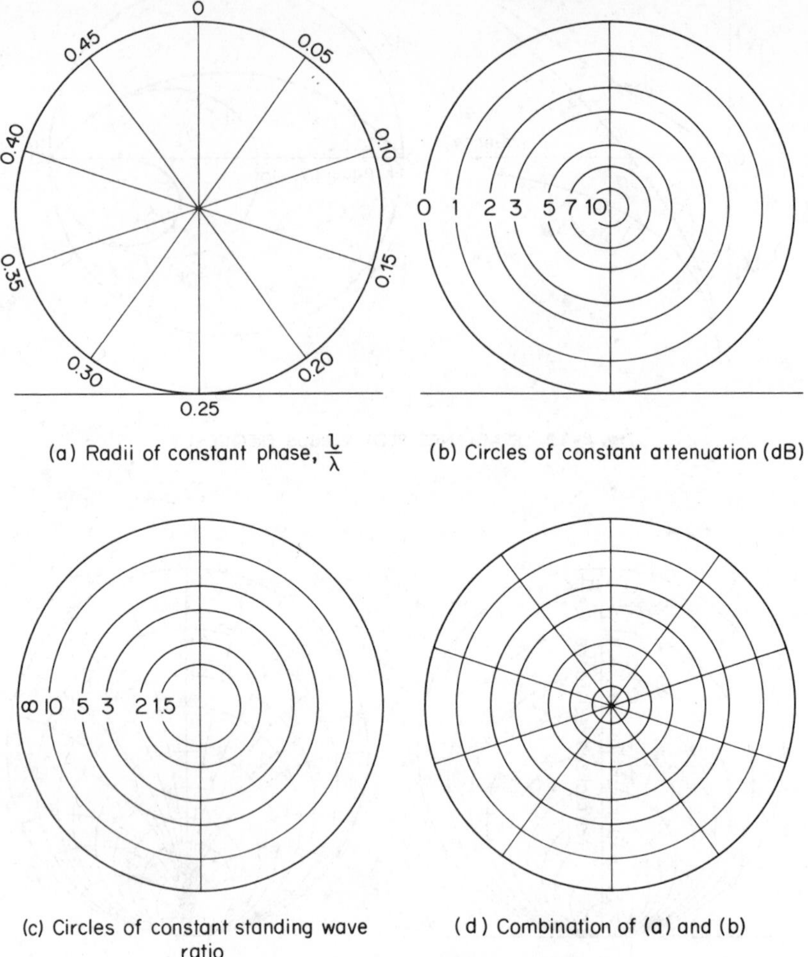

Fig. 8-14 BREAKDOWN OF SWR, PHASE, AND ATTENUATION

Next, the angle circles (labeled in degrees) in the XY curves of Fig. 8-10 become, in the Smith chart, straight lines radiating from the center and numerically labeled on a circular scale around the border of the chart.

In summary, the basic Smith chart is made up of circles of constant resistance [Fig. 8-13(a)], arcs of constant reactance [Fig. 8-13(b)], and a combination of the two [Fig. 8-13(c)]. Next, we use, but do not find, plotted on the main charts (1) radii of constant phase [Fig. 8-14(a)], (2) circles of constant attenuation [Fig. 8-14(b)], and (3) circles of constant impedance and standing-wave ratio [Fig. 8-14(c)]. (d)

An example of plotting technical values is inscribed on the conventional engineer's working diagram shown in Fig. 8-15. Here we establish the normalized load impedance,

$$\frac{Z_r}{Z_0} = 1.3 + j0.75$$

by moving along the 1.3 resistance circle to the reactance point $j0.75$ (see ⊙). A circle through this locating point, with its center at $1 + j0$, passes through the $R = 2$ point and $R = 0.5$ on the horizontal scale, yielding $\rho = 2$ and also giving the locus of sending impedances in r and jx terms for this particular load.

Now, drawing a radial line from $1 + j0$ through our given locating point $(1.3 + j0.75)$ and extending it to the circumference, we find that this load is located at 0.18 wavelength toward the generator.

If we desire to find the sending impedance

one twelfth of a wavelength (0.0833) away, we draw a radial line to the 0.2633 wavelength point (0.0833 + 0.18 = 0.2633) (Δ) and find where this line intersects our impedance circle. See (Δ) plot point. This yields a sending impedance

$$\frac{Z_s}{Z_0} = 1.95 - j0.25$$

Another more common problem is to determine an unknown load impedance for which we have measured the standing-wave ratio and the position of a voltage minimum.

Assuming a VSWR of $\rho = 2.5$ plus a voltage minimum falling at 30° toward the load,

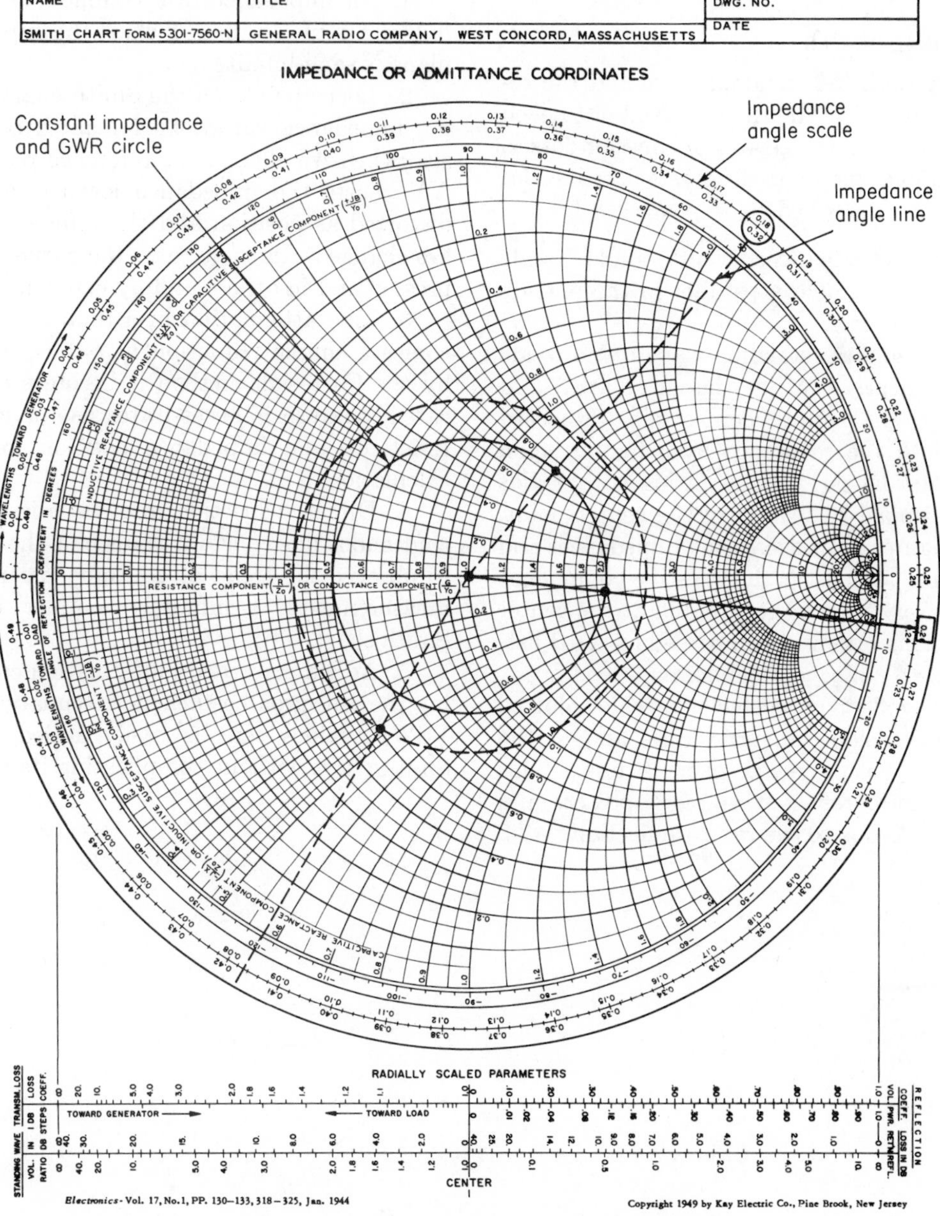

Fig. 8-15 WORKING SMITH CHART FOR DESIGN
(Courtesy of General Radio Co.)

we draw an impedance-locus circle passing through $r = 2.5$ and $r = 2.4$. Then we draw a radial line from the center to the 0.0833 point ($0.08333\lambda = 30°$) on the circumference (Fig. 8-15). Thus at $Z_r/Z_0 = 0.51 - j0.46$, or where the radial line intersects the impedance circle, we have established the desired sending impedance of $Z_s/Z_0 = 0.51 - j0.46$.

ADDITIONAL SCALES

Since both the standing wave and the reflection coefficient are directly related to each other and both related to transmission losses, all three may be plotted on associated scales as herewith described.

1. A reflection-coefficient angle scale around the periphery, parallel to the outside wavelength scale.
2. A voltage standing-wave ratio scale plotted linearly (outside the curve area) and parallel to the main resistance axis. Projected upward to the resistance axis, figures upon this scale deliver voltage ratios with ratios greater than 1 to the left of the center point. These figures show decreasing impedances and minimum voltages as the resistive component moves away from perfect match at the center point. This scale is accompanied by a dB listing next to major voltage points.
3. A reflection coefficient scale plotted linearly (outside the curve area) and to the right of the projected central line. This section gives reflected voltage, power, and losses in dB due to the return reflection.
4. A transmission-loss scale next to and parallel to the VSWR scale. Each of these scales serves the purpose of entering measurement and design factors discussed in Chapters 10 and 11.

SUMMARY

Specifically, the Smith chart has the following advantages:

1. All impedances are confined to within limiting radii. In the XY chart the impedance plane is semi-infinite.
2. Locus points on the Smith chart showing impedance variations remain a constant size when we change the reference point. In XY charts the impedance locus curves are distorted as position and the reference point are changed. In some cases the curve is compressed and in others unduly expanded.
3. Line attenuation in a Smith chart becomes a simple reduction in radius, whereas by the Cartesian method it becomes complicated, because it necessitates use of varying resistance and impedance position for each calculation.
4. Direct inspection and visualization of continuous line constant trials plus the ability to observe the trend of continuously plotted measurements is infinitely simpler in the Smith chart.
5. Added conversion information concerning the reflection coefficient, attenuation, and VSWR values can be easily added to the Smith chart without complicating its format.

Attenuators—Termination—Stubs—Tuners 9

INTRODUCTION

Microwave "fixture" hardware are metallic assemblies which are primarily passive and used only to carry signal or power. These devices attenuate, alter phase, deliberately dissipate power, or otherwise process a signal for measuring or for monitoring.

Specifically, we list them as loads, terminations, attenuators, tuning stubs, *E–H* tuners, and phase shifters—all tailored for the performance criteria listed below. Functionally, to amplify their usage for both low- (signal) and high-power applications we have

1. Attenuators are chiefly used for precise adjustment or in design.
2. Terminations and loads are used for matching and power determination.
3. Stubs and tuners are primarily used for measurement or monitoring.
4. Phase shifters alter the time factor in signal passage.

In addition to the above generalizations, these items may be inserted in either coaxial-line or waveguide systems. We can note that waveguides themselves operate from a cutoff frequency upward and are seldom used for transmission of power at frequencies below 1 GHz.

ATTENUATORS

GENERAL

Attenuators are used to increase the power range of sensitive instruments, to measure gain or loss, and to act as buffers which reduce interaction between instruments. A common use is in providing signal generators with accurate output adjustment for precise measurement.

Type-wise, attenuators may be divided into three categories: (1) fixed pads (2) step-adjustable and (3) continuously variable. The models within each group offer ranges of accuracy and wide band coverage.

The specifications governing attenuator performance are essential facts in selection and application of all models. Key items are:

1. Attenuation: how much reduction from input to output in dB is achieved by a particular model.
2. Characteristic impedance, particularly for coaxial units.
3. Broad-band accuracy: maximum error in the calibrated dial setting over the listed frequency range.
4. Stability: expected change in attenuation over ranges of frequency, temperature, humidity, and power being handled.
5. Insertion loss: residual loss existing

(a)

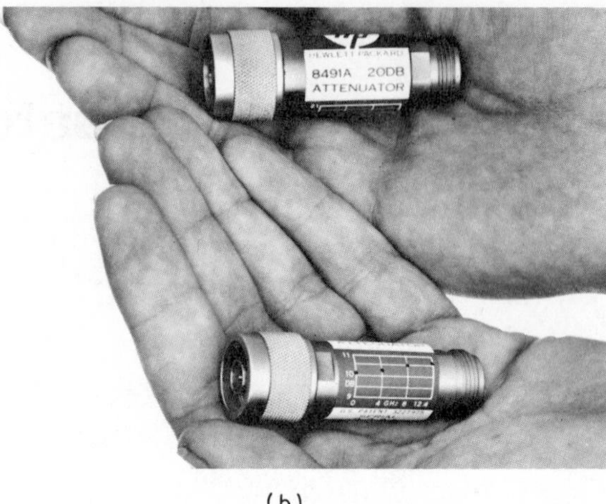

(b)

Specifications

Nominal impedance:	50 ohms
Frequency range:	dc to 8.0 GHz
Standard nominal values:	3, 6, 10 and 20 dB
Maximum deviation from nominal* (Including frequency sensitivity)	
Dc to 4.0 GHz:	±0.3 dB
4.0 to 8.0 GHz	±0.6 dB
Power rating:	
Maximum average:	
Model 5:	3 dB, 15 W; 6, 10 and 20 dB, 10 W Full power to 55° C; derated linearly to 0 W at 125° C
Model 6:	3 dB, 25 W; 6, 10 and 20 dB, 20 W. Full power to 25° C; derated linearly to 0 W at 125° C
Maximum peak:	5 kW for both models at all nominal values
Power coefficient:	$<0.001 \frac{\text{dB}}{\text{dB} \times \text{W}}$
Temperature coefficient:	$<0.0001 \frac{\text{dB}}{\text{dB} \times {}^\circ\text{C}}$
Maximum VSWR:	
Dc to 4.0 GHz:	1.20
4.0 to 8.0 GHz:	1.30
Calibration frequencies:	Insertion loss at dc, 4.0 and 8.0 GHz Data are supplied on calibration form
Construction:	Black anodized, finned body with Type N stainless steel connectors, standard with male or female. Units with both male or both female connectors are available
Weight:	Approximately 5 oz.

*At 25° C (excluding power and temperature sensitivity)

(c) Typical specifications

Fig. 9-1 FIXED ATTENUATOR PADS (Courtesy of Microlab/FXR and Hewlett Packard.)

when the adjustment dial is at zero setting and for the worst case frequencies.

6. VSWR: maximum designed-in value existing at a single calibrated frequency and for worst case under broad-band frequency operation.

7. Resolution: how close the dial-indicated attenuation approaches the true value.

8. Repeatability: how close in repeated cycling the actual attenuation setting remains to the control-knob indicated value.

9. Drift with power: tolerances or variations in true attenuation at different power levels.

Fixed attenuator pads for coaxial lines function in their normal frequency range by use of precision film resistors acting as lossy lines with or without the insertion of lossy dielectric insulation. Figure 9-1(a) pictures a miniature type of fixed coaxial attenuator covering attenuation from 3 to 20 dB. Figure 9-1(b) tabulates the performance specification in models operating up to 18 GHz. Figure 9-2(a) is a set of precision laboratory attenuators intended for extending the range of power meters. Figure 9-1 (b) shows two types of coaxial power dividers, one using reactive attenuation, the other a resistive network. In some waveguide power dividers, division is effected by the use of a directional coupler (see Chapter 5).

Step attenuators for coaxial lines in Fig.

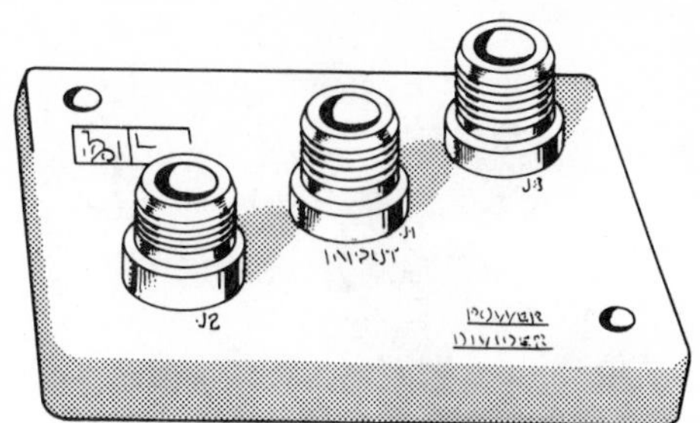

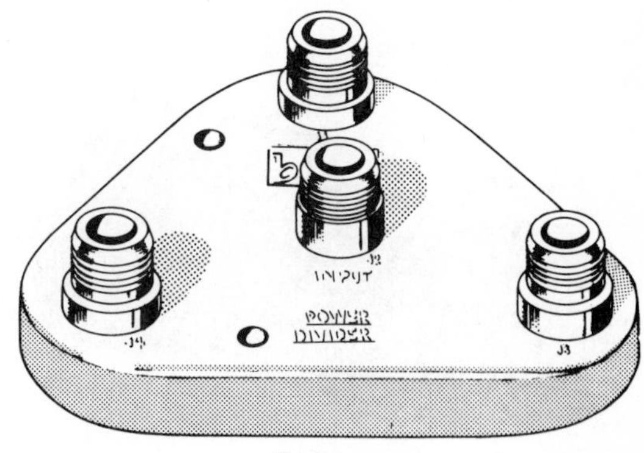

(b) Attenuators

Specifications

Type	Two-way	Three-way
Bandwidth	500-4000 mc	500-4000 mc
Impedance (in ohms)	50	50
Insertion loss (J2, J3 or J4)	3.0 db	3.0 db
Isolation (J2 to J3)	6.0 db	9.5 db
Power ratings:		
Average	100 W	100 W
Peak	10 kW	10 kW
VSWR (max)	1.4	1.4
Connector type*	N female	N female
Ambient temperature range	-50° to +250° F	-50° to +250° F
Finish	Epoxy	Epoxy
Weight	7 oz.	9 oz.
Price	$75.00	$95.00

(c) Power dividers and specification

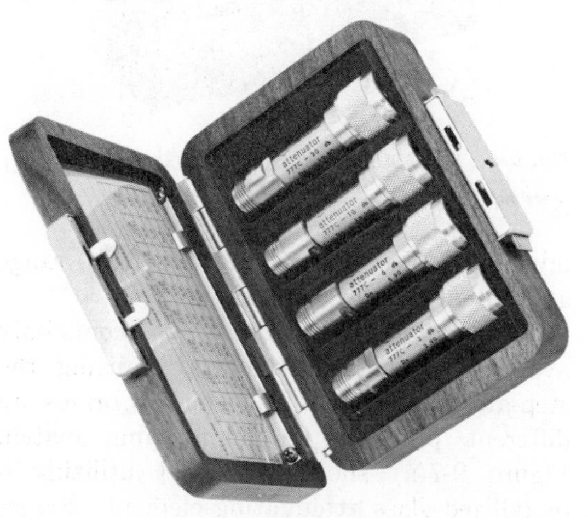

(a)

Fig. 9-2 LABORATORY ATTENUATORS (Courtesy of Microlab/FXR.)

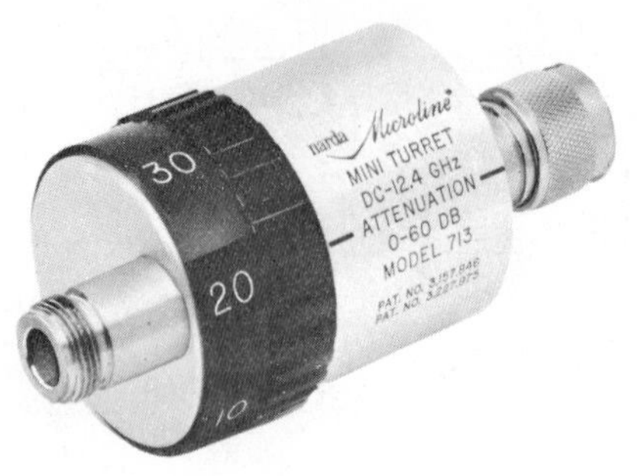

Fig. 9-3 ROTARY STEP COAXIAL ATTENUATORS (Courtesy of Microlab/FXR.)

Fig. 9-4 SLIDING BLOCK SWITCH ATTENUATOR (Courtesy of Kay Electric Co.)

Fig. 9-5 RESISTIVE SWITCH COAXIAL ATTENUATORS (Courtesy of Jerrold Co.)

Fig. 9-6 VARIABLE ARMATURE ATTENUATOR (Courtesy of Merrimac Co.)

9-3 may be rotary-switch actuated for specific insertion of fixed pads, precision cavities linked by sliding-block switches (Fig. 9-4), or a simple series low-frequency rotary coaxial switch using deposited carbon resistors (Fig. 9-5).

A continuously variable wide-band, 50-Ω line attenuator operating a movable armature by means of a gear-reduced micrometer drive is shown in Fig. 9-6. Such models handle as much as 4 W in the low (1- to 8-GHz) range and 2 W in the 10-GHz range.

Waveguide attenuators are commonly available in variable models, omitting the step-attenuation feature, since it brings up different problems in maintaining match. Figure 9-7(a) shows a model utilizing a metallized-glass attenuating element. Figure 9-7(b) and (c) show the use of a rotary vane for continuous power, and Fig. 9-7(d) illus-

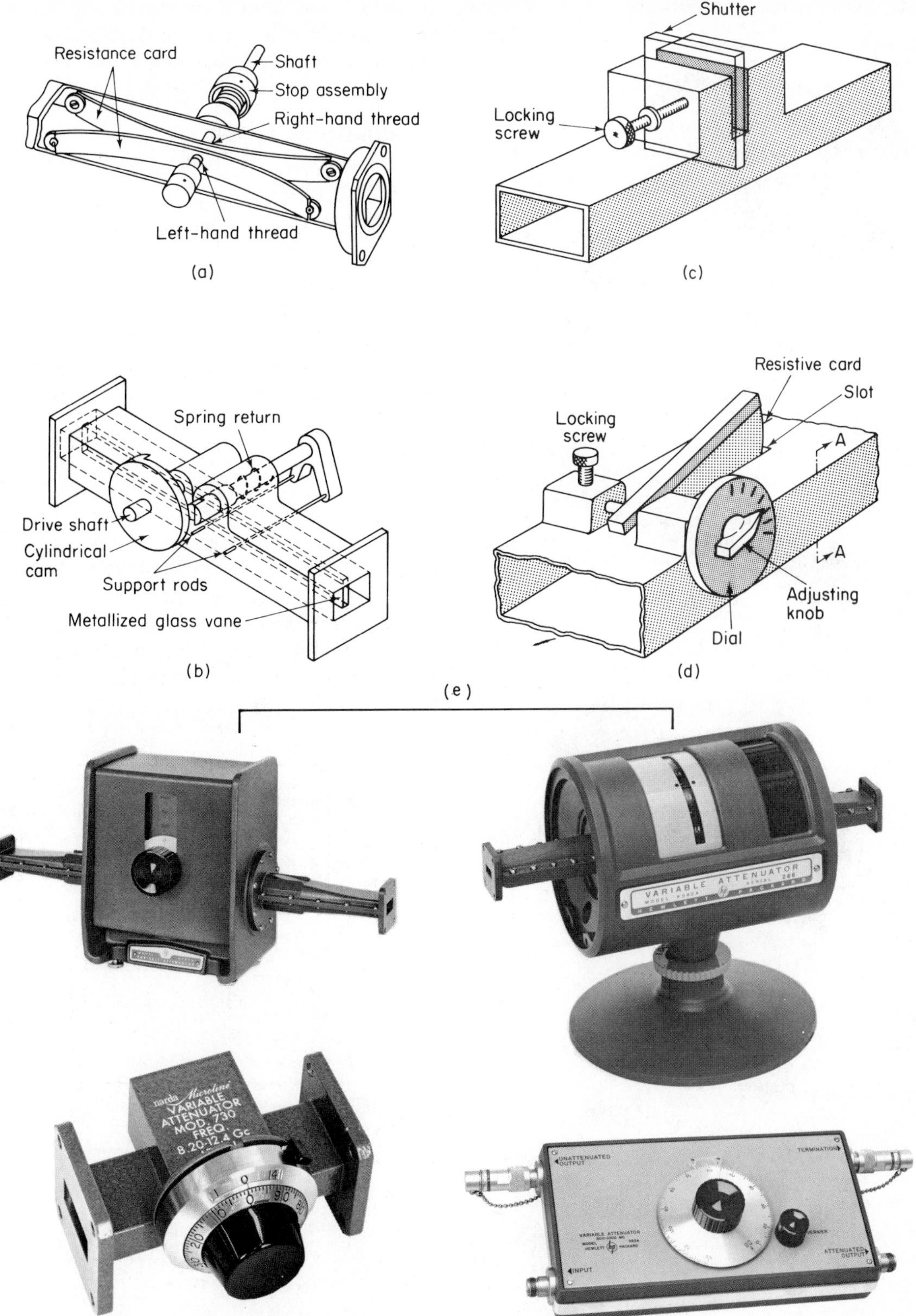

Fig. 9-7 WAVEGUIDE ATTENUATORS (Courtesy of Hewlett-Packard Co.)

trates a broad-band 10-W model with a 2-ft-long directly printed readout scale. Figure 9-7(e) shows commercial models.

TERMINATIONS—LOADS

GENERAL

Coaxial terminations are tailored around resistive elements built directly into the circular structures; Fig. 9-8(a) and (b) show a number of coaxial models and their characteristics. Note the length and provisions for radiation in higher-powered items. Figure 9-8(c) shows power load construction and measurement circuits.

CONSTRUCTION

Figure 9-8(c) shows direct resistive loading by means of a film resistor inserted in and acting as the center conductor. Another direct resistive load used in an rf watt meter is shown mounted in the terminating shell, which includes provision for circulating liquid coolant to dissipate measured power.

Description	Freq. range (kmc)	Power rating (watts)	Max. VSWR	Impedance (ohms)	Length (in.)	Weight (oz)
Low power	0-10	2	1.10	50	1.4	1.6
Medium power	0-4	10	1.10	50	2.0	2.6
High frequency	2-13	10	1.10	50	6.0	3.5
High power	1-13	150	1.10	50	11.1	8.4
Sliding	2-13	10	1.10	50	9.9	4.7
Open circuit	0-13				1.2	1.6
Short circuit	0-13				1.2	1.6

(b) Characteristics

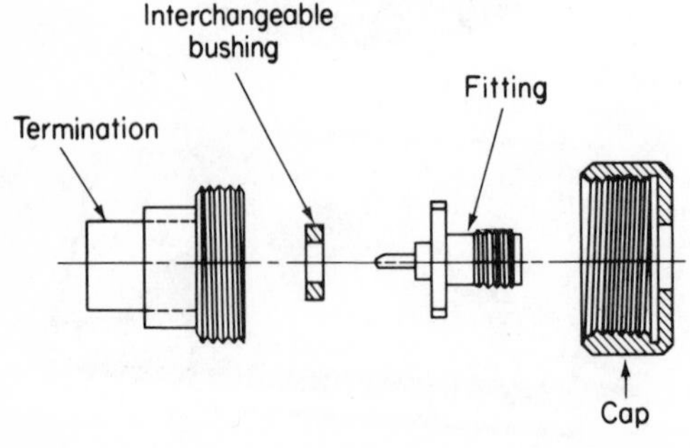

(a) Interchangeable coaxial termination

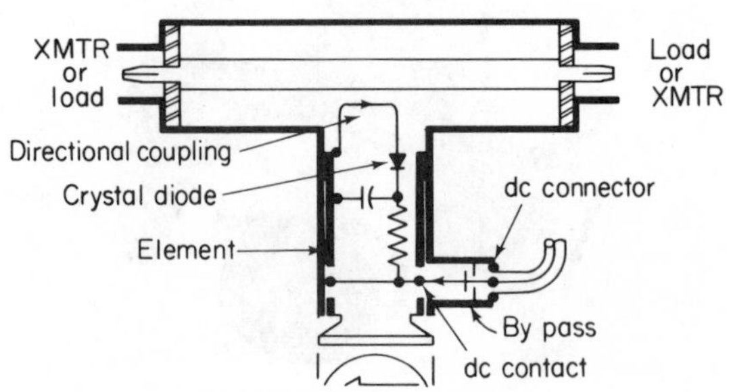

(c) Pour and load termination

Fig. 9-8 COAXIAL RESISTIVE TERMINATIONS (Courtesy of Microlab/FXR.)

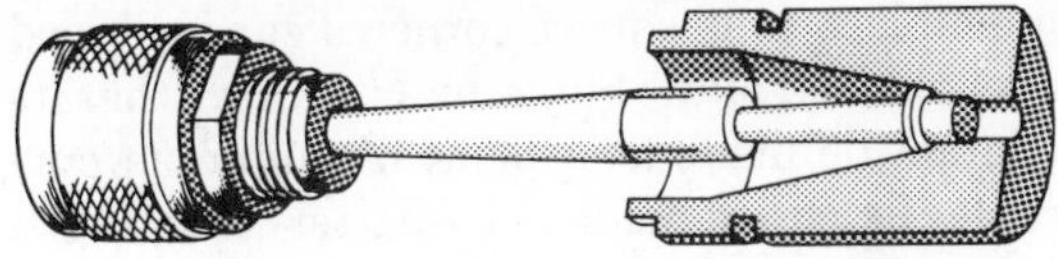

Fig. 9-9 SOLID DIELECTRIC TERMINATION

A solid, dielectric loaded termination is seen in Fig. 9-9. Here the dielectric is tapered for good match and power dissipation.

DESIGN FACTORS

In addition to the usual requirements of acceptable VSWR and power-handling capability, power loads and power terminations must have a connector that will carry the incident power without destroying the VSWR.

Also, there is the choice of whether to use a "dry" resistive, or dielectric termination, or one requiring coolant to dissipate the incident power. Although the former types are available in power capabilities up to 100 W, they are inherently low-powered devices chosen for reliability, small size, and rugged construction with a frequency range for use up through 10 to 20 GHz.

Wet-type terminations use circulating fluid, dielectric or water, to dissipate and to measure output power. Figure 9-10 shows a 50-kW power load using an internal film-type power resistor which is cooled directly by circulating water.

Sliding terminations are variable units, not unlike stubs, which allow for exact overall matching by physical manipulation. Specifically, by such devices we aim to separate the VSWR attributed to the load from that generated by the component or system under test, ensuring that load and component residual VSWR's are not too incompatible.

Fig. 9-10 HIGH-POWER FILM RESISTOR LOAD (Courtesy of Bird Electronics, Inc.)

Figure 9-11(a) shows a sliding-coaxial termination and Fig. 9-11(b) a waveguide model of this device.

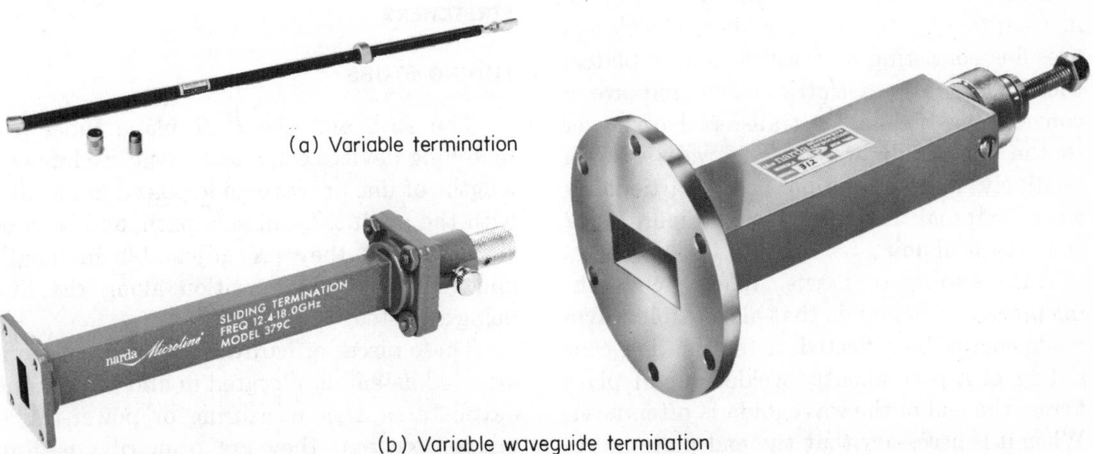

(a) Variable termination

(b) Variable waveguide termination

Fig. 9-11 VARIABLE TERMINATIONS AND SHORTS (Courtesy of Hewlett-Packard Co.)

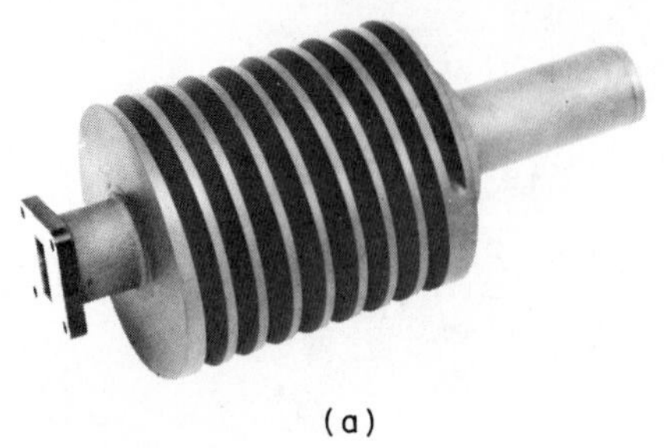

(a)

Microlab series	Dissipative material	Coolant	Fins	Power rating (watts)*	Frequency range (mc)
WA	Lossy plastic	Convection	None	50	1120–8200
WB	Lossy plastic	Convection	Transverse	70	1120–8200
WD	Refractory	Convection	None	150	7500–11500
WE	Refractory	Convection	None	500	1120–40000
WF**	Refractory	Convection	Transverse	—	1120–40000
WG	Refractory	Convection	Transverse	1500	1120–40000
WH***	Refractory	Convection	Transverse	—	1120–40000
WI	Refractory	Convection	Transverse	5000	1700–40000
WJ	Refractory	Convection	Longitudial	750	1120–40000
WL	Refractory	Liquid	None	5000	1120–1800
WM	Water	Water	None	27500	3950–1800
WN	Water	Water	None	55000	5850–1800

*WR 137 version **DA equivalent ***DA load

(b) Characteristics

Fig. 9-12 TYPICAL WAVEGUIDE LOADS AND TERMINATIONS (Courtesy of Microlab/FXR.)

WAVEGUIDE TERMINATIONS AND LOADS

LOADS

Power dissipation in waveguide structures was discussed under impedance matching in Chapter 8. As we remember, the design uses fins consisting of resistive coated plates; these constitute dielectric loads and are a conventional media for transferral of power to the waveguide walls. The walls offer a relatively large radiation area, particularly when external fins are used. Figure 9-12 shows typical units.

LINE SHORTING PLATES. When the matching procedure demands that all possible waveguide energy be reflected, a termination consisting of a permanently welded metal plate across the end of the waveguide is often used. When it is necessary that the end plate be removable, a solid square-cornered cup is slipped over the end of the guide, and to ensure that its contact be made at a point of minimum current, the cup is made one quarter-wavelength long. This construction induces cancelation of the incident and reflected H fields, making contact currents very small, although it is still important to ensure good, smooth, tight contact of the lip of the cup with its adjoining waveguide surfaces.

To ensure exact minimum current we may construct an adjustable plunger in the form of a half-wave channel [Fig. 6-4(a)] which reflects a short circuit across its end where a perfect connection is supposed to exist.

STUBS—*E–H* PLANE TUNERS—LINE STRETCHERS

TUNING STUBS

The stub and the E–H plane tuner are matching devices consisting of short-circuited lengths of line or waveguide placed in parallel with the main transmission path, and so constructed that they are adjustable in length and (effectively) in position along the line being adjusted.

These pieces of hardware are termination-oriented assemblies located in and around the output area of a measuring or power-transmission system; they are primarily used to adjust for perfect match or to eliminate standing waves. The following explanation concerning the assemblies uses nomenclature applying to coaxial lines, since this is where their greatest application lies.

Stubs and tuners are particularly necessary in power transmission and radiating systems where the line may be any number (plus odd fractions) of wavelengths long, and thus the aim is to provide exact local impedance match, say between an antenna and its direct junction with a transmitter. As we know, without fine tailoring high standing-wave ratios reflect variable load upon both the signal or generating source and the transmitter output is thus severely reduced in operating efficiency. These tuning devices when located near the antenna can be more finely adjusted for an optimum of operating efficiency and frequency stability. It might be added that a matching stub markedly reduces the frequency sensitivity of a line having a high standing-wave ratio.

In the operation of a transmitter and radiating-antenna system, a stub is usually placed at a point near the far end of the main line where the conductance is equal to the characteristic conductance of the line; its length is then adjusted by the shorting bar so that the total system susceptance of the point of attachment is zero. The line can then be terminated in its characteristic resistance (or conductance), and the standing-wave ratio between the stub and the transmitter becomes equal to zero.

In an instrumentation setup a similar procedure is followed; (see Chapter 10) discrepancies in mismatch caused by insertion and addition of accessory equipment are "adjusted out," so to speak, by a tuning stub located near the final measurement point.

DOUBLE-STUB TUNERS

Although, for explanation, we have covered the position movable-length adjustable stub for use on an open-wire line, this construction is not suitable for a coaxial line, since a movable stub is difficult to build. On coaxial lines we get around this by using two stubs, usually permanently located $\frac{3}{8}$ wavelength

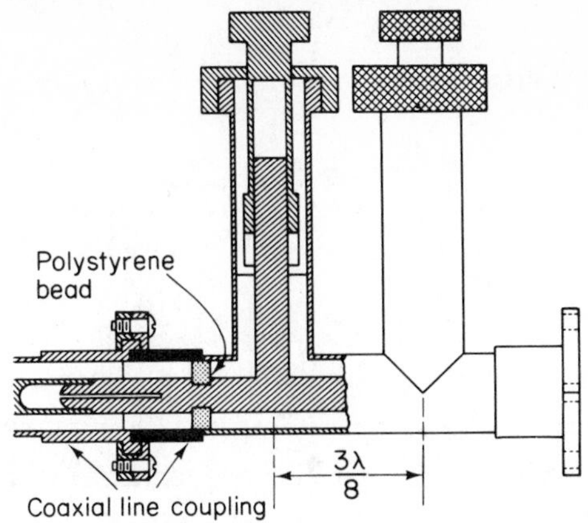

(a) Typical construction

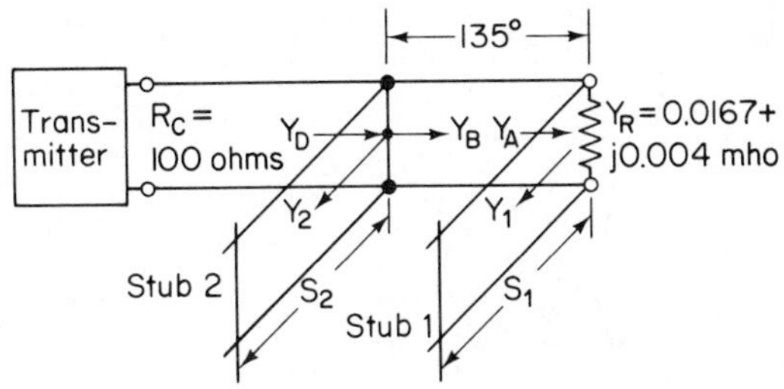

(b) Equivalent circuit

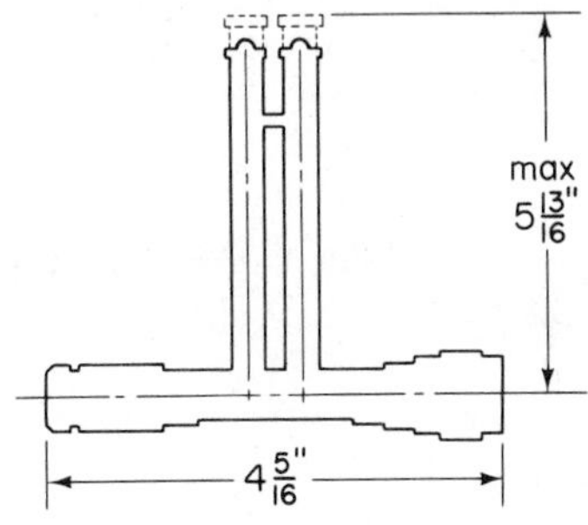

(c) Commercial model

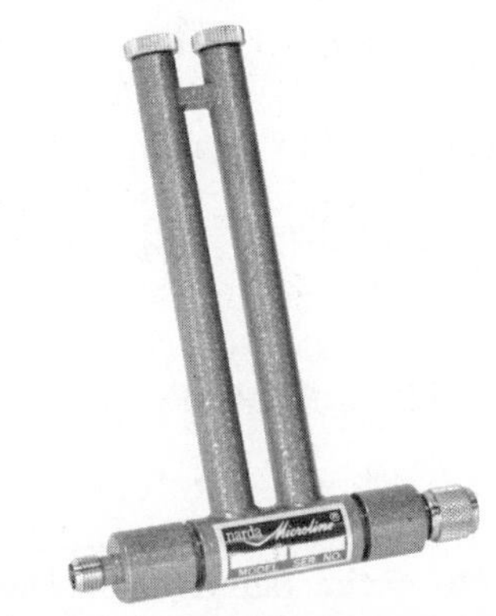

Fig. 9-13 DOUBLE STUB TUNERS (Courtesy of Microlab/FXR.)

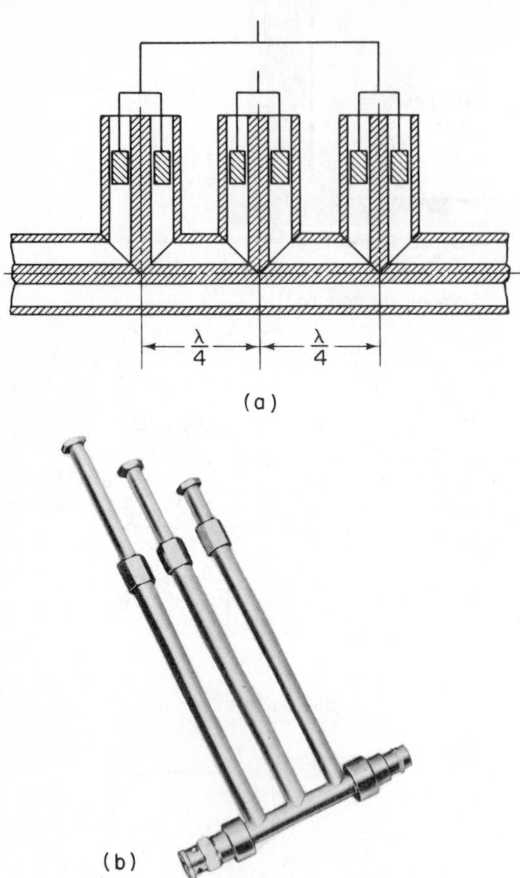

Fig. 9-14 TRIPLE STUB TRANSFORMER (Courtesy of Microlab/FXR.)

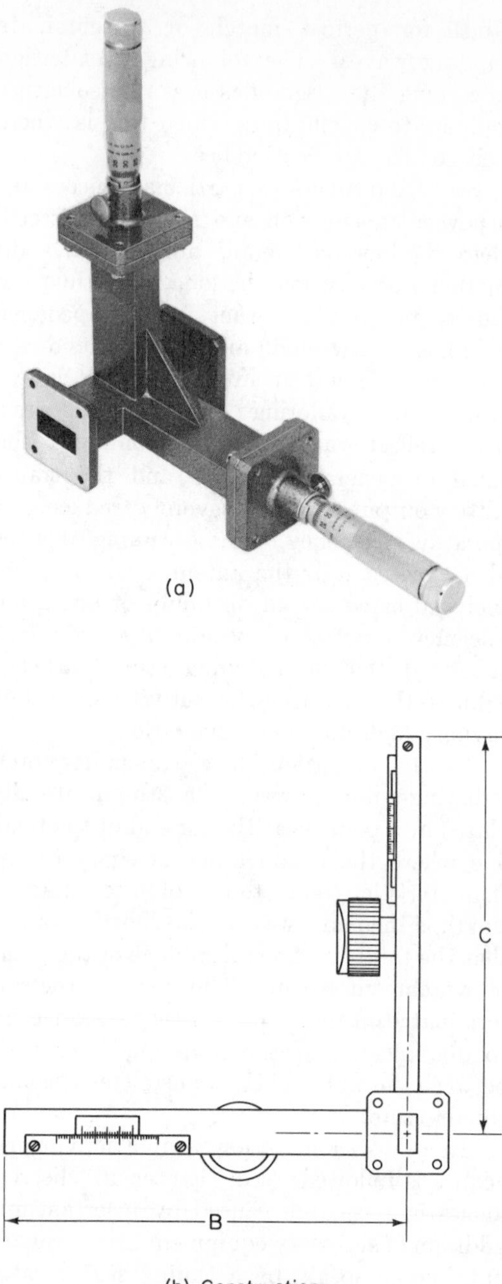

Fig. 9-15 *E-H* PLANE TUNERS (Courtesy of Microlab/FXR.)

apart, each providing its own self-adjustable short-circuited length by means of a movable plug or plunger which makes contact with both the inner conductor and the outside shell.

Figure 9-13(a) illustrates the physical construction; Fig. 9-13(b) the equivalent circuit, and Fig. 9-13(c) two commercial models and their construction. Figure 9-14(a) covers the layout of a triple-stub transformer in which the two end shorting slugs are mechanically ganged. Figure 9-14(b) is a commercial version of this system.

E–H PLANE TUNERS (FIG. 9-15)

These, in effect, are movable short circuits used to match out discontinuities by placement in each of the planes of a waveguide. Assembly-wise, they are straight sections of waveguide with series and shunt tuning arms extending from the center section. Choke-type plungers provide an electrical short circuit in each arm and can be moved to vary the reactance presented at the junction point.

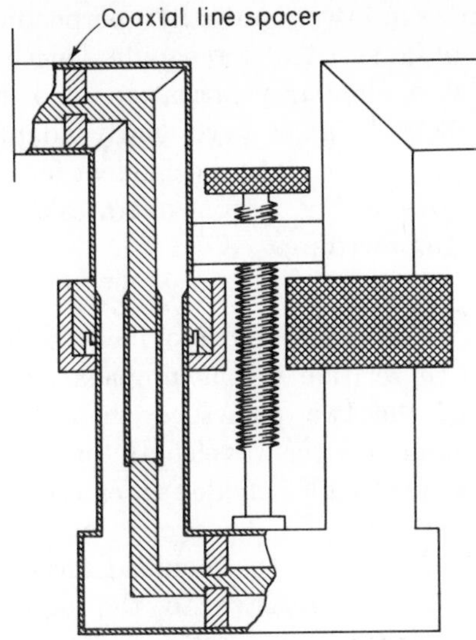

Fig. 9-16 TROMBONE TYPE LINE STRETCHER

LINE STRETCHERS

As noted above, the transmission-line length between transmitter and load (antenna or other) may not be optimum for good impedance match and for power and frequency compatibility. We frequently insert *line stretchers* to remedy this, which, although they do not maintain perfect match between the insertion points, are widely used because of their simplicity and the improvement in power-transfer efficiency. This device is particularly adaptable when matching the high resonant-output circuit impedance of a transmitter through the two series capacitances of the trombone type of line stretcher shown in Fig. 9-16.

PHASE SHIFTERS

This device changes the electrical length of a section of transmission line without changing its physical length and without introducing impedance discontinuities (Fig. 9-17). The mechanism that accomplishes this consists of a disk-shaped dielectric vane eccentrically mounted on a shaft and protruding into the waveguide through a slot. Rotation of the shaft changes the amount of dielectric in the waveguide and hence its effective wavelength.

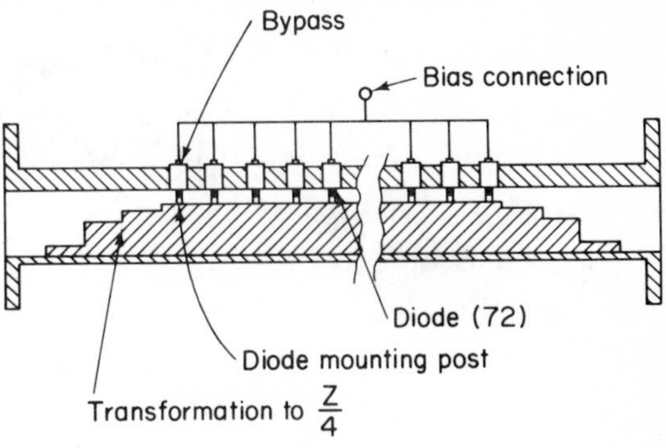

(a) Construction

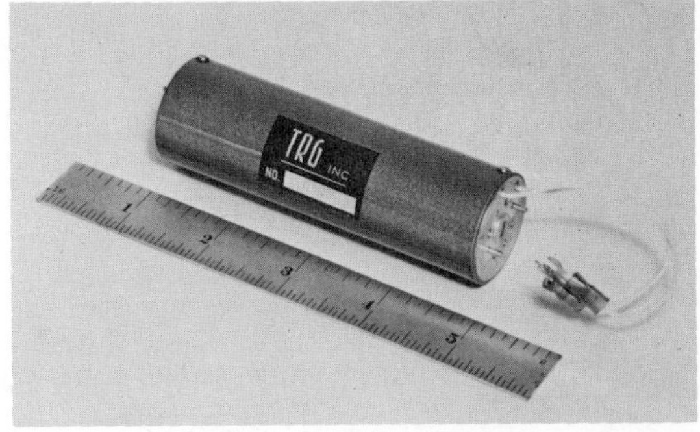

(b) Waveguide phase shifters

Fig. 9-17 PHASE SHIFTER CONSTRUCTION (Courtesy of Microlab/FXR and TRG Inc.)

Circular polarization in phase shifters can be attained in a cylindrical waveguide which contains modifications of a mechanically rotatable line section inserted between its end couplings. By carefully designing the cross section for slightly different propagation

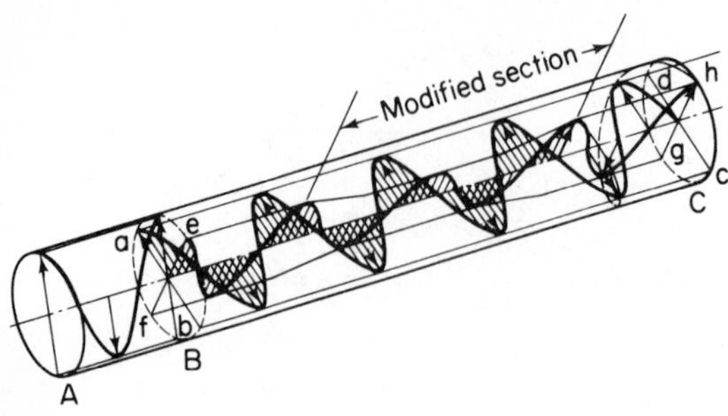

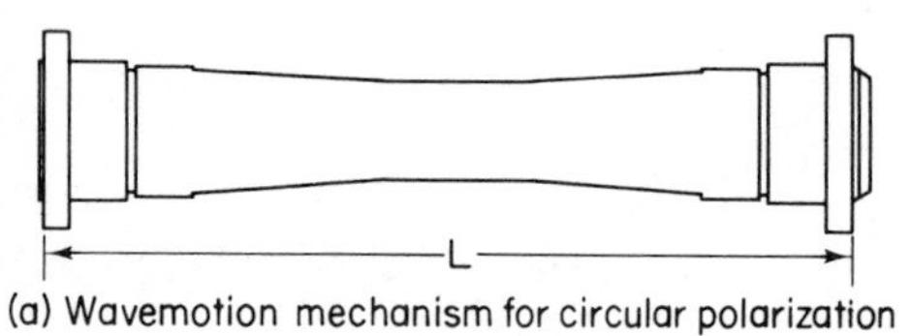

(a) Wavemotion mechanism for circular polarization

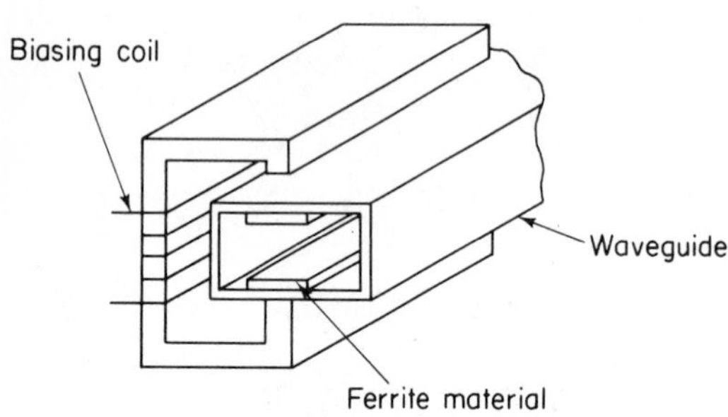

(b) Non-reciprocal unit

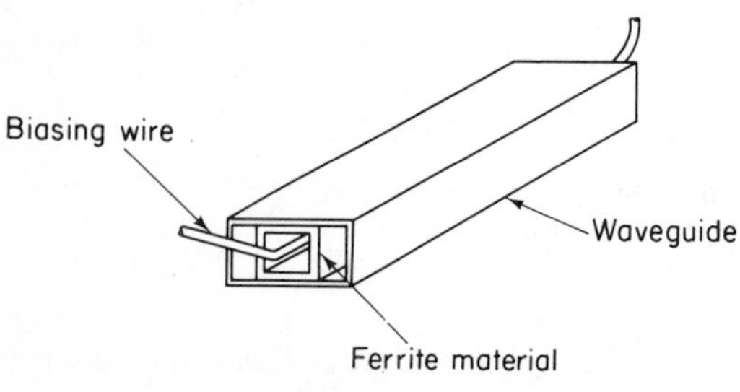

(c) Latching phase shifter

Fig. 9-18 PHASE SHIFTER FOR CIRCULAR POLARIZATION

constants in two mutually perpendicular planes of a selected waveguide length, we arrive at a circular polarization of a linear wave. This design is particularly adaptable to the $TE_{1,1}$ mode. The mechanism for doing this is shown in Fig. 9-18; it operates through two design procedures.

1. The length of the modified section is selected in relation to the propagation constants of the two planes so that there is exactly one quarter-wavelength of the two components of a $TE_{1,1}$ mode as it emerges from the section.
2. The planes of the modified section are oriented at 45° relative to the dominant linearly polarized $TE_{1,1}$ mode wave entering them.

We see this in Fig. 9-18(a) where a linearly polarized wave is passing from *A* to *B*. At *B* the wave is split into two mutually perpendicular components, one in the longitudinal plane and *a*, *b*, *c*, and *d* passing unaffected through the waveguide (by virtue of this plane's waveguide dimensions), and the other *e*, *f*, *g* and *h* altered in phase by passage through section modifications. The result is that from the modified section the two signal components are one quarter-wavelength out of phase. Biased ferrite material disposed within a waveguide can also produce signal phase shift as shown in the nonreciprocal operating units shown in Fig. 9-18(b) and (c).

Figure 9-19 shows how phase shift of polarization can be accomplished in a duplexer by use of transmitter-triggered elements placed in a waveguide. In this particular case, the rectangular waveguide at the antenna end is oriented with its transverse directions 90° from those at the transmitter end. The receiver arm has its electric vector oriented parallel to the electric vector in the waveguide through the received signal which arrives at the antenna.

Both transmitted and received signals are carried in the rectangular waveguide and arrive at the duplexer with a $TE_{0,1}$ mode (that is, with the electric vector at all times to the

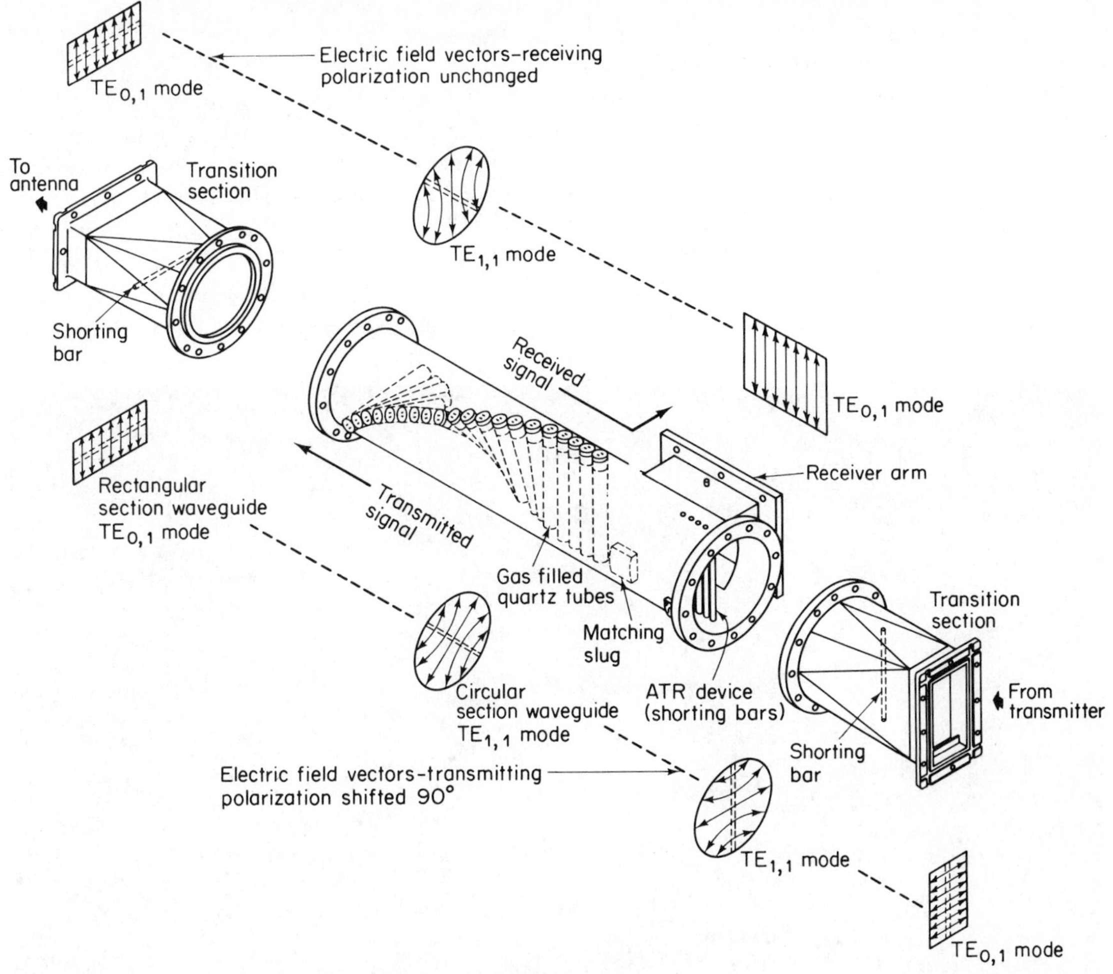

Fig. 9-19 DUPLEXER PHASE SHIFTER

short dimension of the waveguide cross section).

Now, during the send cycle, energy being propagated toward the antenna has its polarization shifted 90° in passing through the duplexer, readily continues in the $TE_{0,1}$ mode through the waveguide to the antenna (see upper vector schedule), and couples very little energy from the transmitter arm into the receiver arm because of the 90° polarization of the E-field vector.

Conversely, when receiving, the input-signal energy in the normal $TE_{0,1}$ mode of the receiver arm passes directly via the circular $TE_{1,1}$ mode and cannot pass through to the transmitter because of the transverse field orientation of the transmitter arm.

This operation is accomplished by a central circular phase shifter section of waveguide located between the end rectangular section, which must of necessity be transition sections in changing their cross section from rectangular to circular and back again to rectangular. The mode phase shifting is accomplished in this cylindrical section by a series of diametrically positioned gas-filled quartz tubes, each with its position successively advanced in angle along the length of the section. When these tubes are fired (in the transmit cycle), they cause a momentary short circuit, change the guide dielectric constant, and angularly shift the phase of the E-field vector. This is done through each of the 16 tubes causing a total of 90° shift in the arrival of the transmitted E-field vector.

Received echo signals do not have enough energy to fire the tubes and, therefore, pass backward through the cylindrical section unchanged. As noted, their polarization is correct for signal entry into the receiver arm; and, in addition, shorting bars located just beyond the receiver input arm prevent their passage to the transmitter.

MILLIMETER WAVEGUIDE TERMINATIONS

All termination-allied units discussed thus far have their counterpart in millimeter wave frequency ranges. Figure 9-20 illustrates typical terminations, shorts, and phase shifters. Some discussion applying to these units will appear under the measurement section (Chapter 10) and in Chapter 11, where millimeter waves are discussed more fully.

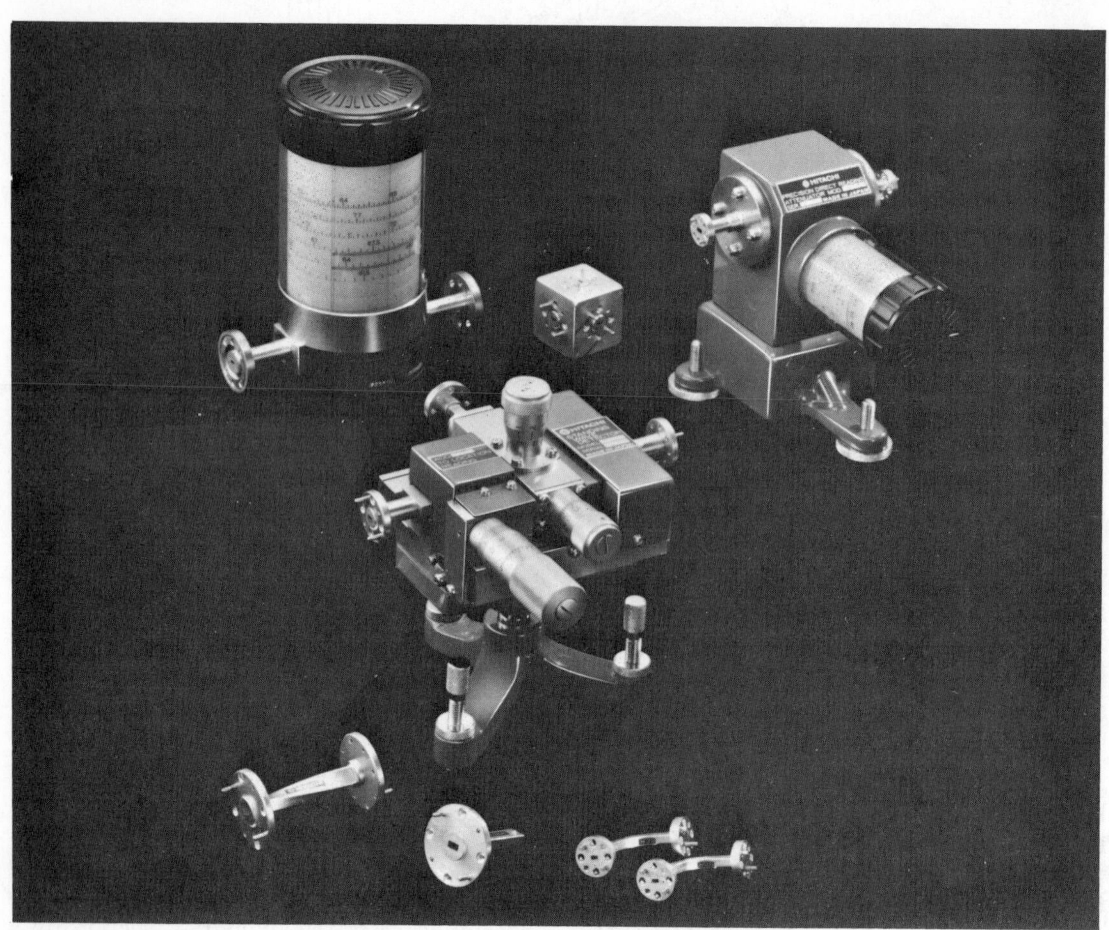

Fig. 9-20 MILLIMETER WAVE TERMINATIONS, SHORTS, AND PHASE SHIFTERS (Courtesy of NARDA Microwave Corp.)

Component Measurement and Equipment 10

INTRODUCTION

Basic measurements in any electronic regime concern (1) amplitude of the voltage–current–power triumvirate, (2) frequency, and (3) component constants. In microwaves in general we cannot directly read out measured voltage and current, so in all except power measurements we must utilize some form of sampling—that is, extracting, detecting, and evaluating some small portion or replica of the voltage–current–power scheme.

Microwave frequency and wavelength are of course directly measurable by means of coupling some frequency-indicating device to the system being measured. *R*, *L*, and *C* constants for microwave components are more obscure and most often expressed in terms of complex impedances. In addition, they must be obtained through a series of VSWR-allied techniques and chart-interpretative procedures (the Smith chart), all of which constitute a measurement procedure of its own which we have included later in this chapter.

Our approach, then, depends upon four major and one minor area.

1. Frequency equipment and measurement techniques. These permeate every measurement because they determine all reactances and to a minor degree guide wavelengths, noise, and other factors.
2. Attenuation. This technique and associated equipment constitute the core of the sampling procedure. It is used throughout all phases of measurement and is subject to combination with the preservation of matched impedances. It is particularly important in power measurements.
3. Impedance. A key measurement, determined and dependent upon existing VSWR's. The matching of impedances is probably the most important generalized procedure peculiar to microwave technology, since it is interdependent upon almost every measurement.
4. Power. High-power measurements are usually calorimetric. Low-power cases utilize an attenuated replica. Both procedures require impedance-matched conditions.
5. Noise. This factor is really a low-power comparison measurement depending upon complex operating conditions, particularly including the preservation of impedance match.

Figure 10-1 summarizes basic measurement sections and shows a number of the interrelationships referred to above. Table 10-1 extends the summary into listing the equipment and techniques commonly used in microwave design and operation. Table 10-2 is a list of the accessories used in microwave operations; it can be referenced to Table 10-1 to amplify the scope of the equipment used.

GENERAL MEASUREMENT APPROACH

ITEMS

All the basic measurements outlined above require setups that have many common com-

TABLE 10-1 MICROWAVE MEASUREMENT REGIME

Type of Measurement	*Source and Type of Signal*	*Measuring Instrument*	*Manner of Measurement*	*Accessories Required*	*General Comments*
Power measurement	CW, Pulsed, Modulated or Noise Equipt. under test	**Calorimeter power meter**	**Temperature rise measured**		**Generally useful only for large powers**
		Bolometer power meter	Measures changes in resistance due to heating	Bolometer mounts, attenuators, bolometer bridges	Useful for powers up to several milliwatts
Frequency measurement	CW, Pulsed or Modulated Equipt. under test Sig. gen.	Cavity wavemeters Slotted sections	Frequency of reaction in system or transmission of maximum power is measured Measure wavelength from VSWR	Attenuators, directional couplers, stub timers, crystal mount and power meter, crystal frequency calibrators	Slotted sections accurate 0.1 to 5 per cent; cavity meters accurate 0.01 to 0.1 per cent; use of crystal reference standards permits accuracies of 0.02 to 0.0001 per cent
Impedance and Admittance	Variable-frequency modulated (CW)	Slotted line, probe, and VSWR indicator	Reflection coefficient measured and impedance plotted on Smith chart	Attenuator, frequency meter	Measurements made at various frequencies; accuracy depends upon slotted line
	Variable-frequency sweep-frequency signal – Sig. Gen.	Impedance plotter	Impedance indicated directly on oscilloscope with chart reticle	Attenuator, frequency meter	Measurements made directly; eliminates need for point by point measurement and plotting
Attenuation	Modulated (CW) Signal Generator	Crystal or bolometer mount and power meter	Substitution method is used–measure tested component against known value of attenuation	Variable attenuator	Not accurate for small values of attenuation because resolution of test equipment is generally too low
Reflection coefficient and VSWR	Modulated (CW) Signal Generator	Slotted line, probe and VSWR indicator	(a) Measure ratio of maximum and minimum field with VSWR indicator, and distance between max. and min. (b) Measure required attenuation to give same maximum and minimum reading on meter	Termination, calibrated attenuator	Method (b) gives more reliable readings when VSWR greater than about 3
Measurement of Q	Modulated Signal Generator	Slotted line, probe and VSWR meter	Measure reflection coefficient at three equally spaced frequencies near resonance then obtain Q graphically on Smith chart	Attenuator, frequency meter	Suitable for low Q measurements
	Sweep-frequency signal Srg. Gen.	Precision wavemeter, crystal mount, oscilloscope	Frequency response is indicated on oscilloscope; 3 dB points are measured with wavemeter and precision attenuator	Precision attenuator, directional coupler	Suitable for low Q measurements
	Pulsed Q Meter	Q meter	Amperes rate of decay of free oscillations in cavity under test with rate of discharge of known capacity into known resistance		Does not require highly stable oscillators, or accurate frequency or attenuation settings; accuracy of measurement ±1 per cent
Frequency spectrum analysis	Signals in frequency band under observation Equipt. under Test	Spectrum analyzer	Frequency band is displayed on screen of oscilloscope; signal amplitudes are displayed vertically, at horizontal positions corresponding to frequencies	Attenuators, directional couplers, etc.	Resolution depends on speed of sweep and bandwidth of analyzer
Noise measurements	Noise level of system and comparison signal of known level Internal	Power indicator	Known amount of signal is added to the input, and noise level is determined from the change in output level	Directional coupler, precision attenuator	Comparison signal which is added must be of same order of level as noise in system; care must be taken not to mismatch input by insertion of test signal
Antenna measurements	CW or modulated Signal Generator	Antenna under test, receiver and power meter	Antenna under test is illuminated by auxiliary antenna, and gain on pattern measured	Standard gain horn, directional couplers, attenuators, frequency meter, crystal mount and power meter	Power required from signal generator depends upon noise level and required dynamic range; servo drive and automatic plotter may be used for pattern measurements
Radar performance	Pulsed Radar under Test	Radar under test	Small portion of transmitter signal is coupled into echo box.	Echo box, directional coupler, attenuator	Simple method of checking some of the basic performance characteristics of radar systems

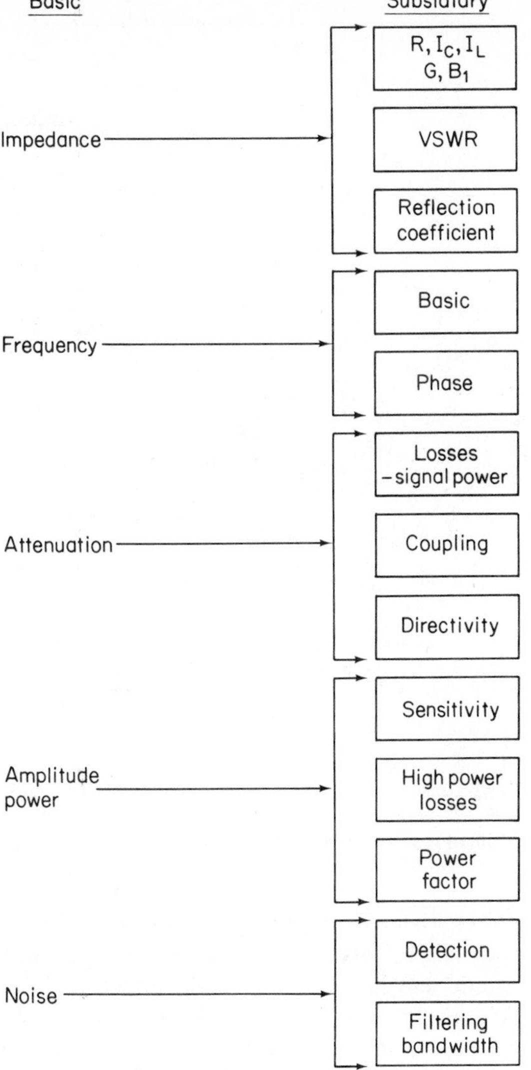

Fig. 10-1 SECTIONALIZED MEASUREMENT APPROACH

ponents. In fact, a common group of microwave components is used to make up the basic measurement setup. All these are universally available and similarly used in microwave technology; likewise, all the basic measurements except noise and power are made with variations of this combination.

Functionally, these components must be assembled so that a measurement can:

1. Apply input signal power to the system within which the component or measurement must be made.
2. Measure the specific frequency of the applied power.
3. Attenuate or regulate the amount of input power being fed to the system.
4. Detect the level of the signal power output at selected points among components being measured.
5. Adjust for impedance necessary for the preservation of matching between electrical components and physical junctions. In other words, this item yields basic information about reflections in parts of the system that are being adjusted or designed.
6. Display signal output and associated control signals beyond the null and maximum indications associated with the VSWR measurement.

BASIC SETUP

Figure 10-2 shows in block form a collection of instruments typical of most measurements. Actually, it is a simplified setup that would be used for frequency measurement and can also serve to guide us in the

TABLE 10-2 MEASUREMENT ACCESSORIES

Attenuators	**Couplers**
Fixed	Directional
Variable	Duo-directional
Precision	Phase shifter
Power divider	**Mounts**
Coaxial	Bolometer
Waveguide	Crystal
Shutter	Mixer
Flap	**Connectors**
Tuners	Coaxial
Double stub	Terminations
Trombone	Adaptors
Line stretcher	Tee
Double slug	Power divider
E-H slide screw	Waveguide shorts
Tuned cavities	**Probes**
Wavemeter	Tuner
Transformer	**Baluns**
Hybrids	**Delay lines**
Magic T	**Filters**
Rat-race	**Coaxial**
Duplexers	**Waveguide**
Diplexers	
Shorting switches	
Semiconductor	
Ferrite	

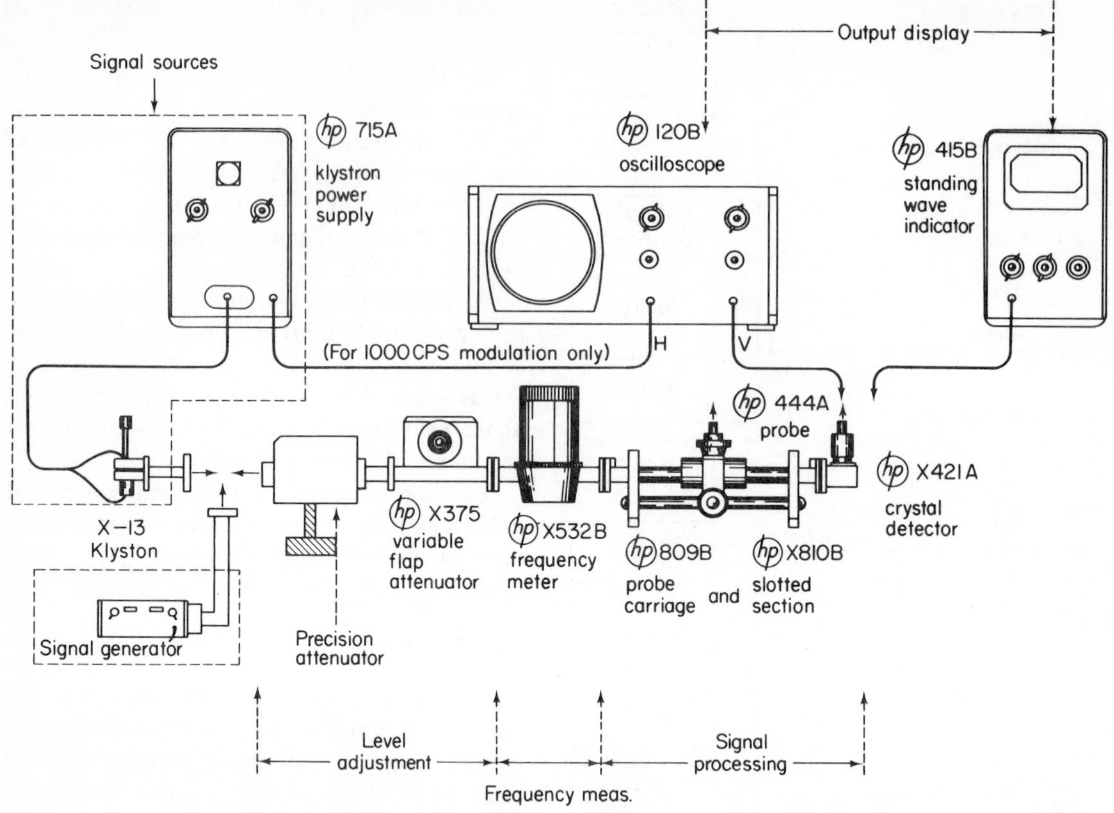

Fig. 10-2 BASIC MEASUREMENT SETUP (Courtesy of Hewlett-Packard Co.)

following. Note that the above items are linked to Fig. 10-1 and will be repeated briefly for equipment layout and again in detail for measurement procedure.

We should, however, note an elementary device used often by itself for specific point-by-point measurement and one that is particularly applicable to our setup; this is the reflex klystron and its associated power supply. The klystron is a low-power signal-generating unit (described in Chapter 7) which, with its mount, is placed at the input end of most basic measurement setups. The combined klystron and its power supply may also be part of an overall signal-generator setup, as shown by the dashed enclosure in Fig. 10-2.

OPERATIONAL AND EQUIPMENT APPROACH

Beyond the brief explanation of the six functional items given above, we shall discover techniques and illustrations in only four main areas: (1) frequency measurement, (2) attenuation, (3) detection, and (4) impedance. We limit our main items because items 1 and 6 above are subordinate and occur or are common to the main sections of interest; their descriptions are inserted at appropriate points. Thus input devices and signal generators plus oscilloscopes appear in all measurements even though they are distinct functional items. A final section on swept-frequency measurements is keynoted because it represents an important section of the entire measurement technology.

INPUT-SIGNAL CONSIDERATIONS

Most modern microwave instrumentation setups use swept-frequency methods, where the input-signal frequency is periodically varied or swept over the frequency range.

This method was made possible through development of the electronically tuned backward-wave oscillator tube. A typical signal generator of this type appears in Fig. 10-3(a).

For point-by-point measurement, a typical signal generator is the HP Model 618-C [Fig. 10-3(b)]. This generator is completely self-contained, has a direct-reading frequency dial covering from 3.8 to 11 GHz, delivers 1 mW of energy into 50-Ω impedance, and has internal modulation for high-sensitivity probe measurements.

For more sophisticated measurements, we would use a generator shown in Fig. 10-3(c), where frequency and attenuation are set on direct-reading digital dials while functions are easily selected by pushbuttons. This unit has two controllable outputs, one for power output up to 10 mW and the other a leveled-down power output to minus 127 dB, the latter being independent of attenuator setting.

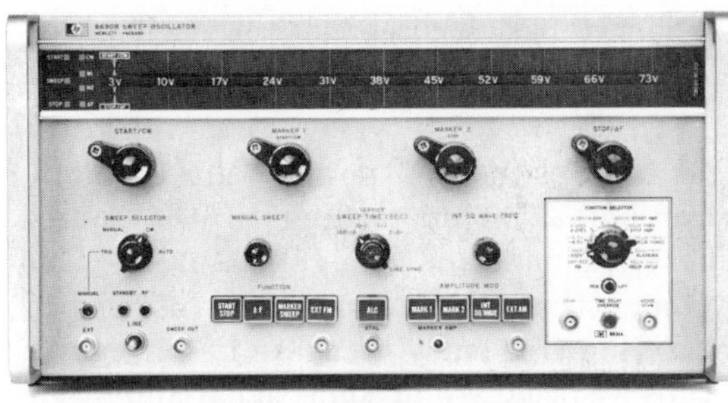

(a)

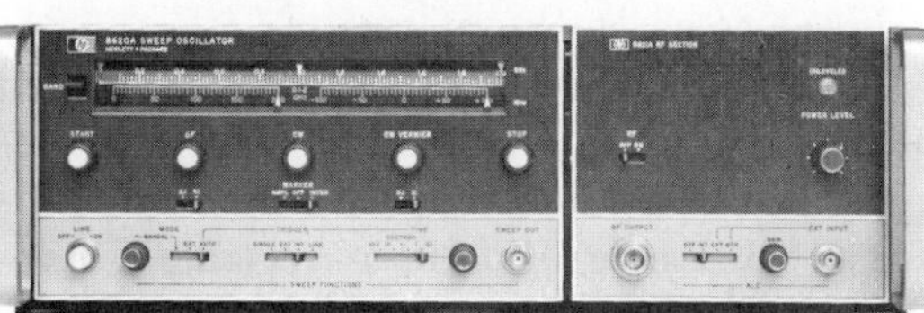

(b)

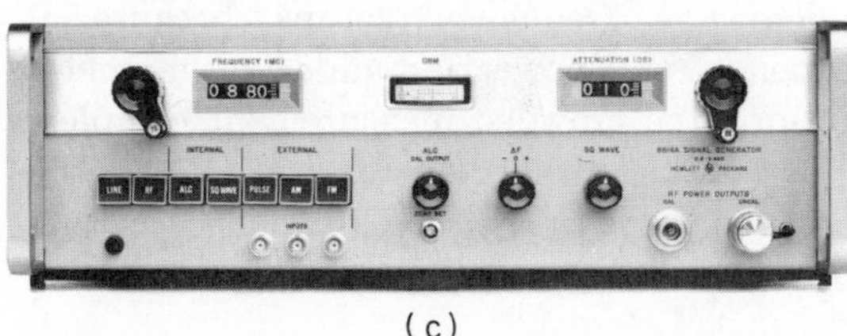

(c)

Fig. 10-3 TYPICAL SIGNAL GENERATORS (Courtesy of Hewlett-Packard Co.)

FREQUENCY

This item logically concerns one of the various types of wave-frequency meters described later. However, their choice, dimensions, and techniques for use extend to a somewhat more complex technology than merely inserting and reading a numerical value.

ATTENUATION

This procedure uses one of the various types of equipment described in Chapter 9. In other words, we measure attenuation by obtaining power ratios before and after adjustment or experiment, or through either if or rf substitution procedures.

DETECTION

This technology concerns all detector–rectifiers and their mounting hardware and circuitry, which must be manipulated, to provide true output indication.

IMPEDANCE

This item centers around the slotted-line, detector, and standing-wave indicator, all used to obtain VSWR data.

DISPLAY

This function consists mainly of oscilloscopic readouts necessary for indicating, inspecting, and interpreting slotted-line VSWR data. Its specific use and adjustment occur under frequency and impedance measurement techniques.

FREQUENCY MEASUREMENT AND EQUIPMENT

GENERAL

Stable, accurate frequency must exist or be established for all microwave systems. We may extract or sample and measure this quantity by

1. Physical wavelength and measurement calculated from distance apart of

minima and maxima voltage points on a slotted-line setup. This method is relatively cumbersome and may not be better than 0.05 per cent accurate. This technique depends upon the fact that a VSWR setup in a transmission line produces nulls every one half-wavelength. If these nulls are detected and the distance between them is measured, the frequency may be calculated. From this information with proper correction, the guide wavelength may be determined.

2. Electronic frequency-counter equipment. These are direct digital lead-out frequency meters, operating up through the GHz frequency range. Their advantage lies in delivery of continuous readable data of reasonable accuracy, but at considerable expense and complexity. Details of this equipment and its operation appear in Appendix H.

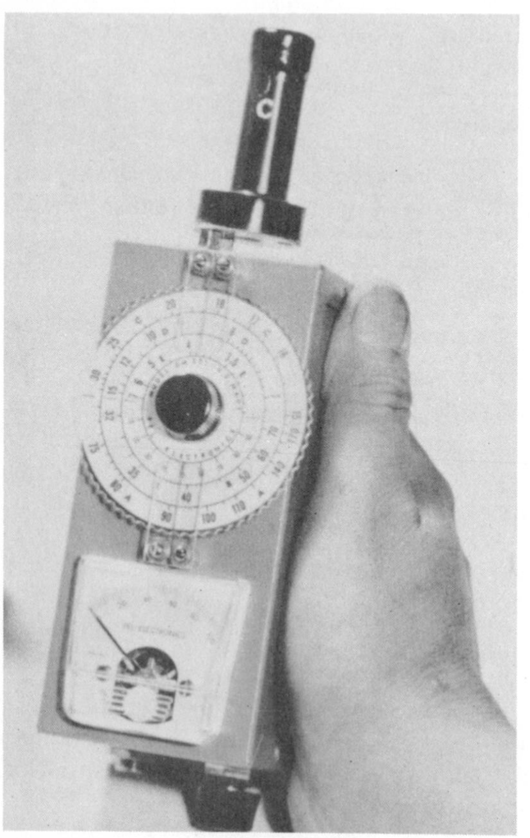

Fig. 10-4 COIL AND CONDENSER WAVEMETER (Courtesy of Pel-Electronics Co.)

3. Heterodyne frequency meters. These meters compare heterodyne "beats" between generated harmonics of a given known frequency with the known microwave signal.

4. Wavemeters. These frequency meters are practical working microwave devices having adjustable, self-resonant, and directly readable outputs. For sampling, these mechanisms must be coupled in some way to the circuit being measured in order to absorb energy for actuating an indicator.

At frequencies below the microwave range, the common wavemeter is a coil-and-condenser, lumped constant, resonant assembly, as illustrated in Fig. 10-4(a).

For microwaves, adjustable, resonant cavities, shown in generalized construction in Fig. 10-5(a), (b), and (c), are used having variable micrometer-type plunger or piston tuning elements. This type of design is used in most of the reaction models described below.

Functionally, wavemeters may be classified in two basic types:

(a) The reaction wavemeter, which utilizes the action of a power "suck-out" by observing a meter reading in the source whose frequency is being measured. This system is the one most commonly used when measuring frequency, because in waveguide setups we usually find some signal-amplitude indicating device that tells us when resonant-induced power suck-out occurs.

(b) The absorption wavemeter, which utilizes the power extracted from the frequency source to actuate its own self-contained indicator, usually a rectifier–microammeter network, a lamp, or an earphone. The absorption wavemeter is uncommon in microwave frequency setups because it expends extra power and adds another meter to an often crowded measurement complex.

OPERATION OF REACTION WAVEMETER

In principle, whether it be for low-frequency or for microwave measurements, this meter uses a circuit equivalent to that in

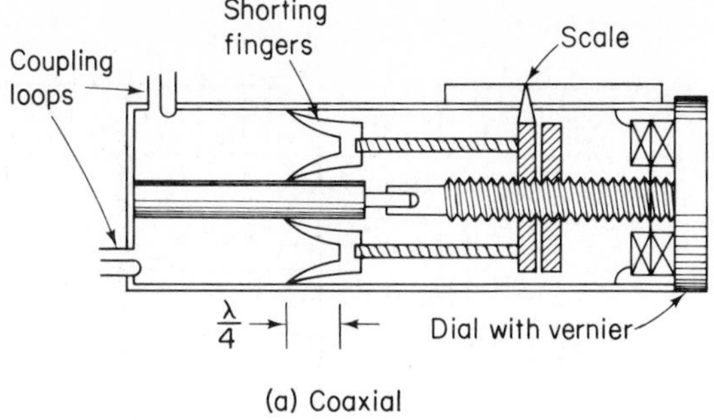

(a) Coaxial

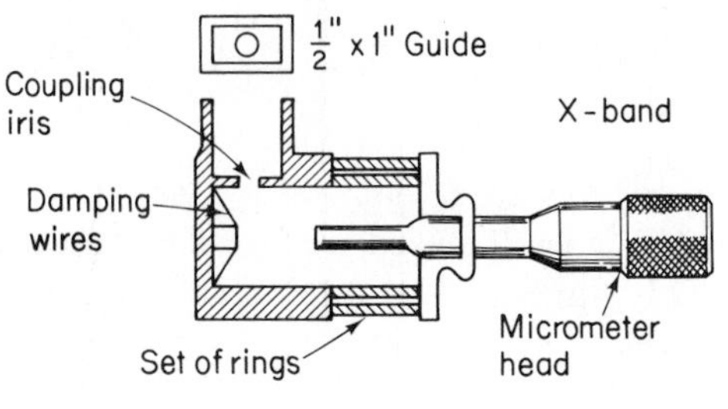

(b) Transitional

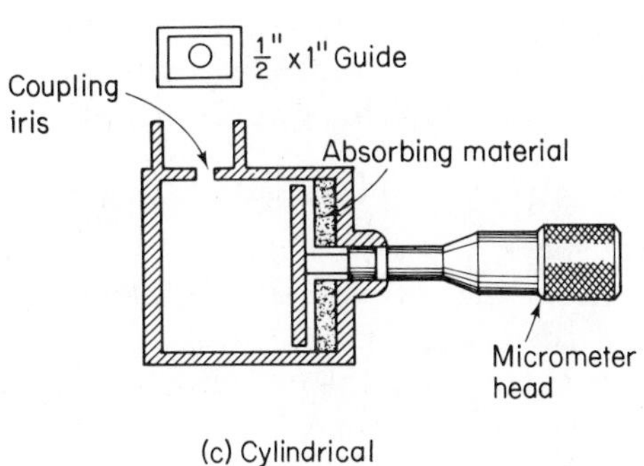

(c) Cylindrical

Fig. 10-5 RESONANT CAVITY WAVEMETERS

Fig. 10-6(a). It consists of a coil L and a variable capacitor C arranged so that when the external coil is loosely coupled to the output coil of the device whose frequency is to be measured, the calibrated capacitor can be tuned until the resonant frequency of the wavemeter is equal to the operating frequency of the device under test.

At this point a dip in the meter indicating output power (say a voltmeter or a crystal detector across the output) will be observed, for, in effect, we are placing a series resonant circuit across the load. This meter-reading dip comes to various minimums (all at the resonant frequency), depending upon the tightness of coupling used between the wavemeter and the circuit under test [Fig. 10-6(b)].

At microwaves, adjustment of the plunger, shorting elements, or disk is equivalent to simultaneously changing both L and C of our tuned circuit in Fig. 10-5; coupling, however, is fixed by the loops extending into the cavity from the external waveguide or by the size of the iris common to the cavity and the waveguide being sampled (Fig. 6-5).

It should be noted that the reaction-type meter has the advantage that when operated slightly off its tuned frequency, its loading upon power transmission is negligible.

COMMERCIAL WAVE-FREQUENCY METERS

Industrial and laboratory instruments have adopted the name *frequency meters* for this equipment because they are almost

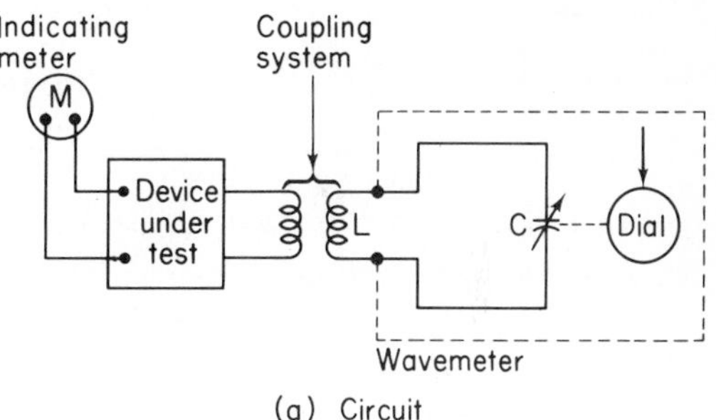

(a) Circuit

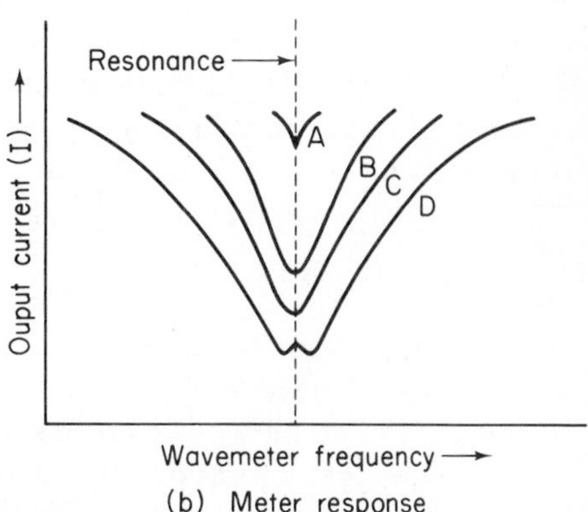

(b) Meter response

Fig. 10-6 REACTION WAVEMETER CIRCUIT AND INDICATOR RESPONSE

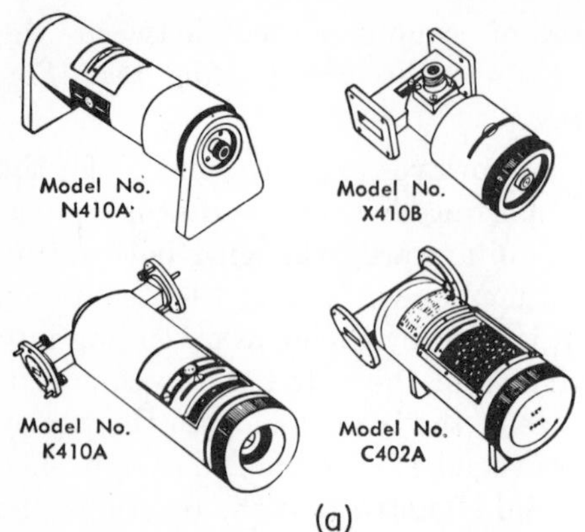

Model No.	Frequency Range (KMc)	Absolute Accuracy (%)	Approx. Q	Waveguide Type RG-( )/U	Flange Type UG-( )/U	Price (F.O.B. Woodside)
COAXIAL TYPES						
N410A	1.00- 4.00	0.10	3000	(3/8" Coax Type N)		$495.00
N414A	3.95-11.0	0.10	500 to 1500	(3/8" Coax Type N)		495.00
WAVEGUIDE TYPES						
*H410B	3.95- 5.85	0.08	8000	49	149A	250.00
*C410B	5.85- 8.20	0.08	8000	50	344	180.00
*W410B	7.05-10 00	0.08	8000	51	51	165.00
*X410B	8.20-12.40	0.08	8000	52	39	150.00
Y410A	12.40-18.00	0.10	4500	91	419	210.00
K410A	18.00-26.50	0.10	4000	53	425	230.00
U410A	26.50-39.50	0.10	3000	96	381	250.00
C402A	5.85- 8.20	0.03	8000	50	344	1275.00
X402A	8.20-12.40	0.03	8000	52	39	1275.00

(a)

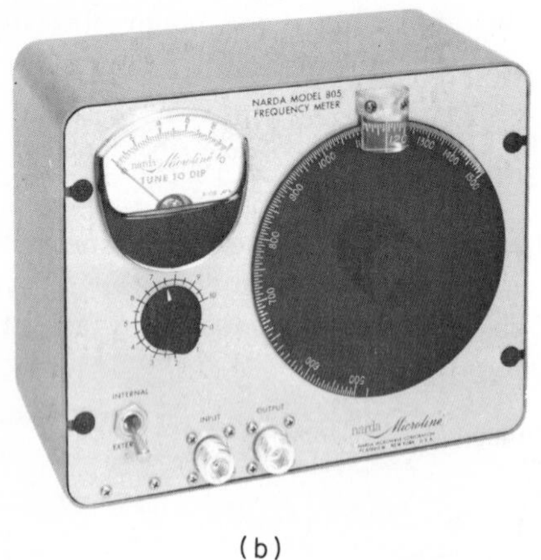

(b)

Fig. 10-7 COMMERCIAL WAVEMETER MODELS (Courtesy of Microlab/FXR and NARDA Microwave Corp.)

universally being made with dial mechanisms which deliver direct frequency readout.

Figure 10-7 illustrates a number of coaxial and waveguide types extending into the GHz range. Most meters have a scale calibrated directly in MHz, eliminating conversion of the micrometer scale divisions into frequency.

MEASUREMENT OF FREQUENCY

Measurement of frequency is a "must" for all setups, since every quantity is determined by the guide wavelength which in turn is a frequency-determined quantity.

For precise measurement or when an independent signal source is being used, a typical measurement setup, as shown in Fig. 10-2, can be used for frequency measurements, since it illustrates most components and their adjustments.

CONTRIBUTORY ADJUSTMENTS

Beyond the routine observation of the reaction-meter dip at resonance, there are four major items controlling and contributing to the typical overall measurement:

1. Adjustment of the signal-generating frequency.
2. Setting of the signal-generator klystron power supply.
3. Connecting the output and modulator indicating oscilloscope.
4. Using the VSWR indicator.

In adjusting a self-contained microwave signal generator, say for dial calibration, we use routine steps. These consist of three main settings: (1) the main dial and vernier, (2) output attenuator setting, and (3) internal modulation adjustment. All these are usually interdependent.

In operating a signal generator that contains a klystron power supply, we are chiefly interested in controlling the applied reflector voltage to obtain correct mode operation. This should be made at the specified or recommended klystron beam voltage and current readable on the accompanying section of the control panel.

The oscilloscope for display in a measurement setup is a visual, dynamic indicator of transmitted power delivered through the system to the crystal detector and of the 1000-Hz horizontal sweep voltages within the main system. Scope operation for a self-contained generator is relatively simple since it uses an integral power supply. In this case, the on-mode klystron operation for delivering maximum power becomes automatic.

A display approximating Fig. 10-8 should appear with correct *H* and *V* scope adjustments. Using this indication of the existence of output power, we proceed to the remainder of the measurements.

VSWR INDICATOR

This equipment is a high-gain, tuned amplifier used to amplify the scope indication by measuring crystal-detector output. See Fig. 10-2, where we change the crystal output lead to the VSWR indicator input.

Since the VSWR indicator is a tuned, high-

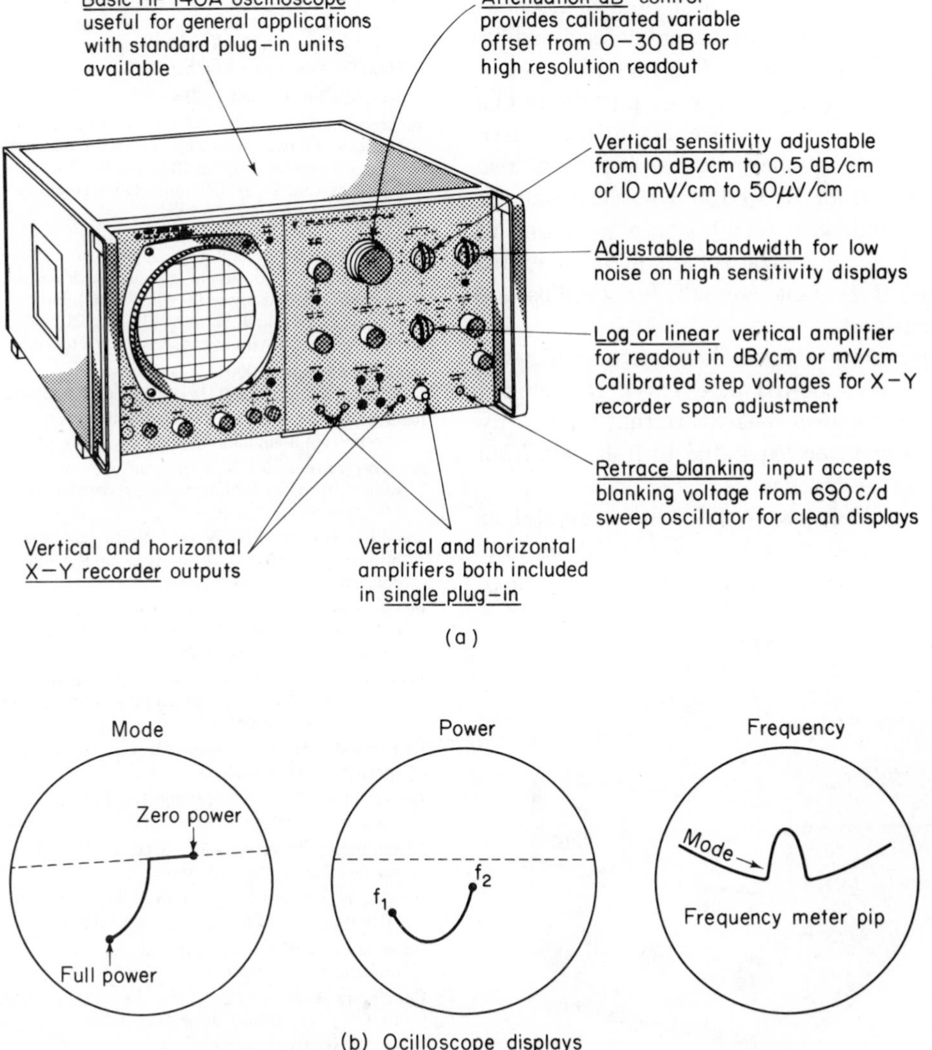
Basic HP 140A oscilloscope useful for general applications with standard plug-in units available
Attenuation dB control provides calibrated variable offset from 0–30 dB for high resolution readout
Vertical sensitivity adjustable from 10 dB/cm to 0.5 dB/cm or 10 mV/cm to 50μV/cm
Adjustable bandwidth for low noise on high sensitivity displays
Log or linear vertical amplifier for readout in dB/cm or mV/cm
Calibrated step voltages for X–Y recorder span adjustment
Retrace blanking input accepts blanking voltage from 690c/d sweep oscillator for clean displays
Vertical and horizontal X–Y recorder outputs
Vertical and horizontal amplifiers both included in single plug-in
(a)
Mode
Zero power
Full power
Power
f1
f2
Frequency
Mode
Frequency meter pip
(b) Ocilloscope displays

Fig. 10-8 OSCILLOSCOPE DISPLAY OF OUTPUT POWER (Courtesy of Hewlett-Packard Co.)

gain amplifier, we can adjust the wavemeter while observing this indicator, watching for a dip in meter reading to pinpoint exact input frequency.

NOTES ON THE *VSWR* INDICATOR (TYPICAL HP MODEL 415-E). This instrument, a basic piece of equipment in microwave measuring circuits, is a low-noise, tuned amplifier–voltmeter, calibrated in dB and VSWR. It is particularly tailored for use with standard 1000-Hz modulated rf signals when used with a square-law crystal detector. It is useful in measuring voltage peaks and valleys, attenuation, gain, or other parameters determined by the ratio of two signals. The tuned frequency is adjustable over a range of plus or minus $3\frac{1}{2}$ per cent for exact matching to the source modulating frequency; its bandwidth is also adjustable from 15 to 130 Hz, the resultant narrow bandwidth facilitating a reduction in noise, while the widest setting accommodates a sweep rate fast enough for oscilloscope presentation.

For precise attenuation measurements, a high-accuracy 60-dB attenuator is included with an expand-offset feature that allows any 2-dB range to be expanded to full scale from maximum resolution.

The unit operates with either crystal or bolometer detectors having high- and low-impedance crystal inputs which are matched, respectively, for source impedances of 50–200 or 2500–10,000 Ω. Precise bias-current circuits (4.5–8.7 mA) feeding into 200 Ω are provided for bolometer operation. Ac and dc outputs are provided for use as a high-gain tuned amplifier or with recorder.

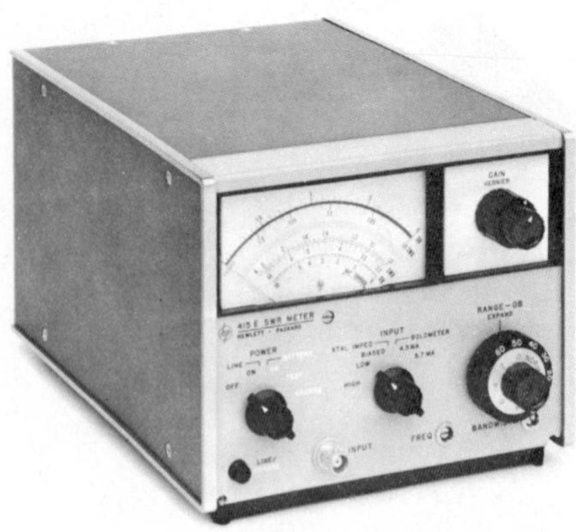

Fig. 10-9 SWR INDICATOR EQUIPMENT (Courtesy of Hewlett-Packard Co.)

Table 10-3 is a tabulated summary of performance characteristics specifically covering the HP Model 415-E. Figure 10-9 shows the actual instrument.

TABLE 10-3 SWR INDICATOR CHARACTERISTICS

Sensitivity: 0.15 μV rms for full-scale deflection at maximum bandwidth (1 μV rms on high impedance crystal input).

Noise: at least 7.5 dB below full scale at rated sensitivity and 130 Hz bandwidth with input terminated in 100 or 5000 Ω; noise figure less than 4 dB.

Range: 70 dB in 10- and 2-dB steps.

Accuracy: ±0.05 dB/10-dB step; maximum cumulative error between any two 10-dB steps, ±0.10 dB; maximum cumulative error between any two 2-dB steps, ±0.05 dB; linearity, ±0.02 dB on expand scales, determined by inherent meter resolution on normal scales.

Input: unbiased low and high impedance crystal (50-200 and 2500-10,000 Ω optimum source impedance respectively for low noise); biased crystal (1 V into 1 kΩ); low and high current bolometer (4.5 and 8.7 mA ±3% into 200 Ω), positive bolometer protection; input connector, BNC female.

Input frequency: 1000 Hz adjustable 7%; other frequencies between 400 and 2500 Hz available on special order.

Bandwidth: variable, 15-130 Hz; typically less than 0.5 dB change in gain from minimum to maximum bandwidth.

Recorder output: 0-1 V dc into an open circuit from 1000 Ω source impedance for ungrounded recorders; output connector, BNC female.

Amplifier output: 0-0.3 V rms (Norm), 0-0.8 V rms (Expand) into at least 10,000 Ω for ungrounded equipment; output connector, dual banana jacks.

Meter scales: calibrated for square-law detectors; SWR: 1-4; 3.2-10 (Norm); 1-1.25 (Expand) dB: 0-10 (Norm); 0-2.0 (Expand); battery: charge state.

Meter movement: taut-band suspension, individually calibrated mirror-backed scales; expanded dB and SWR scales greater than 4¼ in. (108 mm) long.

RFI: conducted and radiated leakage limits are below those specified in MIL-I-6181D.

Power: 115-230 V ±10%, 50-400 Hz, 1 W; optional rechargeable battery provides up to 36 hr continuous operation.

Dimensions: $7^{25}/_{32}$ in. wide, $6^{3}/_{32}$ in. high, 11 in. deep from panel (190 x 155 x 279 mm).

Weight: net, 9 lb (4 kg), 11 lb (5 kg) with battery; shipping, 10 lb (4, 5 kg), 13 lb (6, 3 kg) with battery.

Accessory available: 11057A Handle, fits across top of instrument for carrying convenience.

Combining cases: 1051A, $11^{1}/_{4}$ in. (286 mm) deep. 1052A, $16^{3}/_{8}$ in. (416 mm) deep.

Price: HP Model 415E, $400.

Options: 001, rechargeable battery installed, add $100; 002, rear-panel input connector in parallel with front-panel connector, add $25.

This factor, following frequency and impedance measurements, is of secondary character in that it concerns neither physical-circuit constants (R, L, and C) nor direct values of voltage, current, or field strength. It is an amplitude-oriented ratio factor and is used in almost every operating or evaluative process, either to regulate and adjust measured signals or to determine by comparison in a power system what losses are being encountered.

DIRECT-ATTENUATION RATIO

This is a loss factor for most situations; functionally, it becomes a comparison between two output power readings taken before and after a component has been inserted in a system or before and after applied input power has been changed by a known amount.

As such, this type of measurement would seem to be simple enough, requiring only adjustable signal amplitude or power-measuring equipment, after all precautions for impedance matching and other requirements have been met. It should be noted, however, that in the power-ratio method, some error may enter as a result of the different detector levels used for ratio measurements; detector errors at above 1 mW may rise to 5 per cent. Some of these contributory conditions as they bear upon power measurement are discussed later.

RF SUBSTITUTION MEASUREMENT

This method eliminates detector levels because all output-power readings are taken at a constant value. Here we use a calibrated variable attenuator, adjusting and reading it for some established level after it has replaced the unknown component. When calibrated in dB, the ratio is directly readable.

The measurement procedure follows the setup in Fig. 10-10, where the klystron, power supply, frequency meter, oscilloscope, and standing-wave indicator are located and used with 1000-Hz modulation, as in the generalized setup of Fig. 10-2. This example illustrates how we determine the attenuation calibration of the unknown. Here we use level comparisons on a variable flap attenu-

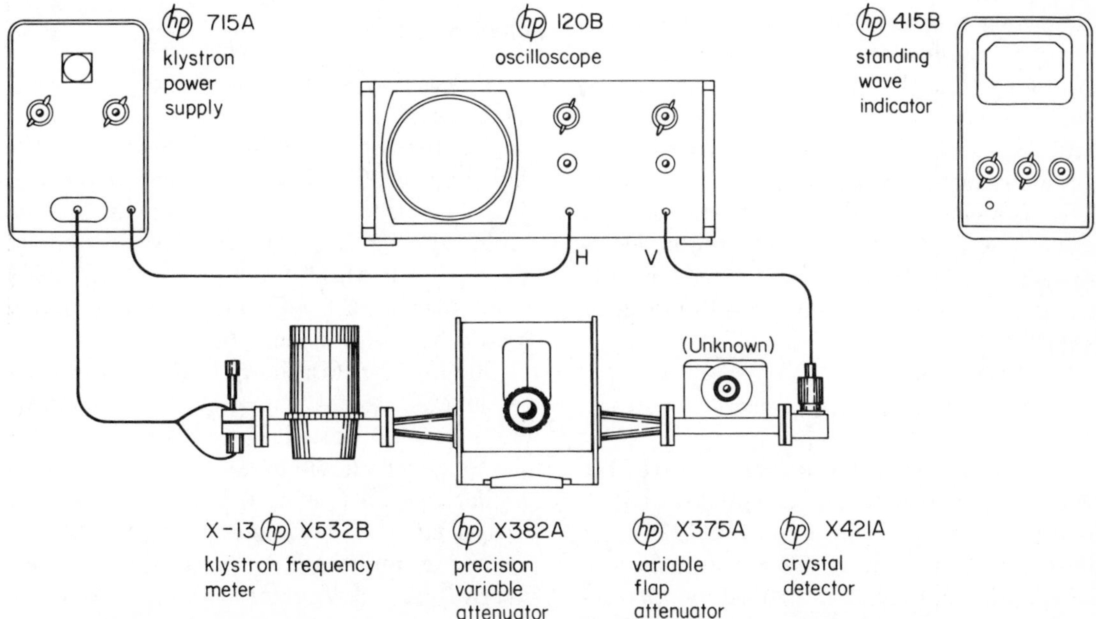

Fig. 10-10 ATTENUATION MEASUREMENT SETUP (Courtesy of Hewlett-Packard Co.)

ator set at predetermined values in conjunction with a standard precision attenuator in the signal path. After using the oscilloscope to obtain approximate operation by disconnecting its lead from the crystal detector and then moving it to the input of the VSWR indicator where final, accurate 1000-Hz amplitude "peaking" is conducted. It is assumed in this illustration that reflections and mismatch are being ignored—hence the deliberate omission of the slotted line used in Fig. 10-2.

Preliminary adjustments beyond those used in conjunction with the standard measurement setup include insurance that the crystal detector signal level should not be over 0.1 mW. For this precaution it is safe to insert another 10-dB attenuator pad between the unknown and the crystal detector and, in addition, to set the precision attenuator for the desired power, given by a reading of 30 dB on the VSWR indicator.

DIRECT POWER RATIO MEASUREMENT

Direct power ratio measurement can give us calibration of the steps on the unknown attenuator by starting at a 0-dB setting and making them coincide with the attenuation settings on the VSWR indicator. To do this we would reduce the VSWR-indicator gain control so that the dial setting is also 0 dB, and then increase the unknown attenuator in 5-dB steps while the VSWR attenuator switch is being reduced the same amount. This move gives coincident "tracking" when there is perfect design of both attenuator units. Deviations in attenuator scale indication from true attenuation appear on the VSWR indicator meter.

RF SUBSTITUTION MEASUREMENT

Rf substitution measurement uses the basic setup of Fig. 10-10. The procedure is to establish and maintain a constant SWR indicator meter reading while the unknown attenuator steps are accompanied by opposite steps in the precision attenuator.

It should be realized that all attenuator steps may not track with true or standard values obtained either from the SWR indicator attenuator or via the precision variable unit. The inconsistencies may arise from design compromises and "end scale" points caused by uneven coupling throughout a given range and other factors. For precise recording of these, attenuator calibration should be done at a number of different frequencies.

DETECTION DEVICES

Microwave detection in a waveguide or transmission line is accomplished by a crystal detector or bolometer. The former is a low-power, square-law field-detecting device employing a solid-state rectifier and actuated by a signal-collecting probe inserted physically into the field area within the waveguide or transmission line being used.

The bolometer is a low-power, signal-extracting, energy-absorption device (using a crystal-type probe) operating chiefly by its property to change internal resistance with temperature rise caused by the heat-dissipating power which it has absorbed from the signal.

CRYSTAL DETECTOR

Physically, a crystal detector is a complete assembly, which includes the crystal holder, the crystal-rectifier element itself, the signal-output fitting, and the probe mounting assembly. This last item usually includes a probe-wire depth-adjustment mechanism, and the hardware that goes along with a tuning stub, that is, the adjustable coaxial tuning section and adjustment knob.

Figure 10-11(a) shows how a crystal detector and coaxial tuning stub are mounted on a coaxial slotted line. Figure 10-11(b) shows closer detail of the hardware for a similar assembly using a mounting that will accept either a bolometer or a crystal element. Adaptation of the probe and accessory hardware to waveguides is shown further on in the waveguide slotted-line assembly.

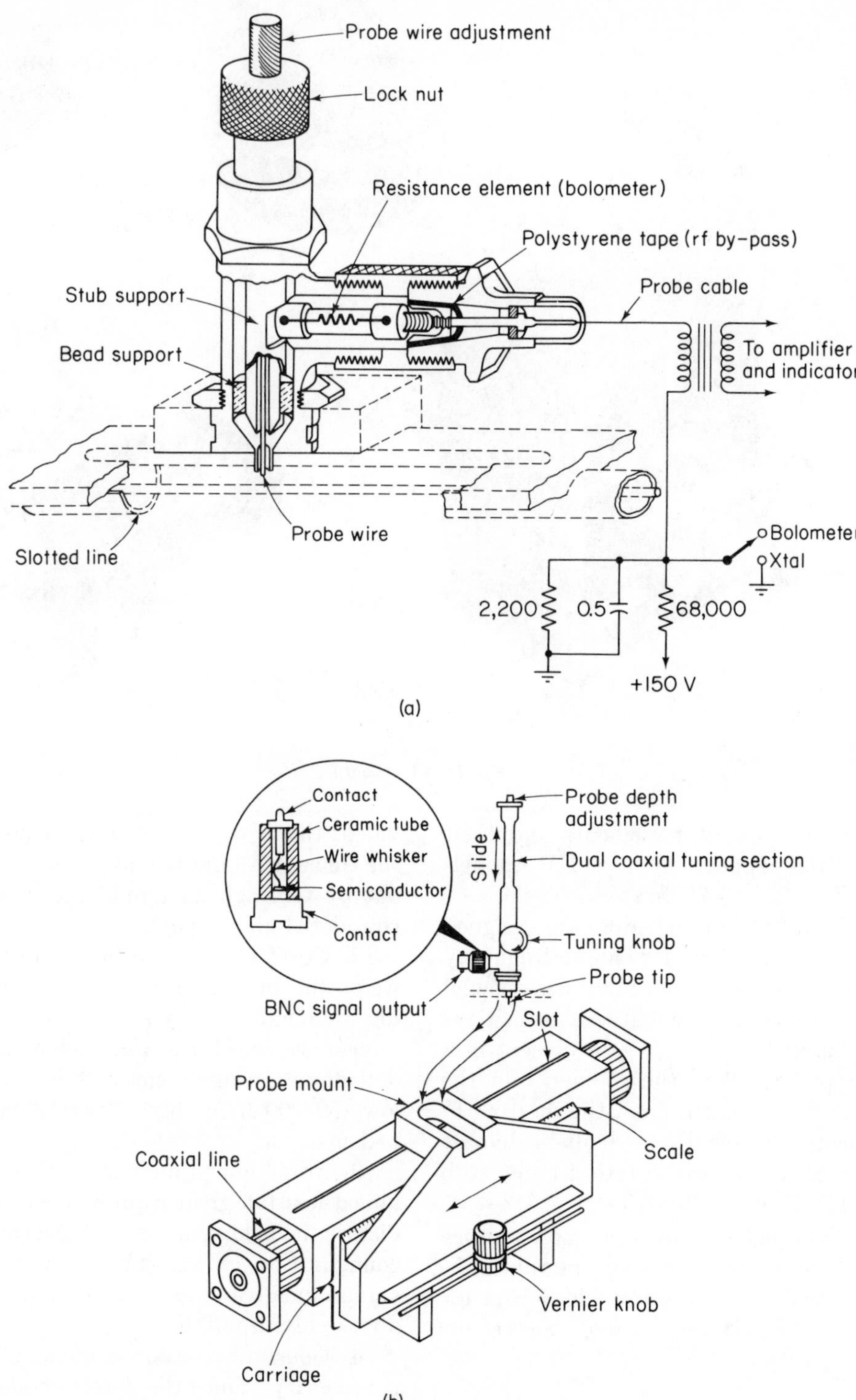

Fig. 10-11 CRYSTAL DETECTOR SETUP AND EQUIPMENT (Courtesy of Hewlett-Packard Co.)

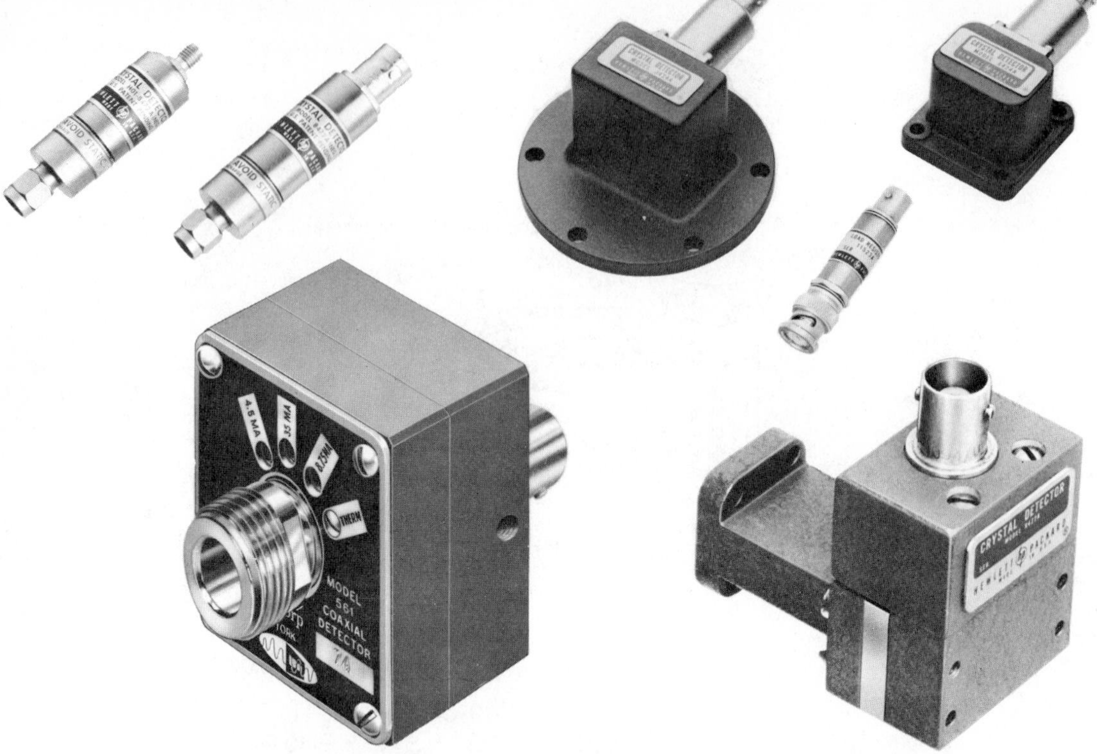

(c) Microwave detectors

Fig. 10-11 (Cont.)

Commercial detector elements and their characteristics appear in the HP models illustrated in Fig. 10-11(c).

Circuits for crystal detectors are designed to operate with 1000-Hz audio-frequency modulation applied to the microwave signal so that the recovered signal can benefit, as we saw above, by the high gain present in a tuned amplifier. Also, the rf energy in the crystal must be isolated from the dc path to the indicated meter by a built-in bypass capacitor or a half-wavelength tuning stub [Fig. 10-11(b)] and Table 10-4.

Biasing the crystal is also common practice when interconnecting with an external indicator; the load circuit also provides means for insertion of a crystal current meter for monitoring output.

In using crystal detectors four precautions must be observed: 1. Protection must be provided; this is usually by a series pad because the element is minute and can be easily burnt out by too high an input (100 mW may be considered high input).

2. Correct load impedance must be used; we do this in connecting the input of a standing-wave indicator by ensuring that the input impedance selector switch should match the detector's specified characteristics, whether low (50–200 Ω) or high (2500–10,000 Ω) impedance.

3. Minimum probe penetration should be aimed for; too great a probe penetration will distort the field symmetry of the electric field configuration and can affect the true maxima and minima readings or may produce too high a residual mismatch.

4. Constant monitoring of crystal currents is necessary. Since the detecting element is a square-law device, it yields a current output proportional to the square of the applied volt-

TABLE 10-4 DETECTOR CHARACTERISTICS

HP Model	Frequency range (GHz)	Frequency resp.[1] (dB)	Low-level sensitivity (mV/μW)	Maximum SWR	RF input	Matched pair available	Square-law load available	Length		Shipping weight	
								(in.)	(mm)	(lb)	(kg)
8471A	100 kHz 1.2 GHz	±0.6; typ ±0.1 over 100 MHz	>0.35	typically 1.3	BNC male	no	no	$2\frac{3}{4}$	70	0.5	0.2
423A	0.01–12.4	±0.2/octave to 8 GHz; ±0.5 overall	>0.4	1.2 to 4.5 GHz; 1.35 to 7 GHz; 1.5 to 12.4 GHz	Type N male	yes[2]	yes[3]	2-15/32	63	1	0.5
8470A	0.01-18	±0.2/octave to 8 GHz; ±0.5 to 12.4 GHz; ±1 overall	>0.4	1.2 to 4.5 GHz; 1.35 to 7 GHz; 1.5 to 12.4 GHz; 1.7 to 18 GHz	APC-7	yes[2]	yes[3]	$2\frac{1}{2}$	64	1	0.5
8472A	0.01-18	±0.2/octave to 8 GHz; ±0.5 to 12.4 GHz; ±1 overall	>0.4	1.2 to 4.5 GHz; 1.35 to 7 GHz; 1.5 to 12.4 GHz; 1.7 to 18 GHz	SMA type male	yes[2]	no	$2\frac{1}{2}$	64	0.2	0.1
S424A	2.60–3.95	±0.2	>0.4	1.35	Waveguide cover flange	yes[4]	yes[3]	2-7/16	62	2	0.9
G424A	3.95–5.85	±0.2	>0.4	1.35		yes[4]	yes[3]	2-1/16	52	1	0.5
J424A	5.30–8.20	±0.2	>0.4	1.35		yes[4]	yes[3]	1-7/8	48	0.5	0.2
H424A	7.05–10.0	±0.2	>0.4	1.35		yes[4]	yes[3]	1-9/16	40	0.5	0.2
X424A	8.20–12.4	±0.3	>0.4	1.35		yes[4]	yes[3]	1-3/8	35	0.5	0.2
M424A	10.0–15.0	±0.5	>0.3	1.5		yes[4]	yes[3]	1	25	0.5	0.2
P424A	12.4–18.0	±0.5	>0.3	1.5		yes[4]	yes[3]	15/16	24	0.5	0.2
K422A[6]	18.0–26.5	±2	≈0.3	2.5		yes[5]	yes[3]	2	51	1	0.5
R422A[6]	26.5–40.0	±2	≈0.3	3		yes[5]	yes[3]	2	51	1	0.5

For all models
Maximum Input: 100 mW peak or average, (8471A: 3 V rms, 4.2 V pk).
Detector element: supplied.

Output polarity: negative (positive output available: for 423A, 8470A, 424A

Output connector: BNC female

1 As read on a 416 Ratio Meter or 415 SWR Meter calibrated for square-law detectors.
2 Frequency response characteristics (excluding basic sensitivity) track within ±0.2 dB per octave from 10 MHz to 8 GHz, ±0.3 dB from 8 to 12.4 GHz, and (8470A and 8472A) ±0.6 dB from 12.4 to 18 GHz.
3 < ±0.5 dB variation from square law up to 50 mV peak output into >75 kΩ; sensitivity typically >0.1 mV/μW.
4 Frequency response characteristics (excluding basic sensitivity) track within ±0.2 dB for S-, G-, J- and H-band units, ±0.3 dB for X-band units, and ±0.5 dB for M- and P-band units, specify
5 Matched pair of unit fitted with square-law loads. Frequency response characteristics (excluding basic sensitivity) track within ±1 dB for power levels less than approx. 0.05 mW; specify
6 Circular flange adapters: 11515A (UG-425/U) for K-band, 11516A (UG-381/U) for R-band.

age, which means that the output meter directly reads power and can increase very rapidly. Sensitive crystals can deliver currents of as high as 20 mW when indicating as little as 30 μW of rf power. Most detectors of this sensitivity must be individually calibrated for absolute power levels.

BOLOMETER

Bolometers are microwave power detectors and may be categorized either as (1) barreters or (2) thermistors. The barreter element (mounted in a fuselike cartridge) consists of a very fine platinum wire, less than 0.0001 in. in diameter. The element is part of a sensitive bridge–amplifier placed in a waveguide so that it comes under the influence of waveguide fields; it becomes heated, produces a self-generated increase in resistance, which is measured by the bridge and indicator, and gives the desired resistance–power relationship.

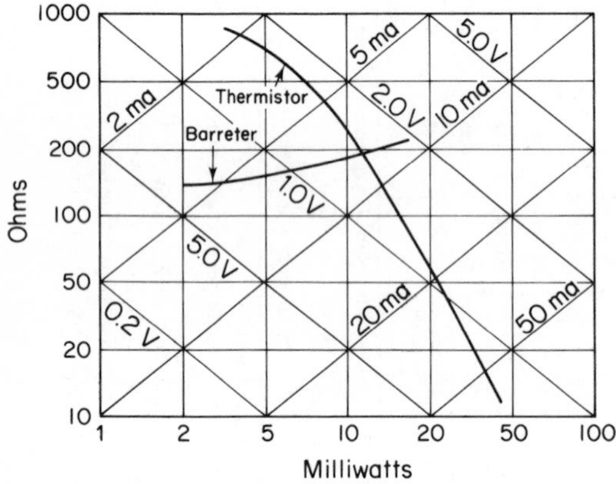

(a) Barreter and thermistor resistance curves

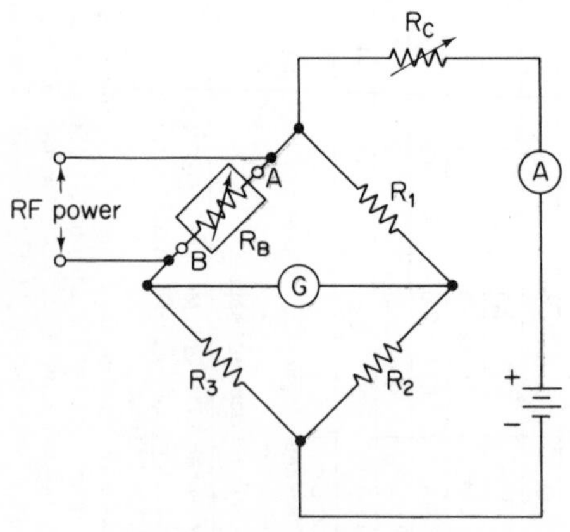

(b) Typical power measurement circuit

Fig. 10-12 BARRETTER AND THERMISTOR RESISTANCE CHARACTERISTICS PLUS MEASUREMENT CIRCUIT

Figure 10-12(a) shows a typical barreter resistance versus milliwatt characteristic curve, plus the bridge circuit used in a typical power-sensing setup.

Film-type bolometers are made from a molecularly thin resistant film based on a strip of glass or mica. These have a more sluggish reaction than the wire type and possess a high burnout rate; their response, however, is more linear.

The thermistor is a bead of semiconductor material placed between two thin parallel wires or between coaxial disks. This unit has negative coefficient (see the characteristic curve in Fig. 10-12), is more sensitive than the barreter, and requires special mount designs to preserve sensitivity and stability under ambient temperature changes. Thermistors are partially self-protected against burnout since their resistance decreases with increased power.

Circuits and use of bolometers. Bolometers can be used with continuous-wave as well as with pulse 1000-Hz signals; in other words, they may be used for detection only or for power measurement. Also, with its longer thermal time constant (100 μsec) some errors may enter.

In 1000-Hz detection, a constant-bias current is applied to a barreter element (see input selector switch on HP Model 415-E and 430-C).

Measurement and installation technique for using bolometers encounter the same preliminary precautions used in crystal-detector measurement setups; the barreter in particular is electrically fragile, and during setup and adjustment the operator should insure that the input selector switches are on maximum setting.

Low-power measurements use barreter or

(a)

HP Model[1]	Frequency range, GHz	Maximum SWR	Operating resistance (ohms)
478A	10 MHz to 10 GHz	1.75, 10 to 25 MHz 1.3, 25 MHz to 7 GHz 1.5, 7 to 10 GHz	200
8478B[2]	10 MHz to 18 GHz	1.75, 10 to 30 MHz 1.35, 30 to 100 MHz 1.1, 0.1 to 1 GHz 1.35, 1 to 12.4 GHz 1.6, 12.4 to 18 GHz	200
S486A	2.60 to 3.95	1.35	100
G486A	3.95 to 5.85	1.5	100
J486A	5.30 to 8.20	1.5	100
H486A	7.05 to 10.0	1.5	100
X486A	8.20 to 12.4	1.5	100
M486A	10.0 to 15.0	1.5	100
P486A	12.4 to 18.0	1.5	100
K486A[3]	18.0 to 26.5	2.0	200
R486A[3]	26.5 to 40	2.0	200

(b) Characteristics

Fig. 10-13 THERMISTOR AND BARRETTER MOUNTS AND CHARACTERISTICS (Courtesy of Hewlett-Packard Co.)

thermistor mounts and hardware pieces, which are substituted for the crystal detector pictured in the basic setup shown in Fig. 10-2. When applied in conjunction with a microwave power meter, refer to the high power applications discussed later in the book. An assortment of thermistor and barreter mounts plus characteristics for both coaxial and rectangular mountings is shown in Fig. 10-13.

IMPEDANCE MEASUREMENT

Microwave electrical circuit constants may be measured by

1. VHF and UHF bridges.
2. Reflectometers.
3. Slotted-line measurement.
4. Vector impedance meter.

VHF AND UHF BRIDGES

Microwave bridge designs are null-voltage comparison systems. They stem from the original four-leg Wheatstone and diamond-arranged circuit, where we apply ac input signal across diagonal terminals and then compare (by a null meter reading) the voltage ratios induced across the opposite diagonal terminals. The null-producing impedance adjustments in the "standard" leg of the bridge deliver an impedance ratio directly related to the unknown leg impedance.

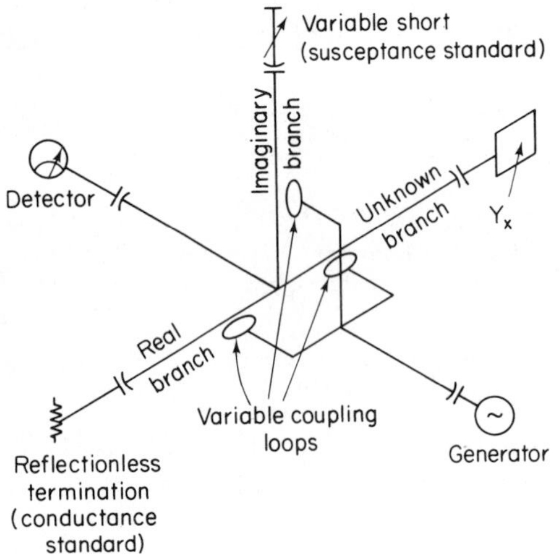

(a) Basic circuit

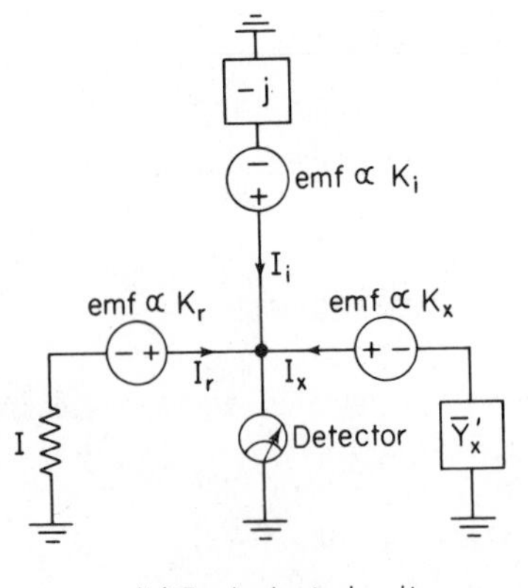

(b) Equivalent circuit

Fig. 10-14 THURSTON ADMITTANCE BRIDGE AND SCHEMATIC

For measuring the admittance of microwave components, the most common variation is the Thurston admittance bridge, shown in Fig. 10-14(a), where three coaxial lines are shown meeting in a T junction. These lines, the unknown branch, the real branch, and the imaginary branch are terminated, respectively, in the unknown admittance, which is a reflectionless load and constitutes a variable short circuit. A fourth branch, perpendicular to the plan of the junction, is connected to the detector.

Signal input is injected into the three admittance branches by three identical but variable coupling loops, excited in parallel by the signal generator. In equivalent schematic form [Fig. 10-14(b)], we see that the injected voltage into each admittance branch is proportional to the three variable coefficients of coupling of the generator to the branches K_x, K_r, and K_i. These branches carry corresponding currents I_x, I_r, and I_i.

Thus the unknown admittance Y_x is the sum total existing at the location of the coupling loops when we look toward the unknown branch, since the emf's are proportional to the three coupling coefficients. Next, we balance the bridge for voltage null by independently adjusting all three of the loops and correctly normalizing Y_x.

After balancing the loops for detector null, the currents add up to zero at the junction; and if we assume that the distance from the loops to the junction is negligible, $I_x + I_r + I_i = 0$; and since $I_x \propto Y_x \times K_x$, $I_r \propto 1 \times K_r$, and $I_i \propto -jK_i$,

$$Y_x = \frac{1}{K_x} - K_r + jK_i$$

where K_r, K_i, and K_x are calibrated indicator dial readings.

So we see that Y_x is proportional to the coupling coefficient K_r and K_i, respectively, plus both being multiplied by $1/K_r$.

Thus, physically, the bridge structure appears as a commercial unit in Fig. 10-15. Each coupling loop is attached to a calibrated indicator located around the perimeter of the structure; the imaginary capacitive susceptance standard is attached directly to the vertical leg while the coaxial unknown admittance extends to the right through a line stretcher or other coaxial fitting.

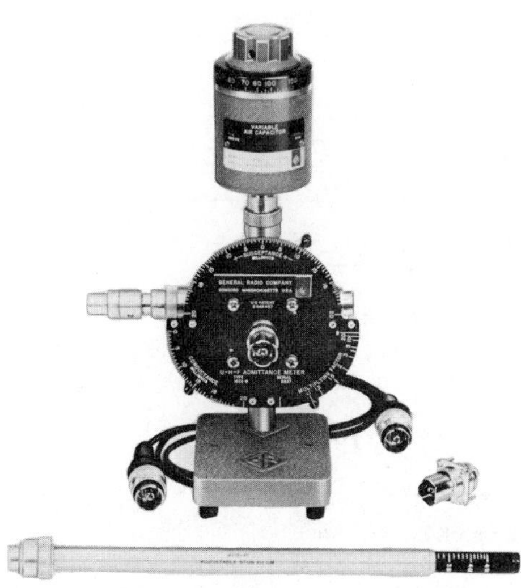

Fig. 10-15 UHF BRIDGE EQUIPMENT (Courtesy of Gen. Radio Co.)

To operate the system we must have the following equipment constituting the setup pictured in Fig. 10-16.

1. Microwave oscillator, GR-1360(b).
2. Standing wavemeter, GR-1234.
3. Heterodyne detector, GR-1249.
4. Line stretcher, GR-874 LKL.

IMMITTANCE BRIDGE

This particular bridge is similar to the 1602 admittance bridge but is tailored particularly for the measurement of transistors, diodes, and two- or four-terminal network parameters. It covers VHF and UHF ranges (25–1500 MHz) and, once the setup is made, it delivers direct readings of forward and reverse complex transfer functions plus the input and output impedances and admittance. Figure 10-17(a) is a skeletonized layout of the system circuitry; Fig. 10-17(b) pictures a commercial unit. Here, again, the circuitry consists of three identical loops fed from a

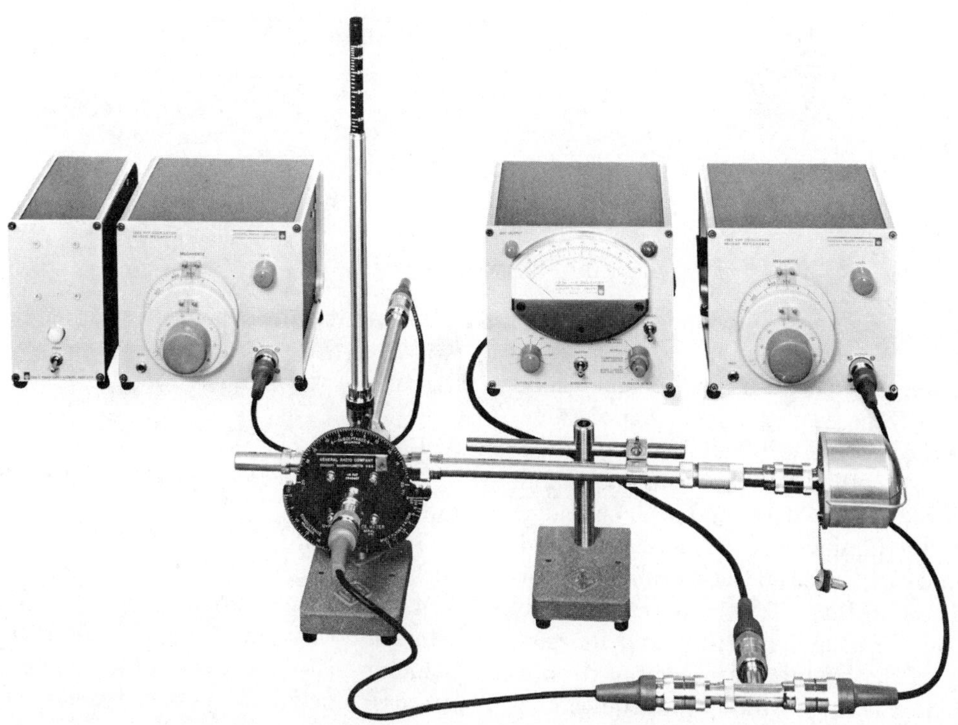

Fig. 10-16 UHF BRIDGE SETUP (Courtesy of Gen. Radio Co.)

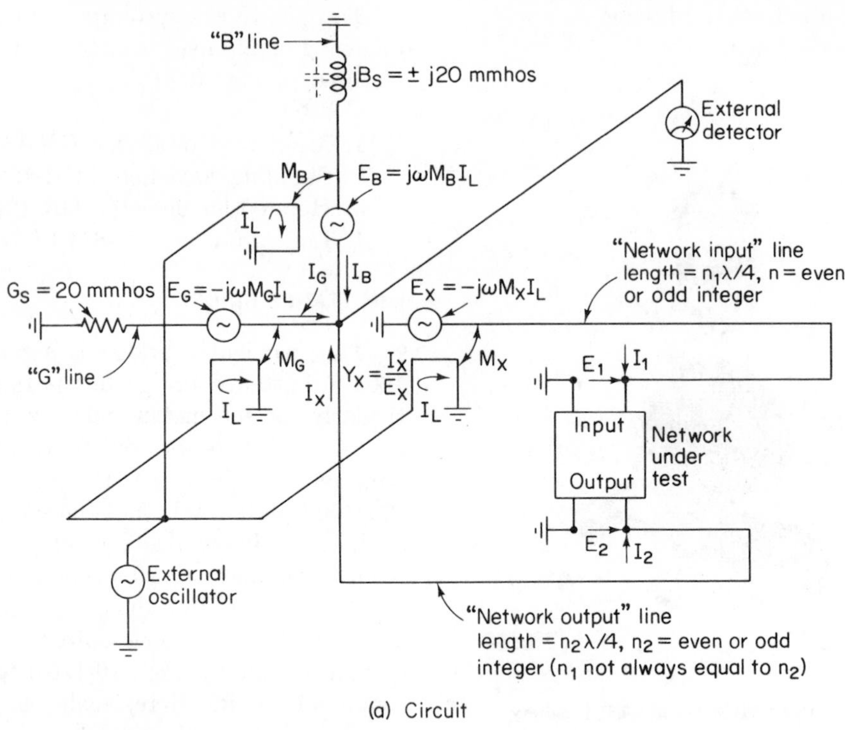

(a) Circuit

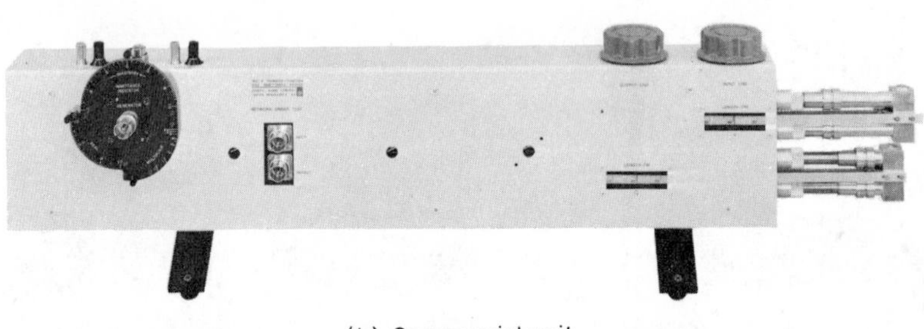

(b) Commercial unit

Fig. 10-17 IMMITANCE BRIDGE LAYOUT AND EQUIPMENT (Courtesy of Gen. Radio Co.)

common signal generator and magnetically coupled to three coaxial lines. One of these is terminated with a self-contained resistance standard, one with a reactance standard, and one with the network to be tested. For bridge balance, the coupling at each loop is adjusted until a null is obtained on the external detector (in which the three lines are terminated). Each loop, its coupling setting, and its relationship to the standard are combined on a calibrated scale from which conductance and susceptance values are directly picked off.

This bridge will measure

1. Transistor "S" parameters.
2. Tunnel diodes—equivalent circuit parameters.
3. General two- and four-terminal networks.
4. Ungrounded components, including resistors and their shunt capacitance; inductors, including self-inductance and resonance; capacitors, including pure capacitance and self-resonance.

REFLECTOMETER

The reflectometer is a device that directly measures the reflection coefficient of a waveguide system. From this ratio of reflected to incident voltages we can calculate the VSWR. Thus,

$$P = \frac{E_{\text{reflected}}}{E_{\text{incident}}} = \frac{\text{SWR} - 1}{\text{SWR} + 1}$$

The reflectometer setup shown in Fig. 10-18 does this by using two back-to-back directional couplers; one samples the forward or incident power passing from the signal to the load and the other samples the reflected power from the load. These two side-by-side 1000-Hz modulated voltages, suitably leveled for square-law operation, are passed through forward and reverse detectors into a ratio meter which delivers direct voltage ratios or reflection coefficient numbers.

Reflectometer accuracy is best at high standard-wave ratios, say SWR = 3 ($\rho = 0.5$), because of coupler directivity, or the inability to distinguish between forward and reverse power flowing in the main arm. This is because the operation is based on the separation of incident and reflected power; when this separation is imperfect, the insufficient directivity will add some main power at an unknown phase to the reflected signal and an error results. We can compensate for this in equivalent coefficient terms by substituting the directivity (in dB) into the return loss equation. Thus, for a reverse coupler directivity of 40 dB, the ambiguity in ρ is plus or minus 0.01.

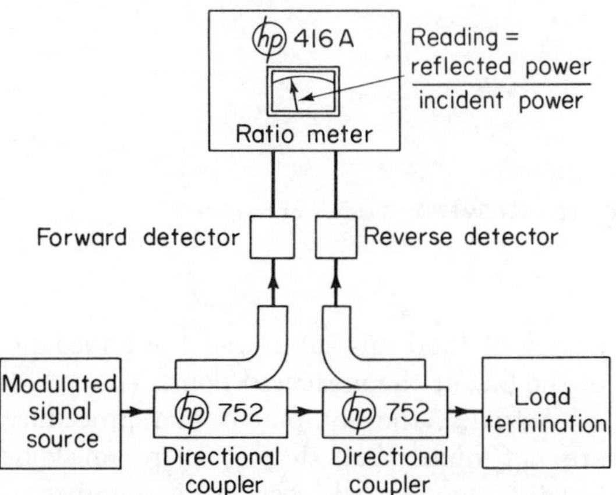

Fig. 10-18 REFLECTOMETER SETUP (Courtesy of Hewlett-Packard Co.)

Also, forward-coupler directivity can contribute to ambiguity, since the directivity signal adds to the incident signal. Here the error is proportional to the power level and may be calculated as

$$\Delta\rho = \pm\rho\left(\log^{-1}\frac{\text{dB}}{20}\right)$$

where dB = coupler directivity

ρ = reflection coefficient of test load

Figure 10-19 illustrates a number of conventional coaxial and waveguide reflectometers and detectors. Figure 10-19(b) illustrates a ratiometer used in associated reflectometry measurements (AP Model 916b).

Modern reflectometer techniques employ swept-frequency procedures involving modification and refinement of the basic method described above. These are described in Appendix I along with further discussion of errors, ambiguities, preattenuation factors, calibration, frequency response, mismatch, detector level, etc.

SLOTTED-LINE MEASUREMENTS

This unit, consisting of a slotted half-wavelength of waveguide or coaxial line plus its movable probe detector, is a universal piece of microwave test gear; it is an indispensable tool for the designer, engineer, or operator. To review, it precisely measures the pattern of standing waves (wave motion cancelation and reinforcement) caused by traveling-wave reflections arising from discontinuities or mismatched loads and terminations in a system.

As we saw in Chapter 8, wave-energy reflections produce physically stationary, measurable voltage maxima and minima along a linear waveguide; these can be detected by inserting a movable probe–detector–amplifier

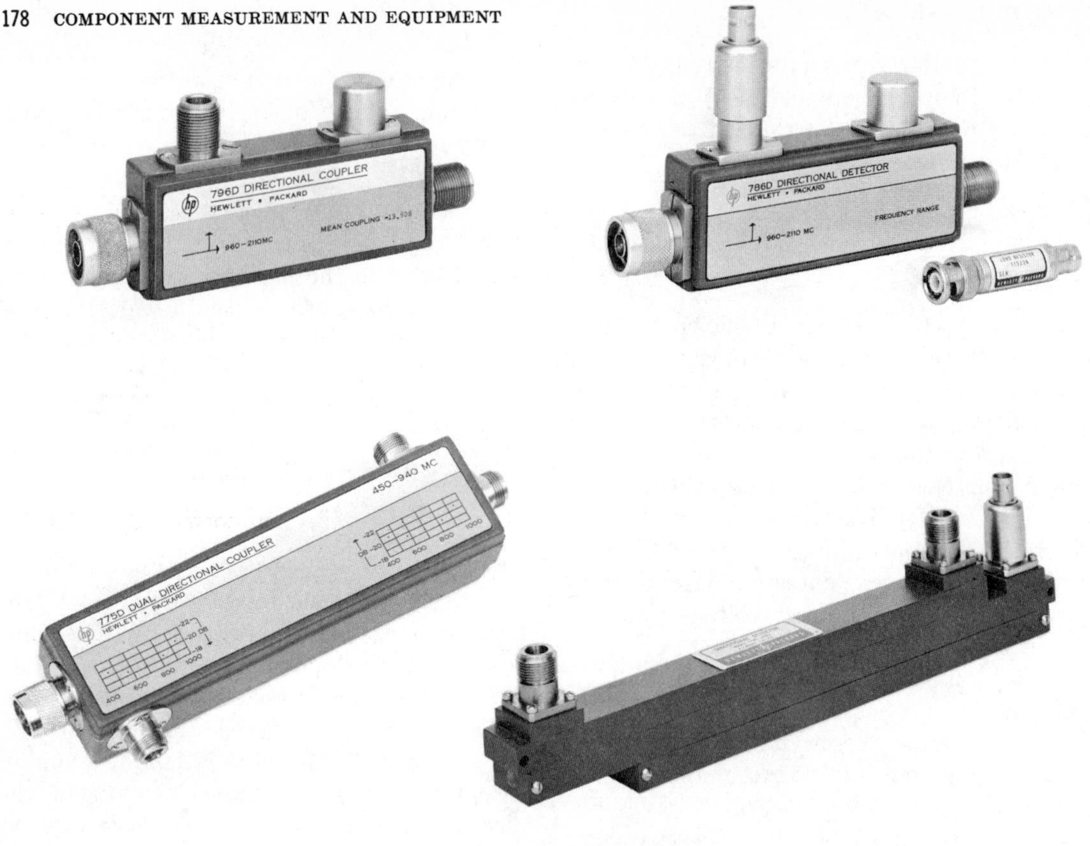

(a) Reflectometer components

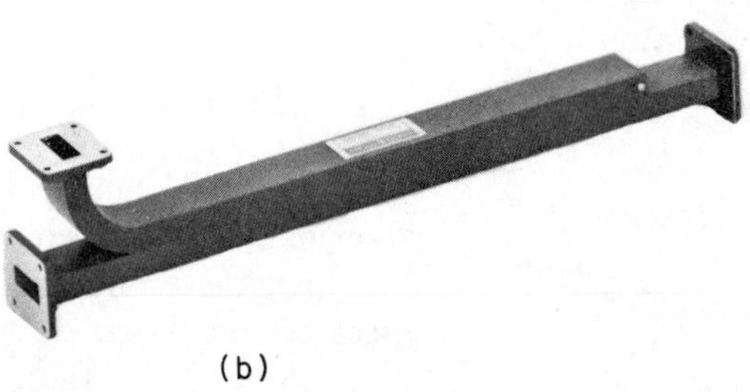

(b)

Fig. 10-19 COMMERCIAL COMPONENTS FOR REFLECTOMETRY (Courtesy of Hewlett-Packard Co.)

mechanism into the slot and positioning for these points.

The voltage ratio of these maxima to minima is the standing-wave ratio—SWR (more often referred to as VSWR). This ratio, as well as the detector's physical position (in terms of waveguide wavelength), tells us the degree of total mismatch and the impedance of the line at the measured point.

Actually, the equipment and procedures are not quite that simple; there must be, besides the slotted section, a number of accessories, and several precautions in design, choice, adjustment, and operation must be observed. A summary of these desiderata and requirements applies to a simple, point-by-point individual measurement that will be discussed below. Each of these form background for intelligent use of the equipment and all lead up to the modern technology of swept-frequency measurements appearing in Appendix J.

MEASUREMENT EQUIPMENT

Figure 10-20 is a block diagram of the equipment supporting a point-by-point VSWR measurement system. In detail, we must have a setup (Fig. 10-21) similar to that used in basic measurements (Fig. 10-2). The main items are

Oscilloscope
V
SWR meter
Tee
Signal source
Variable attenuator
Slotted line
Device
Short circuit (shorting plate)
Square-wave generator

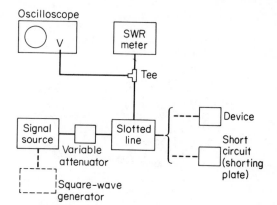

Fig. 10-20 BLOCK DIAGRAM OF VSWR MEASUREMENT (Courtesy of Hewlett-Packard Co.)

1. Square-wave generator. This instrument is typified in the HP Model 220-A, capable of tuning a low-powered klystron oscillator fully on and off at a 1000-Hz rate (Fig. 10-22).

2. Signal-generator source. This instrument is constructed about a klystron-type self-contained power-supply frequency meter and attenuator. Some modern signal generators include all these items within a single package; most units deliver up to a few watts of controlled output energy over a specified frequency range. A generator's output impedance must match or have suitable adaptors for electrically inserting the filter that usually follows it in the hardware setup. Note the performance characteristics concerning stability, output, and harmonic content.

3. Low-pass filter. To reduce false null indications in slotted-line measurements, it is necessary to trap out harmonics and spurious signals in the source that feeds the slotted line.

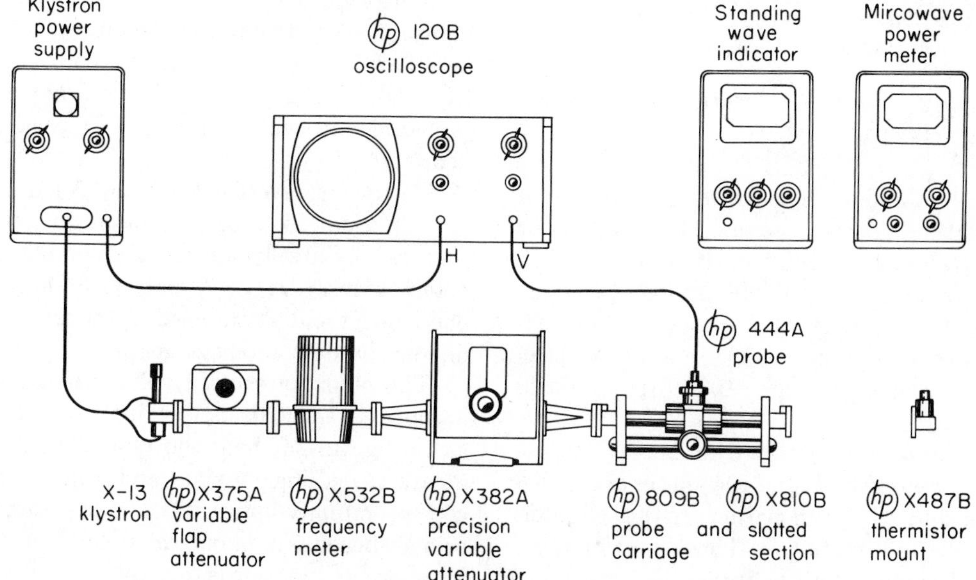

Fig. 10-21 VSWR EQUIPMENT SETUP (Courtesy of Hewlett-Packard Co.)

(a) Squarewave generator

Source impedance: 50 ohms.
Risetime: less than 15 ns.
Overshoot and ringing: less than 5% at 5 volts into 50 ohms
Amplitude
  Model 220A: continuously variable from 0 to − 5 volts into 50 ohms
  Model 221A: continuously variable from 0 to + 5 volts into 50 ohms
Symmetry: variable from approximately 40% to 60%
Repetition rate
  Ranges: from 1 Hz to 10 MHz (8 positions) in decade steps
  Vernier: continuously variable between all ranges
Frequency programming: − 1.2 to − 13 volts applied to external input will program the frequency over selected frequency range
Weight: net, 4 lb (1.8 kg); shipping, 6 lb (2.7 kg)
Power: 115 or 230 volts ± 10%, 50 to 400 Hz, 9 watts
Dimensions: $5\frac{1}{8}$" wide by $3\frac{7}{16}$" high by $11\frac{5}{8}$" deep (131 x 87 x 295 mm)

(b) Characteristics

Fig. 10-22 SQUARE-WAVE GENERATOR (Courtesy of Hewlett-Packard Co.)

These harmonics originate in the process of modulating a signal source, say a klystron or backward-wave oscillator, by causing frequency modulation of the source. In Fig. 10-23(a) we picture a number of low-pass filters covering the usable microwave frequency ranges. Figure 10-23(b) tabulates their characteristics.

4. Detector probes and mounts. The use and characteristics of these pieces were discussed and described earlier in this chapter.

5. Slotted sections. These sections are manufactured for either coaxial or rectangular systems, the former operating up to around 20 GHz, the latter from 4 to 40 GHz. For measuring waveguide systems with a coaxial setup, adaptors exist for making the physical and electrical transition. Figure 10-24(a) shows a coaxial line complete with carriage, slotted line, base plate, and adaptor. Figure 10-24(b) shows carriages and sections for waveguide hardware, and Fig. 10-24(c) shows a number of adaptors for the waveguide-coaxial transition.

Most slotted sections are designed for 50-Ω line impedance, can handle up to 2 W of signal power, and have a residual SWR of 1.02 to 1.04, depending upon the frequency being measured.

6. The SWR indicator described earlier in this chapter translates the detector signal into direct meter indication. It is used when necessary for precision and point-by-point analysis of specific transmission areas.

SLOTTED-LINE APPROACH

Modern microwave instrumentation using oscilloscope, improved display devices, and solid-state components has gone heavily to the use of swept-frequency measurement; however, for explanation, theory, and the basic background of SWR measurement, we advisedly adhere to the point-by-point methods outlined above, which have been used for many years. These items employ a basic approach which, although cumbersome and time consuming, is advisable for novice instruction and will be pursued in brief form below.

To review, SWR is the ratio of maximum to minimum voltage measured at specific physical points by using a movable-probe detector inserted in a half-wave long slot deliberately cut or inserted in series with the line or the device being measured.

The reflection-caused voltage patterns for an open or shorted lossless line would appear as in Fig. 10-25; here the reflections are the greatest, making maxima and minima ratios correspondingly large. Those measuring 10:1 or over become unmanageable and beset with inaccuracies in measurement. If we had access to a number of points on a long lossy

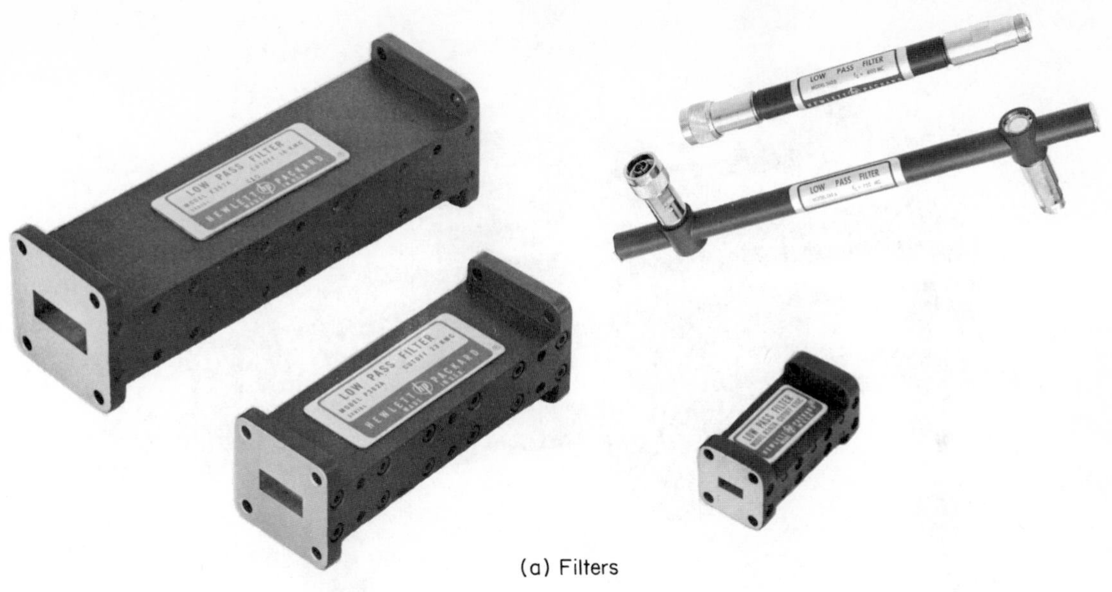

(a) Filters

HP model	X362A	M362A
Passband (GHz)	8.2–12.4	10.0–15.5
Stopband (GHz)	16–37.5	19–47
Passband insertion loss	less than 1 dB	less than 1 dB
Stopband rejection	at least 40 dB	at least 40 dB
SWR	1.5	1.5
Waveguide size, in. (EIA)	$1 \times \frac{1}{2}$ (WR 90)	0.850×0.475 (WR 75)
Length, in. (mm)	5-11/32 (136)	4-15/32 (114)
Shipping weight, lb (kg)	2(0, 9)	1(0, 45)

HP model	360A	360B	360C	360D
Cut-off frequency	700 MHz	1200 MHz	2200 MHz	4100 MHz
Insertion loss	≤1 dB below 0.9 times cut-off frequency			
Rejection	≥50 dB at 1.25 times cut-off frequency			
Impedance	50 ohms through passband; should be matched for optimum performance			
SWR	<1.6 to within 100 MHz of cut-off		<1.6 to within 200 MHz of cut-off	<1.6 to within 300 MHz of cut-off

HP model	Passband frequency (GHz)	Max. passband insertion loss	Rejection band attenuation			
			Below passband		Above passband	
			Frequency (GHz)	Attenuation	Frequency (GHz)	Attenuation
8430A	1 to 2	2 dB	≤0.8	≥50 dB	2.2 to 20	≥45 dB
8431A	2 to 4	2 dB	≤1.6	≥50 dB	4.4 to 20	≥45 dB
8432A	4 to 6	2 dB	≤3.5	≥50 dB	6.5 to 20	≥45 dB
8433A	6 to 8	2 dB	≤5.5	≥50 dB	8.5 to 20	≥45 dB
8434A	8 to 10	2 dB	≤7.5	≥50 dB	10.5 to 17	≥45 dB
8435A	4 to 8	2 dB	≤3.2	≥50 dB	8.8 to 20	≥45 dB
8436A	8 to 12.4	2 dB	≤6.9	≥50 dB	13.5 to 17	≥45 dB

(b) Characteristics

Fig. 10-23 LOW AND PASS BAND FILTERS (Courtesy of Hewlett-Packard Co.)

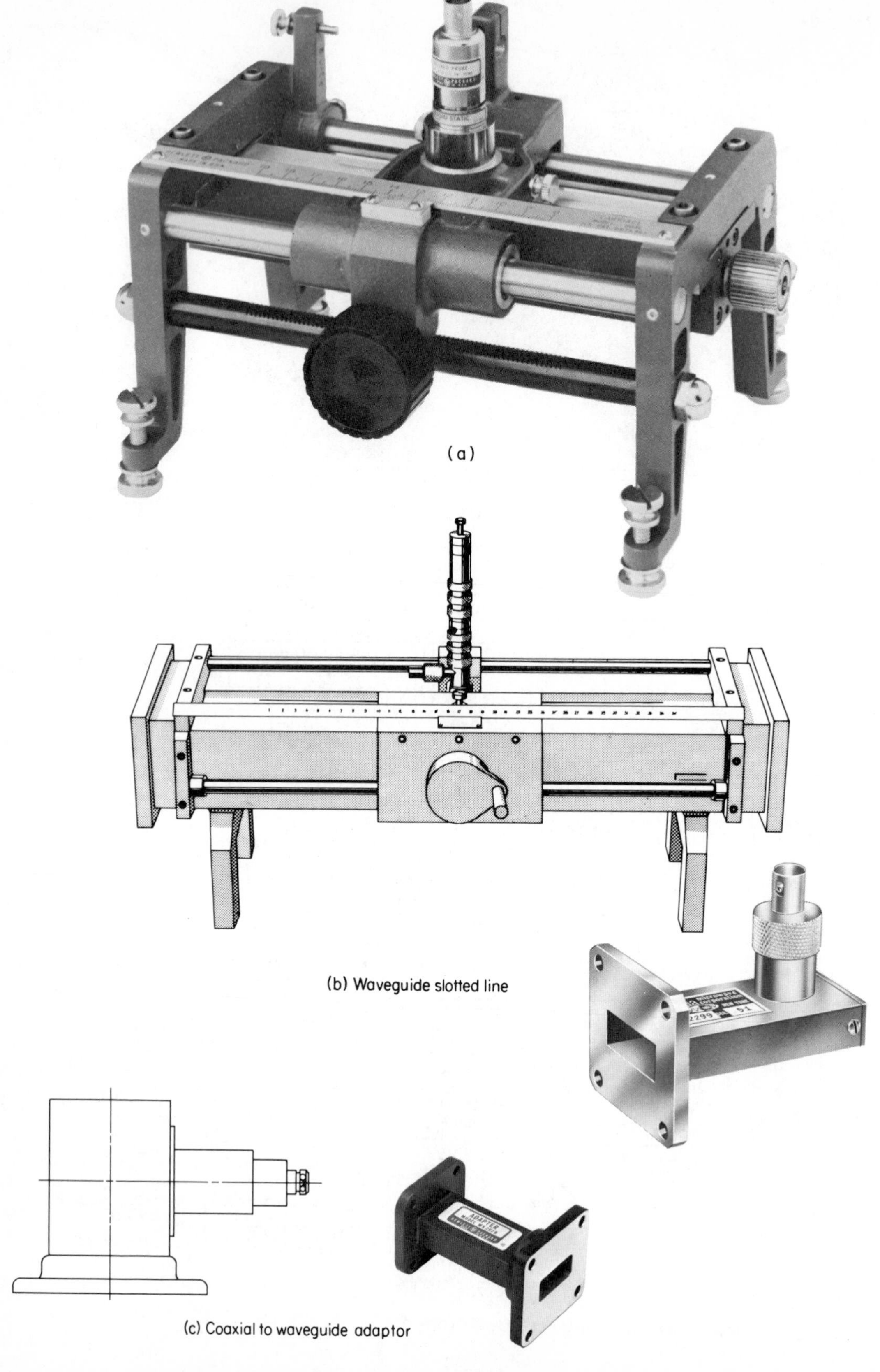

(a)

(b) Waveguide slotted line

(c) Coaxial to waveguide adaptor

Fig. 10-24 SLOTTED LINE AND ADAPTER ASSEMBLIES (Courtesy of Hewlett-Packard Co.)

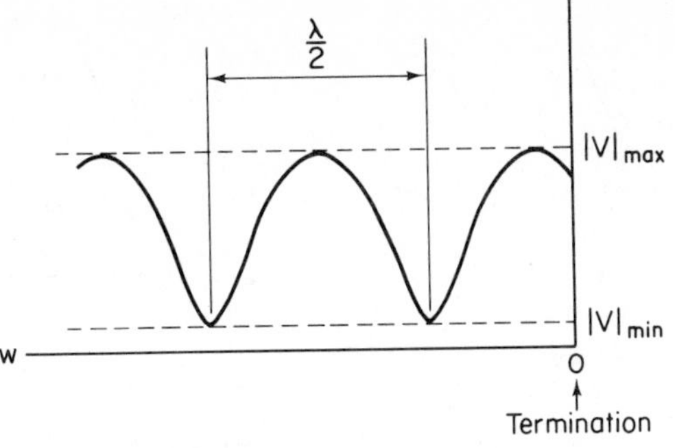

(a) Lossless conditions

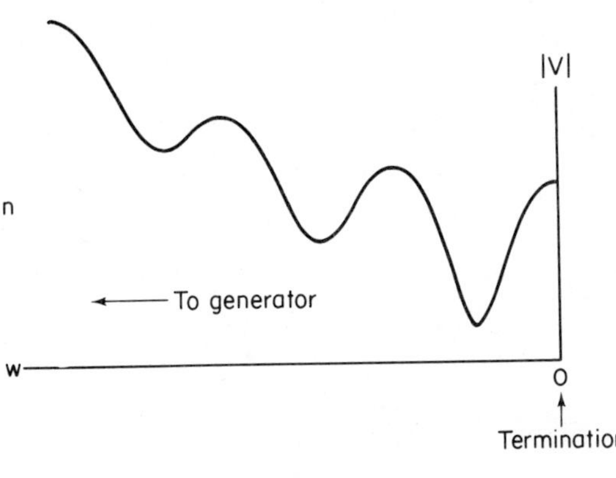

(b) Lossy line

Fig. 10-25 VOLTAGE AMPLITUDE PATTERNS FOR SLOTTED LINE

line and if we could measure a number of pairs of maxima and minima, their amplitudes would decrease with progressive positions along the wave travel, and the pattern would resemble Fig. 10-25(b). Most waveguides have negligible losses, so our SWR and reflection measurements do not include them.

Note that the length of a single standing-wave period, the distance between minima and maxima, is half a wavelength, that is, half the wavelength of a traveling wave. This is because the phases of the forward and reflected waves shift in opposite directions when referenced to changing positions on the line so that the phase angle between the forward and reflected voltages shift by 360° in just half a wavelength. See Chapter 8 for an illustration of the reflection phenomenon.

Now, if we place terminations on the line, approaching its characteristic impedance, the reflections lessen and the maximum–minimum ratio grows smaller. Figure 10-26 shows this for reflection coefficients for the load and standing-wave ratios of 5.7:1 and 1.35:1.

Next, the position of the maxima–minima points with respect to the termination point determines the impedance of the load—mind you, not the characteristic impedance Z_0 (which acts like a pure resistance) but the angular, $R + jX$ value referred (or normalized) in the numerical terms of Z_0. We do this by first converting the physical position of the minimum nodal point into angular degrees (θ the termination angle) and then converting this angle (by the Smith chart)

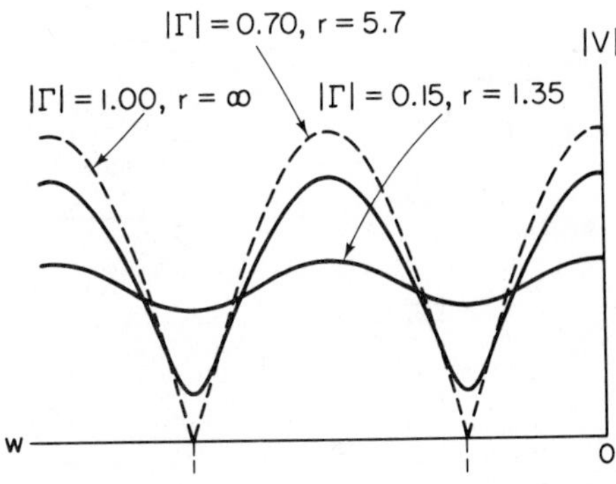

(a) Approaching impedance match

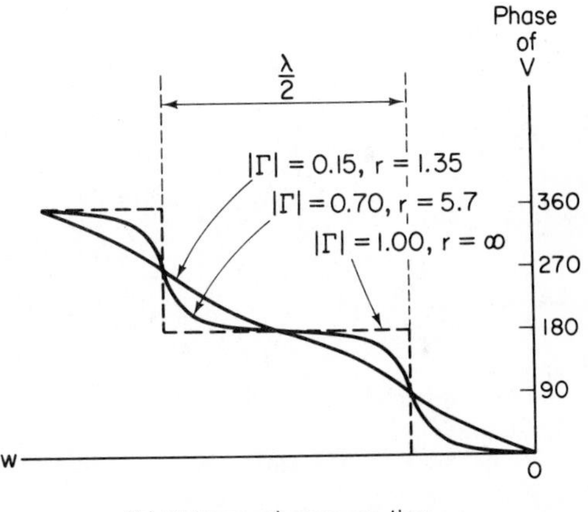

(b) Voltage phases on line

Fig. 10-26 VOLTAGE AMPLITUDE PATTERN FOR MATCHED LINE

R < Zc
R > Zc
(a) Capacitive
(c) Small resistive load
(b) Inductive
(d) Large resistive

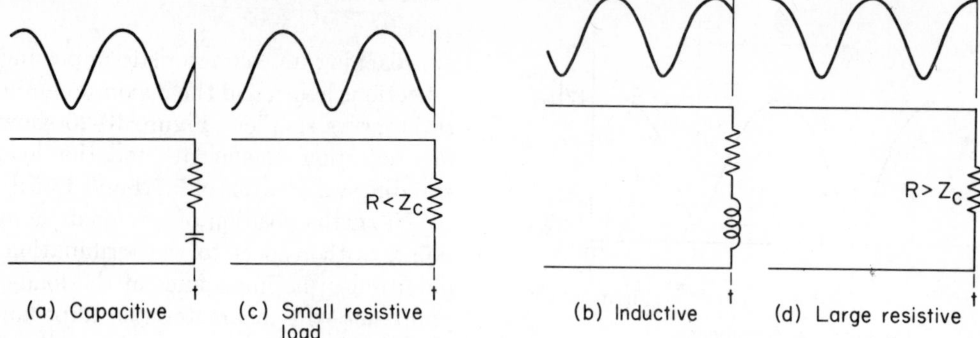

Fig. 10-27 VOLTAGE PATTERN SUMMARY

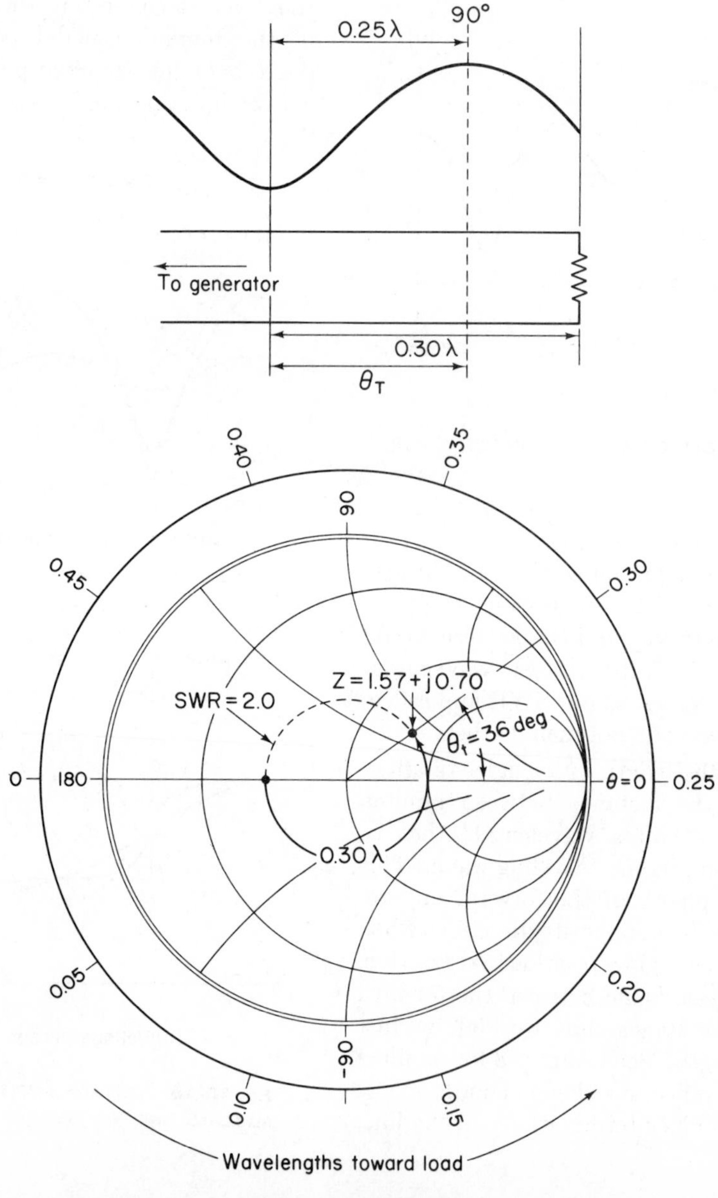

90°
0.25λ
To generator
0.30λ
θT
0.40
0.35
90
0.45
0.30
Z = 1.57 + j0.70
SWR = 2.0
θt = 36 deg
0
180
θ = 0
0.25
0.30λ
0.05
0.20
-90
0.10
0.15
Wavelengths toward load

Fig. 10-28 SMITH CHART EXAMPLE — LINE IMPEDANCE

into $R + jX$ components in terms of the characteristic impedance.

This is logically and correctly executed because the load at the end of the line is at a minimum nodal point, as are all the other nodal points, each half-wavelength (180°) progressing forward along the line toward the generator.

We see this physical-angular positioning for four basic conditions in Fig. 10-27: (1) for inductive, (2) for capacitive, and (3) for resistive loads which are smaller and larger than the characteristic impedance. Note that in (1) the inductive load is larger (say $\theta = +60°$), in (2) the capacitive load is less than Z_c ($\theta = -60°$), and in (3) the resistive load is smaller than Z_c, being at the minimum nodal point, while the resistance at d is larger than Z_c.

Repeating the Smith-chart procedure for a condition where the minimum is at 0.3 wavelength and the SWR = 2, we find the terminal impedance by referring to Fig. 10-28.

Here, starting at the last minimum, $\theta_t = 180°$ on the SWR = 2 circle, and progressing around the chart toward the load until we come to a 0.3-wavelength radial line, our desired impedance becomes the point where this radius intersects our SWR = 2 circle, or at $Z = 1.57 + j0.70$. Normalizing this figure to 50Ω, we would find that our load, as physically measured, would be $Z = 78.5 + j35$.

Comparing the linear and Smith-chart variations of impedance with θ_t in (1) and (2), we see that at first, going from the minimum point, the imaginary component is capacitive (negative) and does not affect the maximum voltage until it has gone enough positive (reactive) to subtract from the resistive voltage component, which by that time is also decreasing.

For SWR's higher than 10 there is possibility of measurement error, because probe coupling must be increased for good voltage minimum reading. This can produce deformation (and consequent errors) of the field pattern at the maximum reading, plus the fact that detector characteristics change at higher levels. If errors occur, we use either the double-minimum-power or the calibrated-

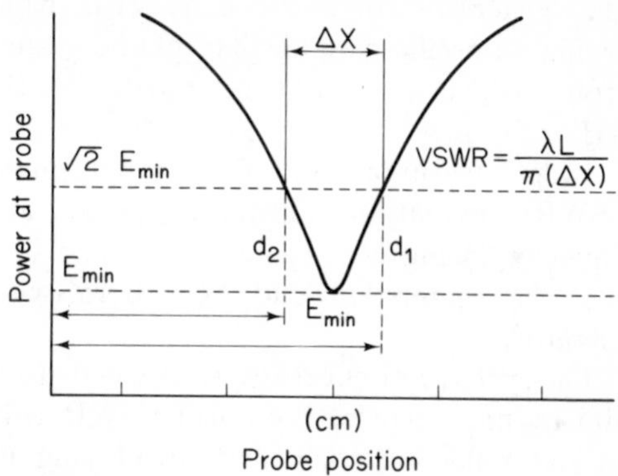

Fig. 10-29 DOUBLE MINIMUM IMPEDANCE LINE

attenuator method. The latter, for safety, attenuates the maximum-voltage level when comparing with minimum voltage readings.

In operating the double minimum power system, we measure the electrical distance (in wavelength) between double power points on each side of the voltage minimum (Fig. 10-29), after first determining the effective waveguide length. Then

$$\sigma_2 = \frac{\lambda_g}{\pi(d_1 - d_2)} = \frac{\lambda_g}{\pi \Delta x}$$

where σ_2 = SWR of load

λ_g = guide wavelength (in centimeters)

Δx = distance between twice minimum power points

Determining the guide wavelength consists of measuring the distance between successive minima when the slotted section is short circuited and doubling this distance. These distances are not the d_1 and d_2 used in measuring the double minimum power points.

SWEPT-FREQUENCY MEASUREMENTS

A swept-frequency microwave measurement system is a moving, dynamic method of periodically varying, say at 60 Hz, the input-signal generator frequency of a setup while viewing an oscilloscope display of the system output, which is also being varied at the 60-Hz rate.

Swept measurements are commonly made of SWR attenuation, impedance, power, and frequency, using the basic setup described below, and centered around the modern sweep generator.

The system, in effect, gives a speeded-up, point-by-point plot of the output SWR voltage and reflection coefficient, displaying instantaneously all maxima and minima across the band being swept. The resulting scope display is an envelope or smear of all the output variations where we obtain a final measurement by observing and calibrating the point of minimum envelope height.

The key function in microwave sweep-generator operation is designed around a system known as output-signal leveling. This development in generator technology stemmed from the backward-wave oscillator tube, the PIN diode applied to attenuator design, and finally to the application of these two in conjunction with an automatic leveling circuit (ALC).

This system is particularly aimed at producing constant generator output over a wide frequency range; the procedure is particularly necessary in the reflection-sensitive area of microwave transmission, where it is necessary to reduce power peaks, cancel source mismatch, and preserve uniform output over flat frequency response. All these items, incidentally, eliminate the use of a ratio meter when making reflectometer measurements.

Figure 10-30 shows a skeletonized operation of ALC in conjunction with a sweep oscillator; the system utilizes feedback of detector voltages produced by reflections present in a directional coupler. These feedback voltages are passed through a leveling amplifier and apply to the BWO–PIN attenuator combination. The loop cancels reflections generated in the directional coupler and can be set at a predetermined power level for proper display and calibration amplitude; so output signal is constant, which eliminates, as we noted, the need of a ratio meter in a reflectometer measurement.

For a slotted-line VSWR measurement we would use a setup similar to Fig. 10-31(a). After adjustment for leveled and calibrated output, the typical envelope display would appear as in Fig. 10-31(b), where the VSWR at A-A is 1.06 and at B-B is 1.06. Slotted-line measurements may be subject to some accuracy limitations, owing to the residual SWR and probe-penetration problems discussed previously. In these cases generator output filters are used.

Reflectometer impedance measurements can deliver calculated SWR's and are simpler in some respects than slotted-line methods. A typical impedance-measuring setup is sketched out in Fig. 10-32 ; here one of the crystals in the dual directional coupler is used

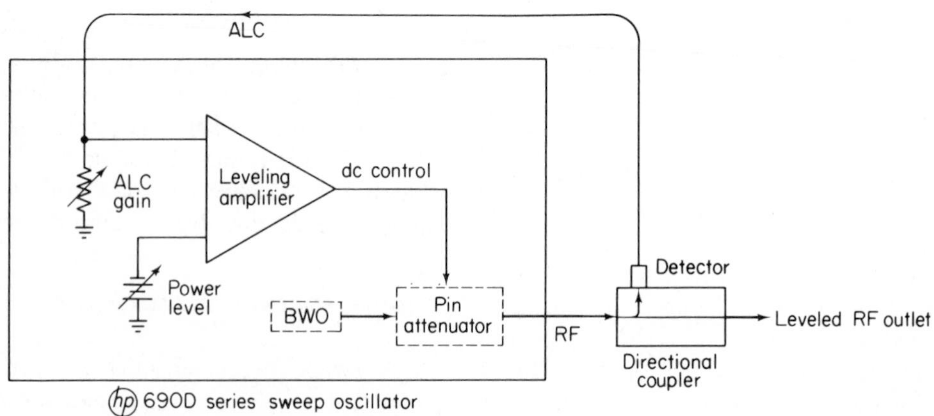

Fig. 10-30 ALC IN SWEPT FREQUENCY MEASUREMENT

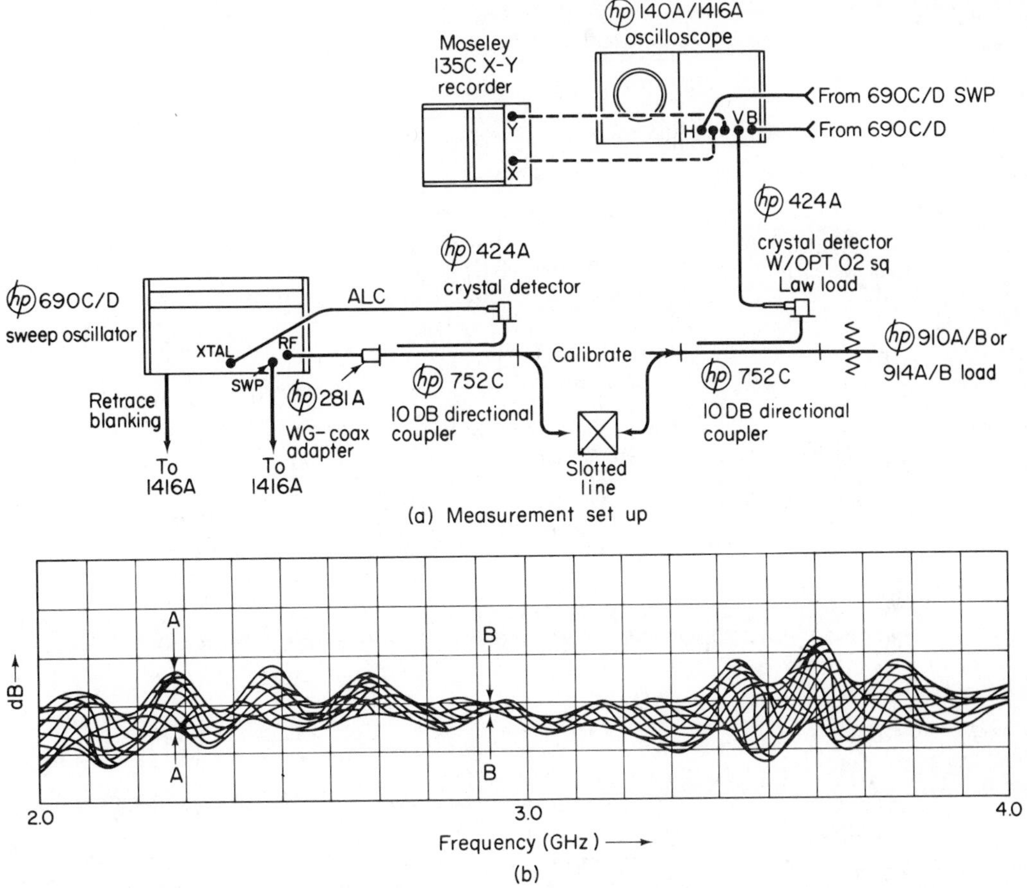

Fig. 10-31 SWEPT FREQUENCY MEASUREMENT OF VSWR (Courtesy of Hewlett-Packard Co.)

as a source for the ALC feedback voltages. Note two other provisions: (a) a coax-waveguide adaptor is necessary to fit the generator output to the directional coupler input, and (2) a calibration load and sliding short circuit are included to adjust the sweeper power level and the scope vertical gain and position settings for compatible deflection and output levels. The sliding short is used to determine 100 per cent reflection from which return loss may be measured.

Reflectometer outputs can be rectified while sweeping the band being measured, and an attenuator–SWR indicator combination inserted (Fig. 10-33). This enables us to pre-insert specific values of return loss in the reverse arm detector and to use an *XY* recorder for plotting differences in mismatch error signals. This setup is also used for determining dielectric constants.

The obvious steps appear from the above outline when measuring attenuation and determining frequency; insertion of either or both types of meters in a test setup will yield the significant output variations—level differences for attenuation changes, plus output dips for frequency coincidence.

Low-power swept-frequency measurements are in some cases preferred to fixed-frequency methods, commonly used at high outputs. We find particular adaptation in amplifier gain measurements, antenna frequency response, and design factors necessary in thermistor mount insertion. These will be briefly mentioned in Chapter 11, where power measurements are discussed in more detail.

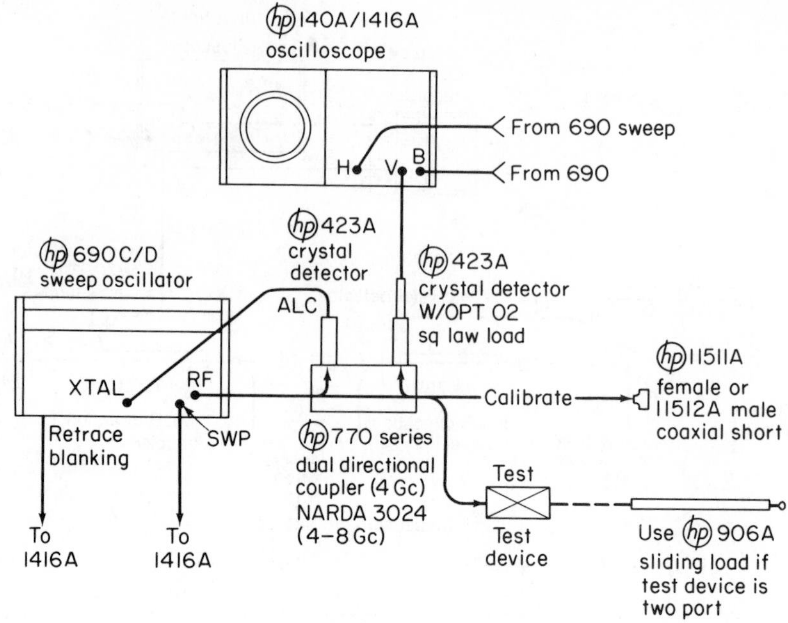

Fig. 10-32 SWEPT REFLECTOMETER MEASUREMENT (Courtesy of Hewlett-Packard Co.)

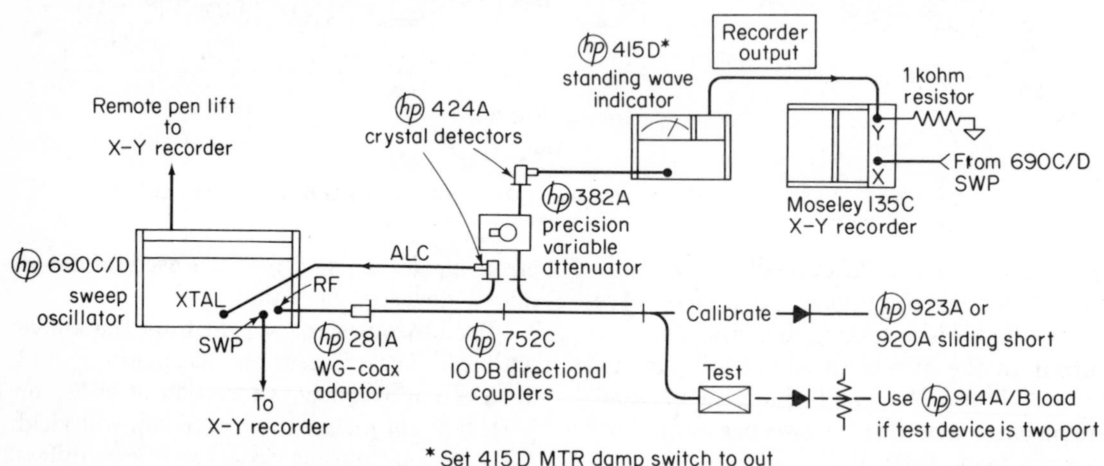

Fig. 10-33 RF PREINSERTION IN SWEPT FREQUENCY MEASUREMENT (Courtesy of Hewlett-Packard Co.)

Microwave Power and Noise Measurements 11

POWER

GENERAL

Power and noise deal fundamentally with extremes of energy, the former mostly in the wattage region from milliwatts to kilowatts, and noise in microwatts.

Power measurements are fundamentally needed in generating and transmitting devices; secondarily, we use them in loss measurements in coaxial lines and waveguides where sensitivity is sometimes at a premium and where, consequently, we call for sensitive power- and signal-detecting devices.

Thus our treatment first deals with high-powered units and then with very sensitive devices, say 10 mW and less. In the former, direct voltmeter power-proportional readings are mostly used, plus calorimetry systems in the highest power ranges. Low-powered units are functionally either heat sensing or calorimetric.

HIGH-POWER WATTMETERS

Power wattmeters from 2 W to 2 kW are inherently direct-reading absorption-type devices where rf wattage readings are delivered as a measured voltage across a known, completely dissipative impedance. Higher power units, 2 to 50 kW, use calorimetry described below.

Load resistors and termination (mostly 50 Ω) are used in these systems illustrated in Fig. 11-1(a). Table 11-1 lists a wide coverage of *Termaline* load-resistor types. Note that models up to 5 kW are air cooled with liquid and air dielectric.

A typical wattmeter that includes the load termination is seen in Fig. 11-2(a). Figure 11-2(b) shows a calorimeter-type meter and the associated water-cooled load. Figure 11-3 illustrates two-directional wattmeters which simultaneously measure forward and reflected power; these instruments are convenient and useful in monitoring power efficiency while, say, a transmitter is operating. Direct reading of forward and reflected power also delivers the SWR figure.

CALORIMETRY

Calorimetric wattmeters are systems in which final power output utilizes the rate of temperature rise in a thermally isolated body of material of known heat capacity, and are these categorized as dry types or flow types. In the former case we may use dry-loaded calorimeters, which consist of a coaxial line filled with a high-loss dielectric. The second type uses circulating water, oil, or ammonia gas as the dissipative agent. Indirectly it is common practice to sense temperature rise through a resistive load immersed in the calorimetric fluid.

In such circulating systems we may calculate power from equation (1). Here the calorimetric fluid is caused to flow so that power (the rate of energy) dissipated P by the volumetric flow rate F, the specific heat of the

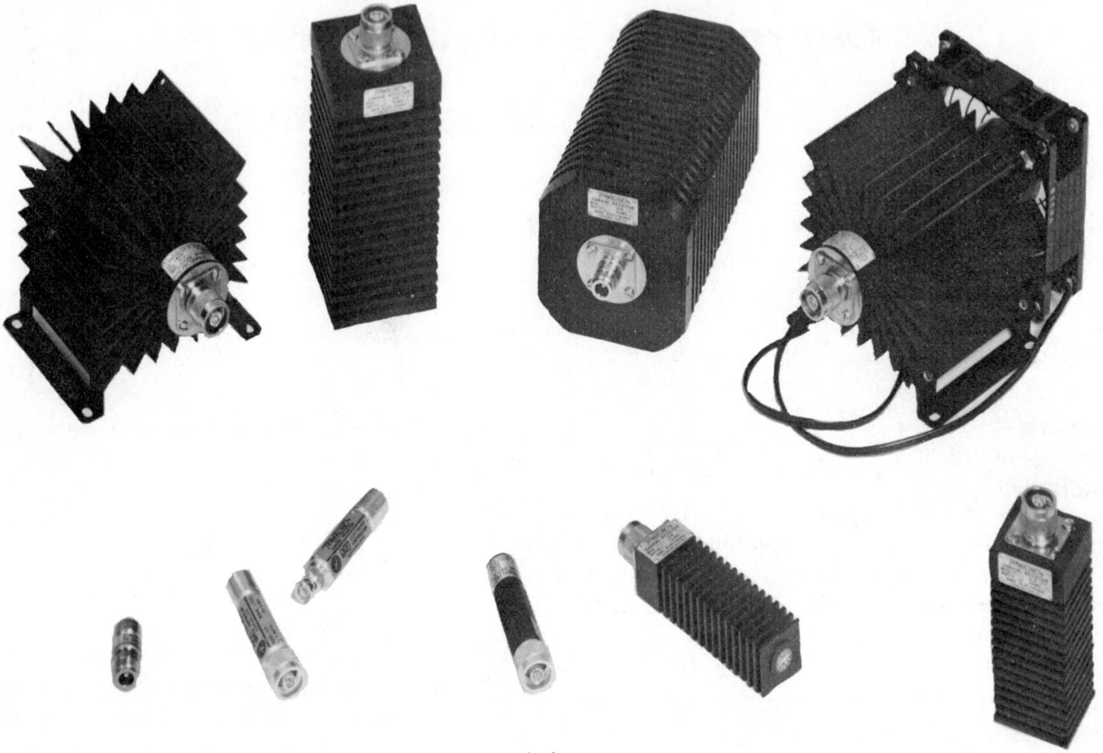

(a)

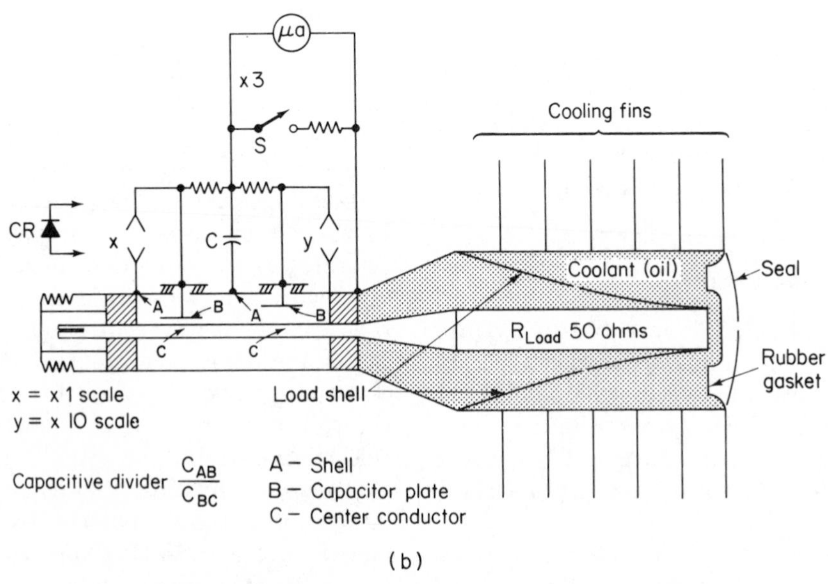

(b)

Fig. 11-1 LOAD RESISTORS AND TERMINATIONS (Courtesy of Bird Electronics, Inc.)

TABLE 11-1 TERMALINE LOAD RESISTORS

Max. Power	*Frequency Range*	*Max. SWR*	*Input Connector*	*Model*
	AIR COOLED			
5w	0-4kmc	1.2	N/M or F	80
5w	0-4kmc	1.2	C, BNC, TNC, M or F	80
20w	0-2kmc	1.2	N/F	80A
50w	0-4kmc	1.2	QC$^{(1)}$	8130
80w	0-4kmc	1.2	N/F	81B
150w	0-4kmc	1.2	QC$^{(1)}$	8135
500w	0-2.5kmc	1.25	QC$^{(1)}$	8201
1000w	0-1kmc	1.1	QC$^{(2)}$	8251
1000w	0-2.5kmc	1.25	QC$^{(2)}$	8833
1000w	0-2kmc	1.25	$1\frac{5}{8}$" EIA Flag.	8813
1200w	0-1kmc	1.1	$3\frac{1}{8}$" Unflg.	887**
1200w	0-2kmc	1.25	$3\frac{1}{8}$" EIA Flg.	888
2500w 5000w*	0-2kmc	1.25	QC$^{(2)}$	8890
2500w 5000w*	0-1kmc	1.1	$3\frac{1}{8}$" EIA Flg.	8891
2500w 5000w*	0-1kmc	1.1	$1\frac{5}{8}$" EIA Flg.	8892
	WATER COOLED			
2500w	0-2.5kmc	1.25	QC$^{(2)}$	8230
5000w	0-2.5kmc	1.25	QC$^{(2)}$	8246
7.5kw	0-500mc	1.3	$3\frac{1}{8}$" Unflg.	8781
7.5kw	0-500mc	1.3	$3\frac{1}{8}$" EIA Flg.	8783
15kw	0-500mc	1.1	$3\frac{1}{8}$" EIA Flg.	8740
25kw	0-500mc	1.1	$3\frac{1}{8}$" EIA Flg.	8750
25kw	0-500mc	1.3	$3\frac{1}{8}$" Unflg.	502**
25kw	0-500mc	1.3	$3\frac{1}{8}$" EIA Flg.	5025
50kw	0-500mc	1.3	$6\frac{1}{8}$" Unflg.	890**
50kw	0-500mc	1.3	$6\frac{1}{8}$" EIA Flg.	8903
	ATTENUATORS (AIR COOLED) 30 dB or 40 dB			
500w	0-1kmc	1.1	QC$^{(1)}$in, $^{(1)}$ out	8325
2000w 4000w*	0-1kmc	1.1	QC$^{(2)}$in, $^{(1)}$ out	8329

* Continuous power rating with BA-88 Blower Accessory　　**51.5 ohms.

Fig. 11-2 TYPICAL COMMERCIAL WATTMETERS (Courtesy of Bird Electronics, Inc.)

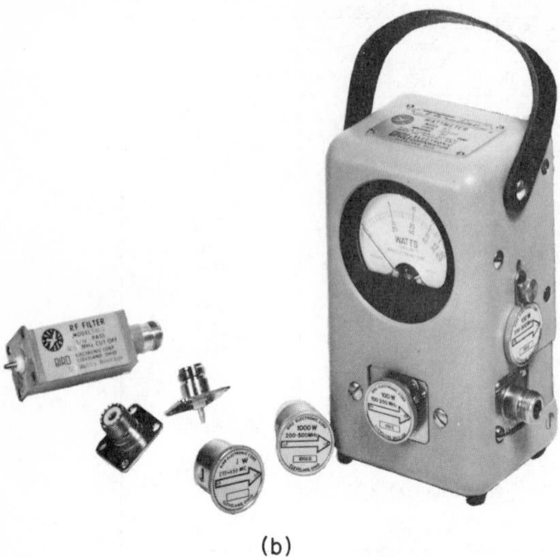

(b)

(a)

Fig. 11-3 TWO-DIRECTIONAL WATTMETERS (Courtesy of Bird Electronics, Inc.)

fluid C, its specific gravity S, and the fluid's temperature rise T can be expressed as

$$P = F \times C \times S \times T \qquad (1)$$

The basic setup for such types of measurement is shown in Fig. 11-4.

Most calorimetric meters use substitution methods for final power readout Fig. 11-4(a). 11-4(b) illustrates and schematically describes an R.F. power bridge using a balanced barretter to sense the rt input; note the comparison system used in Fig. 11-4(c).

Although direct calorimetric measurements are inherently high-power operations, precise efficient meters covering the range from 10 mW to 10 W are widely used. The HP Model 434-A is an excellent example of these types and is shown in Fig. 11-5(c). This meter uses a self-balancing bridge which has identical temperature-sensitive gauges (one in each leg), a high-gain amplifier system, an indicating meter, and two load resistors—one for known input and one for comparison power. The input sensing resistor and one gauge are located for good heat transfer close to the input so that heat energy is carried to its gauge by the oil stream, which unbalances the bridge as a start for the initial step in the measurement cycle. See Fig. 11-5(a) (b).

The unbalanced signal created is amplified and applied to the comparison load resistor, which is located near (thermally close) the other gauge. The heat generated in the comparison load resistor is carried to its gauge by the oil stream which automatically rebalances the bridge.

The panel indicating meter thus measures power supplied to the comparison load needed to rebalance the bridge, this is equivalent to the rf power applied to the input load.

This system has good accuracy because the resistance characteristics of the temperature gauges are the same as are the heat-transfer characteristics of the load; thus the meter can be calibrated directly in input power.

Furthermore, the volumetric flow of oil through the two heads is identical, since all elements in the oil flow system are in series. Differential temperature between oil entering the two heads is eliminated by bringing the oil in each head to the same temperature, by passing it through a parallel-flow heat exchanger.

Thus microwave power dissipated in the rf load resistor is matched by dc heat-produced power in the comparison head, resulting in an output reading.

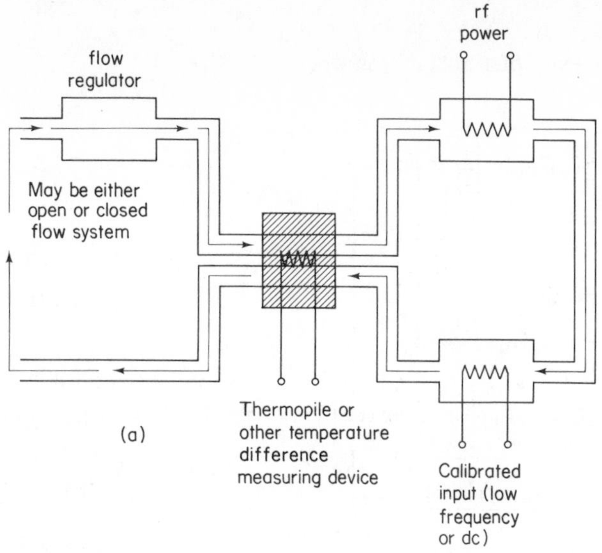

rf
power
flow
regulator
May be either
open or closed
flow system
Thermopile or
other temperature
difference
measuring device
(a)
Calibrated
input (low
frequency
or dc)

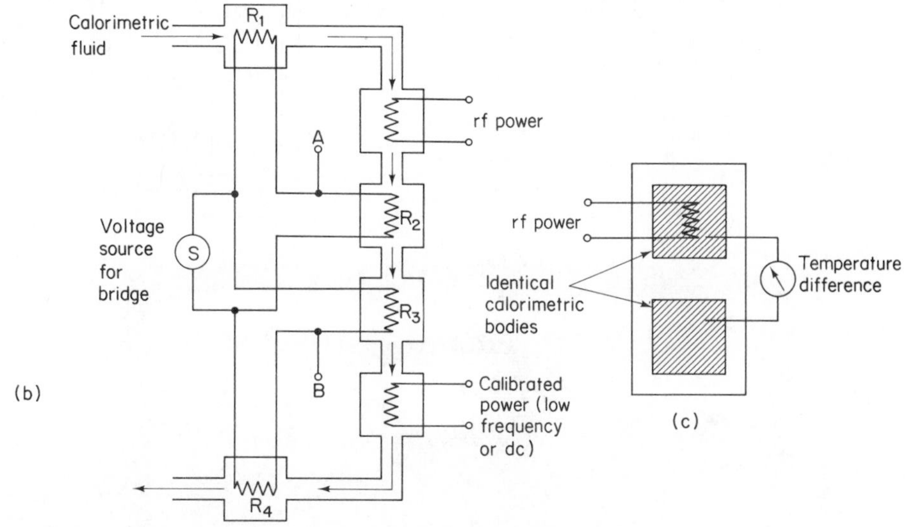

Calorimetric
fluid
R_1
rf power
A
Voltage
source
for
bridge
S
R_2
R_3
B
(b)
Calibrated
power (low
frequency
or dc)
R_4
rf power
Identical
calorimetric
bodies
Temperature
difference
(c)

narda Microline
TEMPERATURE COMPENSATED
PRECISION RF POWER BRIDGE
MODEL 445
POWER
OFF
ON
INPUT
MOUNT
RESISTANCE
200 Ω
100 Ω
R
PRESET
THERMISTOR
OPERATING
RESISTANCE
VERNIER
POWER = $\frac{1}{4R}$ [2Vb VΔ−VΔ²]
VΔ
Va
Vb
VΔ = Vb−Va
RF BRIDGE
REFERENCE BRIDGE
GND

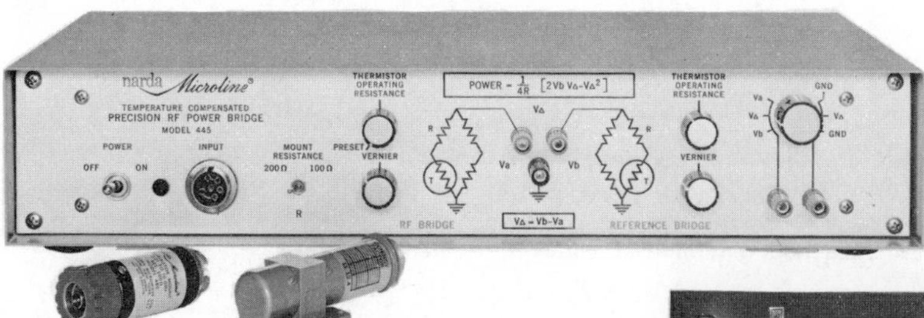

(d) RF Power Bridge and mounts

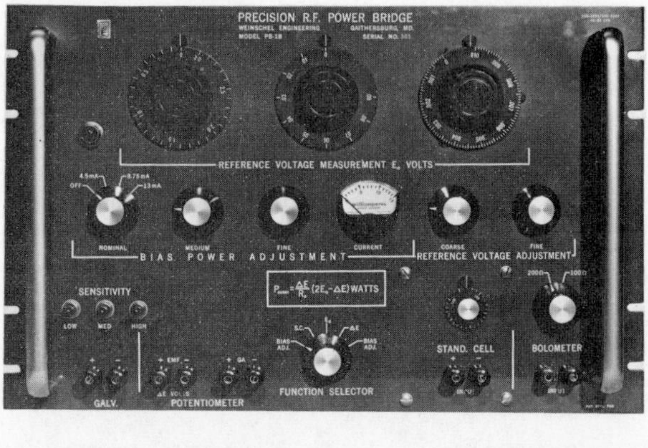

PRECISION R.F. POWER BRIDGE
WEINSCHEL ENGINEERING
GAITHERSBURG, MD.
MODEL PB-1B
REFERENCE VOLTAGE MEASUREMENT E_r VOLTS
OFF
4.5mA
8.75mA
13mA
NOMINAL
MEDIUM
FINE
CURRENT
COARSE
FINE
BIAS POWER ADJUSTMENT
REFERENCE VOLTAGE ADJUSTMENT
SENSITIVITY
LOW
MED
HIGH
$P = \frac{\Delta E}{R_c}(2E_r - \Delta E)$ WATTS
200Ω
100Ω
FUNCTION SELECTOR
STAND. CELL
BOLOMETER
GALV.
POTENTIOMETER

Fig. 11-4 BRIDGE-POWER SENSING CIRCUITS AND RF POWER BRIDGE (Courtesy of Narda Microline and Weinschel Engineering Co.)

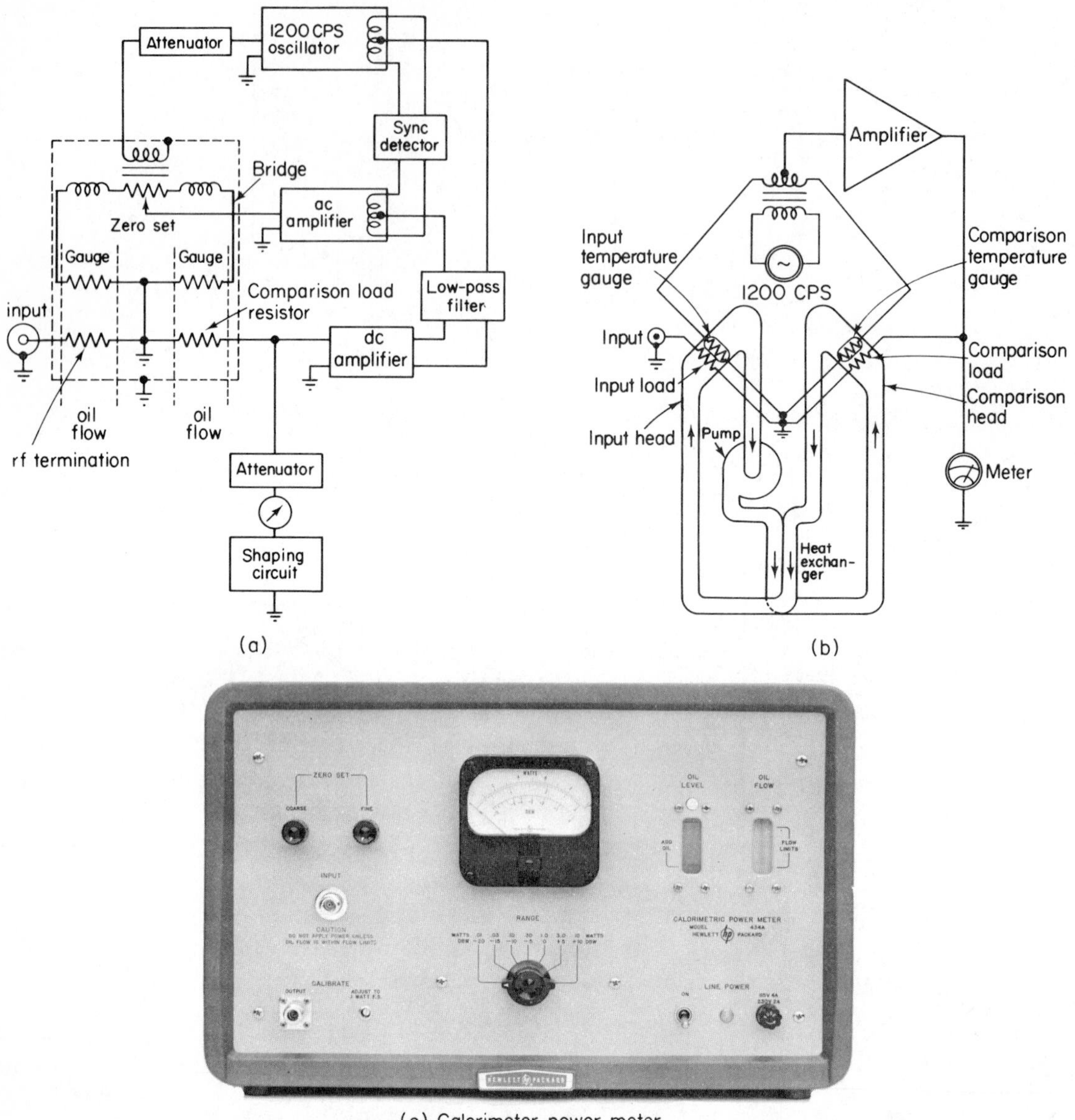

(a) (b)

(c) Calorimeter power meter

Fig. 11-5 CALORIMETRIC WATTMETER (Courtesy of Hewlett-Packard Co.)

LOW-POWER-RANGE WATTMETERS

These specifically deal with micropower resistive sensing devices, say up to 10-mW average power. They are categorized under the type of sensor being used: (1) bolometers, either barreters or thermistors, and (2) thermocouples.

BOLOMETERS. As noted in Chapter 10 the bolometer may be a pure resistive positive-coefficient element (the barreter) or the thermistor, a negative-coefficient semiconductor. The principle of operation is based on the electronic processing of power dissipated in a resistive sensing element, causing a corresponding change in the element's resistance.

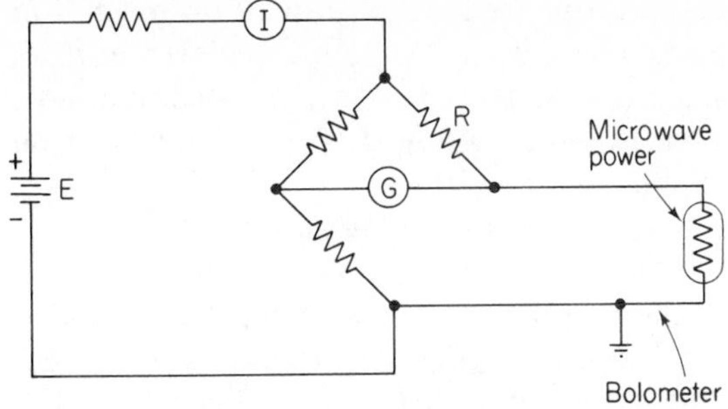

(a) Basic schematic

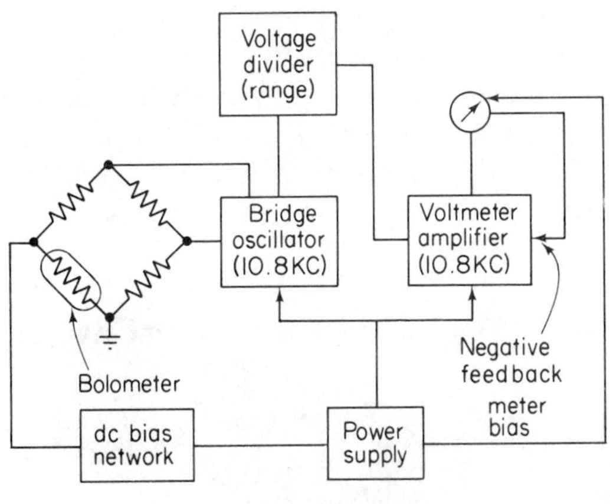

(b) Balancing system

Fig. 11-6 AUTOMATIC BRIDGE SETUP

By proper construction the heating effect of direct current on an element can be made to equal its microwave power dissipation.

A simple Wheatstone-bridge arrangement is shown in Fig. 11-6(a). This circuit operates on a battery–galvanometer combination with resistances so chosen that dc battery operation produces normal balance. When microwave power is applied to the sensing element, its resistance changes, the bridge becomes unbalanced, and balance is restored by decreasing the battery power, the amount of decrease becoming a direct measure of the rf power.

Refinements shown in Fig. 11-6(b) are used to eliminate tedious adjustments where the bolometer element is brought to its predetermined operating resistance in the absence of rf power by simultaneously applying dc and af power. In operation the circuit is so arranged that audio power is automatically removed when rf power is applied. The amount of power removed is amplified and displayed on the indicating meter and equals the true power substitution.

Additional refinements appear in the HP Model 430C, where ac substitution and balancing (except null balance for zero adjustment) are automatic. Figure 11-7 illustrates the automatic balancing produced by feedback; here the feedback from the amplifier produced by unbalance is positive for one

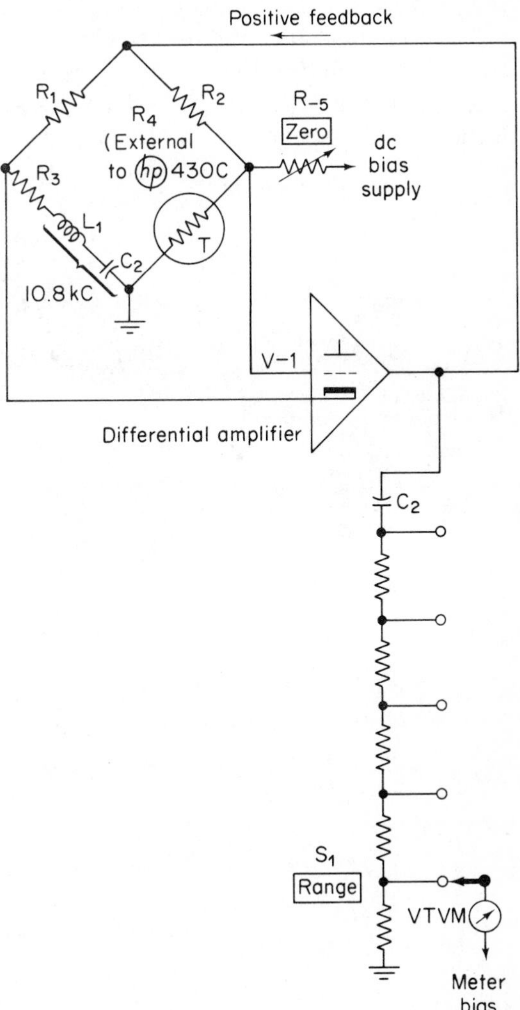

Fig. 11-7 BALANCING SYSTEM IN AUTOMATIC BRIDGE

side of the bridge and negative for the other. The positive feedback is temperature sensitive, depending upon the resistance of the thermistor, so any increase in feedback power, for example, decreases the feedback factor. The negative feedback on the other side of the bridge is, however, frequency sensitive and reaches a minimum value at the resonant frequency of the tuned circuit *L*-1, *C*-1. Now, at the start, when the thermistor is cold and with high resistance, the positive resistance is high; this causes oscillation in the tuned circuit and delivers power to the thermistor, reducing its resistance until the positive feedback becomes smaller and eventually exceeds the negative feedback by a small amount. Note that this shifting of feedback is the critical condition, because both the negative and positive feedback are small (as a result of bridge unbalances) and particularly critical, since the degree by which the positive feedback exceeds the negative is set by the gain of the amplifier circuit. Small differences are thus accompanied by high amplifier gain.

Then, if dc or microwave power is applied to the thermistor, its resistance decreases and the positive feedback is automatically decreased, which causes *L–C* oscillations to decrease by an amount exactly equal to the input power and thus gives automatic balance.

Power readings on the VTVM scale must necessarily be set to a known reference condition compatible with the audio power and applied rf. This is done by applying a suitable amount of dc power from a bias system so that the af power is reduced to a conventional reference value.

Bolometer mounts for either coaxial or

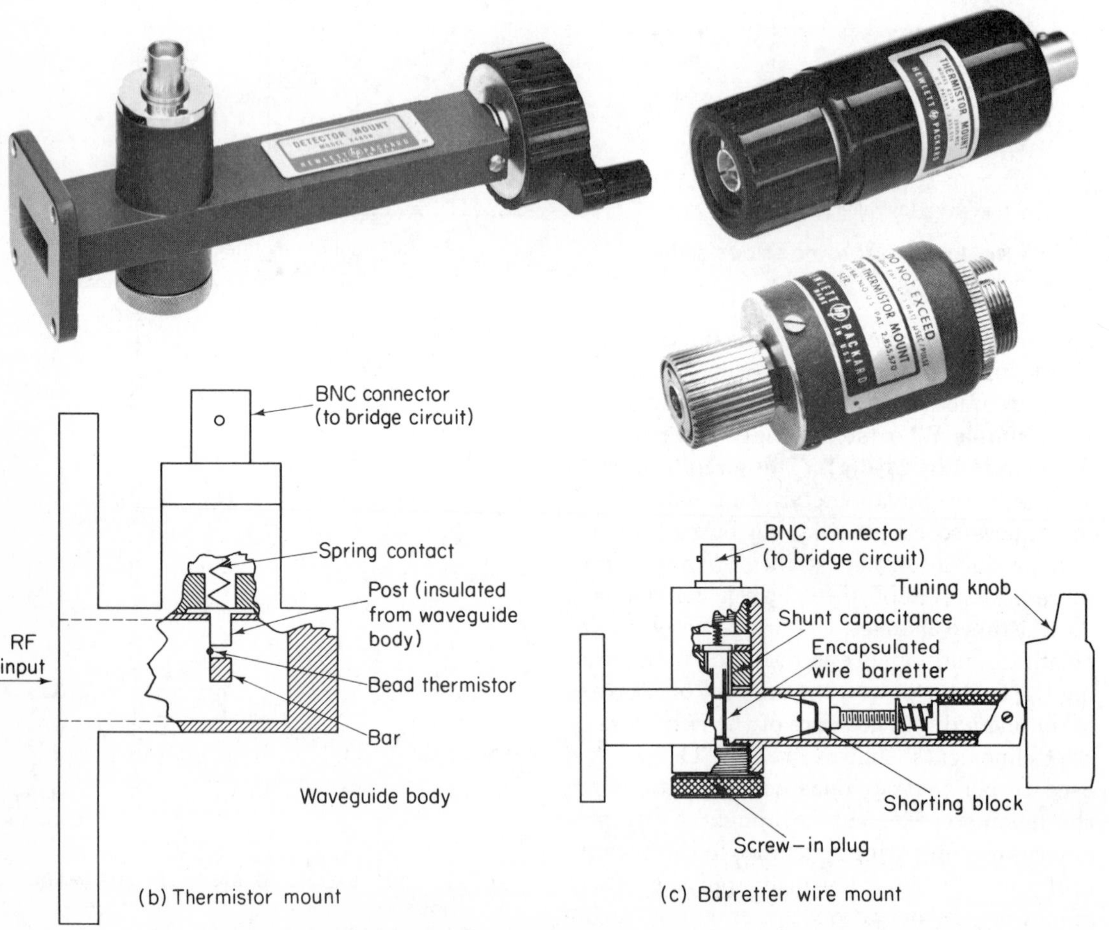

Fig. 11-8 BOLOMETER TYPES AND CONSTRUCTION (Courtesy of Hewlett-Packard Co.)

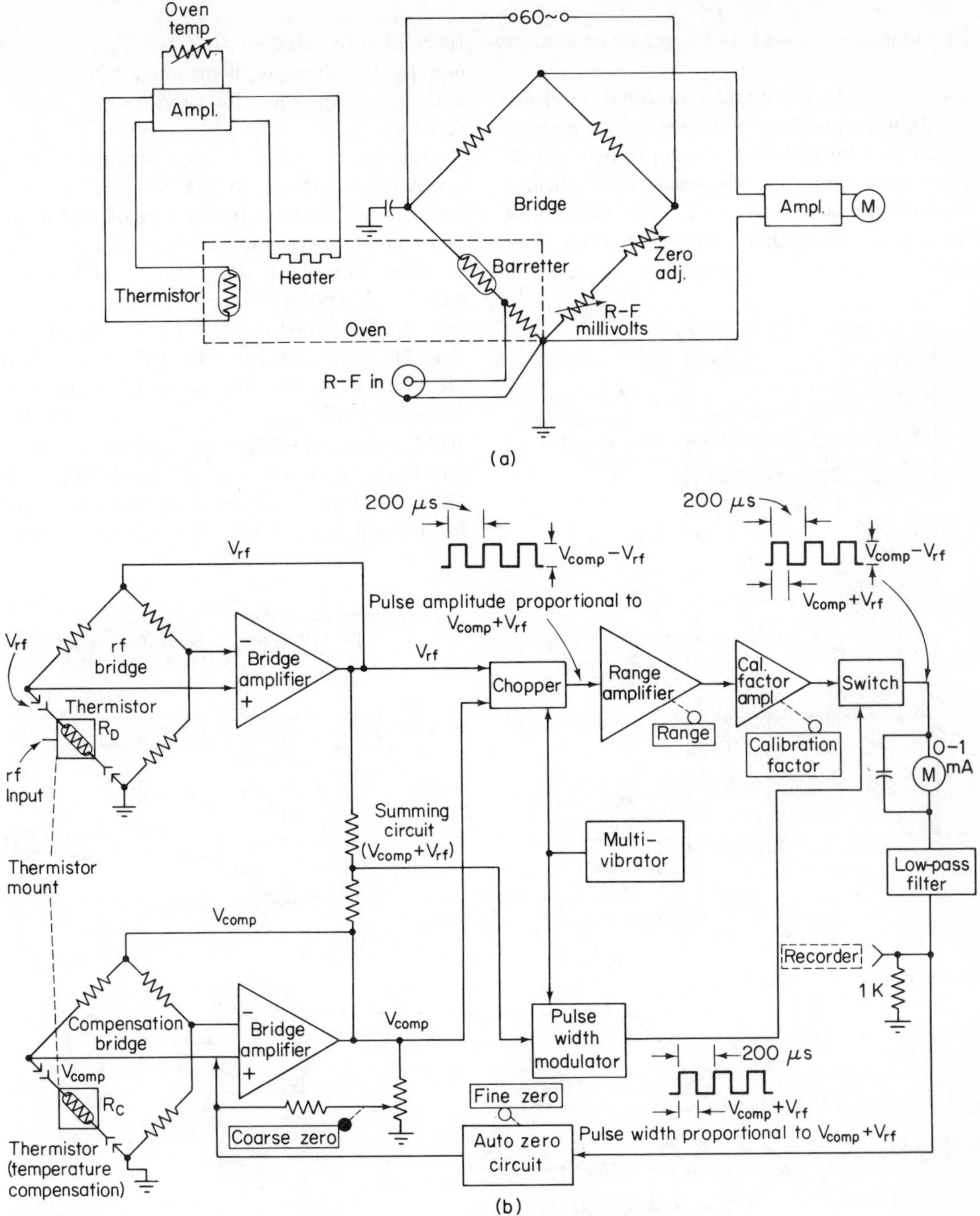

Fig. 11-9 DUAL BOLOMETER BRIDGE CIRCUIT

waveguide structures must be carefully designed so that they will extract maximum power from the transmission signal, present an optimum of impedance mismatch, have low insertion loss, be physically and thermally rugged, and allow a minimum of power leakage around the mounting hardware. Various designs may have fixed tuning, some are tunable, and others have broad-band response. Figure 11-8(a) shows a number of types commonly used with the HP Model 430-C power meter. Figure 11-8(b) shows internal construction of two types.

Thermistor bolometers within themselves

are unable to distinguish ambient environmental temperature changes from applied power and for this reason can produce imbalance variations in high-sensitivity bridge measurements. This calls for thermistor mounts using two thermistor elements in close thermal proximity so that both are affected equally by changes in ambient temperature and can compensate for each other's characteristics.

Likewise, a dual-bridge arrangement is necessary, as shown in Fig. 11-9. Here each element has its own bridge circuit and automatic power meter for measuring items similar to those described above under HP Model 430-C. A typical coaxial arrangement for the dual-thermistor system is shown in Fig. 11-10(a) and (b). In this setup, four identical elements are actually used—two detection units appearing in series with the 10-kHz bias and bridge and mounted in a comparatively massive thermal block [Fig. 11-10(b)], and two compensating units mounted in the cavity associated with the overall struc-

(a)

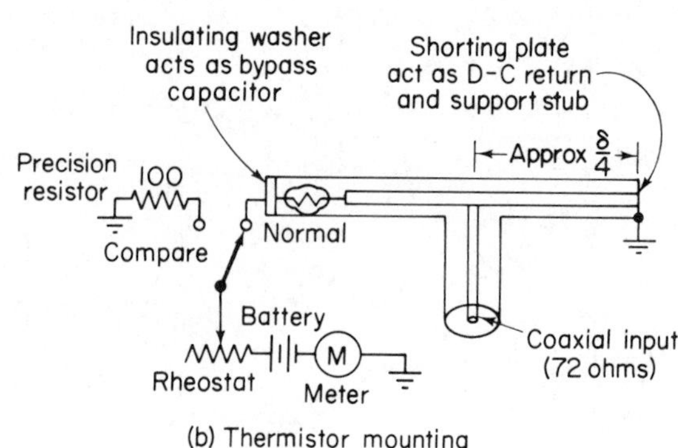

(b) Thermistor mounting

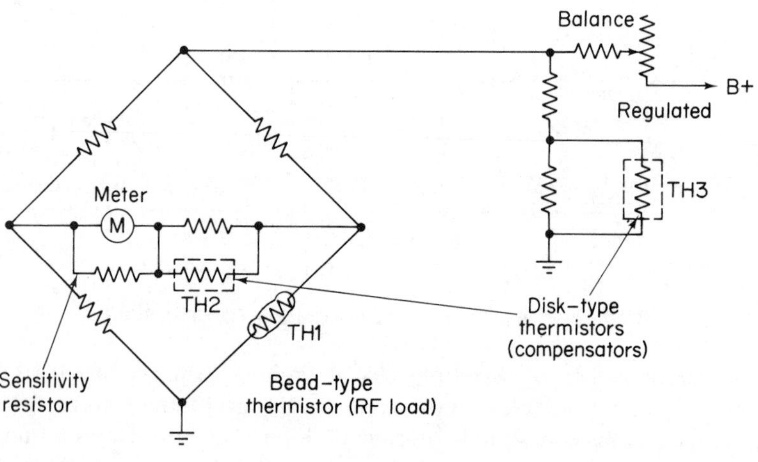

(c) Simplified circuit

Fig. 11-10 COAXIAL MOUNTING FOR BOLOMETER BRIDGE

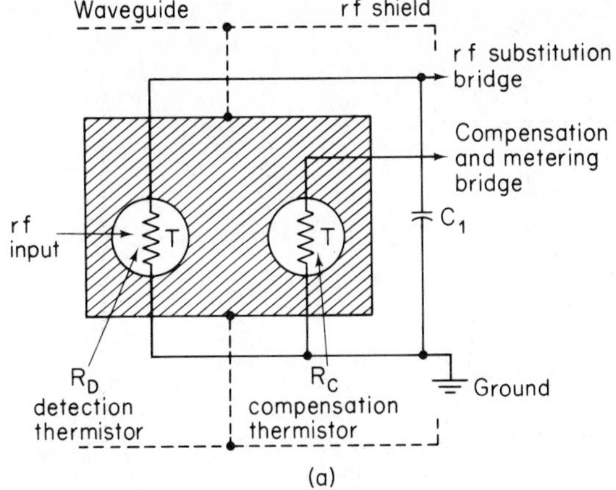

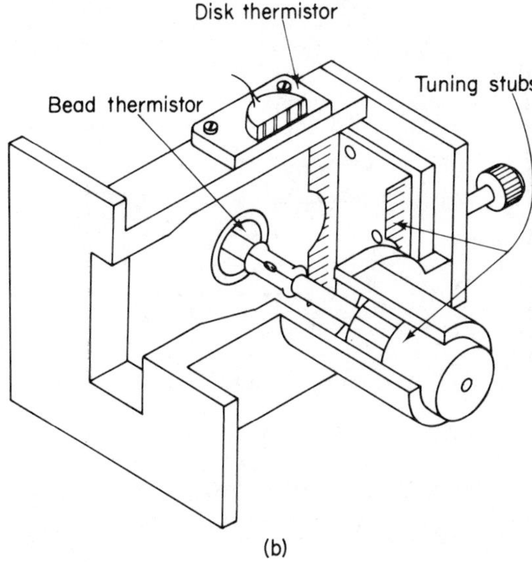

Fig. 11-11 THERMISTOR CIRCUIT AND MOUNT IN WAVEGUIDE (Courtesy of Hewlett-Packard Co.)

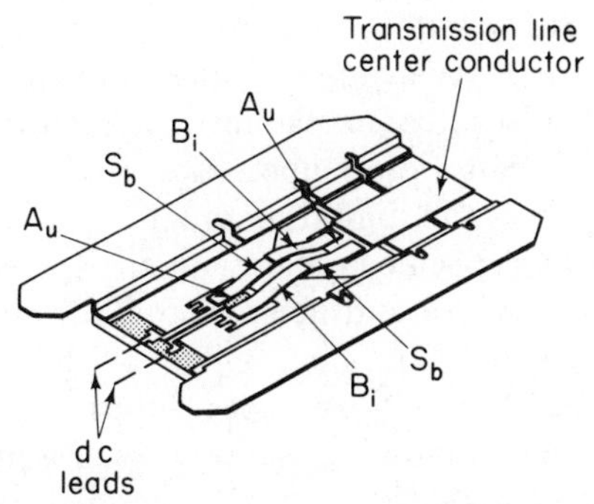

Fig. 11-12 STRIP-JUNCTION THERMOCOUPLE AND MOUNT

ture. In a single-pair waveguide thermistor mount, accuracy and SWR are improved by the post-and-bar arrangement shown in Fig. 11-11(a) and (b).

THERMOCOUPLE POWER METERS. These are essentially fitted for low power ranges comparable to thermistor models; in older high power applications, changes in heater resistance induced mismatch, skin effects produced frequency sensitivity, and susceptibility to burnout made these units critical to manipulate.

New developments at milliwatt power levels eliminate most of these objections and capitalize on the inherent sensitivity of the thermocouple system of sensing, that is, the use of a simple dc millivolt meter to indicate power output.

Thermocouple elements now develop thermal power indication by abutting thin films of antimony and bismuth deposited on thin resistive strips. The resistive strips are part of the series center conductor in a coaxial line and generate the heat necessary to develop the power-proportional thermocouple voltage.

Figure 11-12 illustrates the construction of the strip-junction thermocouple element. Since these units are subject to variations in ambient temperature, overload, and aging, internal-calibration provisions must be included in thermocouple power meters. Figure 11-13(a) is a photograph of this equipment and its accompanying thermocouple mount (b).

Peak power meters are convenience adaptations of fundamental thermocouple–bolometer meters. They may use appropriate circuitry or signal processing to measure the following:

1. Average power—duty cycle usually by calorimeter.
2. Direct-pulse power—scope display manipulation.
3. DC-pulse power comparison.
4. Notch wattmeter readings. The notching process involves gating a reference CW power source "off" during the interval that

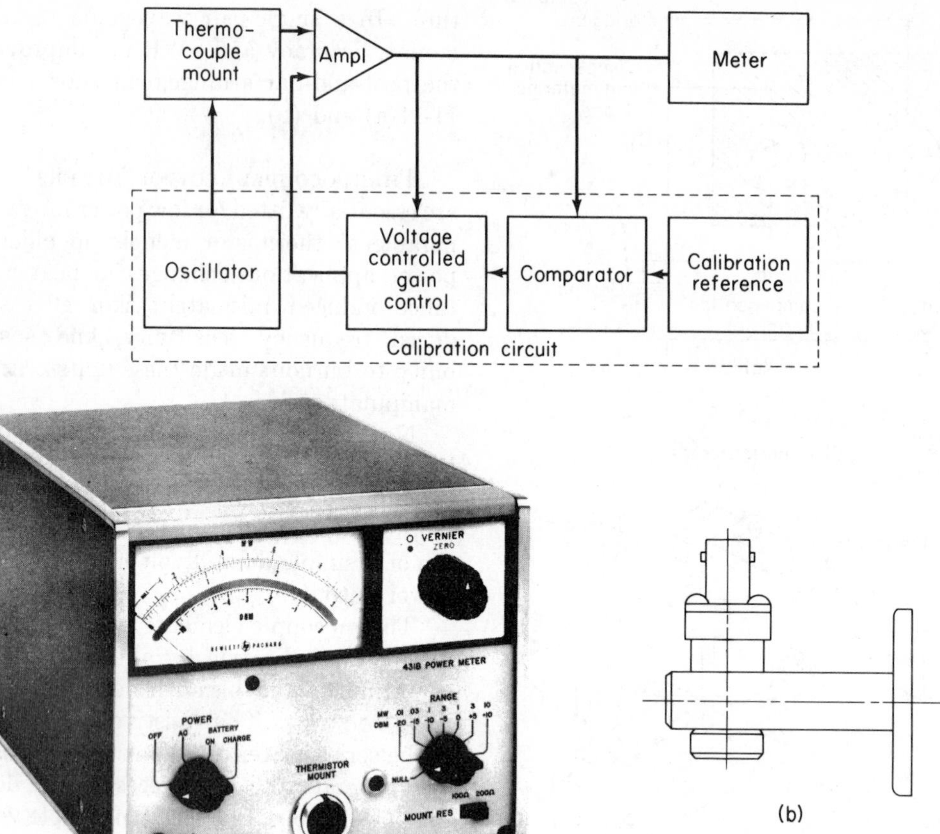

Fig. 11-13 POWER METER AND CALIBRATION CIRCUIT
(Courtesy of Hewlett Packard.)

the pulse source in test is "on." Thus the reference output is "notched" at the same rate, duration, and amplitude as the pulse output of the device in test.

5. Barreter integration–differentiation—circuit modifications necessary as a result of the long thermal time constant of some barreters.

NOISE MEASUREMENT

GENERAL

In microwave receivers three key characteristics are closely linked together and specified for performance. They are overall gain, bandwidth, and noise figure. Built-in amplification, or pure gain, is practically boundless, but the amount that is usable is limited by the inherent noise and necessary bandwidth; too much noise will swamp the information riding upon the modulation or cause errors in digital modulation. Thus gain is always specified along with the corresponding bandwidth and noise factor. Briefly, it is a figure of deficiency telling by how many decibels the self-generated internal receiver noise (contributed in the process of amplification) exceeds the normal thermally existing noise at the receiver input. Here we may have noise coming from the conventional resistive input impedance, say 50 or 100 Ω, reflected by the antenna loading coupled into the receiver terminals, the noise presented by a crystal mixer across the input to an if strip, or noise-producing influences from oscillator injection signals.

NOISE CALCULATIONS

These all stem from temperature conditions, for in any conductor, random electric motion due to thermal agitation produces a voltage, e_n, within the conductor. This motion or power, expressed in watts, is generated by the square of the individually generated electron noise voltages across the generator resistance R_n. Thus

$$P_n = \frac{e_n^2}{R_n}$$

Now, these noise voltages produce output signals containing frequencies randomly distributed throughout the entire rf spectrum. So when amplified by a receiver of a given bandwidth, B, these signals will produce output only in proportion to its bandwidth. Thus, accounting for the absolute temperature, the bandwidth, and the value of the equivalent output resistor, R,

$$e_n^2 = 4kTR_nB$$

where k = Boltzmann's constant ($1.37 \times 10_{-23}$ W-sec/°K)

T = temperature in degrees absolute (Kelvin scale $T = 290°$ at ambient of 20°C)

B = bandwidth (in hertz)

and if the generating resistor is matched to a load, power across it is

$$P_n = kTB$$

This input noise power is amplified in a receiver and picks up noise power along the way—in other resistive amplifier units, in transistor electrodes, etc. So the total noise of an amplifier output consists of two parts, the minimum inherent input noise, N_1, multiplied by the amplifier gain, plus the internally contributed noise power.

If we measure the output of our noisy amplifier, we come up with a combined power, N_1; we then compare it to the minimum input noise power and derive a noise factor:

$$F = \frac{N_1}{kT_0BG} \quad \text{and} \quad N_1 = F(kT_0BG) \qquad (2)$$

But we want an expression for the amount of excess noise added by the receiver itself, N_r. Now, N_r naturally is the difference between the measured noise and the inherent calculated noise minimum, so

$$\begin{aligned} N_r &= N_1 - kTBG \\ &= F(kTBG) - kTBG \\ &= (F - 1)(kTBG) \end{aligned}$$

Inherent receiver noise thus centers around a simple power-meter output measurement, N_1; contributory factors are the determination of the receiver bandwidth and gain (by measurement or specified), plus calculations using specified temperature and Boltzmann's constant.

Measurement of N_1 may be made with a CW noise generator and a conventional power meter; however, if design changes are being made, the bandwidth and gain adjustments for each measurement may be time consuming. The conventional procedure is to connect the noise generator and note the noise power level with the generator turned off. Then turn the noise generator output up until the power output level is double the initial "off" generator output level. The increase in generator power calculated from the increased microvolt level allows us to calculate the noise figure.

BASIC EXCESS POWER NOISE MEASUREMENTS

Differential or excess noise power output measurements using input stimulation by

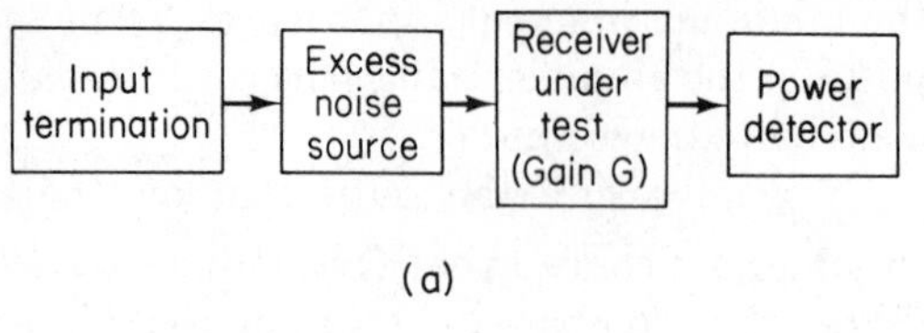

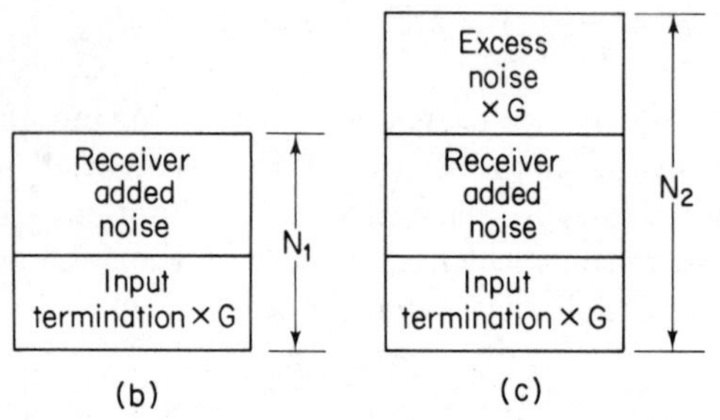

Fig. 11-14 NOISE MEASUREMENT BREAKDOWN

calibrated noise generator sources simplify the whole technique surrounding noise factor measurements. This is because by improving two noise temperature signals (the normal ambient and that of the signal-generator source), gain bandwidth factors, being common, do not enter final calculations. The method is illustrated in Fig. 11-14; it consists simply of taking two measurements: N_1, with the excess-noise source cold, and N_2, with the excess-noise source fired. We can picture the noise readings under these conditions by breaking down the noise components as in Fig. 11-14(b) and (c).

The final ratio, N_2/N_1, is

$$\frac{N_2}{N_1} = \frac{\begin{array}{l}(\text{terminal noise} \times G) \\ + (\text{internal receiver noise}) \\ + (\text{signal gen. excess noise} \times G)\end{array}}{\begin{array}{l}(\text{terminal noise} \times G) \\ + (\text{internal receiver noise})\end{array}} \tag{3}$$

and can be developed using relationships of equation (2), yielding

$$F = \frac{T_2 - T_0}{T_0} \frac{1}{(N_2/N_1) - 1} \tag{4}$$

which converted to logarithmic notation equals

$$F_{\text{dB}} = 10 \log \frac{T_2 - T_0}{T_0} - 10 \log \left(\frac{N_2}{N_1} - 1\right) \tag{5}$$

The noise factor in dB finally equals

$$F_{\text{dB}} = 15.2 - 10 \log \left(\frac{N_2}{N_1} - 1\right) \tag{6}$$

Specific temperature ratios are derived from the temperature of argon-gas tubes typically used in noise-generator equipment. The key ratio stems from the power temperature differences, $(T_2 - T_0)/T_0$, which stems back to the basic noise power $P_n = k(T_2 - T_0)B$. Since the temperature of the argon-gas source is approximately 10,000°K, $(T_2 - T_0)/T_0$ becomes a ratio of 33.3, accounting for the 15.2 dB in equation (3).

DOUBLE-POWER EXCESS-NOISE MEASUREMENT

This measurement uses the conventional setup of Fig. 11-13 and adjusts input so that N_2 is equal to twice N_1. Equation (5) becomes

$$F_{\text{dB}} = 10 \log \left(\frac{T_2 - T_0}{T_0}\right) - 10 \log 1$$

$$= 10 \log \left(\frac{T_2 - T_0}{T_0}\right) \tag{7}$$

Adjusting inputs for $N_2 = 2N_1$, the final noise factor will be 15.2 dB minus the attenuator setting necessary to adjust for $N_2 = 2N_1$.

AUTOMATIC EXCESS-NOISE MEASUREMENTS

As outlined above, we can use the basics of measuring and comparing N_1 and N_2 noise voltages to automatically and continuously measure the noise factor. Figure 11-15(a) illustrates just this sort of equipment employing switched-pulse and signal-comparison processes to deliver a direct value of F on a panel meter.

The device operates according to the skeletonized diagram shown in Fig. 11-15.

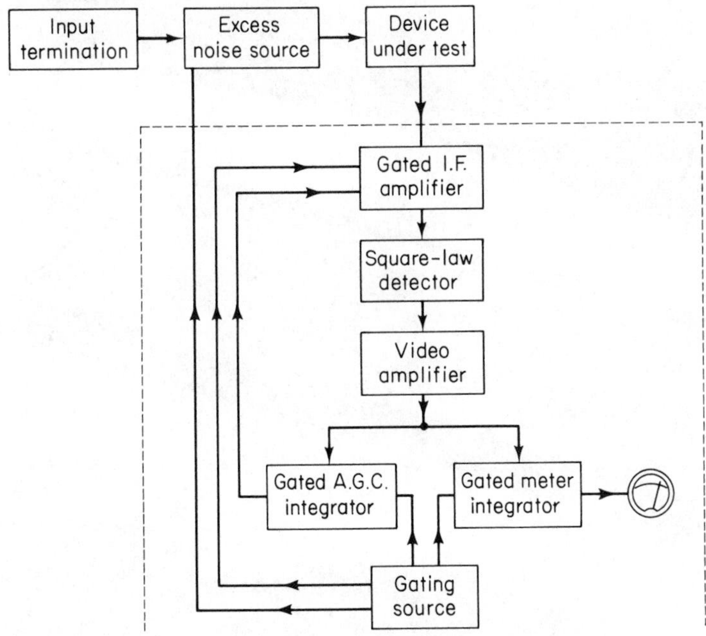

Fig. 11-15 AUTOMATIC EXCESS NOISE MEASUREMENT

Briefly, the device depends upon the periodic insertion of a known excess noise power at the input of a device being tested, amplifying, leveling, and detecting the power levels of two pulse trains (one at N_1 and the other at N_2 level). Measurement of the ratio of these two levels is displayed directly in terms of F_{dB} on the scale of the output meter.

Referring to Fig. 11-15 for more details, the operation consists of

1. Modulating the noise generator on and off with a pulse wave form.
2. Mixing and heterodyning the noise with a basic microwave signal and again heterodyning to a convenient if—say 30 or 60 MHz.
3. Amplifying the noise-modulated 30-MHz amplifier signal.
4. Gating the noise signal if amplifier on and off.
5. Detecting and further amplifying to produce two-level pulse trains.
6. Integrating the pulse train first through an AGC network and then through the meter network.
7. Applying the AGC to the tuned amplifier so that signal leveling is obtained for all conditions.
8. Displaying directly in dB the integrated pulses on a meter movement whose deflection is proportional to the ratio of the noise powers.

ACCESSORY NOISE EQUIPMENT

A wide variety of noise sources is commercially available covering all VHF, UHF, and microwave frequencies, plus some models specifically operating at noise meter if's—30 and 60 MHz.

Figure 11-16 shows three of these noise sources adapted through fittings to coaxial device terminations. All models use the argon tube installed directly in the waveguide.

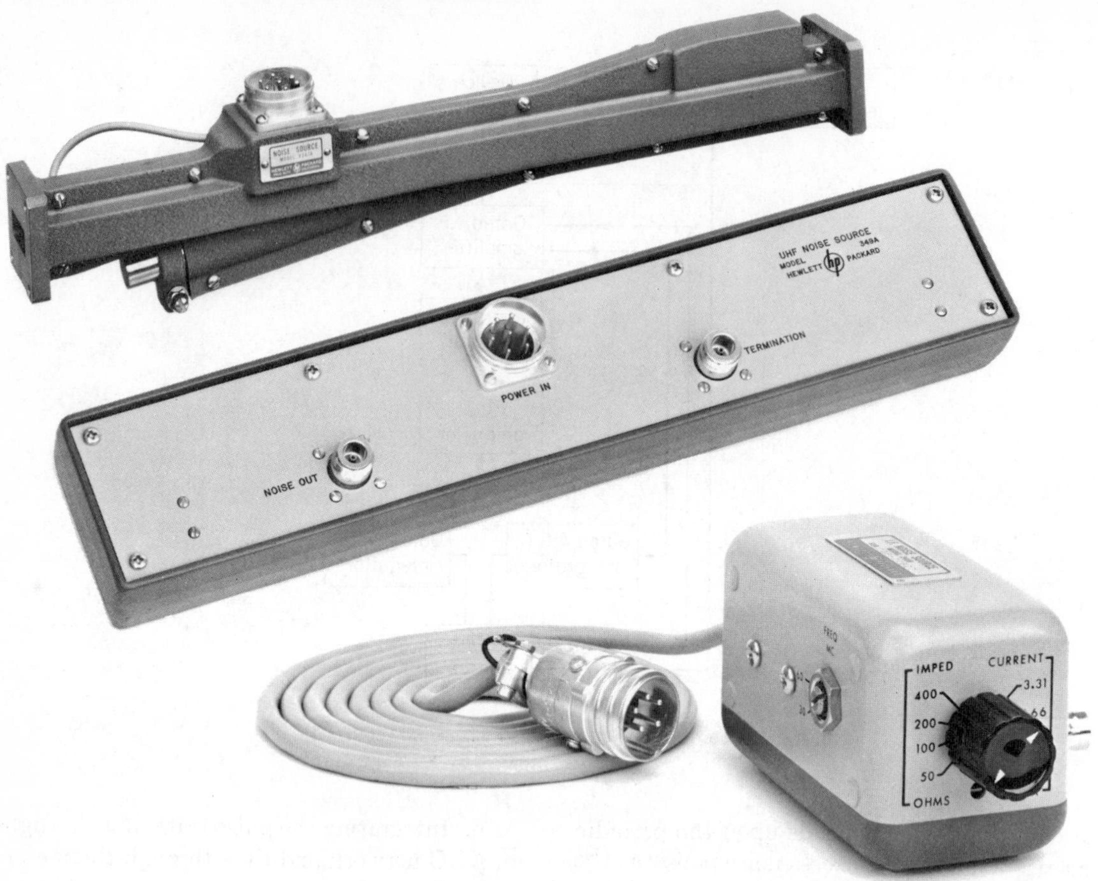

Fig. 11-16 COMMERCIAL NOISE METER AND SOURCES (Courtesy of Hewlett-Packard Co.)

Solid-State Components and Equipment 12

INTRODUCTION

Within a span of ten years the microwave regime has in general paralleled solid-state technological progress. It has been aided chiefly by microminiaturization plus the invention and development of new active and passive devices. In addition, IC construction has made workable unbelievably complex systems and support equipment.

Summarizing, this has all been possible because of four major factors.

1. Reduction in component size with higher frequencies, more minute self-resonances, smaller waveguides, etc.
2. New functional devices. New bulk and molecular-determined principles and modes of operation. Diodes, ferrites, lasers, etc.
3. System design advances in radar, communication, and data processing due to integrated-circuitry improvements and entry into LSI.
4. Wider electronic applications in power devices for industrial heating, biological sciences, instrumentation, space communication, radar, and telemetry.

SOLID-STATE MICROWAVE COMPONENTS

More specifically, the following essentially solid-state devices are inherently peculiar to microwave operation:

1. Oscillators and sources.
2. Amplifiers.
3. Harmonic generators.
4. Mixers.
5. Switches.
6. Delay lines.
7. Attenuators.
8. Couplers.

See also Fig. 12-1.

The discussion and description of these areas are partially combined in the following textual treatment and of necessity may include some overlap within the basic categorizations of 1 and 2 listed above.

CONVENTIONAL OSCILLATORS, AMPLIFIERS, AND HARMONIC GENERATORS

Microwave amplifier and oscillator circuits are constructed alike; except for the Gunn effect, the operating mechanism in a specifically tailored microwave transistor must be wedded mechanically to resonant circuits. In fact, we might schematically consider an amplifier circuit where the transistor is being squeezed into the gap between two 50-Ω lines, as shown in Fig. 12-2. A practical version of this is the 2.45-GHz oscillator using the quarter-wave stub of a coaxial line as one resonant circuit [Fig. 12-3(a) and (b)].

Another tuned-circuit arrangement is seen in Fig. 12-4(a) and (b), where deposited thin-film areas are shown directly integrated with a

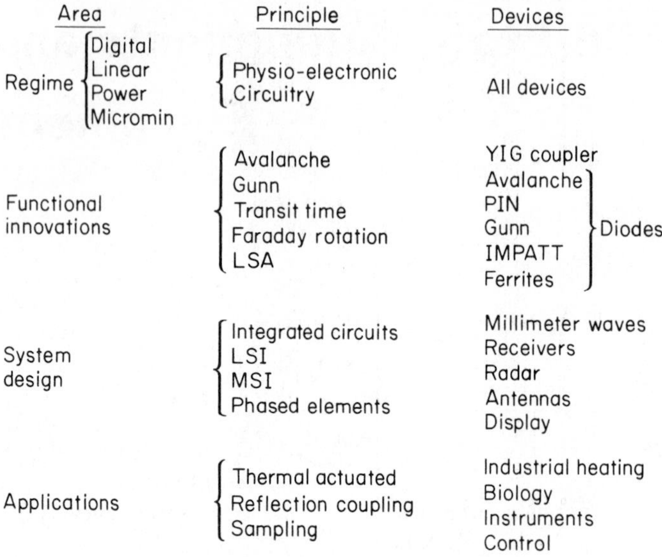

Area
Principle
Devices
Regime
Digital
Linear
Power
Micromin
Physio-electronic
Circuitry
All devices
Functional innovations
Avalanche
Gunn
Transit time
Faraday rotation
LSA
YIG coupler
Avalanche
PIN
Gunn
IMPATT
Ferrites
Diodes
System design
Integrated circuits
LSI
MSI
Phased elements
Millimeter waves
Receivers
Radar
Antennas
Display
Applications
Thermal actuated
Reflection coupling
Sampling
Industrial heating
Biology
Instruments
Control

Fig. 12-1 SOLID-STATE MICROWAVE DEVICES

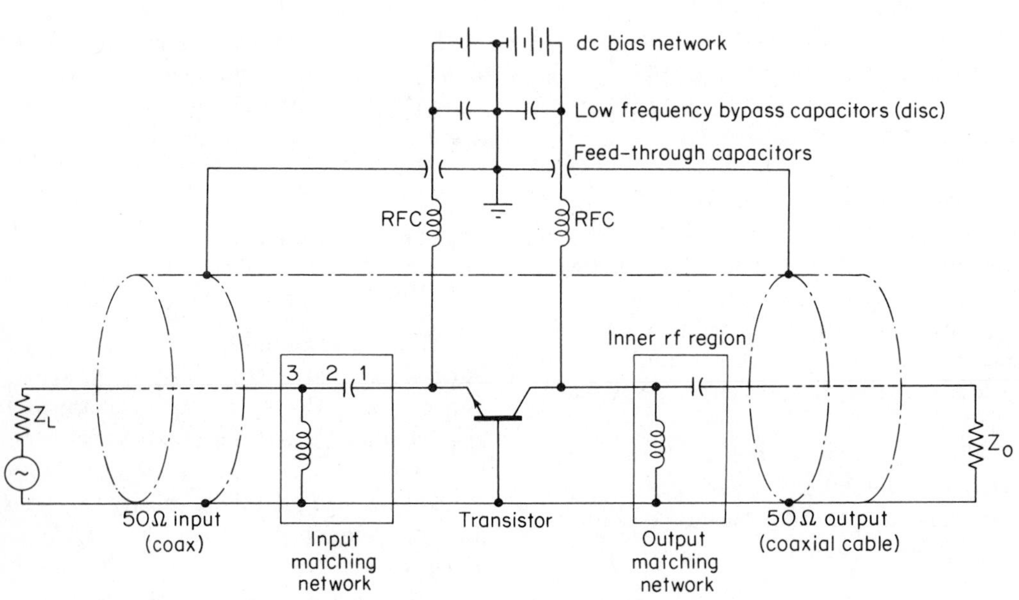

Fig. 12-2 TRANSISTOR COAXIAL LINE AMPLIFIER

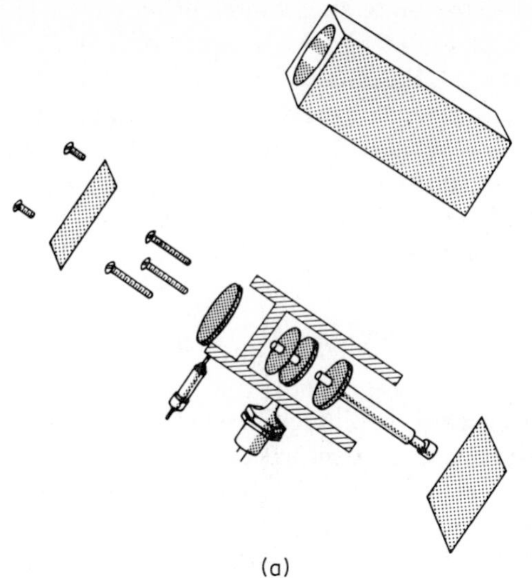
(a)

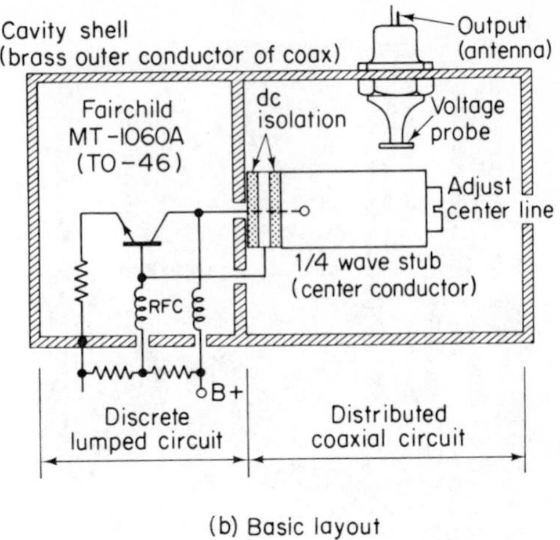

(b) Basic layout

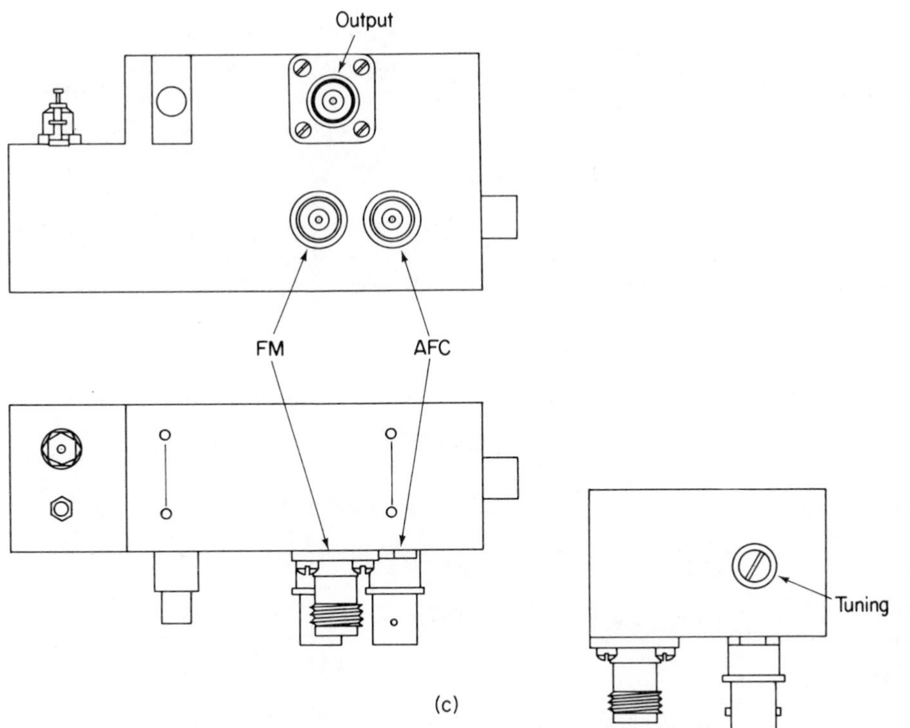

(c)

Fig. 12-3 COAXIAL STUB-TUNED OSCILLATOR STRUCTURE

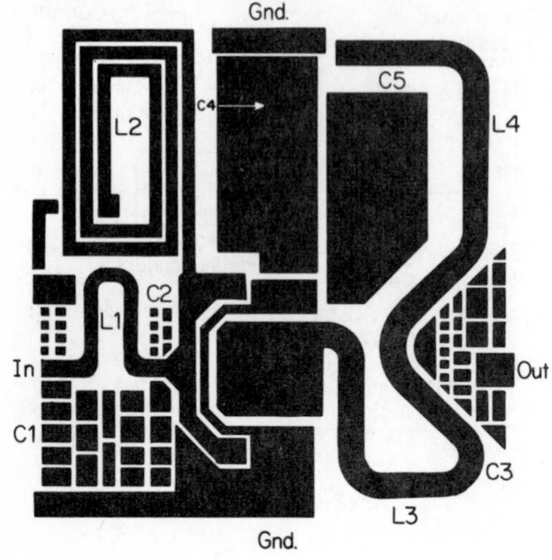

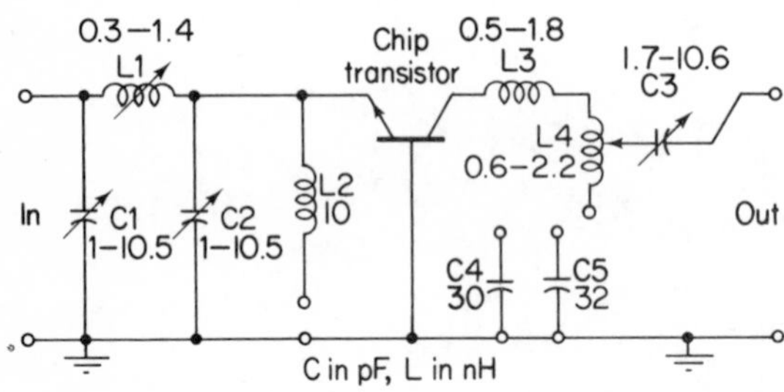

Fig. 12-4 THIN FILM POWER TRANSISTOR AMPLIFIER (Courtesy of RCA)

transistor on a single monolithic chip. This unit has relatively high power output (5 W) in the 2-GHz range.

Transistors used in such GHz assemblies are specifically tailored units with low input and output electrode capacitance and a f_T or cutoff typified in the tabulated units shown in Table 12-1.

VARACTOR SOURCES

The varactor diode is a voltage-variable, back-biased, *PN* junction whose fundamental operation depends upon the electronic control of the capacitance of its specially constructed depletion layer. They thus provide solid-state tuning of circuits hitherto mechanically impossible.

We use varactors for microwave generator sources by capitalizing upon their particularly sharp nonlinear characteristics, which are selected and heavily driven so that they generate pronounced harmonics. In a typical circuit arrangement, successive harmonic multiplication, seen in Fig. 12-5, is obtained by using cascaded varactor doublers which deliver operating power from the last cavity stage.

Physically, with integrated circuit construction and using microstrip-thin-film elements, we would have a unit similar to that shown in Fig. 12-6.

TABLE 12-1 MICROWAVE TRANSISTORS (Courtesy of Raytheon Co.)

L- AND S-BAND POWER TRANSISTORS

Transistor Designation	*Package Type*	*Frequency (Avg.)*	Power *(Min.)*	Gain *(Min.)*	*Efficiency (Typ.)*
LS2501	Stripline	2.0 GHz	1.0 W	5 dB	30%
LS1610	Stripline	1.0 GHz	10.0 W	6 dB	60%
LS1605	Stripline	1.0 GHz	5.0 W	6 dB	60%
LS1604	Stripline	1.0 GHz	4.0 W	6 dB	60%
LS1602	Stripline	1.0 GHz	2.0 W	6 dB	50%
LS1701	Stripline	1.0 GHz	1.0 W	7 dB	50%
LS1501	Stripline	1.0 GHz	1.0 W	5 dB	45%
2N5108A	TO-39 Case	1.0 GHz	1.0 W	5 dB	40%
2N5108	TO-39 Case	1.0 GHz	1.0 W	5 dB	35%
2N4428	TO-39 Case	500 MHz	0.75 W	10 dB	35%
2N3866	TO-39 Case	400 MHz	1.0 W	10 dB	40%
2N3553	TO-39 Case	175 MHz	2.5 W	10 dB	60%

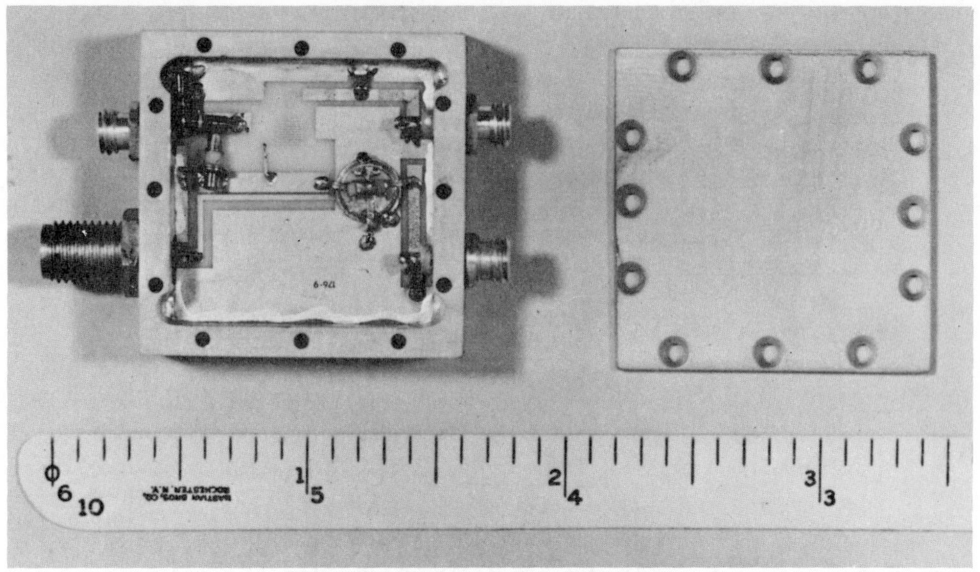

Fig. 12-5 VARACTOR DOUBLER ASSEMBLY

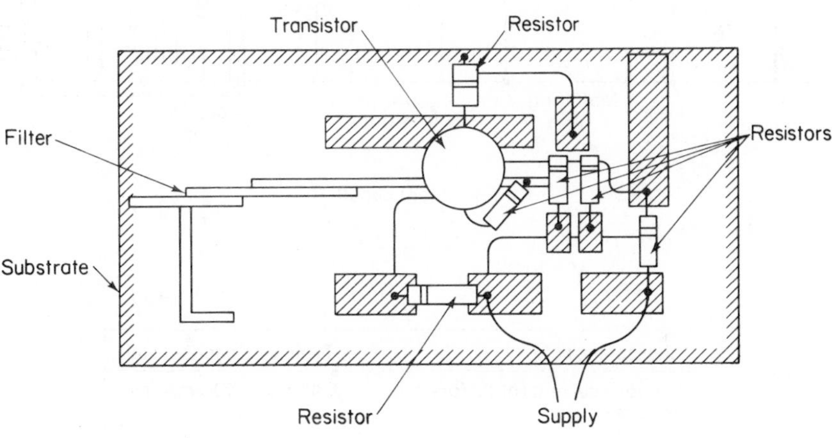

(a) Typical microstrip construction

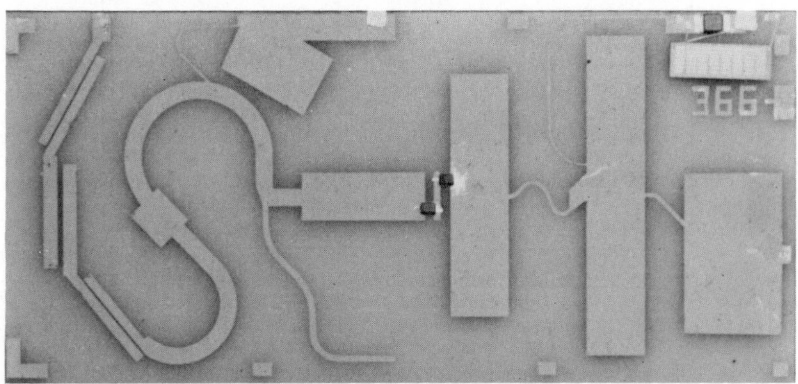

(b) Microstrip, thin film varactor

Fig. 12-6 INTEGRATED MICROSTRIP CONSTRUCTION (Courtesy of Texas Instruments.)

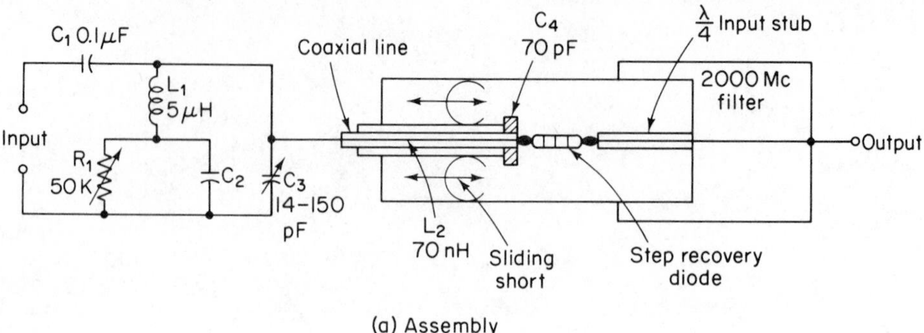

(a) Assembly

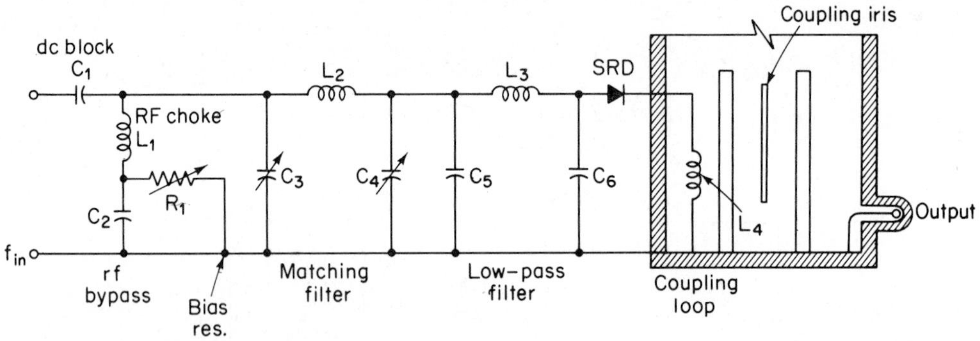

(b) Equivalent circuit

Characteristic	Typical	Specified limit
Forward current (at 1V forward voltage)	645 mA	600 mA min
Breakdown voltage (at −10μA reverse current)	82 V	65 V min
Capacitance	6.6 pF	8.0 pF max
Series resistance (at 50 mA forward current)	0.3 ohms	——
Lifetime, τ	225 ns	100 ns min
Transition time, t_t	400 ps	500 ps max
Cutoff frequency, f_c	160 Gc	——
Lifetime transition time, M	500	——

(c) Performance data for typical step-recovery diode (hpa-0241 step-recovery diode)

Fig. 12-7 STEP-RECOVERY DIODE CAVITY ASSEMBLY

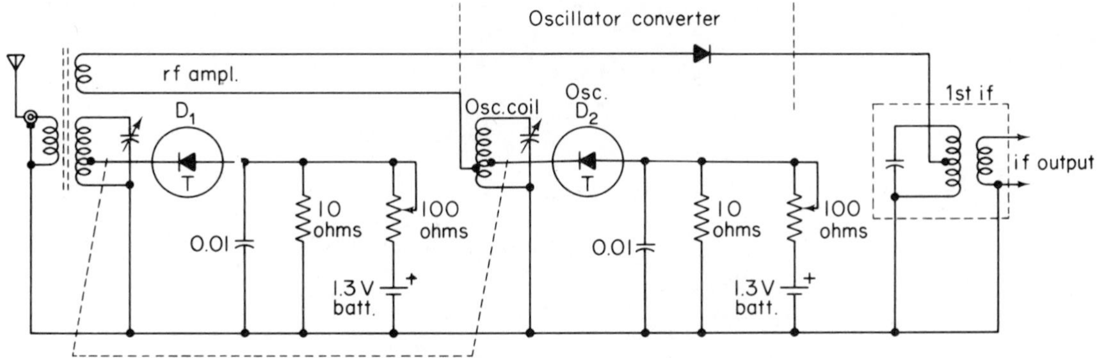

Fig. 12-8 TUNNEL DIODE MICROWAVE AMPLIFIER

STEP-RECOVERY-DIODE SOURCES

The step-recovery diode is a special type of graded junction diode. It employs doping, which is deliberately arranged for the rapid release of maximum forward-current stored charge when reverse bias is applied, and for equally rapid reestablishment of charge for forward conduction. Unlike the capacitance variation or tuning feature of the varactor, its capacitance variations with bias are minor but provide excellent harmonic generation from rapid transition. Figure 12-7 illustrates the hardware, circuitry, and diode characteristics for a step-recovery-diode-generated cavity source of microwave energy.

TUNNEL-DIODE OSCILLATOR

Tunnel diodes find particular use in many signal circuits because of their high efficiency, low noise, frequency stability, simplicity, and low element capacitance. They are, however, low-powered devices delivering only a few milliwatts in the GHz region.

A typical microwave amplifier and oscillator setup, Fig. 12-8 , directly compares the front-end amplifier and oscillator of a commercial receiver with the same circuit constructed with tunnel diodes. The simplicity of the latter shows up in the fact that it uses about one half the number of components.

GUNN-OSCILLATOR SOURCES

The Gunn effect (discovered by J. B. Gunn) is embodied in a single-structured crystal assembly which oscillates by itself at microwave frequencies. The assembly consists of a gallium arsenide sandwich made of intermediate resistivity material placed between two other layers of low-resistivity material [Fig. 12-9(a)]. If the pellet is mounted in a tunable microwave cavity, self-oscillation is sustained similar to the negative-resistance mode of the tunnel diode. This phenomenon exists in the *Q*-band and higher (30–40 GHz). Figure 12-9(b) and (c) illustrate an assembled unit and the equivalent circuit of a practical oscillator. Figure 12-10 illustrates four commercial assemblies.

AVALANCHE-DIODE SOURCES

These diodes are microwave frequency devices using the negative-resistance phenomenon to sustain oscillation. They generate CW energy by operating under avalanche conditions, where secondary emission is induced by reverse-biased *PN* junction breakdown. Similar to but higher powered than the tunnel diode, they can be categorized as follows: (1) Read or *PNIN*, (2) *PIN*, and (3) Impatt–avalanche.

The Read or *PNIN* diode is a four-layer device using an added third layer (*I* layer) of

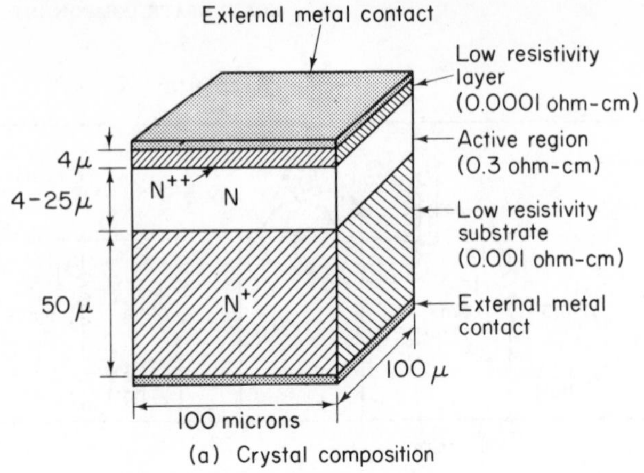

(a) Crystal composition

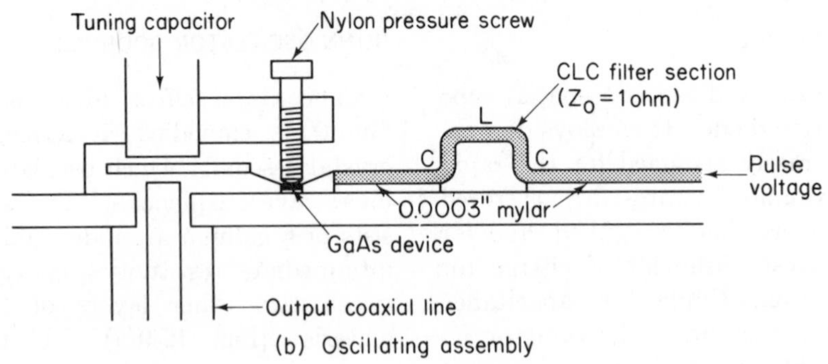

(b) Oscillating assembly

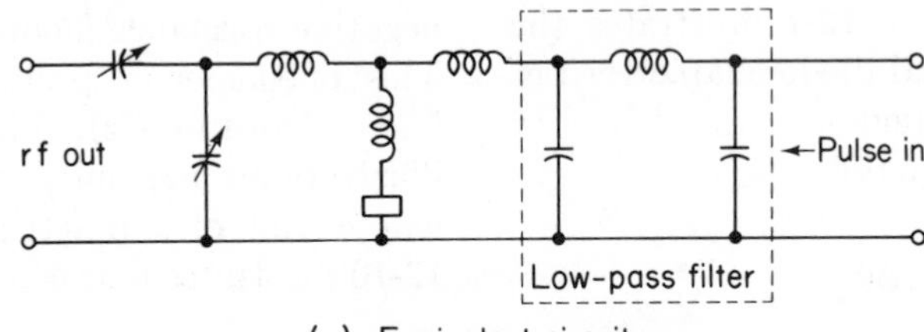

(c) Equivalent circuit

Fig. 12-9 GUNN OSCILLATOR CIRCUIT AND CONSTRUCTION

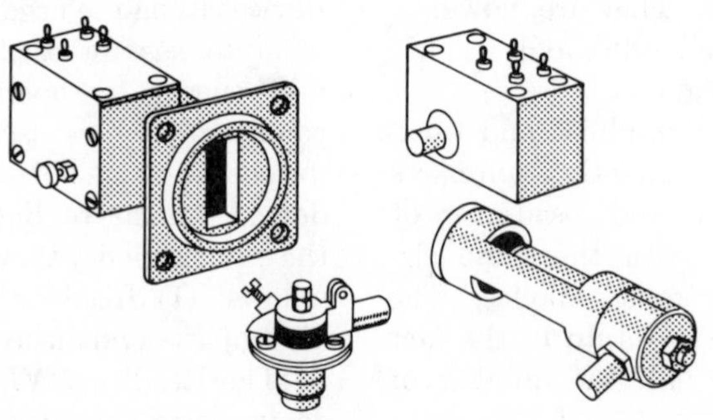

Fig. 12-10 COMMERCIAL GUNN OSCILLATOR ASSEMBLIES (Courtesy of Watkins-Johnson Co.)

intrinsic normal conducting material. The first two layers constitute a junction using steeply graded doping concentration. The middle two junctions together form a layer of low conductivity adjoining the fourth heavily doped N region.

When bias is raised from the reverse leakage condition, an avalanche field is created so that as the overall diode current rapidly increases, a negative resistance is produced which allows the electrons to pass readily through the middle I and N regions as in a tunnel diode. The combined avalanche and tunneling effects can produce hundreds of milliwatts of power in the 10- to 50-GHz range.

IMPATT–AVALANCHE DIODE

The word "impatt" is a mnemonic-functional description in which the letters stand for the following: IMP for "impact," A for "avalanche," T for "transit," and the second T for "time."

The impatt–avalanche diode is a PN junction which, in effect, is a simplification of the Read diode described above. The junction has its impurities so graded and controlled (as in the tunnel diode) that it exhibits negative resistance in the avalanche condition. In this sense it resembles two Read diodes placed back to back, since there is essentially an intrinsic region produced by the doping on each side of the junction.

Figure 12-11(a) shows an enlarged section of a typical avalanche diode describing the epitaxial layer over the N substrate. Over this is diffused a thin layer of P-type impurity creating an avalanche zone next to the main P_+ diffused layer. Figure 12-11(b) shows the impurity profile spread over an expanded section of the junction.

In the central, high-field, avalanche region, electrons circulate outside the P terminal (each with a finite time delay). Microwave generation of circulating electrons within the adjoining layers is aided by the simultaneous presence of negative resistance material, which then exists.

Thus avalanche-transit time junctions have a frequency-dependence, negative resistance characteristic which depends upon the ionization time and the particle movement of secondary emission within the P layer itself; the device is, however, not dependent upon the quantum-mechanical tunneling effects present in a tunnel diode.

Avalanche diodes are, however, somewhat noisy, as a result of internal electron gyrations; they just about reach the noise level present in klystron oscillators. Physically and mechanically they may be combined with a cavity or rely upon fixed tuning. Figure 12-11(a) and (b) show typical assemblies. Avalanche diodes are constructed mostly of silicon (there are some GaAs impatt diodes) and so are somewhat simpler to manufacture than the conventional Gunn diodes and can therefore be included in silicon-based integrated-circuit assemblies.

The power possibilities of avalanche construction are seen in the kilowatt, 1-GHz unit shown in Fig. 12-12(a), where five single diode assemblies Fig. 12-12(b) are stacked in series within a single 1N23 rectifier package. Construction and characteristic for a single unit are shown in Fig. 12-12 (c).

DIODE DETECTORS, MIXERS, MODULATORS, AND LIMITERS

The coaxial mixer and detector are fundamental components in most microwave equipment, particularly in measurement devices where test-drive impedances are usually transformed into coaxial line at 50-Ω levels. As we have seen, they are used in phase detection, power monitoring, and frequency-doubling circuits. Furthermore, almost all microwave receivers employ one or more of these elements.

Characteristically speaking, these nonlinear rectifying devices base their merit upon (1) broad-band properties, (2) conversion loss, (3) noise temperature, and (4) impedance. Their output video signal (dc to 10 MHz) is generated in the load and is directly propor-

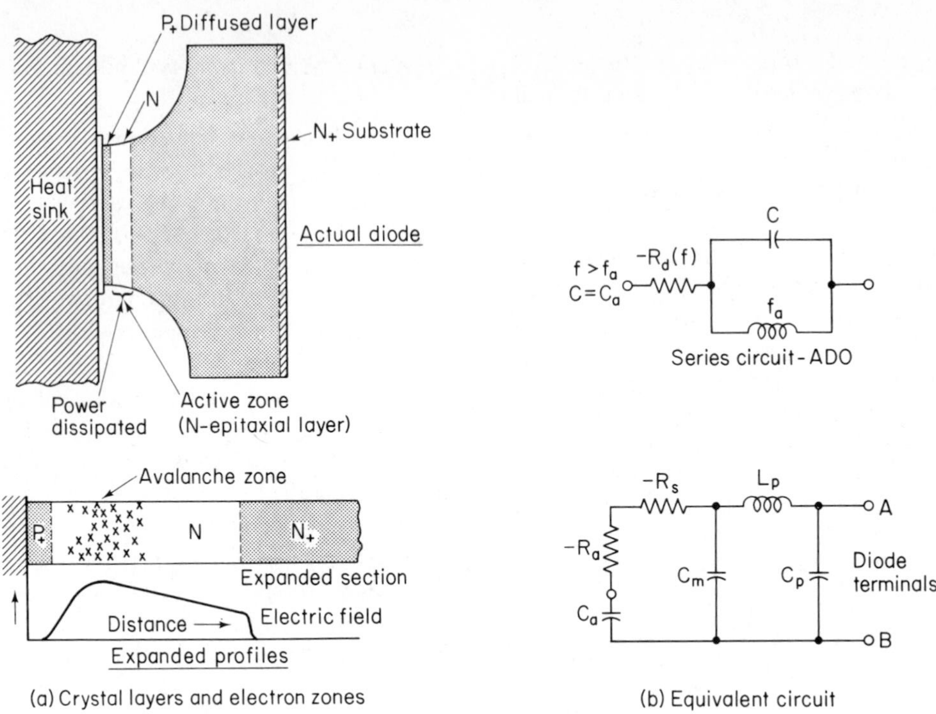

(a) Crystal layers and electron zones

(b) Equivalent circuit

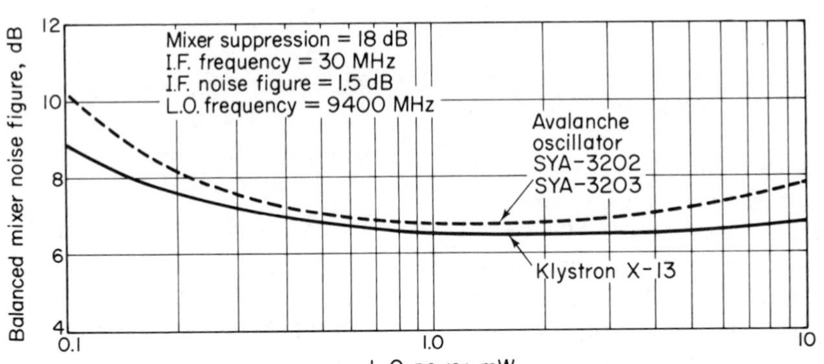

Noise figure of low-noise avalanche diode oscillator as compared to klystron in X-band local oscillator application.

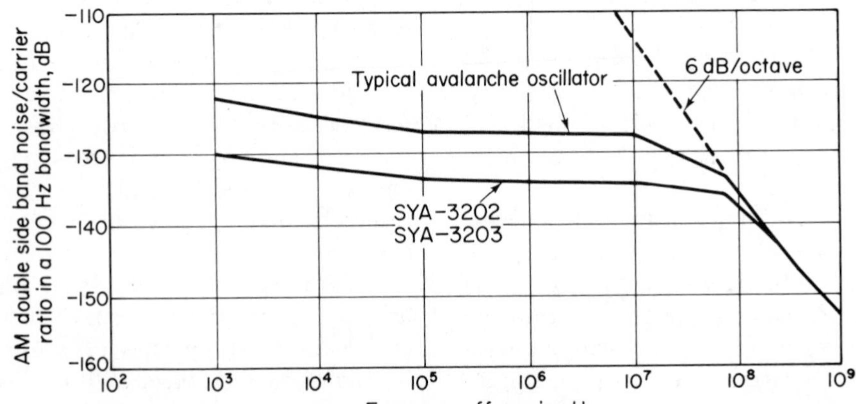

AM noise-to-carrier ratio of a typical avalanche diode oscillator and Sylvania's low-noise avalanche diode oscillator.

(c) Noise characteristics

Fig. 12-11 AVALANCHE DIODE ASSEMBLY CIRCUIT AND NOISE CHARACTERISTICS

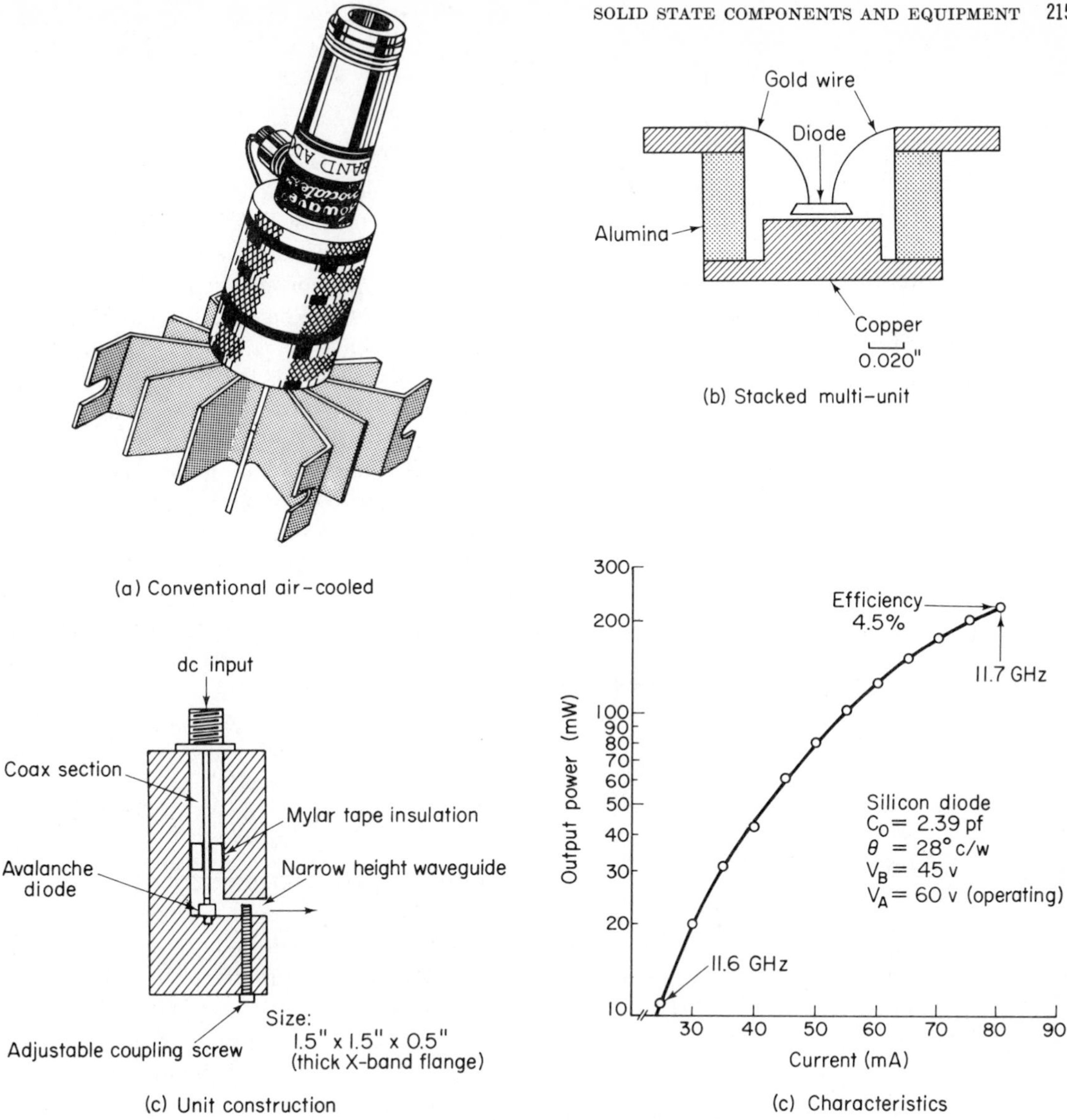

Fig. 12-12 AVALANCHE POWER DIODE UNITS
(Courtesy of Microwave Associates.)

tionate to the square of the small-signal driving voltage. Excluding the bolometer thermally actuated system of detection, these elements can be one of three types:

1. Point-contact diodes.
2. Backward (tunnel) diodes.
3. Schottky barrier or hot-carrier diodes.

POINT-CONTACT DETECTORS

In its early development stages the point-contact diode was the only reliable type of detector available for microwave video detection, and many manufacturers still use them where their particular characteristics fit an application.

At zero bias the point-contact-diode sensivity is relatively high, but its compatible video resistance becomes too great for reasonably high video bandwidth. Also, if the bias is increased, the video load resistance decreases while the noise becomes greater. Thus the point-contact diode's excess noise limits its power-sensitivity capabilities if the lower

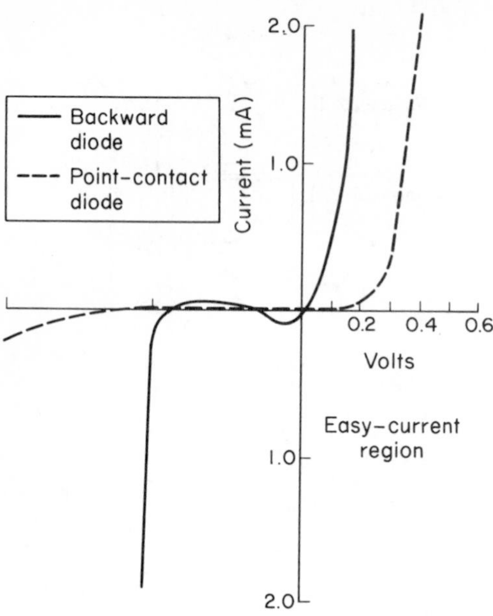

Fig. 12-13 BACKWARD DIODE CHARACTERISTICS

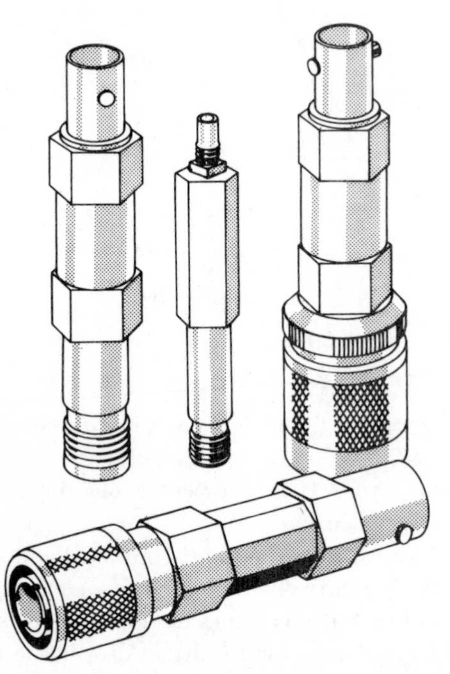

Fig. 12-14 COMMERCIAL BACKWARD DIODES (Courtesy of Microwave, Associates)

cutoff frequency of the associated video amplifier is less than 10 kHz. Here the "$1/f$" noise swamps the "white" noise.

BACK DIODES

The backward diode supports "easy" or heavy current flow in the negative bias or backward direction. Referring to Fig. 12-13, if we picture its current–voltage in an upside-down position, we have a conventional diode characteristic.

Physically, the backward diode is a special tunnel diode (employing quantum-mechanical tunneling) with the knee of the negative avalanche point shifted up to the zero bias position; in this sense it is a Zener diode at the origin point.

Compared to the tunnel diode (see comparable characteristics in Fig. 12-8) it has higher voltage–current sensitivity, lower rf impedance, and excellent frequency versus temperature stability. Also, its $1/f$ noise characteristics are superior to the point-contact detector. Their commercial packaging is similar to other coaxial-mounted detector units (Fig. 12-14).

SCHOTTKY (HOT CARRIER) DIODES

In the hot-carrier-diode concept, or metal-to-semiconductor junction, current flow is supported by majority carriers (mostly electrons) which, when injected at forward (hot) bias, exist at a relatively higher energy level than that of the normal free electron—hence the name "hot carrier."

In operation, the energy barrier at the metal-to-conductor surface (the Schottky barrier) is decreased by forward bias and allows majority carriers to be injected into the metal at high energy levels (above that of free electrons). Here they give up their excess energy very rapidly (lifetime of less than 20 psec) and are accompanied by no flow of minority carriers in the reverse direction. Thus, under signal conditions when the bias is reversed, there is a rapid response, since energy storage at the junction is very low.

All this means that noise levels are cor-

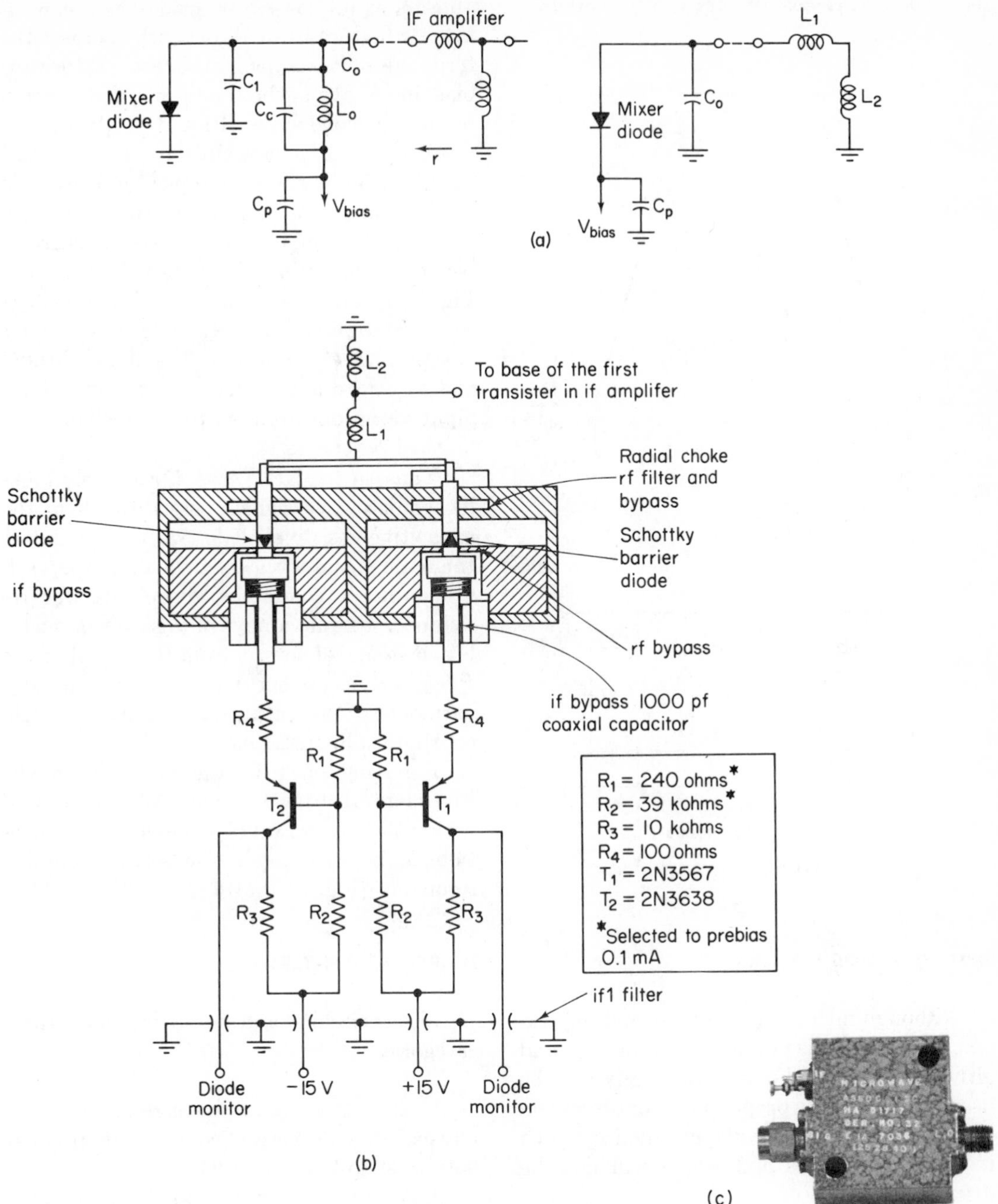

Fig. 12-15 BALANCED SCHOTTKY DIODE MIXER SYSTEM (Courtesy of Microwave Associates.)

respondingly low because recombination turmoil is sharply reduced while conversion efficiency is increased. Thus hot-carrier diodes are widely used in microwave mixer circuits despite their moderate barrier capacitance and series resistance.

Figure 12-15(a), (b), and (c) illustrate the schematic, layout, and physical assembly for a complete broad-band Schottky diode mixer system. The mount uses two dually balanced mixers and the schematic shows how external dc bias is applied to the crystals themselves.

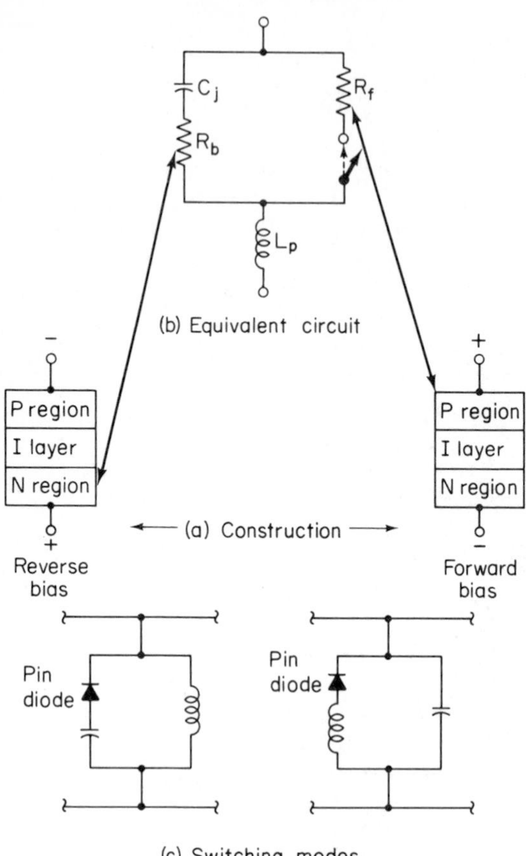

Fig. 12-16 PIN DIODE CONSTRUCTION

SWITCHES—MODULATORS

Although microwave switches and duplexers have been discussed in Chapter 6 and although the *PIN* diode categorically may be used as a switch, its proper place for discussion is with the Read and avalanche diodes, which it closely resembles and which will now be discussed.

PIN DIODE

This is strictly a microwave device; it consists of a thin slice of intrinsic, high-resistance conductor (the *I* layer) abutted by a heavily graded low-resistance junction of P_+ material on its input side, and by a heavily doped N_+ junction on its output side (Fig. 12-16). Thus, in its reverse-biased condition, the structure appears at microwave frequency as a capacitance, high-resistance network across the terminals of its companion circuit. At forward bias, under avalanche conditions, the *I* region becomes flooded with holes and electrons, thus changing from a capacitor to a conductor. Simultaneously, the series capacitor–resistance has become greater, and the diode forward resistance decreases with increasing current. See the schematic representation of this in Fig. 12-16(a). The device accordingly acts as a dc-controlled microwave switch under abrupt bias changes of sufficient amplitude, and as a modulator when proper signal-level input variations are used to cause changes in forward resistance.

Thus, in the latter case, if relatively high-level microwave signals pass through in the capacitive–resistive mode, we can modulate the main signal with low-frequency bias variations of the conductive diode's parallel resistance. The advantage of this type of modulation stems, of course, from the fact that the *PI* junction neither rectifies nor generates harmonics, presenting a relatively high, constant-value resistance.

Referring back to the switching mode, Fig. 12-16(a) (b), we see how, when associated with other microwaves components, transition can be easily made from series to parallel resonance [Fig. 12-16(c)].

MICROWAVE FILTERS

Filter circuits and units fall into three categories:

1. Strip-line filters—depending upon the physical dimensions and construction peculiar only to strip-line construction.
2. Monolithic crystal filters—built after solid-state IC construction using quartz crystals and film-deposition processes.
3. YIG resonator units—using the gyrometric resonance phenomenon of a small sphere of yttrium–iron–garnet composition.

STRIP-LINE FILTERS

Microstrip or strip-line construction is a design technique closely allied to integrated

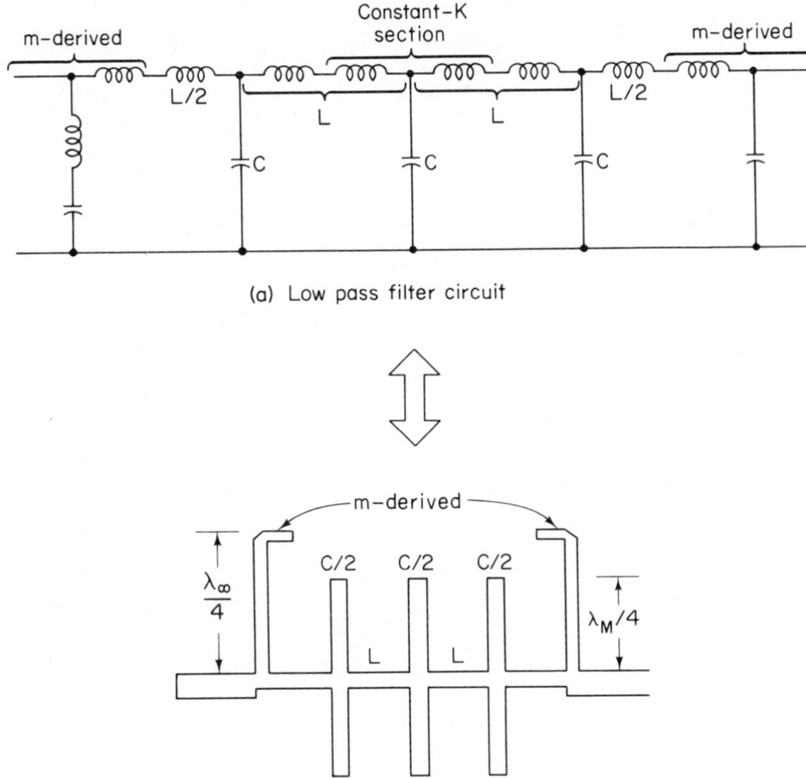

Fig. 12-17 STRIP-LINE LOW PASS FILTER ARRANGEMENT

circuits because it employs thin-film deposition in forming a transmission-line conductor (see discussion later in this chapter).

Elements in a strip-line filter are actually sections of strip transmission line, tailored specifically in size, length, or proximity to each other to give low, high, or band-pass electrical characteristics. Basically, we can picture the elements as small, rectangular, flat, metal areas located for correct electrical value and coupling in two basic configurations, either end coupled or side coupled.

Figure 12-17(a) shows a configuration where the inductor sections and shunt capacitor elements are compatible with the schematic of a low-pass filter [Fig. 12-17(b)].

High-pass and band-pass configurations usually consist of the half-wavelength end-coupled or side-coupled elements shown in Fig. 12-18(a) and (b). Figure 12-18(c) illustrates a combination of these two on a commercial assembly.

If we utilize the directional-coupler strip-line assembly as a directional filter [Fig. 12-19(a)], an array of these can be combined into the strip-line multiplexer assembly shown in Fig. 12-19(b).

MONOLITHIC CRYSTAL FILTERS

When we stripwise integrate (by thin-film construction) the interconnection of a number of ceramic-quartz-wafer-crystal resonators, say in tandem, by proper coupling, we have produced a monolithic filter. These IC-linked assemblies are limited in size and frequency by crystal fragility, seldom being designed for higher than 10- to 20-MHz operation. Figure 12-20 illustrates three typical assemblies.

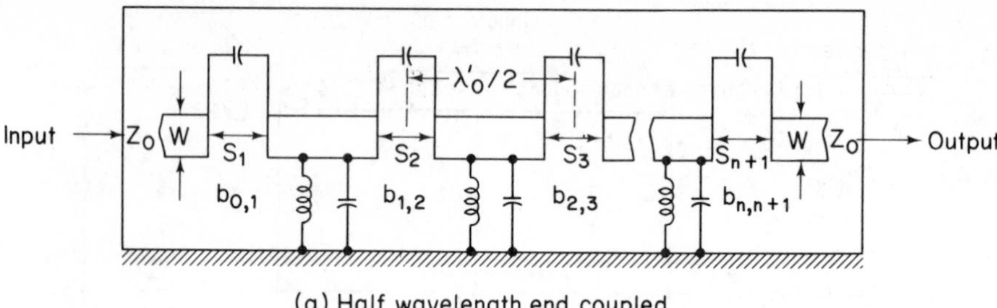

(a) Half wavelength end coupled

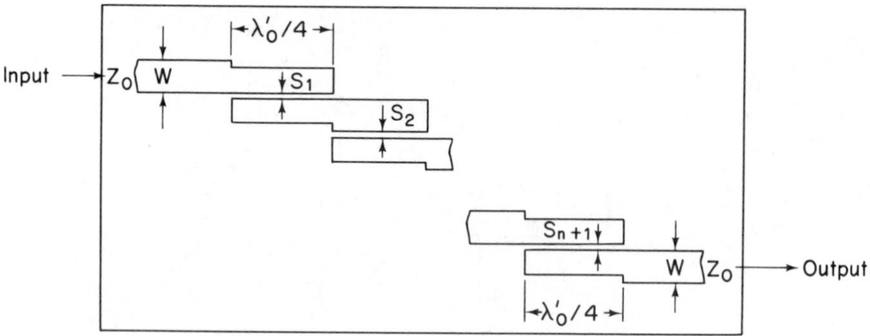

(b) Half wavelength side coupled

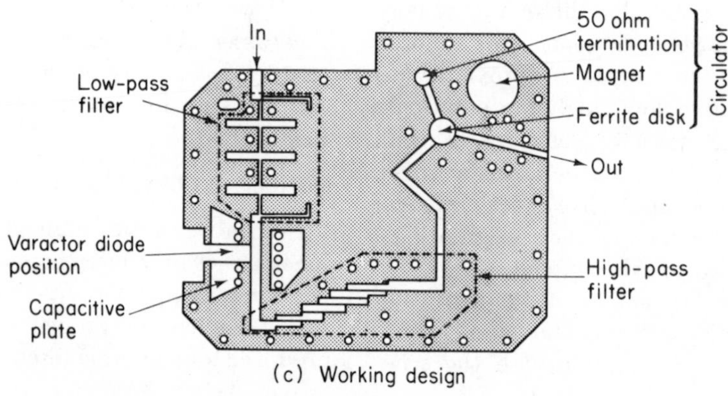

(c) Working design

Fig. 12-18 STRIP-LINE HIGH PASS FILTER ASSEMBLY

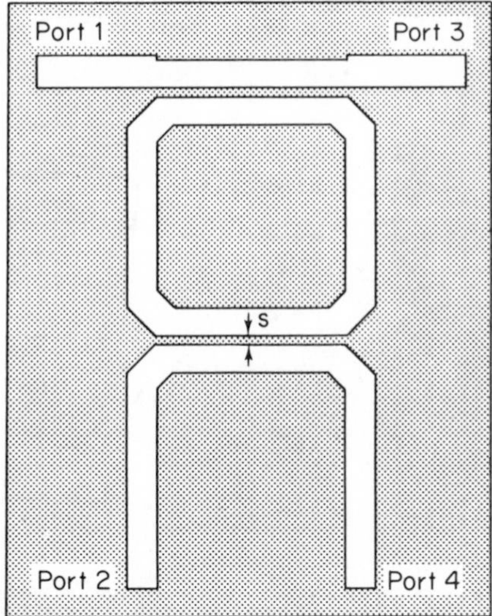

(a) Directional coupler

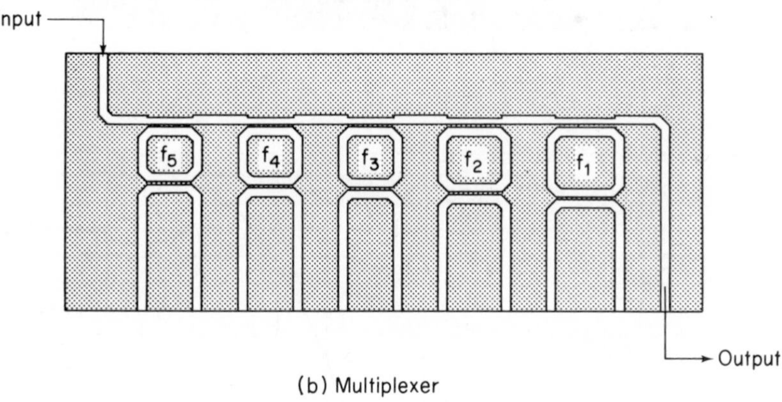

(b) Multiplexer

Fig. 12-19 STRIP-LINE DIRECTIONAL COUPLER AND MULTIPLEXER

YIG-RESONATOR TUNING UNITS

As noted above, the word YIG is an abbreviation for yttrium–iron–garnet; a YIG resonator is a polished sphere a few hundredths of an inch in diameter resting in the center of two coupling loops whose axes are perpendicular to each other, the whole assembly subjected to a magnetic field in a plane mutually perpendicular to the two coupling loops. The whole mechanism exhibits a high Q resonance at a frequency called the *gyromagnetic resonance frequency*, which is tunable simply by changing the strength of the applied magnetic field [Fig. 12-21(a)].

Thus the device is a coupling network operating between 1 and 4 GHz but not at the gyromagnetic resonance frequency. Each YIG unit has a definite bandwidth determined by the ratio of the sphere diameter to the coupling-loop diameter, so a number of successive units placed frequency-wise adjoining each other (out–input–output–input, etc.) form a tandem band-pass filter with high at-

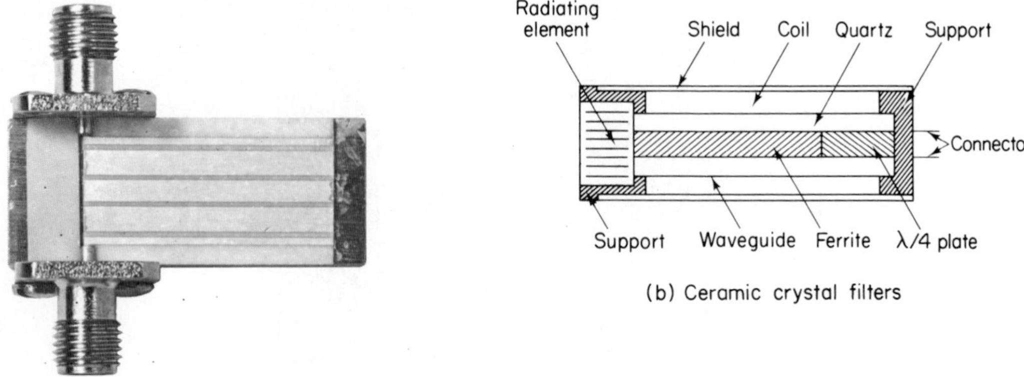

(a) Strip line filters

(b) Ceramic crystal filters

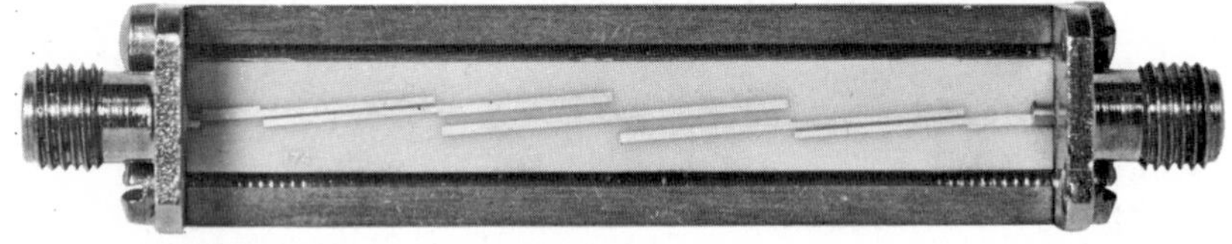

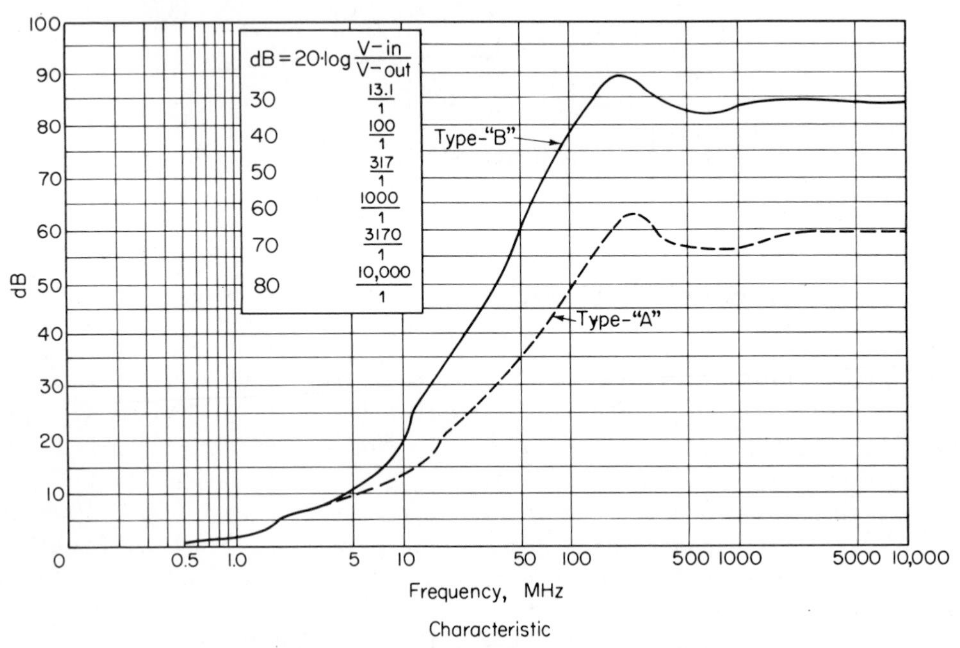

(c) Ferrite-ceramic

Fig. 12-20 FERRITE CRYSTAL FILTER ASSEMBLIES (Courtesy of Texas Instruments.)

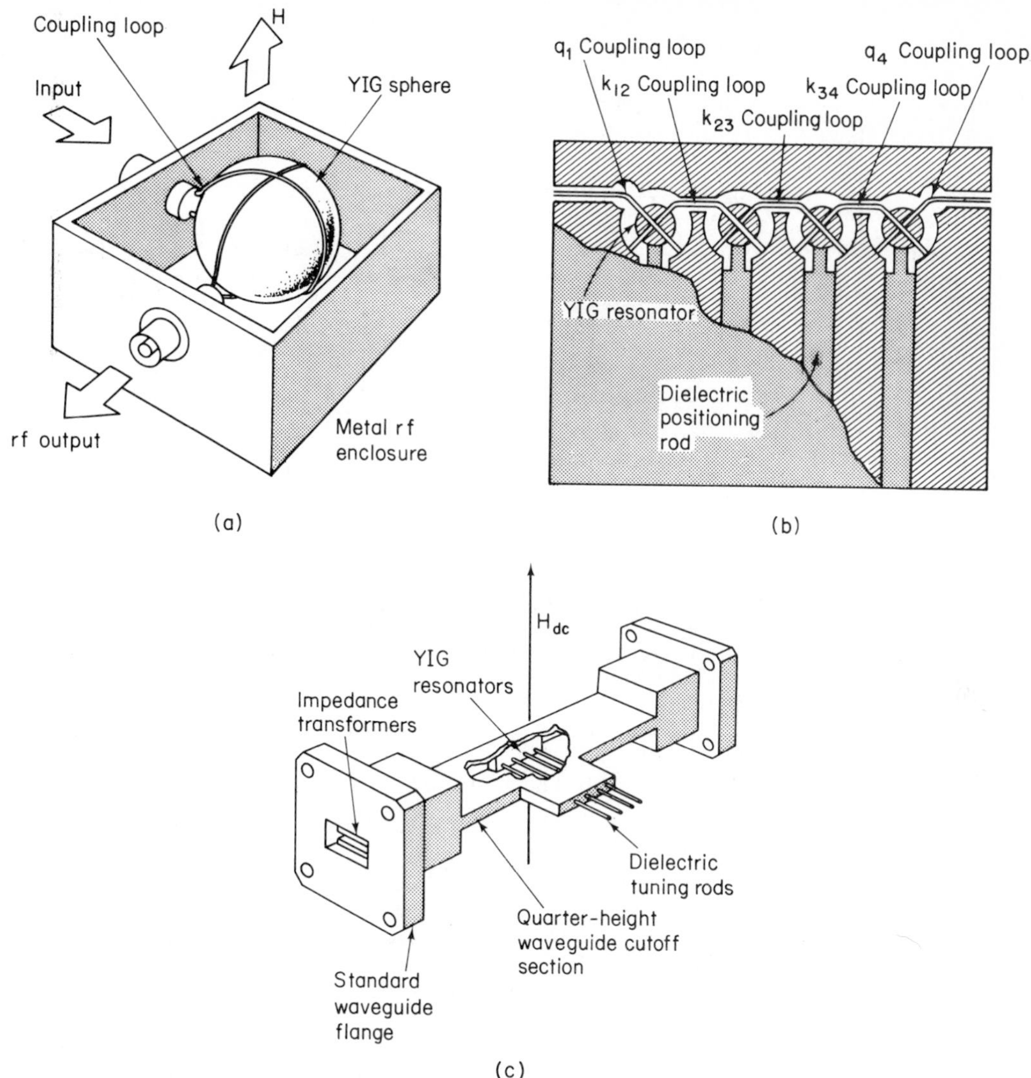

Fig. 12-21 YIG CONSTRUCTION FILTER AND WAVEGUIDE MOUNTING

quencies, but spaced so that the overlap between internal adjoining units gives flat passband response.

Figure 12-21(b) illustrates the tandem placement of four resonator YIG units in a typical filter. Figure 12-21(c) is a similar unit inserted in the narrowed section of a waveguide.

Other applications use the YIG in oscillator circuits, particularly in sweep circuits. See the tuning arrangement in Fig. 12-22.

STRIP-LINE CONSTRUCTION

Conventional strip-transmission-line construction is shown in exaggerated form in Fig. 12-23; it consists of a center conductor guiding the electrical energy between two conducting ground planes. In actual form the center conductor is usually a thin strip (or film) deposited on a dielectric, covering the bottom ground plane and omitting the top half of the ground plane.

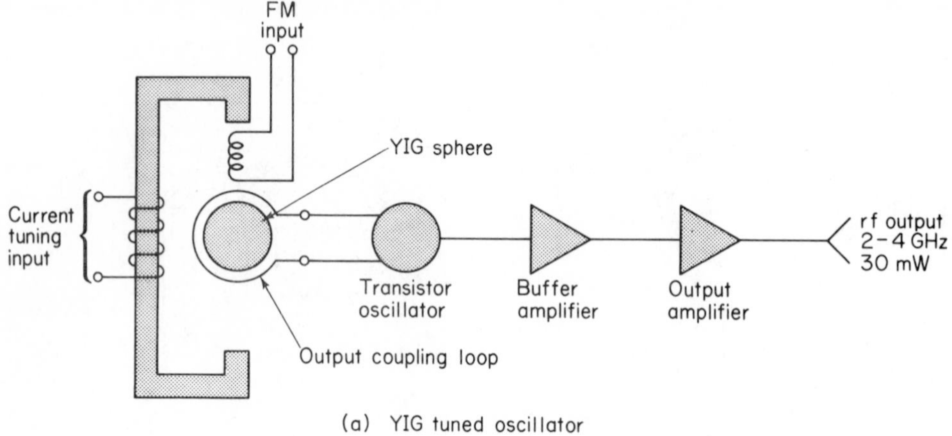

(a) YIG tuned oscillator

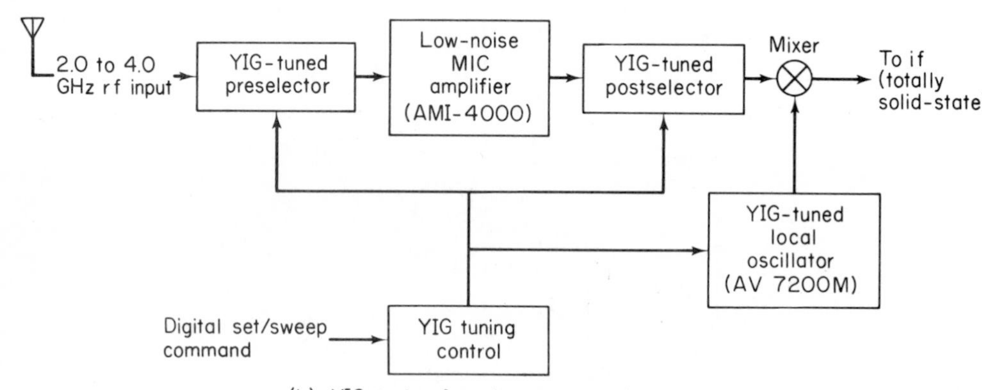

(b) YIG tuning for microwave receiver

Fig. 12-22 YIG TUNING CIRCUIT

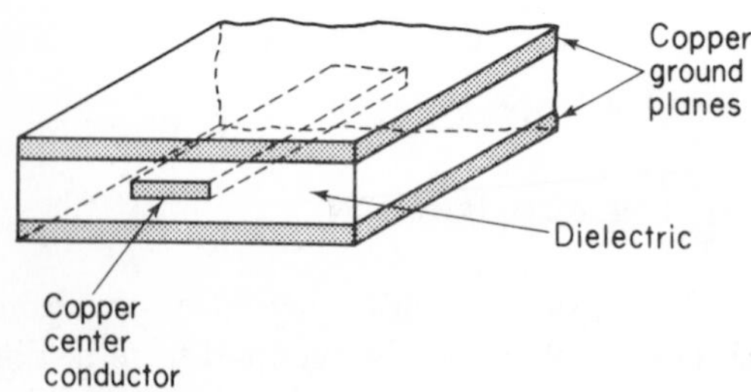

Fig. 12-23 BASIC STRIP-LINE CONSTRUCTION

This "halved" structure is a legitimate transmission structure, for if we take a parallel open-wire line with the field existing as a transverse electromagnetic wave and insert an infinite metal sheet horizontally halfway between the conductors, as in Fig. 12-24, the field pattern will not be disturbed and transmission exists just as if the remaining wire were in position. Actually, propagation is normal since the insertion of a ground plane forms a virtual image of the second wire.

Now, open microstrip produces radio-frequency leakage, so in lines of any length the second ground plane is necessary. Practically, strip transmission line is built using two sheets of copper-clad dielectric—with a $\frac{1}{16}$- to $\frac{1}{8}$-in.-thick insulating layer. To make a complete assembly, the center conductor is formed by chemically etching away the copper on a doubly clad board and abutting it to make the single sandwich structure

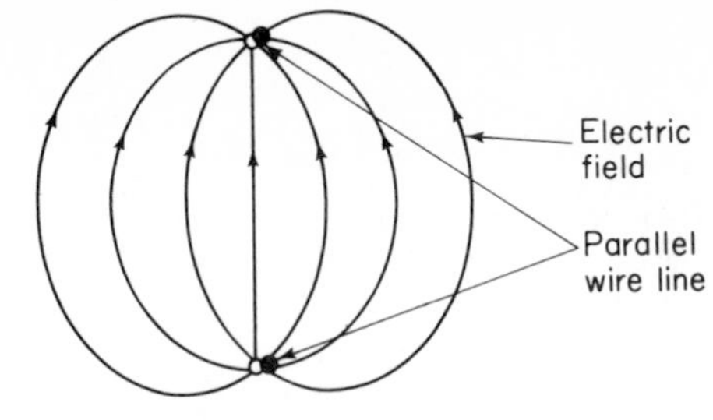

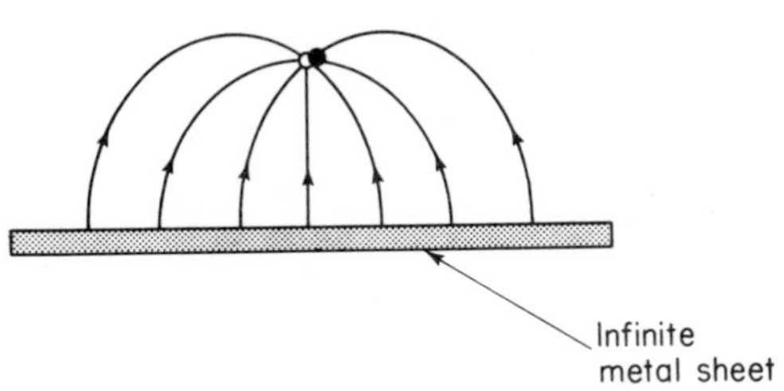

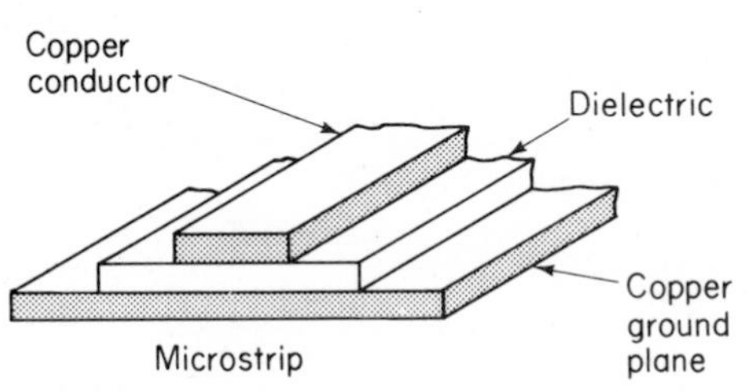

Fig. 12-24 STRIP-LINE HALVED STRUCTURES

shown in Fig. 12-25(a). To give strength and uniformity, the structures are usually clamped together with metal pressure plates. Other configurations [Fig. 12-25(b)] show double-thickness center conductor when using two doubly clad construction sheets. Figure 12-25(c) shows a coupled strip transmission line.

STRIP-LINE COMPONENTS

These have probably more application in integrated-circuit structures than in straight transmission lines. Passive and active semiconductor areas can be deposited, embedded, or diffused into sections of the line.

Also, in the GHz region the actual physical lengths of sections become quarter-wave and half-wave units which fit directly into filters and coupler structures. A good example of this is the strip-base filter structure illustrated earlier in Fig. 12-19.

In designing simple strip-line and strip-line-constructed components, a mathematical and technological regime exists, paralleling that used in coaxial and waveguide systems. Only a few fundamental characteristics can be discussed.

Design Factors [Fig. 12-26(a)]

Characteristic Impedance

Here

$$Z_0 = \frac{60}{\sqrt{\epsilon_r}} \ln\left(\frac{4b}{\pi d_0}\right) \qquad \Omega$$

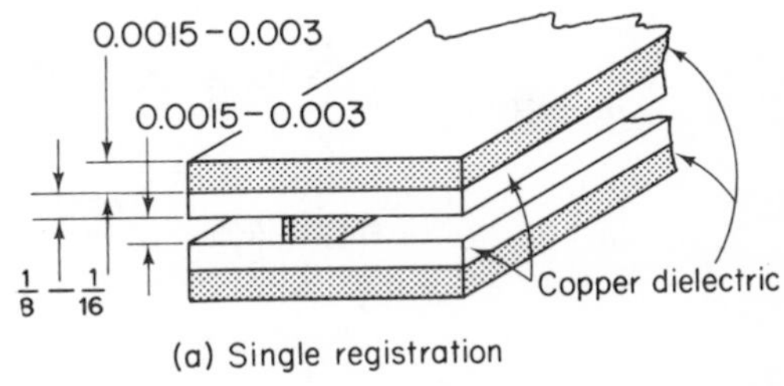

(a) Single registration

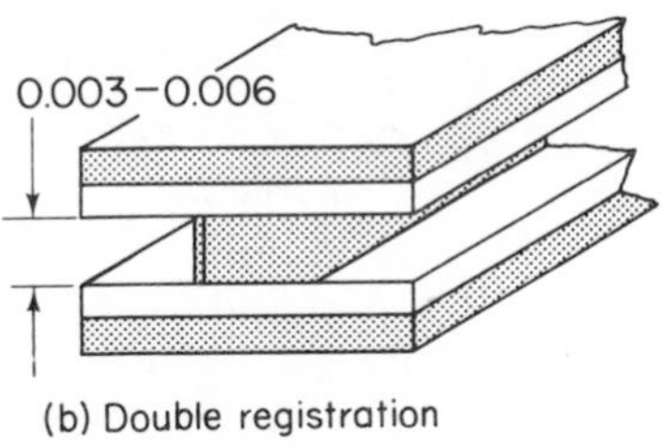

(b) Double registration

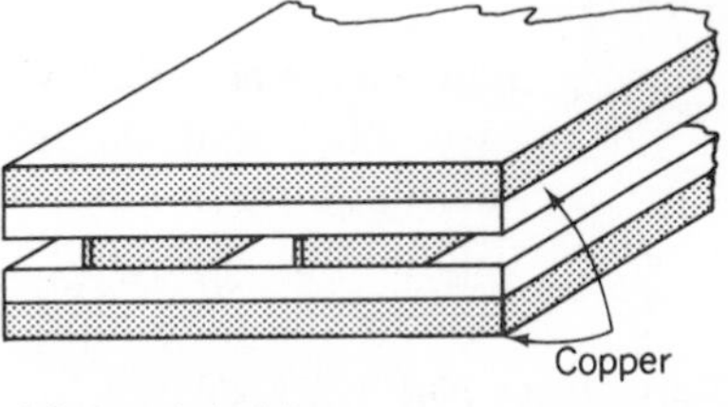

(c) Coupled strip

Fig. 12-25 STRIP-LINE SANDWICH

$$Z_0 = \frac{60}{\sqrt{\epsilon_r}} \ln\left(\frac{4b}{\pi d_0}\right) \text{ ohms}$$

ϵ_r = Relative dielectric constant

d_0 = Determined from graphs[2]

$W/b < 0.35$

(a) Characteristic impedance

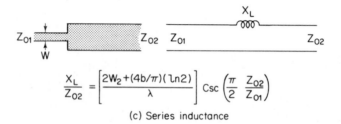

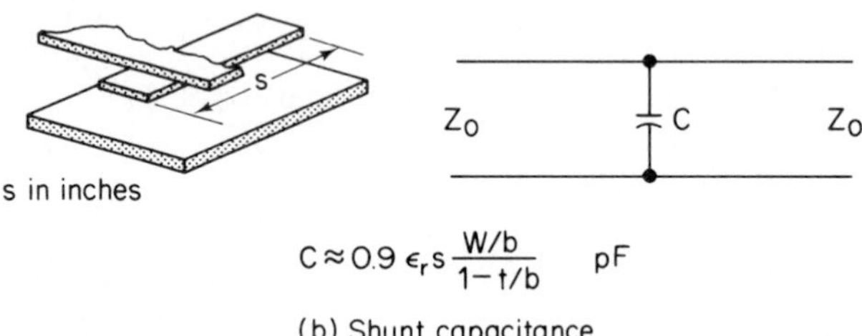

$$C \approx 0.9\, \epsilon_r s \frac{W/b}{1 - t/b} \quad \text{pF}$$

(b) Shunt capacitance

$$\frac{X_L}{Z_{02}} = \left[\frac{2W_2 + (4b/\pi)(\ln 2)}{\lambda}\right] \csc\left(\frac{\pi}{2}\, \frac{Z_{02}}{Z_{01}}\right)$$

(c) Series inductance

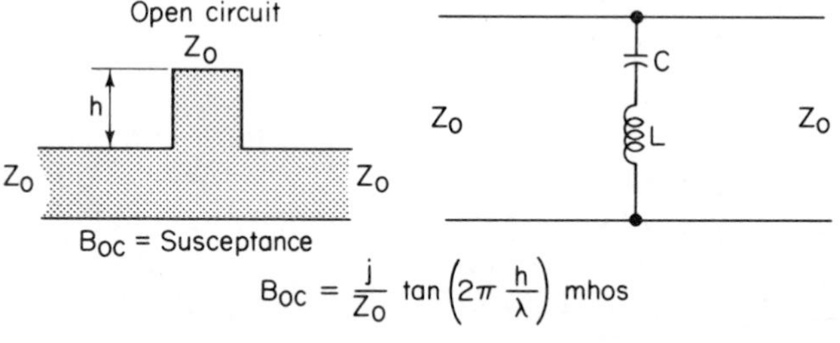

B_{OC} = Susceptance

$$B_{OC} = \frac{j}{Z_0} \tan\left(2\pi \frac{h}{\lambda}\right) \text{ mhos}$$

(d) Series-resonant shunt circuit

Fig. 12-26 STRIP-LINE DESIGN FACTORS

where ϵ_r = relative dielectric constant

d_0 = ratio of height to width of conductor

w/d = 0.35 in.

Shunt Capacitance [Fig. 12-26(b)]

$$C = 0.9\epsilon_r \times S \times \frac{w/b}{1 - t/b}$$

Series Inductance [Fig. 12-26(c)]

$$\frac{x_L}{Z_{02}} = \left[\frac{2w_2 + (4b/\pi)(\ln 2)}{\lambda}\right] \csc\left(\frac{\pi}{2} \times \frac{Z_{02}}{Z_{01}}\right)$$

Series-Resonant Shunt Circuit [Fig. 12-26(d)]

$$B_{0c} = \frac{1}{Z_0} \tan\left(2\pi \frac{h}{\lambda}\right) \qquad \text{mhos}$$

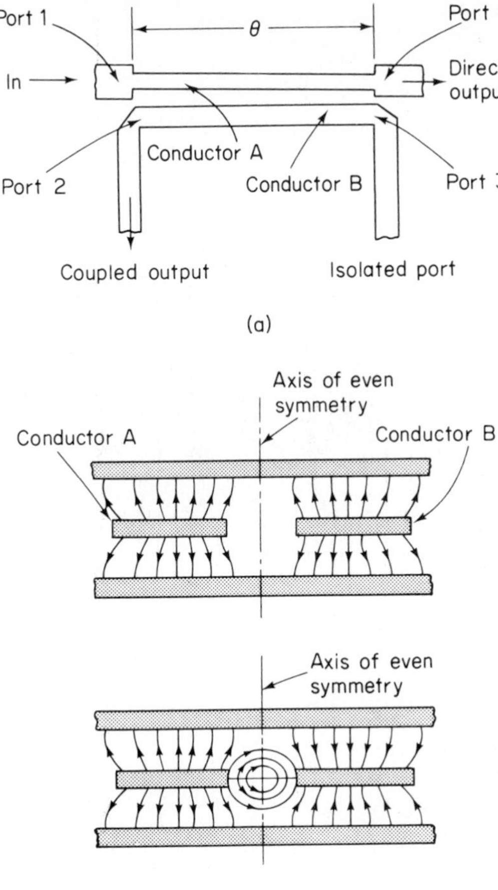

Fig. 12-27 STRIP-LINE DIRECTIONAL COUPLER

STRIP-LINE DIRECTIONAL COUPLER

This unit is illustrated in Fig. 12-27(a), where the even and odd modes of the conductors are supported as shown in the electric-field distribution drawn in Fig. 12-27(b).

Microstrip hybrids, mixers, dividers, and amplifiers are commercially available. Figure 12-28 illustrates a number of typical assemblies.

MICROWAVE RECEIVERS

Microwave components are most typically combined into receiver assemblies; this requires integrated circuitry—hence the coined abbreviation MIC, microwave integrated circuitry.

The whole process of design and manufacture of MIC's naturally originates with solid-state technology; it developed generally as a two-step process, (1) the microminiaturization techniques and operations, and (2) the combining and interconnecting of microminiature components in two circuits. This latter phase of MIC technology centers around the microstrip fundamentals discussed above.

Figure 12-29(a) summarizes this wedding of IC and microwave techniques in generalized form, showing most receiver circuit components grouped on a mounting plate and interconnected by strip conductors. The components are grouped in a hybrid assembly typified by

1. Etched-line and discrete resistors.
2. Thin-film units deposited by vapor plating or other methods which deliver passive semiconductors, resistors, and capacitors.
3. Monolithics—deposited or diffused on active semiconductor substrates. This includes transistors and diodes plus some types of resistors and capacitors.

Figure 12-29(b) gives more operational specifics of a hybrid layout for use in a transceiver circuit.

More specifically, we see all these techniques practically incorporated by the Texas Instrument Co. in an X-band receiver shown schematically and physically in Fig. 12-30(a). This entire unit is 2 by 3 in. and is built on a slab of ceramic. The thick- or thin-film strip line is first located either by deposition or etching to form a five-chip device—transistors, diodes, and monolithic circuits. These are then attached and interconnected by parallel gap ultrasonic welding or thermocompression bonding.

The Texas Instrument Company has used literally hundreds of MIC units assembled in an all solid-state, X-band, electronically scanned, phased array radar. This system is called MERA, for modular electronics for radar applications, and is used widely for terrain following, ground mapping, terrain avoidance, and air-to-ground arranging.

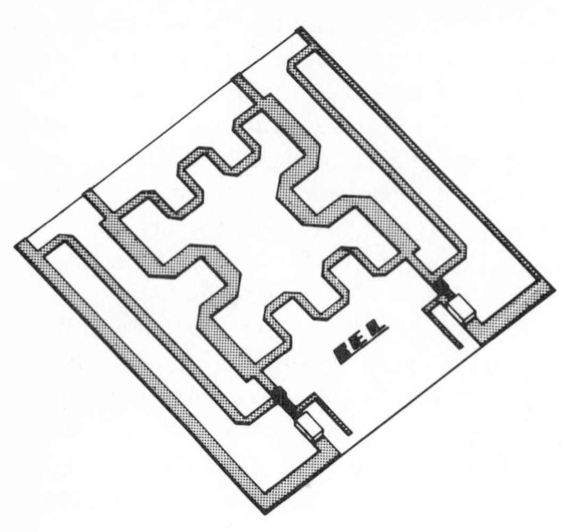

(a) Balanced "L" band mixer

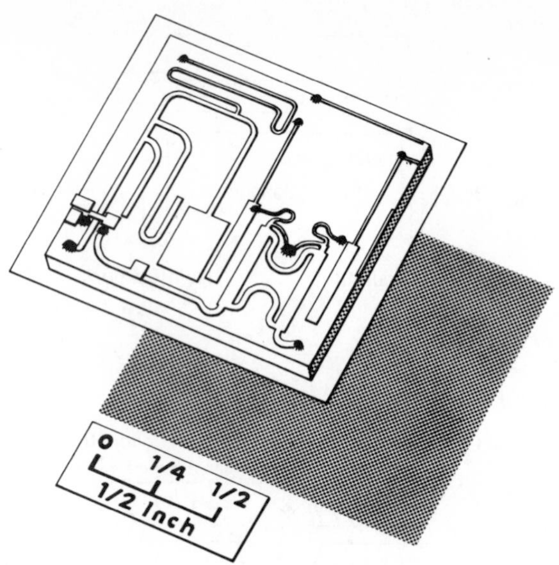

(c) Typical "S" band amplifier

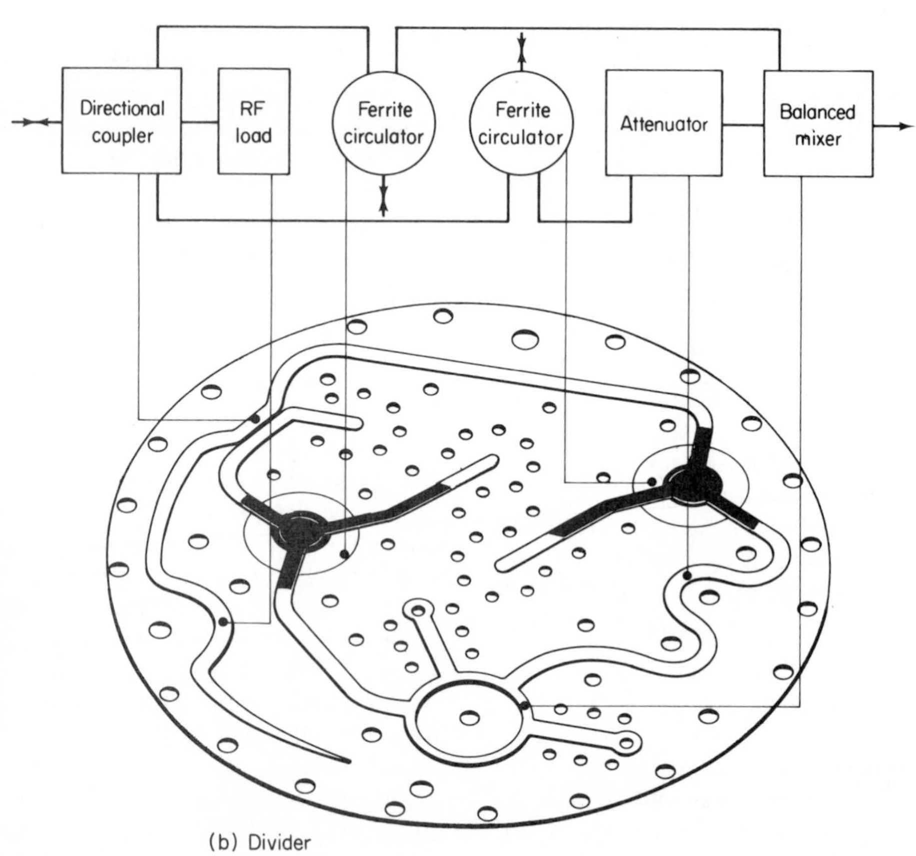

(b) Divider

Fig. 12-28 COMMERCIAL STRIP-LINE ASSEMBLIES

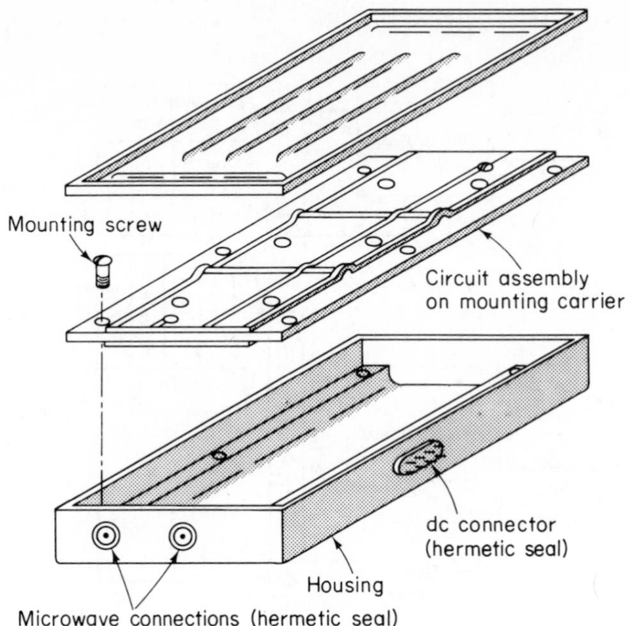

(a) Entire IC package

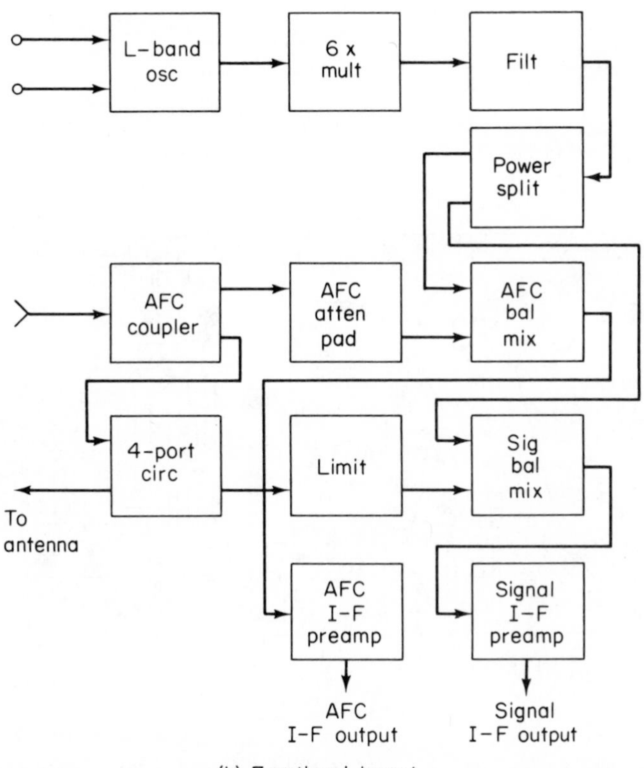

(b) Functional layout

Fig. 12-29 STRIP-LINE RECEIVER ASSEMBLY

rf module No. 1
rf module No. 2
rf module No. 604
Support structure
2.25/ 2.125 GHz
500 MHz
rf distribution manifold
2.25/ 2.125 GHz
Signal generator
2nd LO 440 MHz
Δ if
Σ if
Receiver/ processor
Control
To other units
28 VDC
115 VAC
Rack mounted DC power supplies
MRI video
Σ video
Δ video
Range
strobe
tf/range computer
tf commands
Range analog range lock-on
Row φ
Column φ
φ Reset
Module gate
Synchronizer/ scan computer
Selected video
System zero time
Scan timing to other units
Indicator
Template video

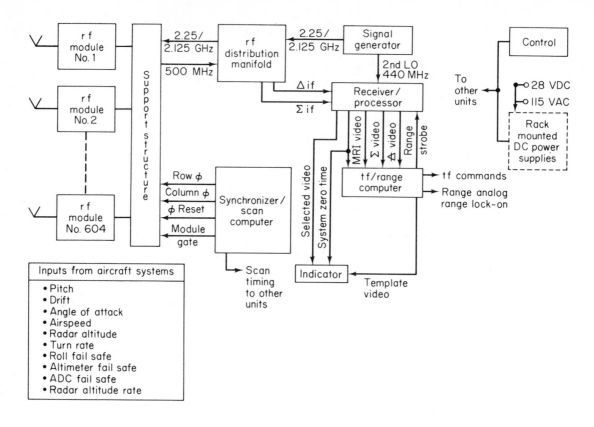

Inputs from aircraft systems

- Pitch
- Drift
- Angle of attack
- Airspeed
- Radar altitude
- Turn rate
- Roll fail safe
- Altimeter fail safe
- ADC fail safe
- Radar altitude rate

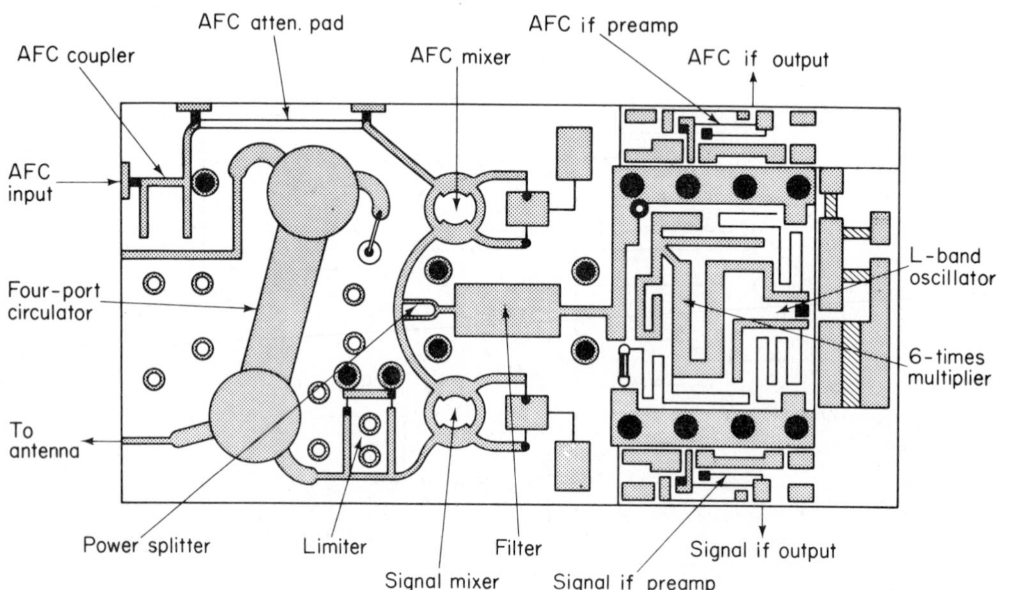

Fig. 12-30 MERA LAYOUT AND RECEIVER MODULE (Courtesy of Texas Instrument Co.)

Millimeter Waves, Masers, and Parametric Amplifiers 13

INTRODUCTION

Millimeter waves are merely an extension of the microwave spectrum; specifically, this region encompasses frequency from 30 to 300 GHz (10–1 mm). Gigahertz systems, residing between microwaves and optoelectronics, have practical performance advantages in radar, communication, tracking, navigation, mapping, and computing. Physically, for instance, microwave antenna beam widths of 1° are achievable, while high-GHz-system beam widths may be measured in seconds.

Information rates, being dependent upon a 10° of carrier bandwidth for millimeter waves, may be around 200 megabits per second for a 2-GHz (S-band) system and 10 gigabits for a 100-GHz system. This performance of course must be weighted against practical advantages in lower-frequency microwave systems—lower attenuation under foggy conditions, link operational flexibility, terrain factors, etc., particularly in all-weather operation over fairly long ranges.

Then again, millimeter waves offer a number of advantages in low-loss space transmission. At 100 GHz a 12-in. dish would give the same directivity and gain obtainable in a 50-ft dish operating at 2 GHz; furthermore, the waveguides, components, generators and other equipment offer size and weight reductions of 100:1 or more.

Probably the greatest drawback in millimeter-wave communication is the attenuation of radiated energy within the earth's atmosphere. Transmission is seriously affected by energy absorption within molecules of oxygen and water vapor. At sea level this makes

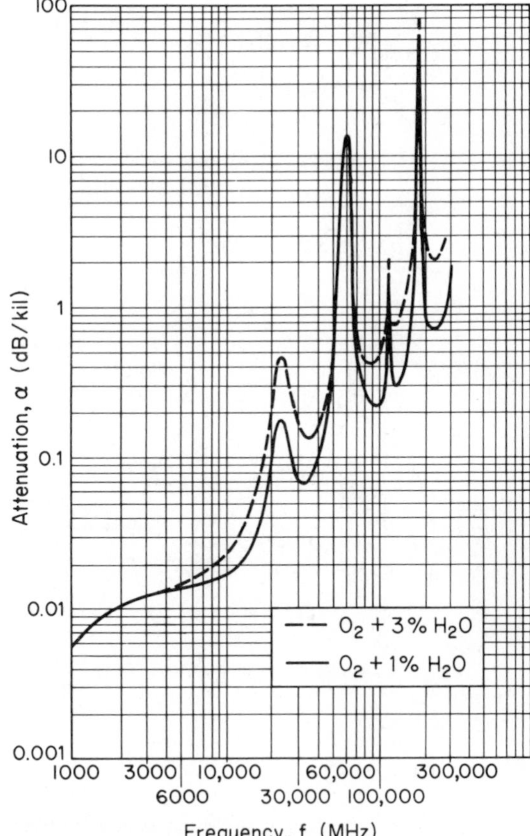

Fig. 13-1 MILLIMETER WAVE TRANSMISSION ATTENUATION CURVES

communication more or less impractical, but above 30,000–40,000-ft, where water vapor has disappeared, there are "windows" or frequency areas where millimeter wave attenuation is less than other microwave frequencies.

Figure 13-1 illustrates this difference between sea-level and high-altitude attenuation; the inference is obvious, then, that millimeter communication and radar systems between satellites or between high-flying aircraft, or between the latter and satellites, possibly offer a 100:1 advantage. Incidentally, millimeter wave signals successfully and definitely penetrate the plasma sheath developed when a spacecraft is on its reentry phase through the earth's atmosphere.

In the course of millimeter wave development, power-generating sources and receiving equipment are the most obvious avenues, both involving extension by scaling down of existing microwave devices plus a number of new approaches utilizing electron-orbital phenomena.

The scaling-down process encounters dimensional difficulties because inaccurate tolerances in tuning cavities and at magnetic gaps produce losses and nonuniform performance. A 6-in. cavity (for 1-GHz operation), for instance, must be 0.06-in. at 100 GHz. Furthermore, as frequency increases, efficiency decreases, and to produce hundreds of watts

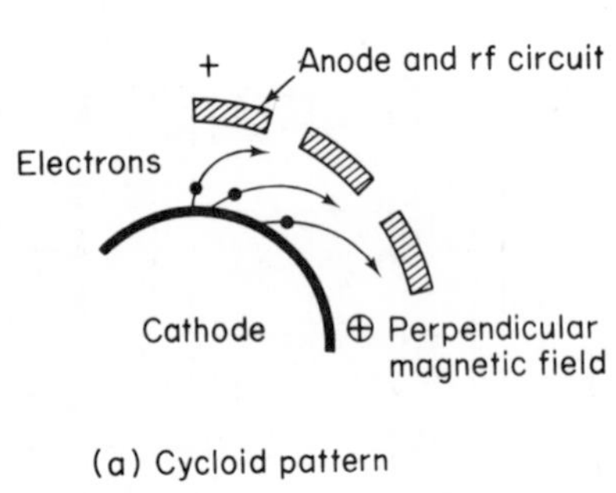

(a) Cycloid pattern

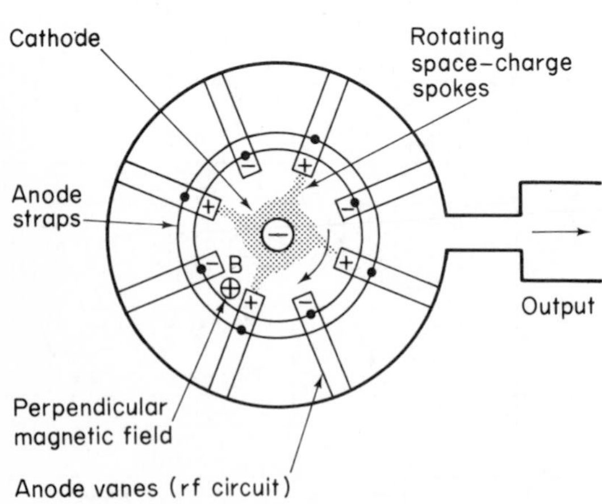

(b) Electron cloud forming relating space charge

(c) Commercial units

Fig. 13-2 MAGNETRON ELECTRON PATHS AND COMMERCIAL UNITS
(Courtesy of Raytheon Co.)

at moderate frequencies requires power supplies of impractical weight and size.

In millimeter receivers the problem becomes one of noise; mixing and amplifying by use of masers, parametric amplifiers, and associated devices are partial answers and will be discussed below.

MILLIMETER WAVE GENERATORS

Two approaches dominate the field of millimeter generators:

1. Scaled-down electron-tube oscillator–generators.
2. Special electron devices.

ELECTRON-TUBE OSCILLATORS

Electron-tube oscillators for millimeter waves come under two general categories:

1. Cross field (magnetrons).
2. Linear beam, under which are included klystrons, traveling-wave tubes, and backward-wave oscillators.

MAGNETRON. Magnetron design for millimeter waves aims in general at miniaturizing the structures described in Chapter 7. To review, the designs aim to physically reduce the mechanism of the dc magnetic field parallel to the cathode axis in order to confine the curved electron trajectories as they pass between the potentially different anodes and cathode elements [Fig. 13-2(a)]. The cycloidal electron paths caused by this crossing of the dc electric and magnetic fields generate a rotating space charge which interacts energy-wise with rf standing waves between the tips of the anode vanes [Fig. 13-2(b)]. This regularly occurring action produces oscillatory energy at the output terminals of the rf structure.

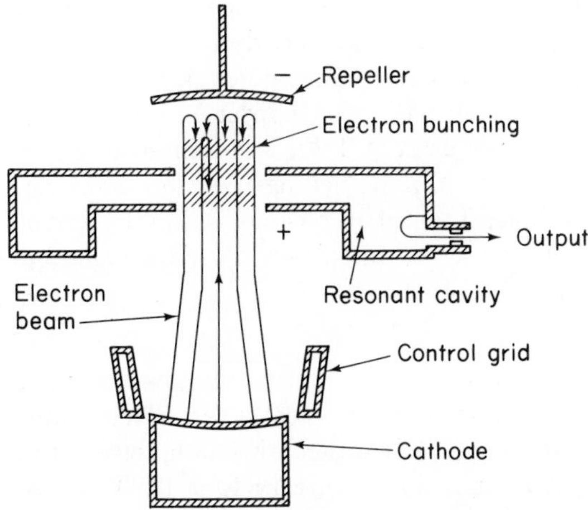

Fig. 13-3 SINGLE-CAVITY KLYSTRON

For power capabilities in millimeter magnetrons, designers use high-emission cathodes plus an increased number of anode vanes. As in low-frequency units, the magnetron offers higher power output than linear-beam tubes, offering units as high as 50- to 100-kW power output in the 50- to 100-GHz range.

KLYSTRON. Klystrons for millimeter-wave generation are mostly reflex structures. Extending the description in Chapter 7, millimeter wave design must of necessity produce a compact mechanism for a simple single-cavity structure (Fig. 13-3). Here we use a cathode-generated electron beam, accelerated by a positive cavity potential; the beam is thus velocity-modulated by rf voltages in passing through the cavity gap. The electron drift is terminated by the repeller, where the negative potential causes the electron to reverse direction and pass through the gap a second time (in the opposite direction); in so doing they become bunched and are prevented from returning on a third trip by proper shaping of the electrodes. The oscillator feedback, plus the bunching action and its interaction with the rf field, produces output energy. Tuning is accomplished by adjustment of the repeller voltage or by physically altering the cavity dimensions through a flexible diaphragm forming part of the top wall of the cavity.

Figure 13-4(a) illustrates the klystron action in a two-cavity amplifier structure. Here the basic conversion of power-supply energy to rf output power occurs through interaction of the electron beam with the field

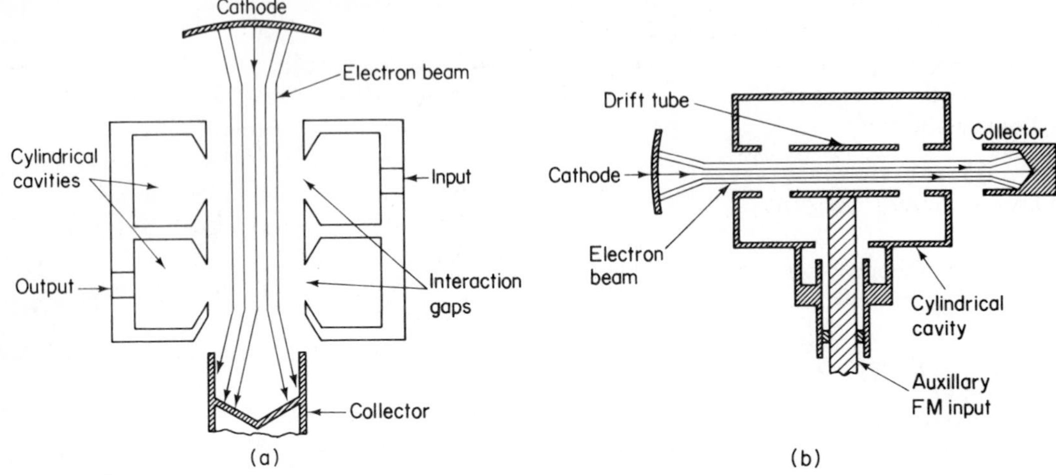

Fig. 13-4 TWO-CAVITY KLYSTRON ACTION

of the surrounding rf circuit. In this way the electrons are exposed to the rf field across the gap of the upper cavity and are accelerated or retarded depending upon what time they enter the gap. This induces bunching or velocity modulation of the electrons during their passage through the drift region existing between the two cavities. The bunched beam induces rf currents as it enters the second cavity, resulting in output power. This design can also be converted to an oscillator by providing proper amplitude and phase feedback.

A simplified type of klystron of increased efficiency uses two tightly coupled cavities [Fig. 13-4(b) is (essentially a single cavity)] with the electron beam acting as a "floating" drift tube.

Traveling-wave tube. Traveling-wave tubes for the lowest millimeter wave region present mechanical design problems over those encountered in magnetrons and klystrons. These physical limitations appear in the basic structure of the rf helix and delay lines, where an accelerated electron beam is surrounded by an rf wave having a strong field component in the direction of the beam travel. Fig. 13-5 shows the basic structure. Operation aims at making the velocity of the beam and the rf wave nearly the same so that the feedback can be arranged and oscillation can be induced (see Chapter 7). Millimeter wave operation of TWT's is limited to under 100 GHz. The device has become, however, a powerful factor in receiver design by virtue of its low-noise operation at frequencies between 20 and 60 GHz.

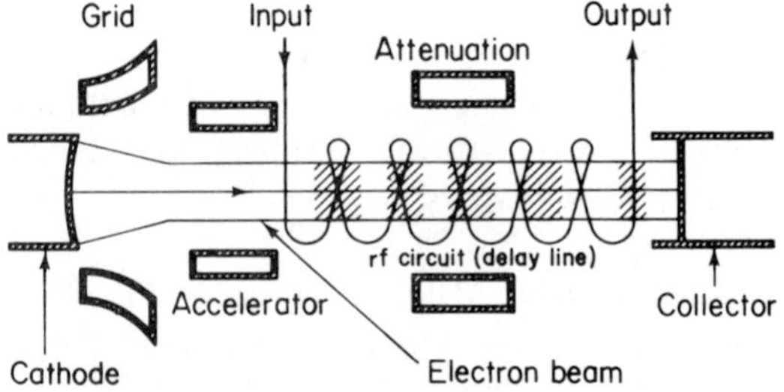

Fig. 13-5 BASIC TWT STRUCTURE

O-type backward-wave oscillators (commonly called O-BWO's). These tubes have design advantages in millimeter wave operation in that, contrary to a TWT, whose structure they closely resemble, the electron beam and the rf energy wave travel in op-

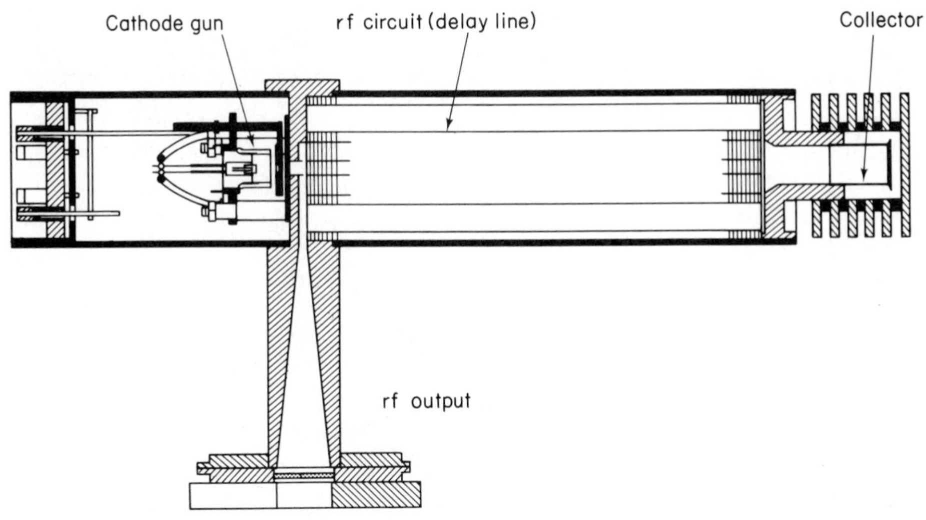

Fig. 13-6 O TYPE BACKWARD WAVE OSCILLATOR

posite directions. This allows easier physical manipulation of the feedback loop, as we observe in the O-BWO structure in Fig. 13-6. Here the rf energy is led out at the gun end of the line while the collector end is terminated in the line's characteristic impedance. Feedback is arranged by making the velocity of the electron beam synchronize with the phase velocity of the rf line acting as a slow-wave circuit so that oscillation starts once the beam current exceeds a certain level.

The millimeter wave O-BWO also has the advantage of using purely electronic tuning by the simple procedure of varying the beam-accelerating voltage or, in other words, changing the beam velocity; so synchronism and feedback conditions can be moved from one frequency to another without physical manipulation. In spite of the mechanical design problems of scaling down an O-BWO's structure, good solid millimeter wavelength structures are possible with careful fabrication of the slow wave section.

Delivering sizable rf powers in the 50- to 100-GHz range (hundreds of watts), the O-BWO promises to constitute the backbone of millimeter wave generating devices. We can note this trend in the tabular rundown of millimeter tube types in Table 13-1.

SPECIAL MM-TUBE GENERATORS

A number of design variations have been developed as outgrowth of the above basic generator types.

1. Monotron. This is a modified floating drift-tube klystron, essentially omitting the drift-tube function. Basically, it bunches electron energy during a long transit path across the cavity gap. Although simple to build, it possesses complexities of operation that have made it commercially none too popular.

2. Reflectron. This tube combines the operating principles of the monotron and the O-BWO. Figure 13-7 shows a cross section of this device.

3. Laddertron. The laddertron is a version of the reflectron, a tunable single-cavity klystron using a "flat" electron beam passing between two parallel "ladders" which provide coupling gaps through which the rf fields of the cavity can interact with the electron beam.

4. M-type BWO. This combines field-crossing techniques with a backward-traveling wave. Figure 13-8 shows the construction of the delay line in conjunction with the

TABLE 13-1 MILLIMETER TUBE CHARACTERISTICS

General	*Sub-types*	
Magnetrons (crossed-field)	Conventional Amplitron M-BWO Inverted Coaxial ┌— Reflectron —┐	Stabilitron Bitermitron Monotron
Klystrons	Multi Cavity Single Cavity Reflex	Laddertron
TWT's	Conventional Low Noise Type	
O'BWO's M'BWO's	Conventional └— Reflectron —┘ Scalloped Beam	
Special tubes	Tornudotron Ubitron Cerontley Radiator	Rebatron Laddertron

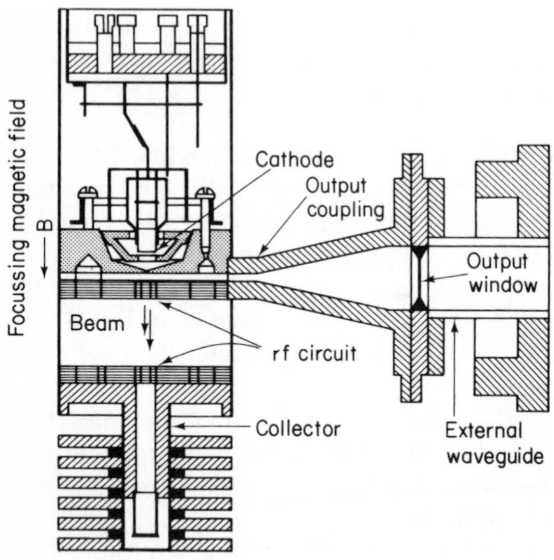

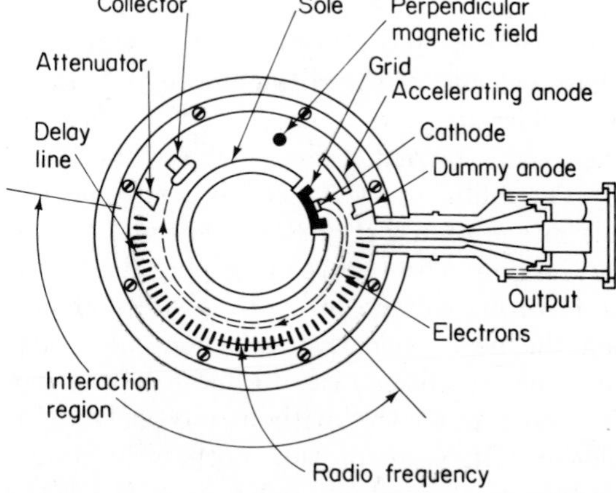

Fig. 13-7 REFLECTRON CONSTRUCTION

anode–cathode–cavity system common to a magnetron.

5. Amplitron (see Chapter 7). This device is a broad-beam, cross-field amplifier using a reentrant electron stream. The reentrant electrons interact with the backward wave of a nonreentrant rf structure. Efficiency and

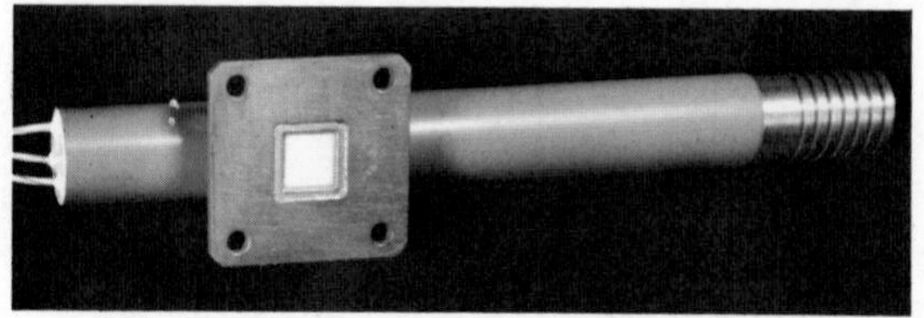

Fig. 13-8 M TYPE BWO (Courtesy of Watkins-Johnson.)

simple operation are claimed for this device (see Table 13-1).

SPECIAL MILLIMETER WAVE GENERATORS

Generating millimeter waves is an electron-orbital phenomenon whether it be through the use of an electron beam, as in the devices described above, or by some direct use of the electron's orbiting characteristics. All these devices rely upon shocking, pulsing, bunching, or pumping electrons or clouds of electrons from one energy level to another; this transition is accompanied by release or generation of useful output energy. A few devices coming under this category are

1. Tornadotron.
2. Ferrite-based structures.
3. Semiconductor diodes and varactors.
4. Bunched-beam devices.
5. Maser devices.

TORNADOTRON

The Tornadotron injects and traps a cylindrical column or cloud of electrons within an rf cavity. The electrons are then subjected to an rf field at right angles to the magnetic field. This causes the electron in the cloud to be pumped into an orbital motion and to produce radiation or power output at some submillimeter frequency.

FERRITE GENERATORS

Ferrite generators utilize an electron's ferromagnetic resonance within the bulk substance itself; electrons have a spinning characteristic similar to a gyromagnet and can be caused to precess by simultaneously applying a dc plus an rf magnetic field. This dual application delivers the pumping action, which alters the electron precession and generates power at some millimeter wavelength. Useful operation of this device has produced tens of watts up into the hundred GHz region.

SEMICONDUCTOR DIODES

These are bulk semiconductor devices and produce millimeter wave energy in either of two methods: (1) by harmonic generation or input-frequency multiplication using varactor diodes, and (2) by self-internal oscillation embodied in the Gunn oscillator, the impatt, and LSA oscillators.

VARACTOR MULTIPLIERS

Varactor multipliers are independent of voltage-capacitance relationships. They rely for efficient operation upon the sharp transition of their V-I characteristic to deliver an output that is rich in harmonics. At microwaves their physical dimensions enhance performance, and silicon and gallium arsenide or point-contact construction are commonly used. In gigahertz ranges, equipment must be tailored with diode total inductance value (case included) of fractions of a nanohenry. Figure 13-9(a) shows typical construction of three types of units in this range.

Microwave circuit assemblies with high multiplication usually employ an output filter for most efficient selection of compositely developed harmonics. Figure 13-10(a) shows four resonant-circuit multipliers (2x, 4x, 8x, and 10x) feeding a varactor through a sliding, quarter-wave choke. The varactor embedded in the first section of the filter operates under high input impedance, charge-controlled conditions for best efficiency; the choke also isolates output circuits from the idler-multiplier circuits. Figure 13-10(b) shows the equivalent circuitry. Output multiplication of 18x is the result of selecting the sum of idler harmonics of 2x products and 8x products.

A strip-line varactor doubler operable in the low gigahertz range is shown in Fig. 13-10 (c). Three strip transmission lines, A, B, and C are compatibly tuned, forming bandpass filters A and B at the fundamental frequency and B and C at the second harmonic frequency (about 1.6 GHz). The doubling varactor is at

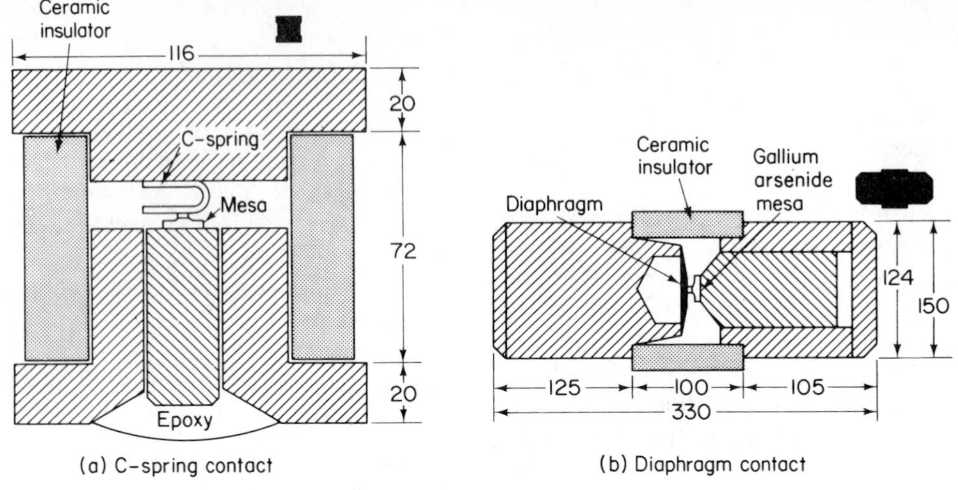

(a) C-spring contact

(b) Diaphragm contact

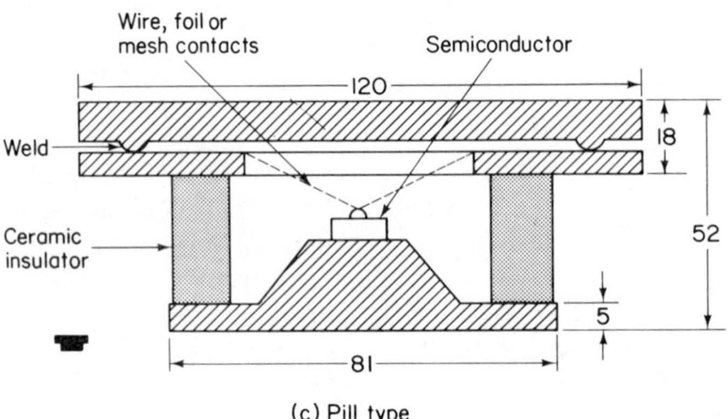

(c) Pill type

Fig. 13-9 VARACTOR CONSTRUCTION

the end of the B filter. The assembly, on a 1/16″ Teflon glass fiber strip 1″ by 2½″ by $\frac{3}{4}$″ delivers one watt of power at 45 per cent efficiency and less than 1 per cent spurious response.

Gunn diodes and impatt devices were treated in Chapter 7.

Figure 13-11 summarizes power output characteristic of common solid-state oscillator devices in the gigahertz range.

LSA Diode. The LSA diode (limited space-charge accumulation) uses gallium arsenide crystals specifically processed by doping, thickness, etc. The crystal operates by so proportioning the driving signal that space charge is prevented from accumulating and distorting the field within the junction area while maintaining sufficient overall field amplitude across the diode. The circuit is arranged to automatically quench, so to speak, the accumulation of space charge and allows oscillation to persist at an exceedingly efficient mode. Increased power at higher frequencies (44–88 GHz) than possible in Gunn oscillators has been obtained. Figure 13-12(a) and (b) show a conventional circuit and mounting arrangement for the LSA oscillator assembly.

BUNCHED ELECTRON-BEAM DEVICES

These devices operate by various means of periodically varying the cross section of an electron beam during its trajectory. This

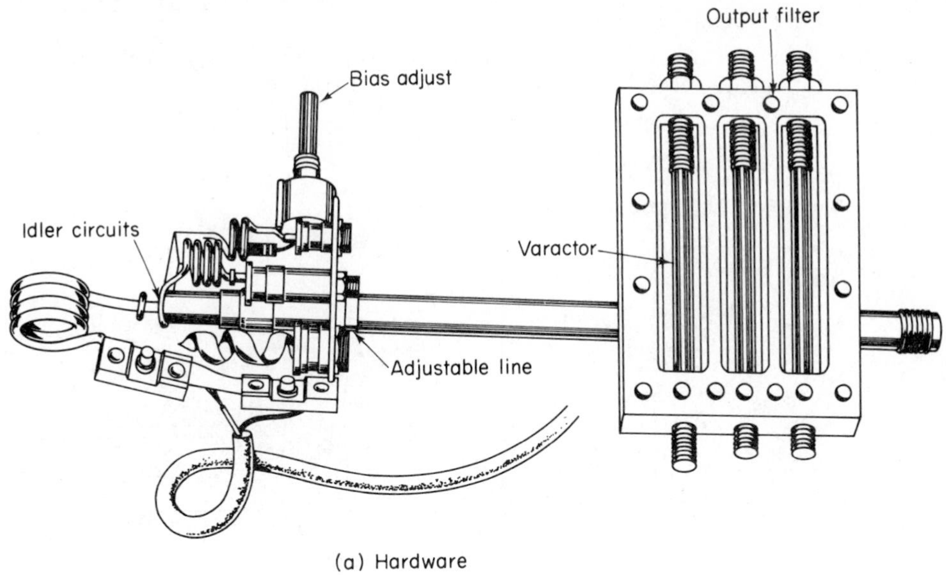

(a) Hardware

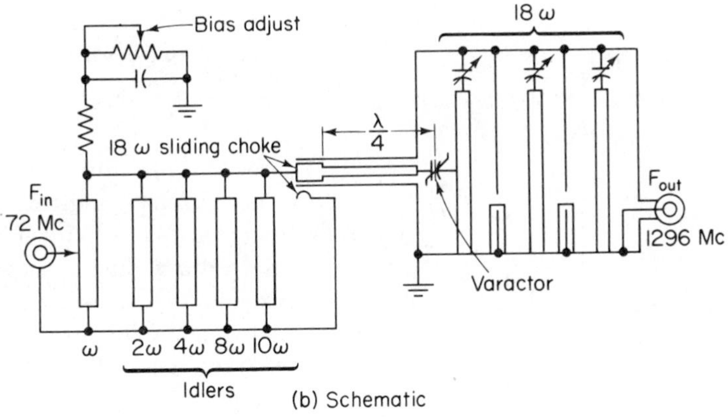

(b) Schematic

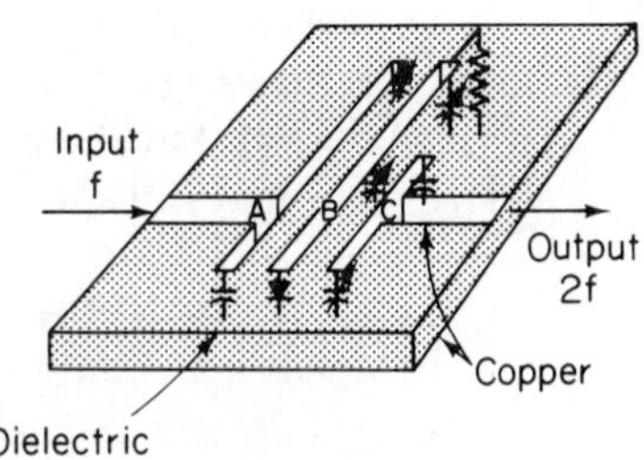

(c) Strip transmission line type

Fig. 13-10 VARACTORS IN MULTIPLIER CIRCUITS

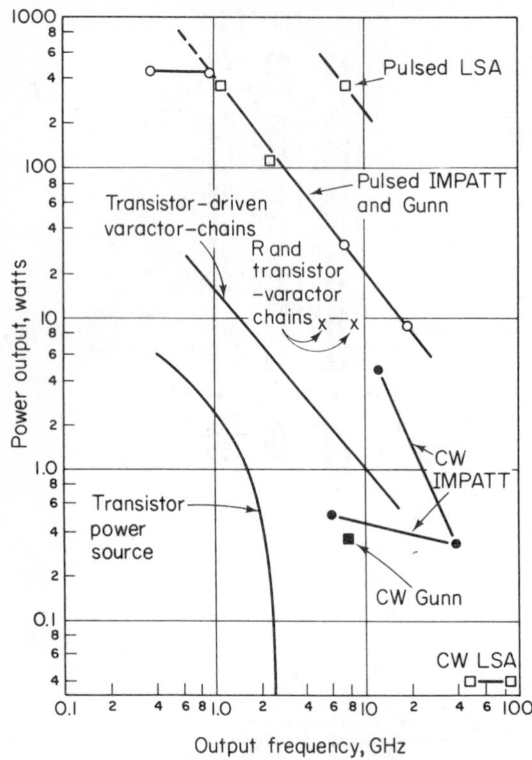

Fig. 13-11 GIGAHERTZ OSCILLATOR SOURCES

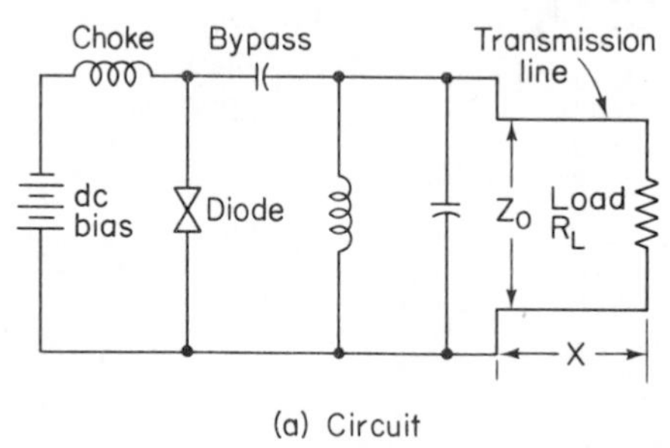

(a) Circuit

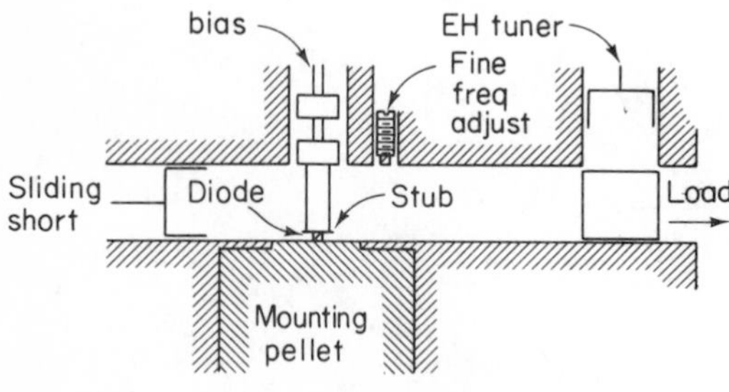

(b) Construction

Fig. 13-12 LSA OSCILLATOR CIRCUIT AND MOUNTING

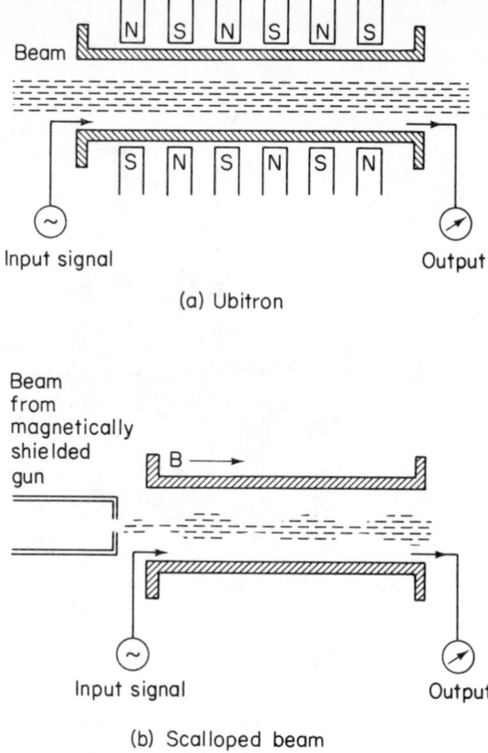

Fig. 13-13 UBITRON AND SCALLOPED BEAM TUBES

procedure produces interaction between electron groups, resulting in electron bunching and, ultimately, in the generation of millimeter wave output energy. Summarizing the various types, we find them classified by the name of the inventor, a trade name, or by the characteristic of the primary generated beam.

Ubitron (undulating beam.) This structure is basically an amplifier using a cylindrical beam injected axially into a rectangular waveguide propagating the $TE_{1,0}$ mode, and so arranged that the beam is made to undulate by means of a periodic, transversely directed magnetic-field pattern [Fig. 13-13(a)]. Peak power of 150 kW at 55 GHz has been obtained.

Scalloped-beam amplifier. The scalloped-beam amplifier uses the natural periodic undulations in the cross section of an electron beam emanating from the shielded gun; in addition, the beam travels inside a waveguide of circular cross section and functions because of the interaction between the fundamental space component of the scalloped beam and the forward-traveling $TM_{1,0}$ mode of the waveguide [Fig. 13-13(b)]. With kilovoltages accelerating the beam, the device exhibits about 10 per cent efficiency, delivering peak power at 40 GHz.

SWEPT-BEAM DEVICES. The Bermutron acts essentially as a harmonic amplifier. It uses a high kV electron beam transversely swept at a GHz rate by passing it through the strong electric field present in its coaxial cavity. Actually, the beam scans along a slotted section of a rectangular waveguide at a rate such that the injected electrons stay in phase with the traveling $TE_{1,0}$ wave.

CERENKOV RADIATOR. The Cerenkov radiator passes a lightly bunched electron beam, velocity modulated at some frequency below the millimeter band, within close proximity of a dielectric material (usually through a hole). Radiation of energy from the beam into the dielectric takes place if the beam velocity is greater than the speed of light in the dielectric; in other words, the transfer of power results when electrons passing through the hole possess higher velocity than the slower electrons passing through the dielectric. This device has limited power capabilities and is allied to a similar device known as the Rebatron. Figure 13-14(a) shows a physical arrangement of components in the Cerenkov radiator.

MAGNETIC UNDULATOR. The magnetic undulator, a prebunched beam device, is similar to the ubitron but functions without interaction onto an enclosing waveguide. Prebunching of the beam delivers extra energy transfer through increased undulation delivered by the periodically transverse magnetic field. [See Fig. 13-14(b)].

MASERS—PARAMETRIC AMPLIFICATION

The maser warrants brief discussion for it is an amplifying and generating device in the rf spectrum overlapping millimeter waves

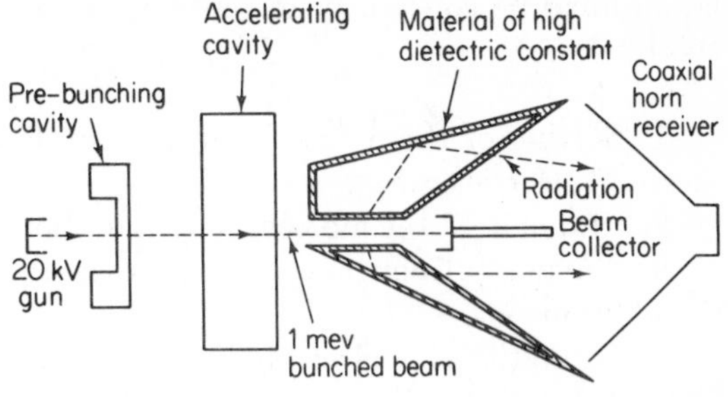

(a) Basic layout

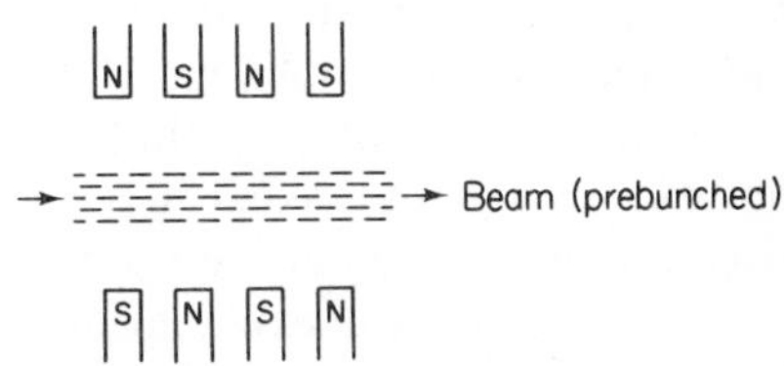

(b) Prebunching in magnetic undulator

Fig. 13-14 CERENKOFF RADIATOR AND THE MAGNETIC MODULATOR

in the region between about 100 GHz and a few thousand GHz.

Maser devices are closely allied operationally to the mechanism described above under bunched-beam generators. They are energy-conversion devices depending upon atomic structures and spinning electrons in their orbits around the nucleus; the operation is dependent upon changing energy levels and uses spin of the individual electrons themselves.

In changing from a lower to a higher level, energy is consumed; conversely, in a generating device, energy is freed when the energy level is reduced. Power is liberated in the form of photon radiation, whose frequency depends upon the energy-level transition. The energy change is externally triggered by shocking or pulsing individual atoms with photons of an electromagnetic wave of the proper frequency. To sustain the conversion process, the higher-energy-level electron population must be constantly replenished; we do this by:

1. Sorting of the molecules in a beam by means of electric or magnetic fields.
2. "Pumping" the molecules by radiation, where the pump frequency is higher than the signal frequency.
3. Excitation by electron bombardment.

The parametric amplifier uses maser processes at the high end of the microwave spectrum, below the range where most gas masers operate. It is a device introducing varactor diodes into the amplifying circuits, thus adding another parameter—hence the name parametric amplifier. It is an energy-delivering device (by pumping the molecules) and in so doing imparts extra, noiseless energy, thus giving the signal a better noise figure.

A simple physical explanation, rather than the specific atomic-electron interactions, will serve to illustrate the action if we visualize a practical analogy of the pumping mechanism. We do this by considering the tuning cavity as partially made up of separate capacitor plates (constituting the varactor's variable capacitance) with the plates being alternately pushed apart and brought together. If, at the instant of maximum voltage across the capacitor plates, they are physically moved apart, their capacitance will be decreased and the voltage across them will increase, as shown in the "jump" in Fig. 13-15(a). If, then, one quarter of a cycle later, at zero voltage across the plates, they are physically moved together to their original position, again no change in capacitance has been produced, since there is no charge on the capacitor at that instant; consequently, there is no electrical energy transfer. Repeating the operation on the negative half-cycle and continuing it from there on, we see in Fig. 13-15 that energy has been added to the signal (and none subtracted as the plates are moved together) as a result of this pumping action applied to the capacitor plates.

Electrically, this process is most efficiently done in the simplest cases by pumping at double the signal frequency. This means the adding of a pumping circuit to the resonant signal circuit of an "idler" frequency—one which is appropriately filtered and adjusted for maximum pumping action so that it does not interfere with the signal frequencies. All this is simply done at microwave frequencies

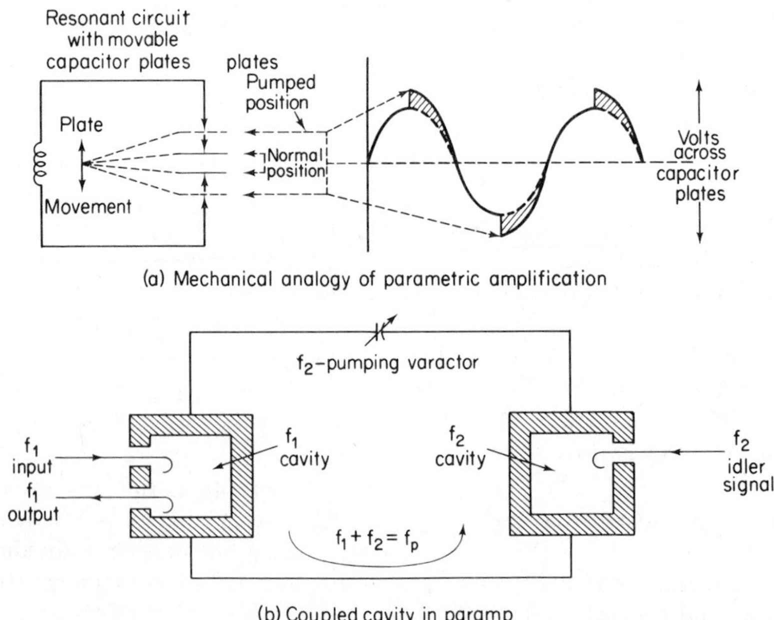

Fig. 13-15 CAPACITOR PLATE PUMPING ANALOGY IN PARAMETRIC AMPLIFIER

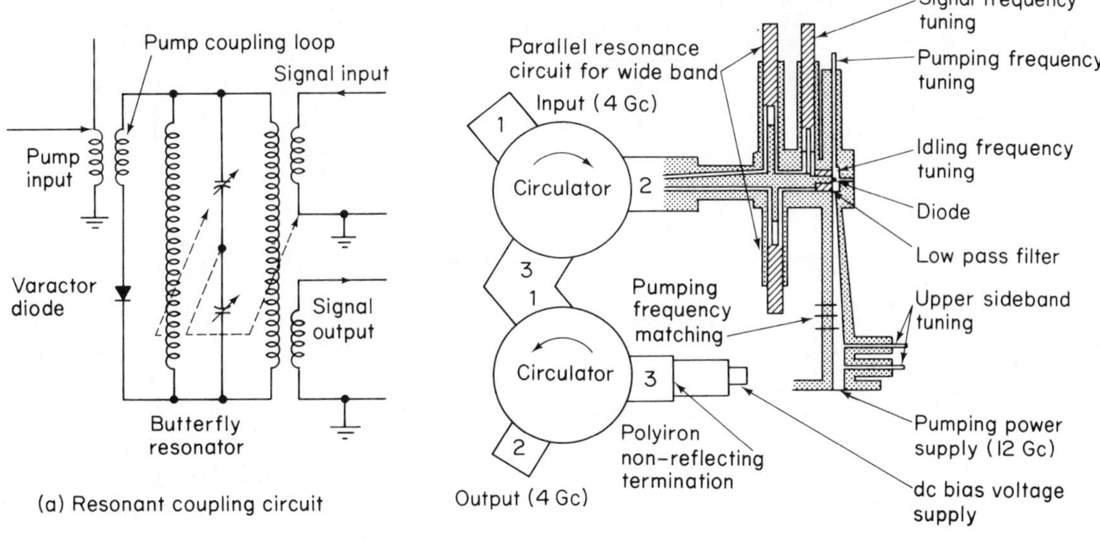

Fig. 13-16 PARAMETRIC AMPLIFIER ASSEMBLY AND SCHEMATIC

by a varactor diode, which is physically small enough to fit directly into a waveguide or a cavity, particularly since it generates within itself second harmonic voltages directly at the pumping frequency and at the expense of no fundamental signal energy.

Circuitwise, the injection of external parametric power into a resonant circuit generates negative resistance, reducing inherent losses and bringing the circuit toward oscillation. Typically, in a parametric amplifier circuit the input signal is enhanced in its own cavity or waveguide and the amplifier output is picked off at another point, say through another port in a microwave circulator. A schematic arrangement of a negative resistance amplifier of the coupler cavity type is shown in Fig. 13-15(b).

Figure 13-16(a) shows the schematic arrangement of a VHF parametric amplifier using a butterfly resonator; Fig. 13-16(b) illustrates a 4-GHz microwave amplifier with varactor-developed triple-harmonic pumping and with the input and output signals separated by microwave circulators.

MILLIMETER WAVE RECEIVERS

GENERAL

Communication in this area is basically dependent upon noise figures in solid-state mixers and amplifiers. Here developments in the former have not progressed as fast as in generating devices, chiefly because such devices become inherently noisier as operating frequency increases. Parametric amplification using gallium arsenide diodes, however, offers excellent opportunity for noise-level improvement.

Traveling-wave amplifiers are another approach to better performance in the 30- to 100-GHz range. Figure 13-17 illustrates the schematic and physical arrangement of a low-noise TWT amplifier delivering noise figures of 10 to 12 dB in the K and R bands. Noiseless signal enhancement exists because of the basic reasons discussed above.

Gas masers in microwave systems can produce low-noise operation in the 30- to 100-GHz range; for example, a ruby crystal using chromium ions has been "masered" in

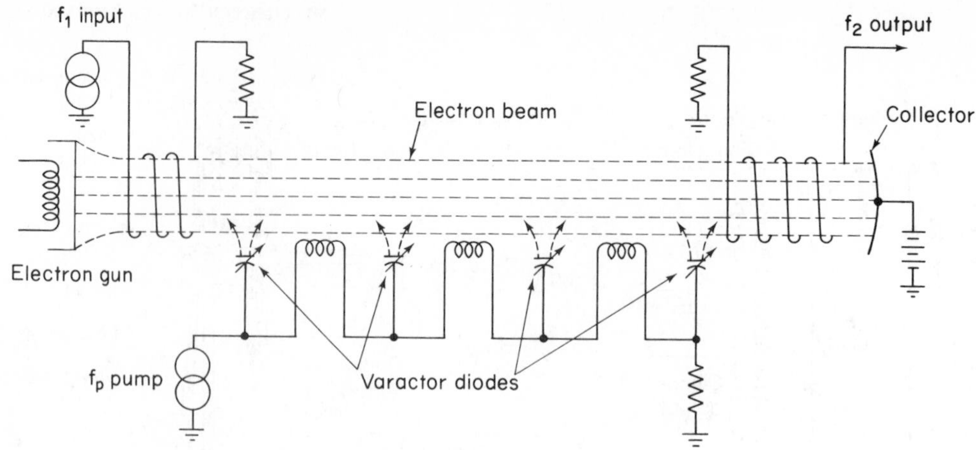

Fig. 13-17 LOW NOISE TWT AMPLIFIER

the 20- to 40-GHz range using pump frequencies from 43- to 47 MHz.

HARDWARE CIRCUITS—CONSTRUCTION

In general, millimeter wave hardware, excluding generating devices, consists of dimensionally scaled-down components of conventional design and contour. The same holds for measurement devices. Figure 13-18 illustrates a conglomerate assembly view encompassing most typical millimeter wave hardware components, including measurement items.

Amplifiers for millimeter wave receivers in the low-GHz region are logically constructed of integrated circuit components. Circulators, avalanche pump oscillators, and varactor diodes making up a typical assembly are seen in Fig. 13-19. The final package weighs 10 oz, is $2\frac{1}{2}$ by $2\frac{1}{2}$ by 1 in. using 0.05-in. polished alumina substrates and chromium–copper gold microstrip lines. The varactor is a high-cutoff-frequency junction diode exhibiting series resonance at the paramp idler frequency; the avalanche pump oscillator is a silicon diode.

Fig. 13-18 MILLIMETER WAVE HARDWARE (Courtesy of Hitachi, Inc.)

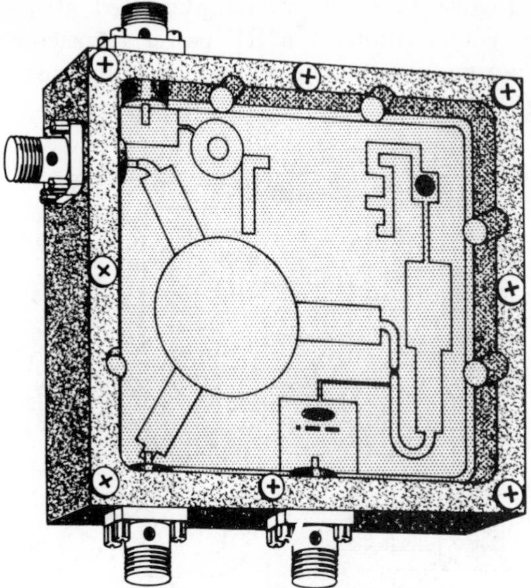

IC microstrip paramplifier

Fig. 13-19 IC MICROWAVE AMPLIFIER

The amplifier has a 17-dB gain and 2.5-dB noise figure; it operates over a 25-MHz instantaneous bandwidth within the frequency range 2.2–2.3 GHz.

Figure 13-20(a), (b), and (c) show similar amplifier units in the low-GHz frequency range.

Amplifier performance specifications

Frequency

PA3940	1.0 – 1.5 GHz
PA3941	1.5 – 2.0 GHz
PA3942	2.0 – 2.3 GHz

Gain (min. across band)

PA3940	6 dB
PA3941	5 dB
PA3942	4 dB

Output power (min.) = 10.0 W at 25° case temp.
± 0.5 dB from −30° C – +70° C

Input impedance = 50 ohms

Output impedance = 50 ohms

Gain flatness ≤ 0.5 dB

dc power 28 V ± 5%

Input and output VSWR ≤ 1.5

Efficiency ≥ 25%

Transistor thermal impedance θ_{jc} 4.0° C/W

Max. load VSWR 3 : 1 all phase angles

Storage temperature −50° C – +130° C

Dimension $1\frac{1}{2}'' \times 3\frac{3}{4}'' \times \frac{3}{4}''$

(a)

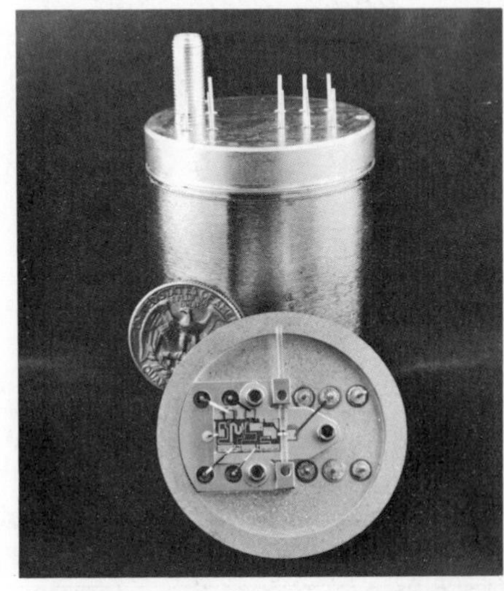

(a)

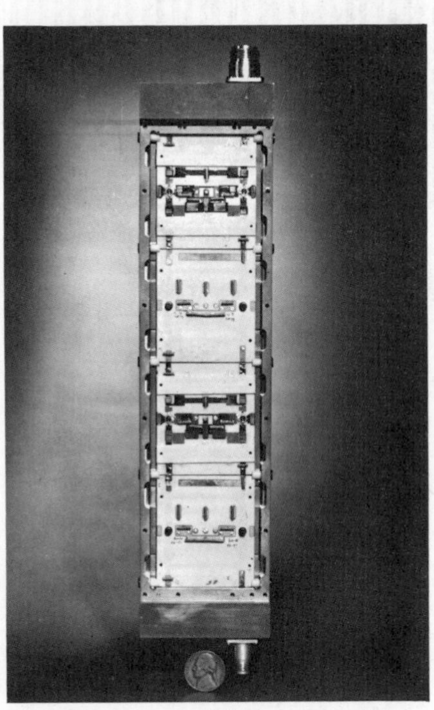

(b)

Fig. 13-20 WIDE BAND AMPLIFIER (Courtesy of Texas Instruments.)

Figure 13-21 shows interdigitated strip-line construction of a RF power transistor delivering a watt at low microwave frequencies.

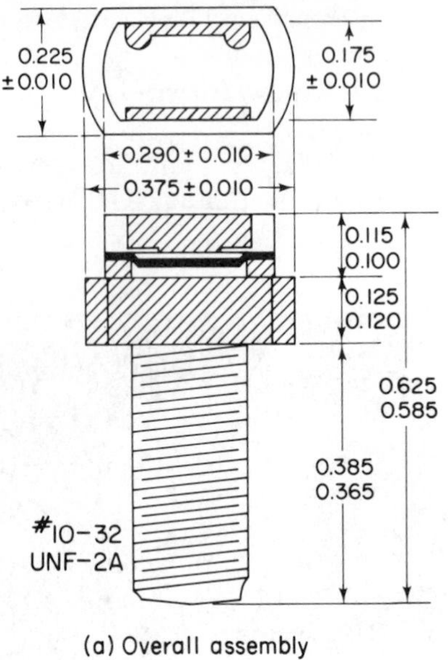

(a) Overall assembly

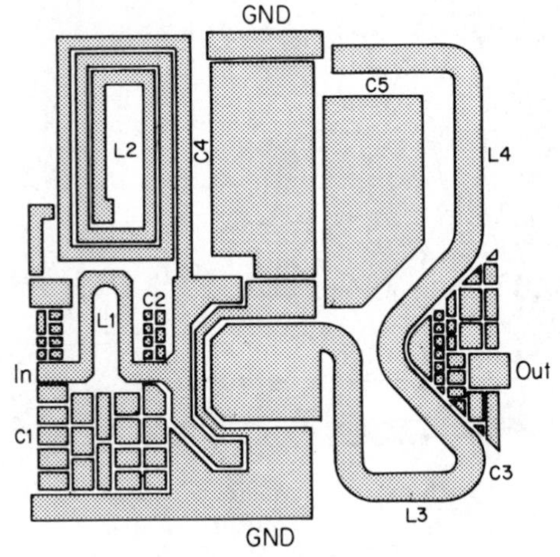

(c) Amplifier construction

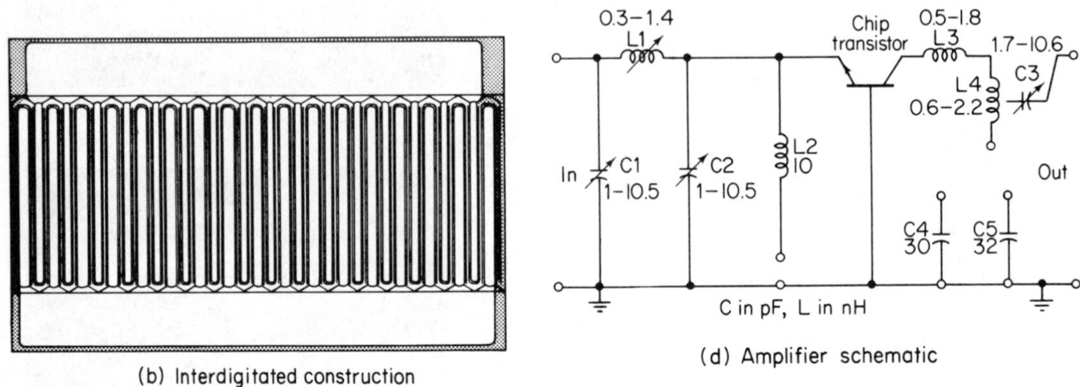

(b) Interdigitated construction

(d) Amplifier schematic

Fig. 13-21 MICROWAVE POWER TRANSISTOR (Courtesy of RCA.)

Microwave Antennas 14

GENERAL

The principles underlying the physical design and performance of microwave antennas differ chiefly from those of long-wave antennas within the scaling-down factors. During design, however, the basic design characteristics for both include five parameters:

1. Impedance.
2. Bandwidth.
3. Gain.
4. Polarization.
5. Pattern.

In general, the reduction in wavelength and size that go along with microwave antennas offers a number of advantages, plus freedom and variations in design parameters; these give flexibility to construction and operation of a considerable number of antenna types. The flexibility, permissive variations, and advantages stem from five factors inherent in short-wave power-radiation characteristics.

1. Extension of basic waveguide structural dimensions into radiating elements.
2. Reflection properties of shorter wavelengths.
3. Inherent lens action.
4. Traveling surface waves.
5. Beam scanning.

Thus our main approach, after discussion and establishment of fundamental antenna action plus associated component disposition, will be a description and the design relationships pertaining to each of these basic factors. Our approach then treats antenna measurement techniques and practical applications.

BASIC THEORY AND ACTION

The basis of radiated antenna energy originates from the magnetic and electric fields originating and surrounding two finite lengths of current-carrying wires driven by an ac generator. As they exist in parallel physical position [Fig. 14-1(a)] and are spread angularly to 45° and then positioned in-line to a complete 180° opening, the fields persist as shown and the current surges back and forth the same as in an ac circuit terminated by a capacitance.

This simulation creates the elemental conditions surrounding an energized dipole or vertical half-wave antenna consisting of two quarter-wave radiating wires. We do this for electrical efficiency because quarter-wave line lengths are naturally self-resonant.

Next, Fig. 14-2 pictures the completed electric and magnetic field lines of this structure in their respective polarities, phases, and amplitudes of generated current and voltage as the energy pulses back and forth from the generator to the antenna elements. This, of course, means that both electric and mag-

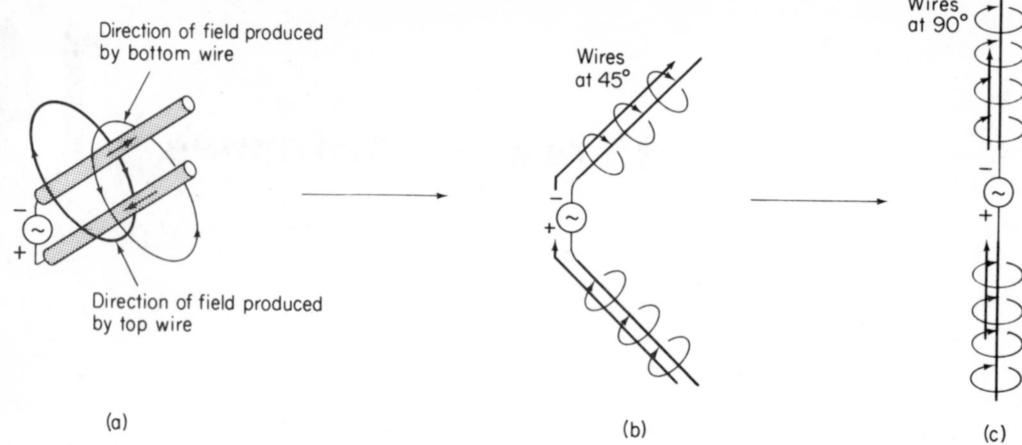

Fig. 14-1 BASIC ANTENNA DIPOLE STRUCTURE

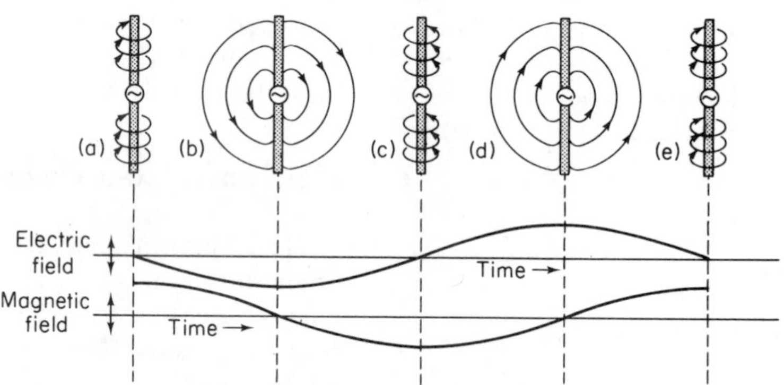

Fig. 14-2 FIELDS SURROUNDING A DIPOLE

netic fields emanate outward with increasing current and voltage, but, in classical terms, deliver no energy since the capacitive load returns its stored energy when voltage and current conditions are reversed.

However, we do know that energy is radiated, and from all evidence this absorption is related mainly to the collapse of the electrostatic field. This means that in the cycles of expanding and collapsing fields some of the field energy never gets returned to the generator and is literally "snapped off," detached, and in effect radiated from the illuminated or excited structure.

Now, physically this happens because the moving electric field actually completes itself, forming discrete bundles, of individual electrostatically enclosed chunks of air space dielectric, which are dynamic energy quanta, traveling outward at the speed of light; these are labeled the radiation field.

All the while, the magnetic field is also expanding (at right angles to the electrostatic field), but its predominant radiation effect is limited to the area directly surrounding the radiator, forming, in effect, a "near" or induction field; this field differs from the space-velocity nature of the charged-particle phenomenon existing in the electric field because there is less "snap off" of magnetic energy. Here we find that the magnetic field collapses completely when the driving current reverses and, in this case, acts conventionally as if the antenna were a resonant circuit. The induction-field strength decreases rapidly, varying inversely as the square of the distance from

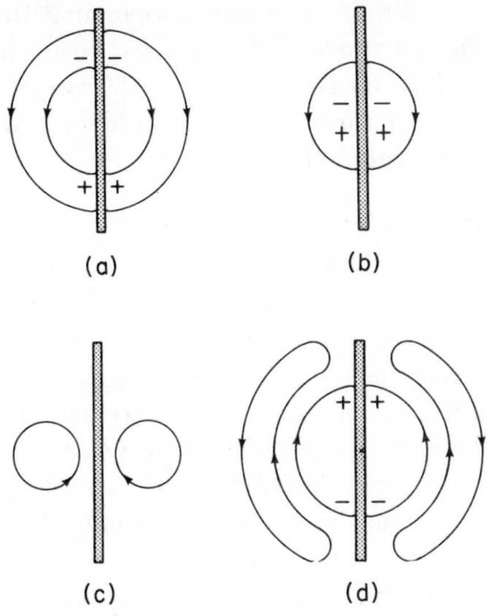

Fig. 14-3 DYNAMIC FIELD MECHANISM

the antenna. The radiation field also varies, but only directly, as the distance from the radiating antenna.

A simple way of visualizing this is to picture the electrostatic field lines of Fig. 14-2(b) and (d) starting to collapse as in Fig. 14-3(a) and (b). Later on, in Fig. 14-3(c), as the unlike charges near the antenna approach each other, one complete electrostatic field line forms, so to speak, a complete path or circuit, enclosing an area of uniformly stressed dielectric. Lines of force farther away from the antenna, however, are kept from collapsing completely before the next charge reversal occurs and are stranded in space, as we see in Fig. 14-3(d).

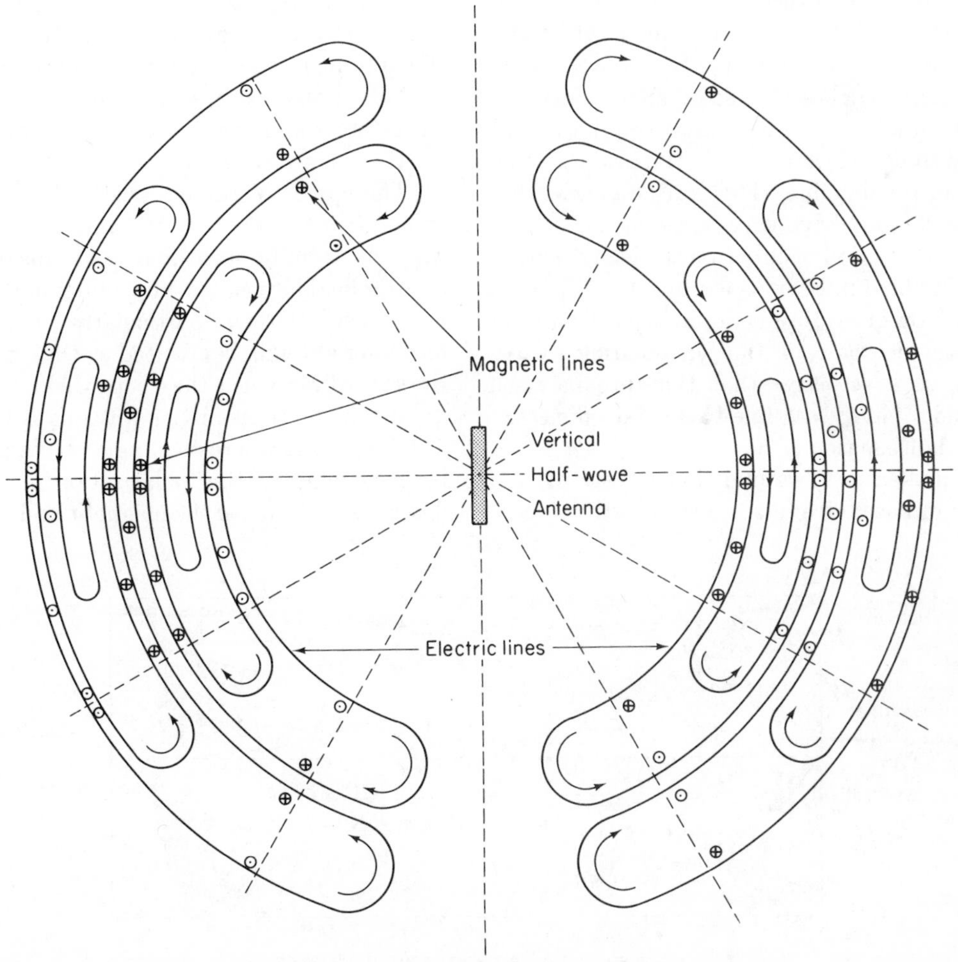

Magnetic lines
Vertical
Half-wave
Antenna
Electric lines

Fig. 14-4 RADIATION OF STRANDED FIELD QUANTA

Thus, picturing the field relationships of this stranded, radiated electromagnetic wave in Fig. 14-4, we see that it consists of both magnetic and electric components and their field lines, each in space quadrature with each other, but also both in time phase with its companion. Note that this radiated and propagated field possesses both magnetic- and electric-field strength lines, each supporting the other, but that the combination is not the induction field (also consisting of magnetic and electric lines). As noted above, the induction field strength drops off rapidly as we progress away from the antenna, reaching a value at one sixth of a wavelength to a value equaling the electric field. Nearer the antenna this radiation continues to dominate, but it decreases directly as the square of initial strength and power while the electrostatic field decreases with distance from the antenna.

From the above we see that for most efficient radiation we must have a radiator that bears some relation to a quarter-wave section of transmission line. This is necessary so that the pulsating drive energy is working with a resonant section of line (for efficiency) and so that its field pulsations can most easily get a "grip" on or attain maximum energy transfer to the dielectric of the surrounding media. This, in effect, says that the antenna itself should have a line-to-media dielectric impedance match.

Furthermore, we find that the strength of field radiated from a given length of conductor varies in direct proportion to the applied frequency. We can see then that a half-wave, 60-Hz antenna, 1550 miles long would be impractical and would in addition produce negligible power.

PROPAGATION—POLARIZATION

The radiation field representing energy radiated from an antenna moves outward in ever-expanding spheres. One (any one) sectional surface of such a sphere is known as a wavefront; it has all its energy components in phase and perpendicular to the direction of energy travel. When the pulsating magnetic and electric lines of force maintain their respective directions in the course of travel, they are said to be linearly polarized. When, with a vertical antenna, the electric lines of force lie in a vertical position, the wave is said to be vertically polarized, that is, at right angles to the surface of the earth. Reorienting the antenna horizontally produces a horizontally polarized wave. Figure 14-5 illustrates horizontal and vertical polarization of the traveling wavefront.

The choice of the electric field as the reference is made because the intensity of the wave is usually measured in terms of the electric-field intensity (volts, millivolts, microvolts per meter), particularly when the receiving antenna is oriented so that it lies in the same direction as the electric field.

When the transmitting antenna is located in an aircraft, where the angle above ground is considerable, the polarization is better described in terms of the magnetic and electric

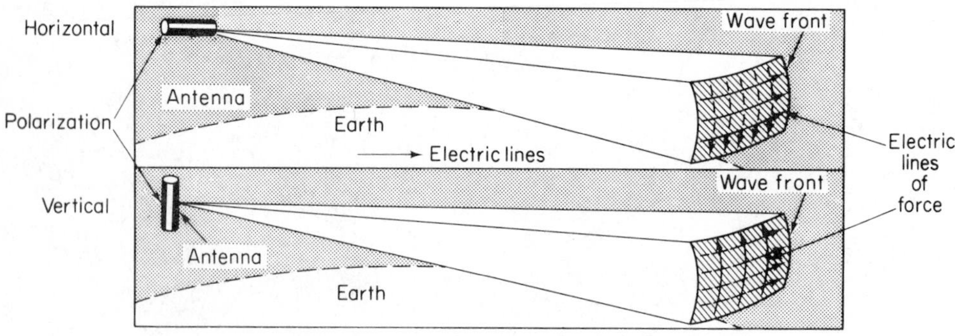

Fig. 14-5 POLARIZATION MECHANICS

fields in a three-dimensional, polar-coordinate system.

In addition, linearly polarized antennas may be circularly polarized—by transmitting two perpendicular, linearly polarized fields which have a 90° phase differential. Here, depending upon the sense of rotation, the circular polarization may be either right- or left-handed.

When antennas radiate energy in a polarization different from that intended, the unwanted radiation is known as *cross polarization*. Deviation from circular polarization exists when the sense of rotation is opposite to that intended and the axial ratio of field components shows ellipsoid characteristics.

DIRECTIVITY—ANTENNA RADIATION PATTERNS

Every antenna, and particularly microwave configurations, possesses some type of directivity. In other words, the electromagnetic radiation encompasses a three-dimensional volume of space with respect to the midpoint of its structure. We describe or measure this directivity by recording the amplitude of the field intensity (or power density) of the radiated energy present on the surface of a sphere with the antenna at its center. In rectangular coordinates we use for abscissas the angular deviation measured from the maximum radiation ordinate (straight ahead) as shown in Fig. 14-6. The ordinates may be in power, voltage, or as a numerical ratio (decibels) taken with respect to an ideal dipole.

The polar-coordinate form of field-intensity record, consisting of field intensity radii, is useful in picturing antenna coverage over a full 360°. Figure 14-6(b) illustrates typical side lobes and back radiation. Antenna patterns may be conventionally calculated or tested for variations by using the unit under test as a receiving antenna and then taking measurements as it is rotated in a particular plane.

Conventional radiation patterns, beyond the omni- or broadcast type, are the beam-shaped configurations summarized below—pencil, fan, and special. Note that the omni- or broadcast-type patterns find use in communication services with the horizontal-plane pattern generally circular in shape, while the vertical-plane pattern may have directivity in order to increase gain.

CARDIOID PATTERNS. Cardioid patterns, which are figure-eight patterns, are the basis of beam construction developed in the designs discussed later. Figure 14-7 illustrates a number of typical patterns for a center-fed dipole using varying lengths. Minor lobes exist because of diffraction between different standing voltage modes existing with specific radiator dimensions.

PENCIL BEAM PATTERNS. Pencil beam patterns are highly directional and are used to obtain maximum gain when the directed energy is confined to as narrow an angular sector as possible. Beam widths in the two principal fixed planes are essentially equal.

FAN BEAM PATTERNS. Fan beam designs are similar to pencil beam patterns except that the overall configuration has a truly fan-shaped beam cross section which is elliptical rather than circular; in other words, the beam width in one plane is broader than the beam width in the other.

SPECIAL BEAM PATTERNS. A number of applications dictate that radiation patterns be specifically tailored for particular coverage. We may find, for instance,

1. The cosecant pattern for providing constant radar-signal return over a range of angles in the vertical plane.
2. Figure-eight patterns for direction finding.
3. Split-beam amd multilobe patterns for special applications.

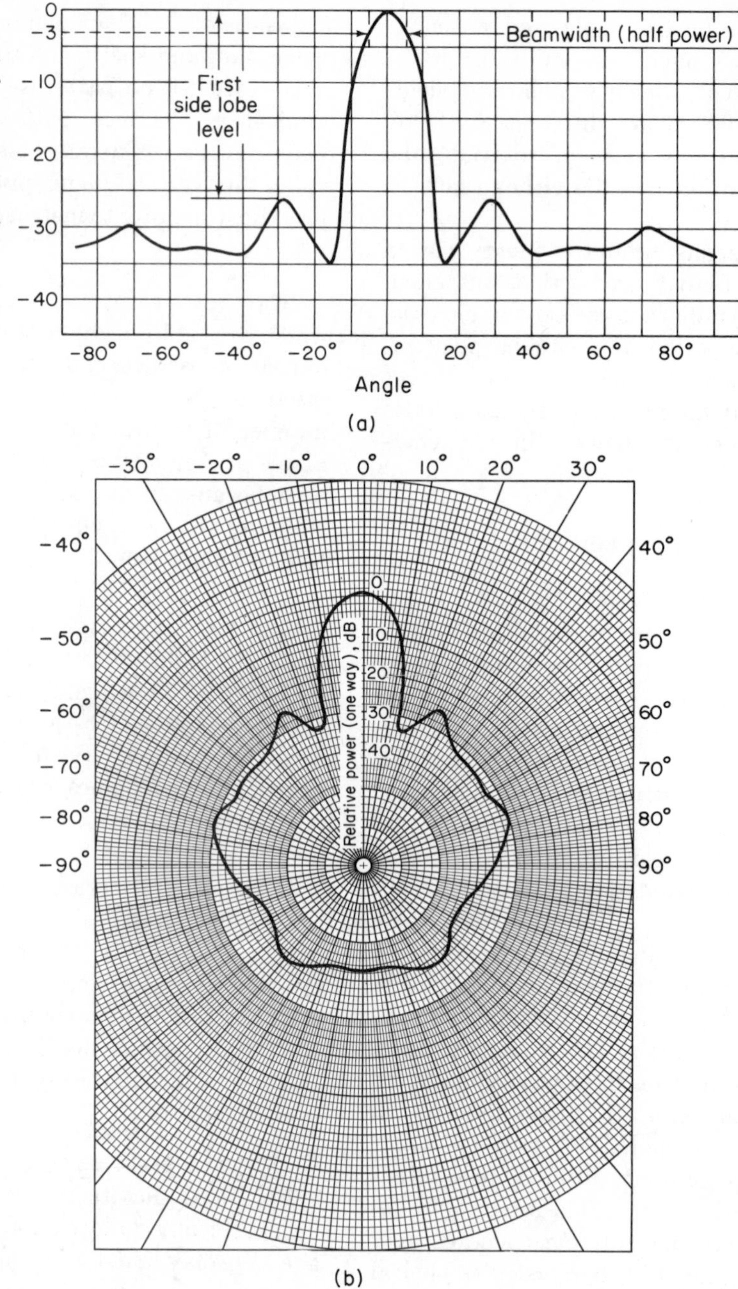

Fig. 14-6 RECTANGULAR AND POLAR FIELD PATTERN

POWER GAIN

Basic antenna gain is a figure of merit closely associated with directivity; it is the ratio (dB's) of an antenna's field intensity (in a given direction) to the field intensity produced (in the same direction) by either an isotropic or a half-wave dipole radiator.

Gain comparisons are usually listed with respect to both the isotropic and to the half-wave dipole antenna; the former is a convenient, ideal (but unattainable) imagi-

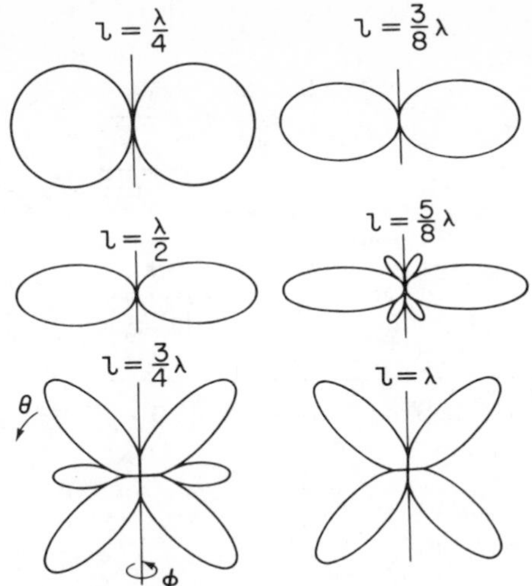

Fig. 14-7 TYPICAL CARDIOID PATTERNS

nary antenna, easily visualized for direct comparisons.

High-gain figures are associated with narrow beam widths and are usually obtained by direct comparison measurements against a standard gain antenna.

IMPEDANCE

To feed, drive, or use an antenna most efficiently, its basic impedance must match the actuating (or receiving) transmission line. Calculation of this quantity is for the most part a specialized operation, so practically we use predetermined or theoretical values from most available data, interconnect the unit with the system, and adjust matching stubs or filters after measurement for minimum VSWR.

We might add that the general procedure for calculating antenna impedance is to divide the total radiated power by the square of the effective antenna current. For short antennas this is a significant quantity, because it enables the engineer to estimate the overall radiation efficiency by separating the radiation component from the loss resistance due to the ground system and that caused by impedance mismatching.

For this computation it is necessary to know the radiation field pattern of the antenna in terms of the current flowing at the point to which the radiation is referred. The total radiated power is then computed by integrating the total power density passing through a sphere surrounding the antenna.

BANDWIDTH

An antenna's bandwidth can roughly be defined as the frequency limits within which it meets most of its performance characteristics.

This is at best only an approximation, for, unlike circuits or component requirements, that can be directly related to frequency, the combination of related factors around which an antenna performs presents a multipronged requirement when linked to specific frequency limits. We must, therefore, when referring to bandwidth, specifically consider frequency-dependent changes in

1. Pattern shape or direction.
2. Gain characteristics.
3. Side-lobe level.
4. Polarization.
5. Impedance.

We may note that in microwave antennas which have inherently small dimensions (i.e., when the linear dimensions are of the order of one half-wavelength or less) the limiting factor is impedance variations. In circularly polarized antennas, the angular polarization characteristics suffer first with frequency variations.

In structures with simple radiation patterns, where a single lobe is predominant, the equivalent Q of the antenna, as determined by its radiation resistance, bears a direct relation to the bandwidth (Fig. 14-6); the higher the Q, the less the bandwidth.

ANTENNA TYPES—APPLICATIONS

Broadly speaking, antennas may be categorized as either under dipole or aperture types. The former stems from the basic radiator described above, is physically (with

Type	Configuration	Z_r ohms	ΔB %	Gain (dB) above Isotrope	Gain (dB) above Dipole
Small diphole		—	—	1.74	−0.4
$\frac{1}{2}\lambda$ diphole /thick tube L/D = 276	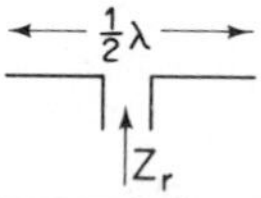 ← $\frac{1}{2}\lambda$ → Z_r	60	34	2.14	0 Reference dipole
$\frac{1}{2}\lambda$ diphole /thick tube L/D = 51	Similar to above	49	55	2.14	0
$\frac{1}{2}\lambda$ diphole cylinder L/D = 10	Similar to above	37	100	2.14	0
$\frac{1}{4}\lambda$ folded diphole	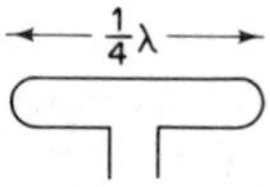 ← $\frac{1}{4}\lambda$ →	6000 (resistive) 260 (av. surge Z)	5	1.64	−0.5
$\frac{1}{2}\lambda$ folded diphole	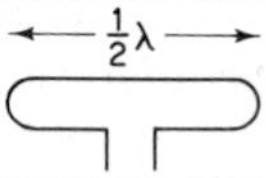 ← $\frac{1}{2}\lambda$ →	300	45	2.14	0
1 λ diphole cylinder	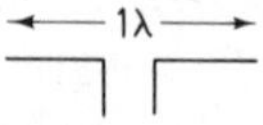 ← 1λ →	150	130	3.64	1.5
$\frac{1}{2}\lambda$ biconical	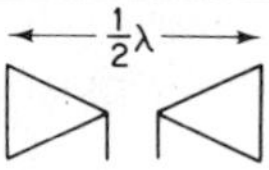 ← $\frac{1}{2}\lambda$ →	72	100	2.14	0
1 λ biconical	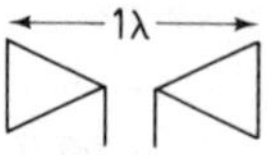 ← 1λ →	350	200	2.14	0
Crossed dipholes or turnstile/ (1 stack)		150	50	−0.86	−3
Turnstile 2 stack $\frac{1}{2}\lambda$ /separation	Similar to above except stacked	—	—	2.14	0
Super turnstile/or batwing					
1 Sections		—	—	2.14	0
3 Sections		—	—	7	4.8
6 Sections		—	—	10.14	8
12 Sections		—	—	13	10.9
4 bay helical		—	—	15.14	13
$\frac{1}{2}\lambda$ diphole and reflecting/sheet	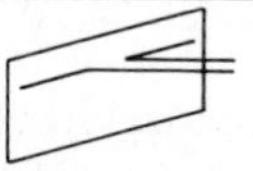	150	20	7.14	5

Fig. 14-8 DIPOLE ANTENNA CHARACTERISTICS

one exception) one half-wavelength overall, and radiates power in some form of lobe pattern. Aperture antennas concentrate, confine, reflect, or otherwise guide the radiated energy so that it appears as though passing an aperture.

Scannable antennas, those that can be adjusted or varied, are a hybrid category using the features of various basic types.

DIPOLE ANTENNAS

Each of these has some particular construction or group of performance characteristics listed in the tabular form of Fig. 14-8. For the most part these designs are physically pictured and placed as shown in the first column and deliver horizontal polarized waves. In addition to their impedance, front-to-back ratios, and gain above an isotrope antenna, their characteristics and applications are covered in the following discussion.

1. Simple half-wave dipole. This design has no rejection for back-path reflections (no front-to-back ratio) and is good for narrow frequency coverage.

2. Dipole with screen reflector. This design increases dipole impedance and improves front-to-back ratio but has narrow frequency coverage.

3. Folded dipoles. These represent a still further increase in dipole impedance.

4. Single bow-tie with reflector. Forward range and front-to-back ratio are increased over a dipole structure.

5. Arrays. These consist of a number of dipole radiators placed side by side or endwise so that individual radiated lobes will reinforce each other, producing increased gain and narrow beams. This is particularly true in the end-fire and broadside arrays.

(a) Log periodic array. This array delivers uniform gain and VSWR over a wide frequency range. Gain is excellent; bandwidth is narrow.

(b) The Yagi array is a collection of closely coupled parasitic arrays where the parasitic element may function as either a director or reflector. Moderate gain with narrow frequency coverage results.

(c) Vertically polarized dipoles. A number of these with moderately low impedance and conventional dipole gain require special configuration to obtain vertical polarization.

APERTURE ANTENNAS

Aperture antennas are classified generally as reflector, horn, lens-actuated, and traveling waves. Figure 14-9 characterizes them and each of the general and sub-types is discussed in the following section. Note that some of the designs are described and related to the types of feed employed.

REFLECTORS. These probably represent more general usage and different types than either lenses or horn types. In general, they generate pencil, fan, or specially shaped beam-radiation patterns and deliver from 10- to 20-dB gain over the isotropic reference. In detail we may have:

1. Parabolic antennas with point source feed: The reflector may be conventionally symmetrical or cylindrical or offset.

2. Cassegrain antennas: This design feeds energy through the rear of the center of the main parabolic surface to a hyperbolic sub-reflector placed at the wave-focusing point. This arrangement avoids the bulky structure of the horn-feed elements and the losses encountered with a small driven element. Similar structures with modified driving systems are the Gregorian and the line-fed-spherical structure.

3. Corner reflectors [Fig. 14-9(c)]: These particularly fitted for narrow apertures and, having flat plane reflectors, are relatively easy to construct. Most usually they use a 90° corner angle with a driving dipole placed between one half and one wavelength away from the apex.

4. Special reflectors include the pill-box, the hourglass, the parabolic torus, the Zeiss cardioid, the hog horn, and the periscope. Each of these bears a direct relationship to the basic types described in the above.

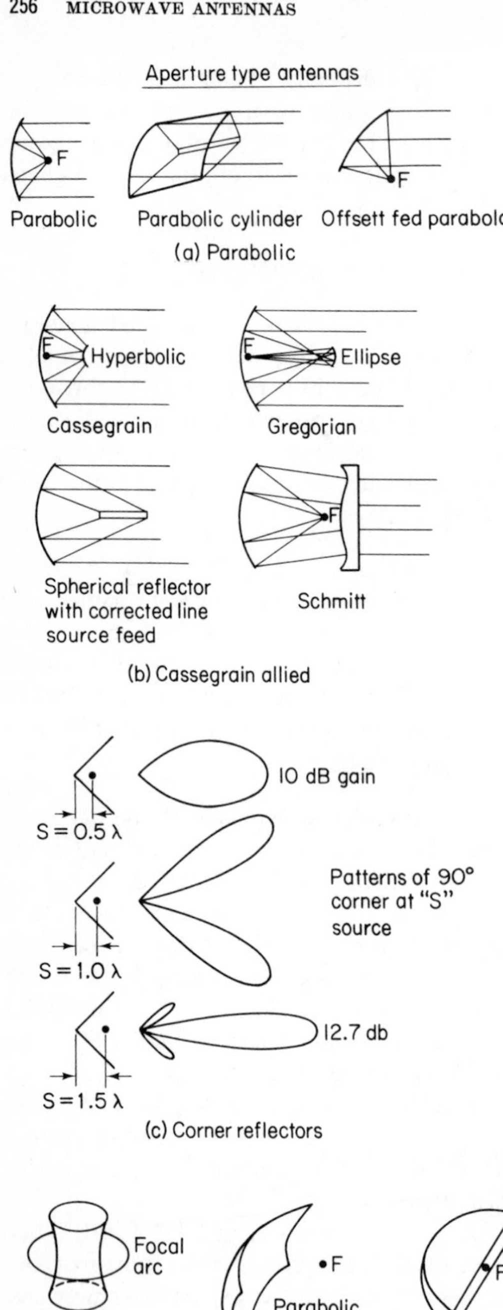

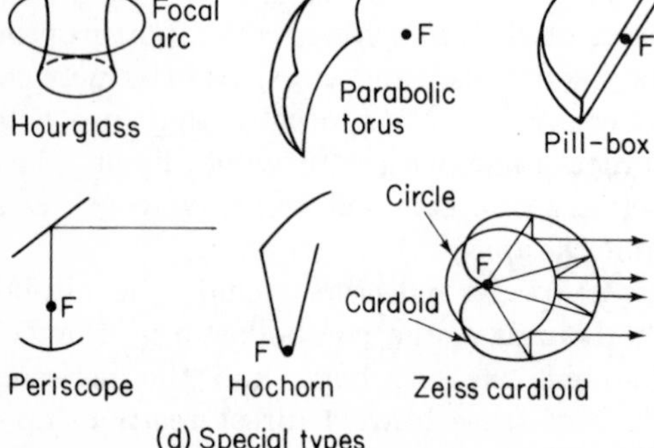

Fig. 14-9 APERTURE ANTENNA CHARACTERISTICS

Horns. These structures combine the feed with the radiation function by merely shaping the output end of the driving waveguide. Designwise, horn dimensions can easily control beam shape, gain, and polarization; they are particularly useful at microwave frequencies because they eliminate feed-to-radiator transition losses (Fig. 14-10).

Lens antennas. Microwave lens-type antennas are the electrical counterpart of optical lenses using dielectric materials instead of glass. In general, they resemble enlarged optical units and "bend" illuminating microwaves for focusing, thus accomplishing the objective performed by reflectors. For specific characteristics they have some advantages, and economically a lens may be cheaper to construct than a reflector since it has less critical dimensions. However, gain is usually 1–2 dB less than an equivalent reflector; on the other hand, its rearward radiation is lower. Figure 14-11 illustrates typical structures.

A special lens–reflector, called the Luneberg lens, has become popular because of its wide response angle. This reflector is a solid dielectric sphere using material whose index of refraction varies with the distance from the sphere's center. It collects energy that falls upon the surface of one hemisphere of the structure, refracting it through the sphere and bringing it to focus at the center of the surface of the opposite hemisphere (Fig. 14-12). If we place a reflecting cap at this focal point, energy will be reradiated in the direction from which it originated. Change in the direction of incident waves shifts the focal point, so that if we place a reflecting metallic cap over part of the back area, the target will respond over a wide range of angles of incidence.

Surface-wave and slot antennas. Traveling waves guided along a surface will radiate power directly from the wave itself. Similarly, slot-antenna radiation is a phenomenon known as leaky-wave propagation.

Traveling surface waves operate because of the discontinuity of the guiding surface, which causes the wave to give out power. The

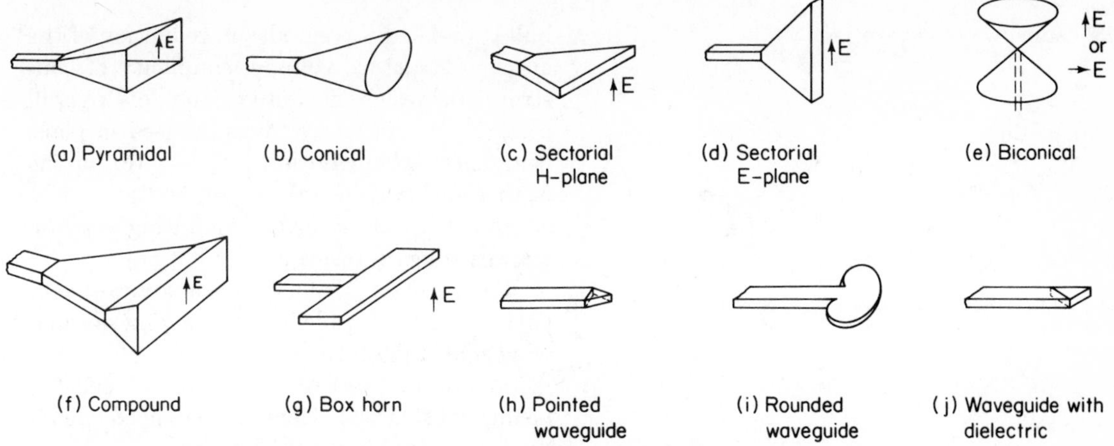

(a) Pyramidal (b) Conical (c) Sectorial H-plane (d) Sectorial E-plane (e) Biconical

(f) Compound (g) Box horn (h) Pointed waveguide (i) Rounded waveguide (j) Waveguide with dielectric

Fig. 14-10 HORN ANTENNAS

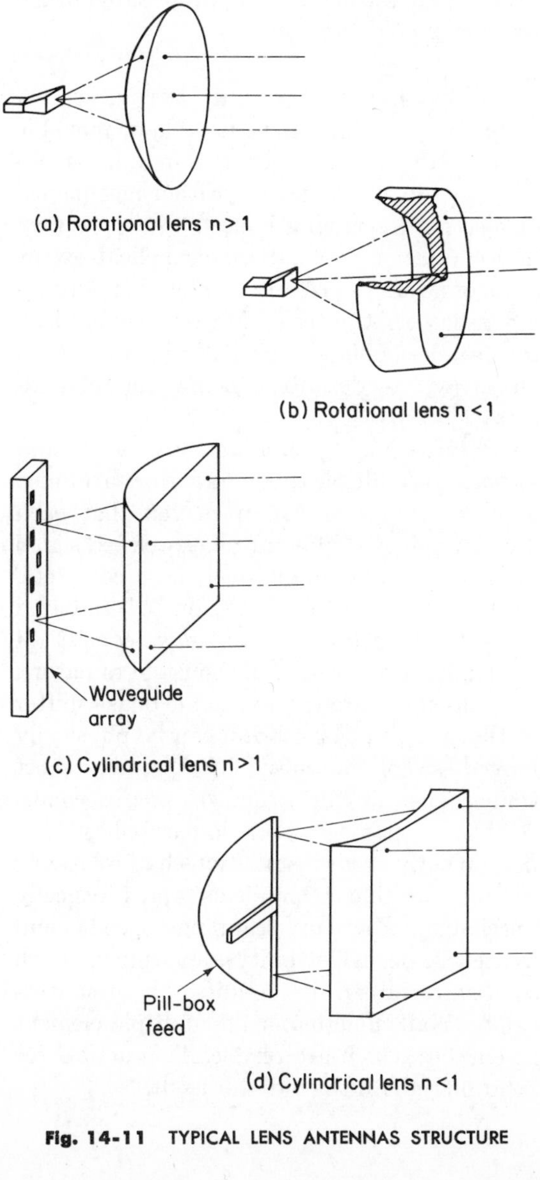

(a) Rotational lens n > 1

(b) Rotational lens n < 1

Waveguide array

(c) Cylindrical lens n > 1

Pill-box feed

(d) Cylindrical lens n < 1

Fig. 14-11 TYPICAL LENS ANTENNAS STRUCTURE

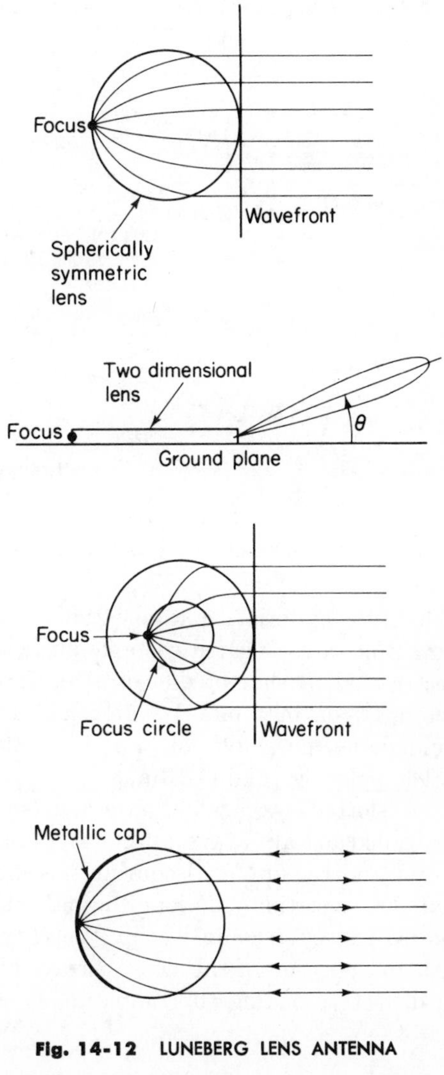

Focus

Spherically symmetric lens

Wavefront

Two dimensional lens

Focus

θ

Ground plane

Focus

Focus circle

Wavefront

Metallic cap

Fig. 14-12 LUNEBERG LENS ANTENNA

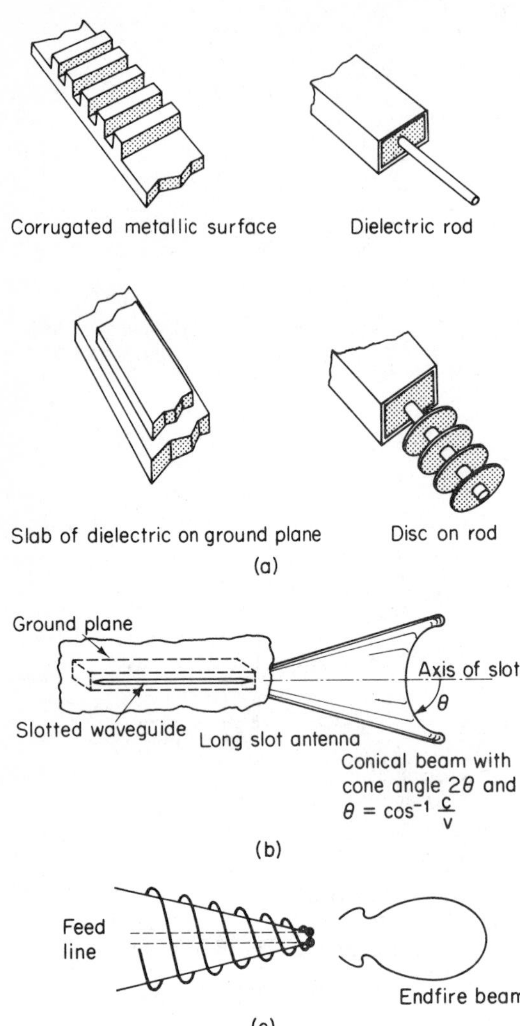

Fig. 14-13 SURFACE WAVE ANTENNAS

surface of the material can take various forms: for example, a corrugated metallic surface, a dielectric rod, a slab of dielectric or ferrite which may or may not be attached to a ground plane, or a disk on a rod of either metal or dielectric [Fig. 14-13(a)].

The slotted waveguide antenna (sometimes called a leaky-wave antenna) radiates energy by virtue of the discontinuities introduced by insertion of longitudinal slots. Radiated energy is controlled by the varying length and position of the slots. Arrays built from dielectric loading offer advantages over bulky air-filled waveguides in reduction of the antenna structure volume, simplicity of construction, design flexibility, and low overall cost. They can many times be used in place of dipoles, large cavities, helices, etc., in the 5- to 10-cm wavelength region and are easily adapted to flush or surface mounting—say on aircraft wings or fuselage.

Figure 14-13(b) illustrates a conical beam pattern made possible by using a long slot in a waveguide radiator.

Figure 14-13(c) shows a special end-fire beam constructed from a balanced helix-diode radiator built in conical form.

Surface-wave antennas are obviously end-fire devices, the beam being propagated in the antenna's axial direction.

Scannable or search antennas. A scannable unit aims at adjustable or movable action of the radiating beam, as in radar tracking, or in direction finders. Either mechanical or electronic scanning is used where we may desire rotation, oscillation, or helical excursions of either the reflector or (with a stationary reflector) the feed. Modern trends lean toward electronic scanning of all elements of an array, as described below for phased-array systems.

Phased-array configurations are constructed of multiple dipole radiators arranged in clusters and so fed or driven that each dipole receives a different phase-related signal in order to make up or construct a complete, electronically scannable beam of radiated energy. A phased array may consist of several hundred to several thousand radiators, each supplied through a variable phase shifter so that the composite beam can be physically moved, swept, or pulsed in extremely short time spans—of the order of microseconds.

The mechanical steerable parabolic dishes used in early radar installations had relatively complex rotating mechanisms, were physically unreliable, slow in scanning speed, and limited in beam flexibility and agility. With solid-state integrated circuits, phase shifters can be small enough to make multiple-element assemblies which are physically practical for location right at the radiator itself.

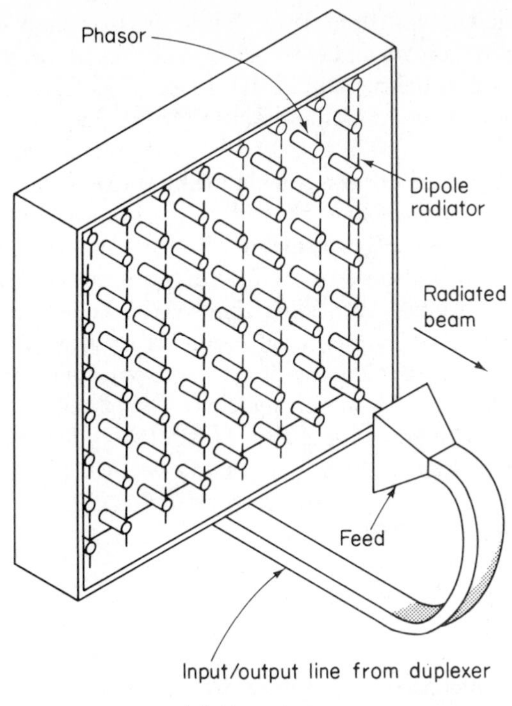

(a) Horn fed

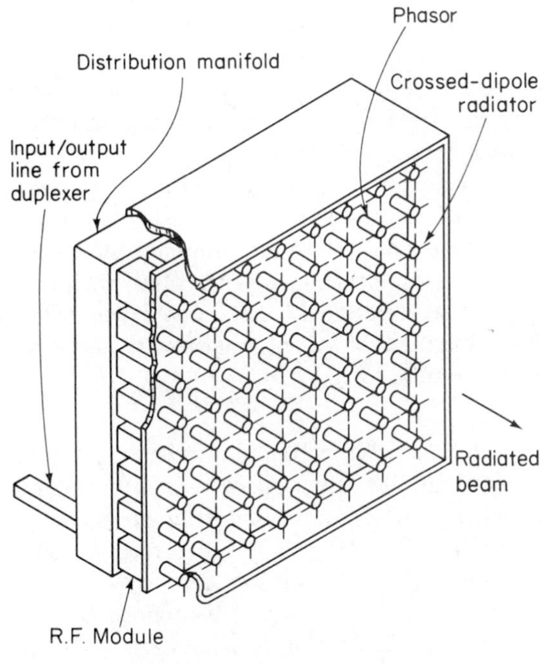

(b) Manifold feed

Fig. 14-14 PHASED ARRAY ANTENNAS

Even though there is increased circuit complexity in multiple phase-shifting networks needed to excite each dipole, the stepup in solid-state reliability far more than compensates for the increased number of circuit elements.

Figure 14-14(a) illustrates a horn-fed reflective array; Fig. 14-14(b) illustrates a similar array fed by driving signals through a distribution manifold which houses the individual phase shifters.

TEST AND MEASUREMENT

Antenna measurement concerns basic characteristics, mainly impedance, gain, and pattern. The subsidiary characteristic of polarization and the phase front of wave radiation will not be covered but a number of the practical considerations of construction will be mentioned.

IMPEDANCE

This characteristic in an antenna is treated in the conventional way described in Chapter 10, where we conduct component measurement via a VSWR measurement upon the desired impedance in order to determine it (with the aid of Smith charts). This determines the distance between voltage or current minimums and also the reference point.

Unlike a simple component measurement, physical conditions affect an antenna measurement, causing errors supplementary to the items listed below. Typical errors arise from:

1. Probe loading of the slotted line.
2. Mismatch of signal source.
3. Nonuniformity of detector characteristics.
4. Harmonics, frequency modulation, spurious signal sources, received signals, etc.
5. Reflections from nearby objects.
6. Coupling from antenna directly into the detector.

Items 5 and 6 are applicable to antenna measurements, for if the antenna under test is not located in a clear area, reflections from various objects will reenter the slotted line through the antenna and cause false measurements. To check this aberration we simply move the antenna one quarter-wavelength and note the change in the VSWR minimum location.

Antenna energy directly coupled to the detector (rather than through the transmission feed line) may be eliminated by lengthening the feed line, surrounding the antenna with absorbing material, or reorienting the antenna itself.

PATTERN

Radiation patterns are, of course, aimed at far-field intensity measurements so that the distance between the stimulating antenna and the antenna under test must be large enough to be physically handled. The test range, or site equipment, must have a transmitting source, the antenna under test, a detector for indicating the magnitude of the received field, and a mount for turning the antenna under test. Figure 14-15 illustrates a typical setup.

The measurement technique assumes that the transmitting array should have as high a gain as possible to eliminate phase-front discrepancies, should be linearly polarized, have a good definite main lobe with negligible minor lobes (at least 20 dB less than the main lobe), and possess a wide beam width so that variations of field strength over the aperture and circle of rotation of the antenna under test will also be negligible (under 5 per cent).

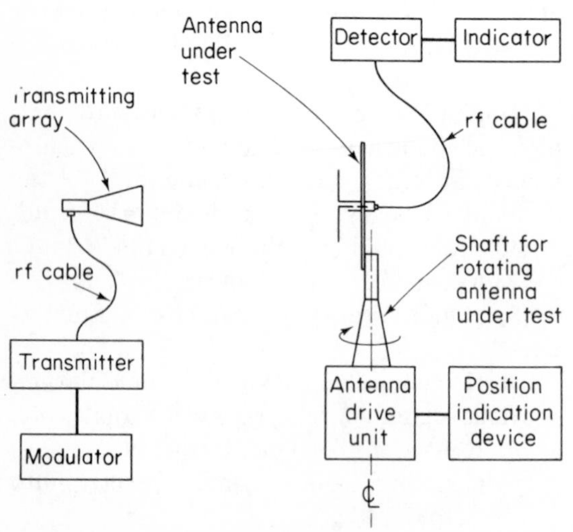

Fig. 14-15 PATTERN MEASUREMENT SETUP

Ground reflection (from vertical polarization) may cause considerable error; remedies for such spurious readings may sometimes be eliminated by mounting the test antenna on a tower or by making the transmitting antenna more highly directional in the vertical plane. Also, various metal fence or reflector structures may be used.

GAIN

As noted earlier, antenna gain comparisons are rated in reference to a lossless isotropic source, that is, a hypothetical source that transmits power in all directions. This device is impossible to construct but is convenient to use in calculating or plotting directivities of different antenna types. For comparison, then, we inspect gains of any antenna in any direction as the power flow per square meter transmitted in that direction compared to the power flow in an isotropic source.

For actual concrete gain figures, however, we must compare measurements on actual test antennas to some reference antenna, usually a dipole (Fig. 14-16). Here we make the actual numerical comparison given below by adjustment of the attenuator in the receiver gain circuits.

1. With the antenna under test and pointed in the direction of maximum signal intensity, some convenient transmitter level is applied, P_1, to give a specified receiver output with attenuator setting A_1.

2. A standard antenna is placed at the distance R from the transmitting antenna and again the attenuator adjusted to A_2 for the same specified receiver output.

3. Gain in decibels is thus A_1 minus A_2. The techniques attendant to the above are

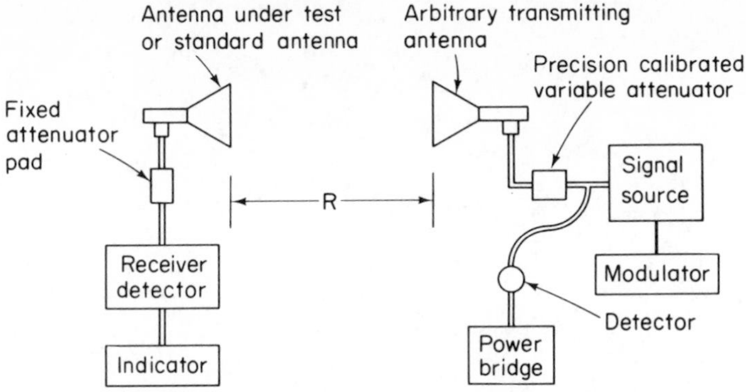

Note: when calibrating two identical antennas, the receiving and transmitting antennas shown are replaced by two identical antennas

Fig. 14-16 DIPOLE GAIN COMPARISON SETUP

practical considerations of placement and adjustment. We must:

(a) Properly space the receiving and transmitting antennas.
(b) Eliminate or account for all ground and object interference.
(c) Orient both setups for maximum reception.
(d) Preserve match when changing antennas.
(e) Maintain operational conditions such as frequency, detection efficiency, attenuator calibration, etc.

The above approach assumes that we know the gain of the standard antenna; precision results thus are only determined by the definiteness and creditable accuracy of the gain of the standard antenna which is being used. Calculations from measurements for this depend upon the physical aperture size, the wavelength, etc.—all involving more specifics than are warranted in this book.

A skeletonized procedure using basic equations and employing two identical antennas spaced at a distance R, with one radiating a power P_t while the other receives a power P_r, is thus

$$\frac{P_r}{P_t} = \frac{\lambda^2 \times G_1 \times G_2}{16\pi R^2}$$

From this equation, if the respective powers, λ and R can be accurately measured (depending upon aperture, current, voltages etc.), the elemental gain is the square root of $G_1 \times G_2$.

RECEIVING ANTENNAS

Receiving and transmitting antennas essentially are theoretically identical. Physically, their differences are mechanical and structural, except in radar, where the same element serves for both transmitting and receiving.

Receiving antennas electrically differ in one important factor, their inherent thermal noise. Just as in high-gain amplifiers, the components constituting a receiving antenna system contribute to the system's inherent noise temperature, which, when combined with the receiver input circuit components, affects the useful amount of signal level for useful communication. Microwave receiving systems are frequently quite noisy; at about 2000° K, noise varies from 5 to 10 dB and part of the noise is contributed by the antenna, which typically runs at about 200° K. Thus the structure's capture hardware, grounding arrangements, aperture excitation, and mismatch must be carefully proportioned.

Added to this are a number of input noises from sea and land, atmosphere absorption, and noise from galactic and radio stars. These factors do not enter into transmitter antenna design, where the chief concern in propagation is the ratio of power between the main beam and the side lobes.

Appendices

Tables of Sines, Cosines, and Tangents A

Angle	Radians	Sine	Cosine	Tangent	Angle	Radians	Sine	Cosine	Tangent
0°	.0000	.0000	1.0000	.0000	**45°**	.7854	.7071	.7071	1.0000
1	.0175	.0175	.9998	.0175	**46**	.8029	.7193	.6947	1.0355
2	.0349	.0349	.9994	.0349	**47**	.8203	.7314	.6820	1.0724
3	.0524	.0523	.9986	.0524	**48**	.8378	.7431	.6691	1.1106
4	.0698	.0698	.9976	.0699	**49**	.8552	.7547	.6561	1.1504
5	.0873	.0872	.9962	.0875	**50**	.8727	.7660	.6428	1.1918
6	.1047	.1045	.9945	.1051	**51**	.8901	.7771	.6293	1.2349
7	.1222	.1219	.9925	.1228	**52**	.9076	.7880	.6157	1.2799
8	.1396	.1392	.9903	.1405	**53**	.9250	.7986	.6018	1.3270
9	.1571	.1564	.9877	.1584	**54**	.9425	.8090	.5878	1.3764
10	.1745	.1736	.9848	.1763	**55**	.9599	.8192	.5736	1.4281
11	.1920	.1908	.9816	.1944	**56**	.9774	.8290	.5592	1.4826
12	.2094	.2079	.9781	.2126	**57**	.9948	.8387	.5446	1.5399
13	.2269	.2250	.9744	.2309	**58**	1.0123	.8480	.5299	1.6003
14	.2443	.2419	.9703	.2493	**59**	1.0297	.8572	.5150	1.6643
15	.2618	.2588	.9659	.2679	**60**	1.0472	.8660	.5000	1.7321
16	.2793	.2756	.9613	.2867	**61**	1.0647	.8746	.4848	1.8040
17	.2967	.2924	.9563	.3057	**62**	1.0821	.8829	.4695	1.8807
18	.3142	.3090	.9511	.3249	**63**	1.0996	.8910	.4540	1.9626
19	.3316	.3256	.9455	.3443	**64**	1.1170	.8988	.4384	2.0503
20	.3491	.3420	.9397	.3640	**65**	1.1345	.9063	.4226	2.1445
21	.3665	.3584	.9336	.3839	**66**	1.1519	.9135	.4067	2.2460
22	.3840	.3746	.9272	.4040	**67**	1.1694	.9205	.3907	2.3559
23	.4014	.3907	.9205	.4245	**68**	1.1868	.9272	.3746	2.4751
24	.4189	.4067	.9135	.4452	**69**	1.2043	.9336	.3584	2.6051
25	.4363	.4226	.9063	.4663	**70**	1.2217	.9397	.3420	2.7475
26	.4538	.4384	.8988	.4877	**71**	1.2392	.9455	.3256	2.9042
27	.4712	.4540	.8910	.5095	**72**	1.2566	.9511	.3090	3.0777
28	.4887	.4695	.8829	.5317	**73**	1.2741	.9563	.2924	3.2709
29	.5061	.4848	.8746	.5543	**74**	1.2915	.9613	.2756	3.4874
30	.5236	.5000	.8660	.5774	**75**	1.3090	.9659	.2588	3.7321
31	.5411	.5150	.8572	.6009	**76**	1.3265	.9703	.2419	4.0108
32	.5585	.5299	.8480	.6249	**77**	1.3439	.9744	.2250	4.3315
33	.5760	.5446	.8387	.6494	**78**	1.3614	.9781	.2079	4.7046
34	.5934	.5592	.8290	.6745	**79**	1.3788	.9816	.1908	5.1446
35	.6109	.5736	.8192	.7002	**80**	1.3963	.9848	.1736	5.6713
36	.6283	.5878	.8090	.7265	**81**	1.4137	.9877	.1564	6.3138
37	.6458	.6018	.7986	.7536	**82**	1.4312	.9903	.1392	7.1154
38	.6632	.6157	.7880	.7813	**83**	1.4486	.9925	.1219	8.1443
39	.6807	.6293	.7771	.8098	**84**	1.4661	.9945	.1045	9.5144
40	.6981	.6428	.7660	.8391	**85**	1.4835	.9962	.0872	11.43
41	.7156	.6561	.7547	.8693	**86**	1.5010	.9976	.0698	14.30
42	.7330	.6691	.7431	.9004	**87**	1.5184	.9986	.0523	19.08
43	.7505	.6820	.7314	.9325	**88**	1.5359	.9994	.0349	28.64
44	.7679	.6947	.7193	.9657	**89**	1.5533	.9998	.0175	57.29

B Four-Place Log Tables

N	0	1	2	3	4	5	6	7	8	9	u. d.
10	0000	0043	0086	0128	0170	0212	0253	0294	0334	0374	4.2
11	0414	0453	0492	0531	0569	0607	0645	0682	0719	0755	3.8
12	0792	0828	0864	0899	0934	0969	1004	1038	1072	1106	3.5
13	1139	1173	1206	1239	1271	1303	1335	1367	1399	1430	3.2
14	1461	1492	1523	1553	1584	1614	1644	1673	1703	1732	3.0
15	1761	1790	1818	1847	1875	1903	1931	1959	1987	2014	2.8
16	2041	2068	2095	2122	2148	2175	2201	2227	2253	2279	2.6
17	2304	2330	2355	2380	2405	2430	2455	2480	2504	2529	2.5
18	2553	2577	2601	2625	2648	2672	2695	2718	2742	2765	2.4
19	2788	2810	2833	2856	2878	2900	2923	2945	2967	2989	2.2
20	3010	3032	3054	3075	3096	3118	3139	3160	3181	3201	2.1
21	3222	3243	3263	3284	3304	3324	3345	3365	3385	3404	2.0
22	3424	3444	3464	3483	3502	3522	3541	3560	3579	3598	1.9
23	3617	3636	3655	3674	3692	3711	3729	3747	3766	3784	1.8
24	3802	3820	3838	3856	3874	3892	3909	3927	3945	3962	1.8
25	3979	3997	4014	4031	4048	4065	4082	4099	4116	4133	1.7
26	4150	4166	4183	4200	4216	4232	4249	4265	4281	4298	1.6
27	4314	4330	4346	4362	4378	4393	4409	4425	4440	4456	1.6
28	4472	4487	4502	4518	4533	4548	4564	4579	4594	4609	1.5
29	4624	4639	4654	4669	4683	4698	4713	4728	4742	4757	1.5
30	4771	4786	4800	4814	4829	4843	4857	4871	4886	4900	1.4
31	4914	4928	4942	4955	4969	4983	4997	5011	5024	5038	1.4
32	5051	5065	5079	5092	5105	5119	5132	5145	5159	5172	1.3
33	5185	5198	5211	5224	5237	5250	5263	5276	5289	5302	1.3
34	5315	5328	5340	5353	5366	5378	5391	5403	5416	5428	1.3
35	5441	5453	5465	5478	5490	5502	5514	5527	5539	5551	1.2
36	5563	5575	5587	5599	5611	5623	5635	5647	5658	5670	1.2
37	5682	5694	5705	5717	5729	5740	5752	5763	5775	5786	1.2
38	5798	5809	5821	5832	5843	5855	5866	5877	5888	5899	1.1
39	5911	5922	5933	5944	5955	5966	5977	5988	5999	6010	1.1
40	6021	6031	6042	6053	6064	6075	6085	6096	6107	6117	1.1
41	6128	6138	6149	6160	6170	6180	6191	6201	6212	6222	1.0
42	6232	6243	6253	6263	6274	6284	6294	6304	6314	6325	1.0
43	6335	6345	6355	6365	6375	6385	6395	6405	6415	6425	1.0
44	6435	6444	6454	6464	6474	6484	6493	6503	6513	6522	1.0
45	6532	6542	6551	6561	6571	6580	6590	6599	6609	6618	1.0
46	6628	6637	6646	6656	6665	6675	6684	6693	6702	6712	.9
47	6721	6730	6739	6749	6758	6767	6776	6785	6794	6803	.9
48	6812	6821	6830	6839	6848	6857	6866	6875	6884	6893	.9
49	6902	6911	6920	6928	6937	6946	6955	6964	6972	6981	.9
50	6990	6998	7007	7016	7024	7033	7042	7050	7059	7067	.9
51	7076	7084	7093	7101	7110	7118	7126	7135	7143	7152	.8
52	7160	7168	7177	7185	7193	7202	7210	7218	7226	7235	.8
53	7243	7251	7259	7267	7275	7284	7292	7300	7308	7316	.8
54	7324	7332	7340	7348	7356	7364	7372	7380	7388	7396	.8

N	0	1	2	3	4	5	6	7	8	9	u. d.
55	7404	7412	7419	7427	7435	7443	7451	7459	7466	7474	.8
56	7482	7490	7497	7505	7513	7520	7528	7536	7543	7551	.8
57	7559	7566	7574	7582	7589	7597	7604	7612	7619	7627	.8
58	7634	7642	7649	7657	7664	7672	7679	7686	7694	7701	.7
59	7709	7716	7723	7731	7738	7745	7752	7760	7767	7774	.7
60	7782	7789	7796	7803	7810	7818	7825	7832	7839	7846	.7
61	7853	7860	7868	7875	7882	7889	7896	7903	7910	7917	.7
62	7924	7931	7938	7945	7952	7959	7966	7973	7980	7987	.7
63	7993	8000	8007	8014	8021	8028	8035	8041	8048	8055	.7
64	8062	8069	8075	8082	8089	8096	8102	8109	8116	8122	.7
65	8129	8136	8142	8149	8156	8162	8169	8176	8182	8189	.7
66	8195	8202	8209	8215	8222	8228	8235	8241	8248	8254	.7
67	8261	8267	8274	8280	8287	8293	8299	8306	8312	8319	.6
68	8325	8331	8338	8344	8351	8357	8363	8370	8376	8382	.6
69	8388	8395	8401	8407	8414	8420	8426	8432	8439	8445	.6
70	8451	8457	8463	8470	8476	8482	8488	8494	8500	8506	.6
71	8513	8519	8525	8531	8537	8543	8549	8555	8561	8567	.6
72	8573	8579	8585	8591	8597	8603	8609	8615	8621	8627	.6
73	8633	8639	8645	8651	8657	8663	8669	8675	8681	8686	.6
74	8692	8698	8704	8710	8716	8722	8727	8733	8739	8745	.6
75	8751	8756	8762	8768	8774	8779	8785	8791	8797	8802	.6
76	8808	8814	8820	8825	8831	8837	8842	8848	8854	8859	.6
77	8865	8871	8876	8882	8887	8893	8899	8904	8910	8915	.6
78	8921	8927	8932	8938	8943	8949	8954	8960	8965	8971	.6
79	8976	8982	8987	8993	8998	9004	9009	9015	9020	9025	.5
80	9031	9036	9042	9047	9053	9058	9063	9069	9074	9079	.5
81	9085	9090	9096	9101	9106	9112	9117	9122	9128	9133	.5
82	9138	9143	9149	9154	9159	9165	9170	9175	9180	9186	.5
83	9191	9196	9201	9206	9212	9217	9222	9227	9232	9238	.5
84	9243	9248	9253	9258	9263	9269	9274	9279	9284	9289	.5
85	9294	9299	9304	9309	9315	9320	9325	9330	9335	9340	.5
86	9345	9350	9355	9360	9365	9370	9375	9380	9385	9390	.5
87	9395	9400	9405	9410	9415	9420	9425	9430	9435	9440	.5
88	9445	9450	9455	9460	9465	9469	9474	9479	9484	9489	.5
89	9494	9499	9504	9509	9513	9518	9523	9528	9533	9538	.5
90	9542	9547	9552	9557	9562	9566	9571	9576	9581	9586	.5
91	9590	9595	9600	9605	9609	9614	9619	9624	9628	9633	.5
92	9638	9643	9647	9652	9657	9661	9666	9671	9675	9680	.5
93	9685	9689	9694	9699	9703	9708	9713	9717	9722	9727	.5
94	9731	9736	9741	9745	9750	9754	9759	9763	9768	9773	.5
95	9777	9782	9786	9791	9795	9800	9805	9809	9814	9818	.5
96	9823	9827	9832	9836	9841	9845	9850	9854	9859	9863	.5
97	9868	9872	9877	9881	9886	9890	9894	9899	9903	9908	.4
98	9912	9917	9921	9926	9930	9934	9939	9943	9948	9952	.4
99	9956	9961	9965	9969	9974	9978	9983	9987	9991	9996	.4

C Decibels versus Voltage and Power

Voltage Ratio (Equal) Impedance	Power Ratio	− dB +	Voltage Ratio (Equal) Impedance	Power Ratio
1.000	1.000	0	1.000	1.000
0.989	0.977	0.1	1.012	1.023
0.977	0.955	0.2	1.023	1.047
0.966	0.933	0.3	1.035	1.072
0.955	0.912	0.4	1.047	1.096
0.944	0.891	0.5	1.059	1.122
0.933	0.871	0.6	1.072	1.148
0.923	0.851	0.7	1.084	1.175
0.912	0.832	0.8	1.096	1.202
0.902	0.813	0.9	1.109	1.230
0.891	0.794	1.0	1.122	1.259
0.841	0.708	1.5	1.189	1.413
0.794	0.631	2.0	1.259	1.585
0.750	0.562	2.5	1.334	1.778
0.708	0.501	3.0	1.413	1.995
0.668	0.447	3.5	1.496	2.239
0.631	0.398	4.0	1.585	2.512
0.596	0.355	4.5	1.679	2.818
0.562	0.316	5.0	1.778	3.162
0.531	0.282	5.5	1.884	3.548
0.501	0.251	6.0	1.995	3.981
0.473	0.224	6.5	2.113	4.467
0.447	0.200	7.0	2.239	5.012
0.422	0.178	7.5	2.371	5.623
0.398	0.159	8.0	2.512	6.310
0.376	0.141	8.5	2.661	7.079
0.355	0.126	9.0	2.818	7.943
0.335	0.112	9.5	2.985	8.913
0.316	0.100	10	3.162	10.00
0.282	0.0794	11	3.55	12.6
0.251	0.0631	12	3.98	15.9
0.224	0.0501	13	4.47	20.0
0.200	0.0398	14	5.01	25.1
0.178	0.0316	15	5.62	31.6
0.159	0.0251	16	6.31	39.8
0.141	0.0200	17	7.08	50.1
0.126	0.0159	18	7.94	63.1
0.112	0.0126	19	8.91	79.4
0.100	0.0100	20	10.00	100.0
3.16×10^{-2}	10^{-3}	30	3.16×10	10^3
10^{-2}	10^{-4}	40	10^3	10^4
3.16×10^{-3}	10^{-5}	50	3.16×10^3	10^5
10^{-3}	10^{-6}	60	10^4	10^6
3.16×10^{-4}	10^{-7}	70	3.16×10^4	10^7
10^{-4}	10^{-8}	80	10^5	10^8
3.16×10^{-5}	10^{-9}	90	3.16×10^5	10^9
10^{-5}	10^{-10}	100	10^6	10^{10}
3.16×10^{-6}	10^{-11}	110		10^{11}
10^{-6}	10^{-12}	120		10^{12}

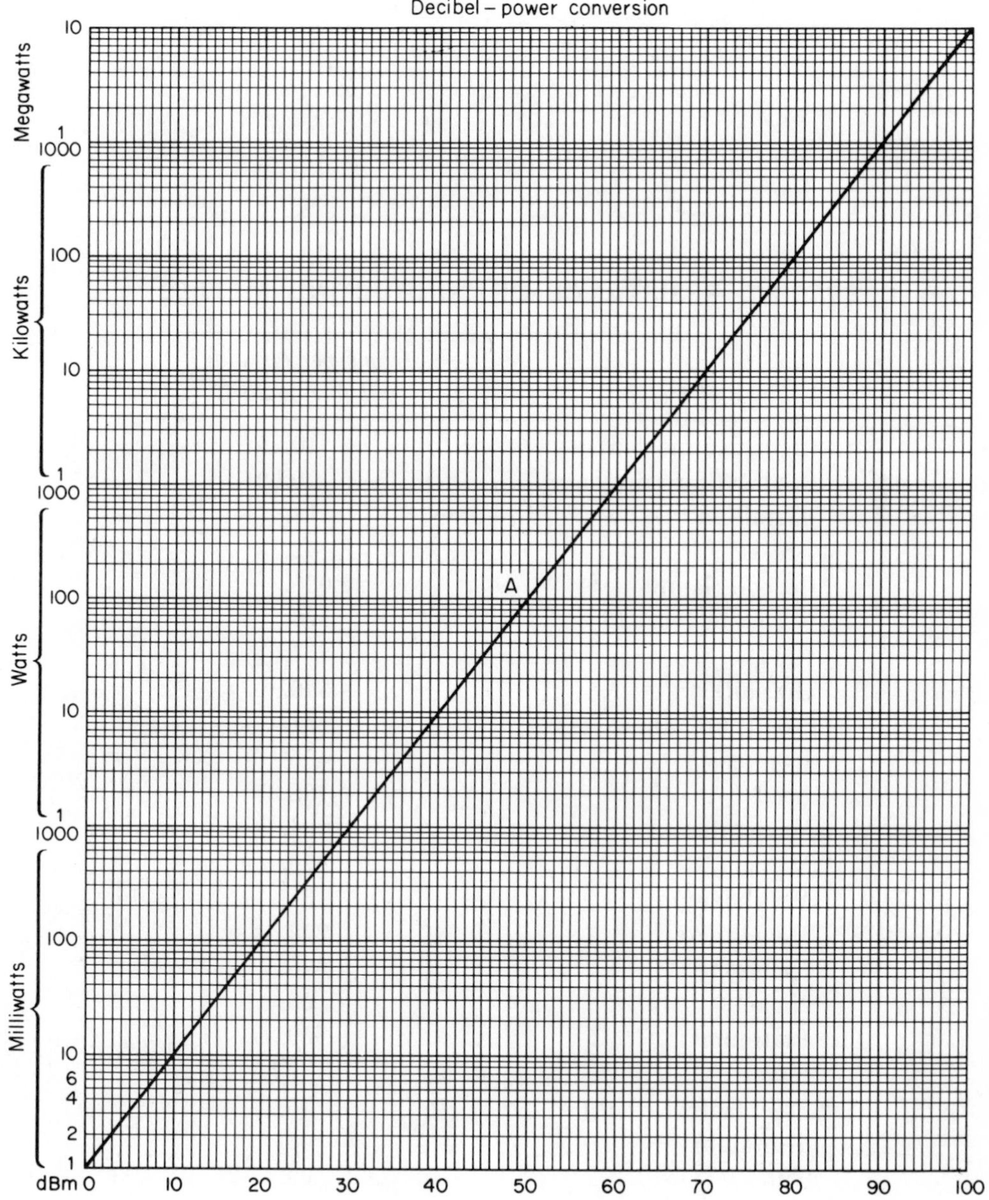

Decibel – power conversion
Megawatts
Kilowatts
Watts
Milliwatts
10
1
1000
100
10
1
1000
100
10
1
1000
100
10
6
4
2
1
dBm 0
10
20
30
40
50
60
70
80
90
100
A

D Specific Material Resistances

SPECIFIC RESISTANCE OF CONDUCTING METALS

Type of metal	*Ohms per circular mil-foot at 20°C*
Copper (annealed)	10.35
Copper (hard drawn)	10.60
Aluminum	17.
Brass	42.
Carbon (lampblack)	22,000.
German silver (18%)	198.
Gold	14.6
Graphite	4,300.
Iron (pure, annealed)	61.
Iron (cast)	435.
Lead	132.35
Manganin	264.
Mercury	576.
Molybdenum	34.
Monel metal	252.
Nichrome	675.
Nickel	60.
Platinum	60.
Silver	9.56
Steel (soft, carbon)	96.
Tantalum	93.
Tin	69.
Tungsten	34.
Zinc	35.

E

Useful Waveguide Formulas

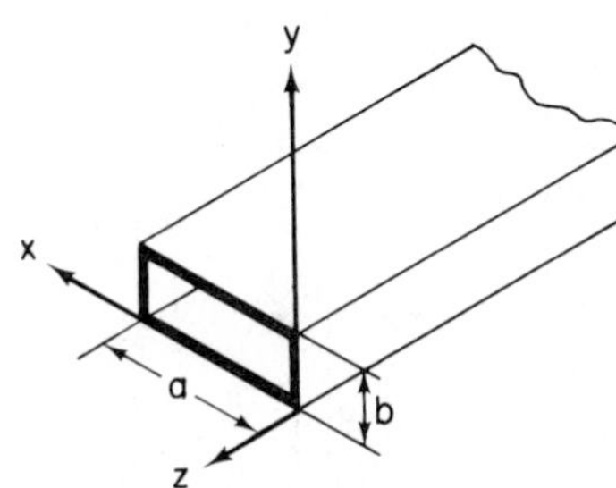

Guide wavelength (λ_g)

$$\lambda_g = \frac{\lambda}{\sqrt{1-(\lambda/\lambda_c)^2}}$$

λ = free-space wavelength
λ_c = cutoff wavelength

Phase constant (β)

$$\beta = \frac{2\pi}{\lambda_g} = \frac{2\pi}{\lambda}\sqrt{1-\left(\frac{\lambda}{\lambda_c}\right)^2} = \sqrt{\frac{\omega^2}{c^2} - \frac{\pi^2}{a^2}}$$

c = velocity of light

Group and phase velocity

$$\frac{\text{Phase velocity}}{\text{Velocity of light}} = \frac{v_p}{c} = \frac{\lambda_g}{\lambda} = \frac{1}{\sqrt{1-(\lambda/\lambda_c)^2}}$$

Cutoff wavelength
(General case)

$$\lambda_c = \frac{2a}{\sqrt{(m^2) + (na/b)^2}}$$

a and b = guide dimensions in drawing
m = first subscript describing mode
n = second subscript describing mode

Cutoff frequency calculations					
Rectangular guide (a, b)		Square guide (a, b)		Circular guide (2a)	
a = 2b		a = b		radius = a	
Mode	Cutoff wavelength	Mode	Cutoff wavelength	Mode	Cutoff wavelength
TE_{10}	2a	TE_{10}	2a	TE_{11}	3.41a
TE_{01}	a	TE_{01}	2a	TM_{01}	2.61a
TE_{20}	a	TE_{11}	1.4a	TE_{21}	2.06a
TE_{11}	0.89a	TM_{11}	1.4a	TE_{01}	1.64a
TM_{11}	0.89a	TE_{20}	a	TM_{11}	1.64a

F

Attenuation in Coaxial Cables

Frequency	Attenuation (db per 100 ft.)					
MHz	RG-58A	RG-55	RG-8	RG-9	3/8" Spirafil	1/2" Spirafil
100	6.20	4.80	2.00	2.00	1.20	0.80
200	9.30	7.00	3.00	2.90	1.75	1.16
400	14.0	10.4	4.50	4.25	2.55	1.70
600	18.0	12.8	6.00	5.40	3.15	2.10
800	21.0	15.0	7.20	6.90	3.70	2.45
1000	24.0	17.0	8.00	7.30	4.20	2.80
3000	45.0	32.0	16.5	15.5	7.50	5.00
5000	83.0	46.0	27.0	23.0		
10,000	247	130		36.0		

G Commonly Used Semiconductor Letter Symbols

Roman Letters

A_f	Band pass of a resonant circuit
agc	Automatic gain control
A_i	Current gain
am	Amplitude modulation
A_v	Voltage gain
bfo	Beat frequency oscillator
C	Constant (when following an equal (=) sign)
CB	Common-base configuration
C_{cb}	Interelement capacitance between base and collector
C_{eb}	Interelement capacitance between base and emitter
C_{cc}	Interelement capacitance between collector and emitter
CC	Common-collector configuration
CE	Common-emitter configuration
cm	Centimeter
C_N	Neutralizing capacitor
CR	Junction diode
C_{sp}	Reflected capacitance
cw	Continuous-wave transmission
E_{IN}	Input voltage
E_{OUT}	Output voltage
fm	Frequency modulation
f_a	Cutoff frequency
f_r	Frequency of resonance
G	Power gain
g_{bc}	Base-collector conductance
g_{be}	Base-emitter conductance
g_{ce}	Collector-emittance conductance
g_{fe}	Forward transfer conductance
g_{ie}	Input conductance
G_m	Maximum power
g_{oe}	Output conductance
g_{re}	Reverse transfer conductance
h	Hybrid
h_{ib}	Input resistance of common-base configuration with output short-circuited
h_{ic}	Input resistance of common-collector configuration with output short-circuited
h_{ie}	Input resistance of common-emitter configuration with output short-circuited
h_{ob}	Output conductance of common-base configuration with input open
h_{oc}	Output conductance of common-collector configuration with input open
h_{oe}	Output conductance of common-emitter configuration with input open
I_e	Total emitter current
I_b	Changing base current (a-c)
I_B	D-c base current
i_c	Changing collector current (a-c)
I_C	D-c collector current
if	Intermediate frequency
m	Turns ratio of a transformer
mA	Milliampere
mV	Millivolt
mW	Milliwatt
N	Number of turns in an indicator
N-type	Semiconductor with donor impurity

NPN Transistor with one P-type and two N-type semiconductors

P-type Semiconductor with acceptor impurity

P Power

PN Combination of N-type and P-type semiconductors

PNP Transistor with one N-type and two P-type semiconductors

Q Selectivity of a resonant circuit

RT Thermistor

r_b A-c base resistance

RC A coupling circuit employing a resistor and a capacitor

r_d A-c collector resistance

r_e A-c emitter resistance

r_{fe} Forward transfer resistance with input open

r_{ie} Input resistance with output open

R_L Load resistor

r_m Mutual resistance

R_N Neutralizing resistor

r_{oe} Output resistance with input open

r_{re} Reverse transfer resistance with input open

S_I Current-stability factor

S_V Voltage-stability factor

T Transformer

TC Time constant

t_f Fall time

t_p Pulse time

t_r Rise time

t_s Storage time

V_{BE} Fixed base-emitter voltage

V_{CE} Fixed collector-emitter voltage

V_g Source voltage

v_{be} Changing voltage (a-c) between base and emitter

v_{ce} Changing voltage (a-c) between collector and emitter

V_{SAT} Saturation voltage

Y Admittance

Z Impedance

Greek Letters

α Alpha; current amplification factor

α_{fb} Forward short-circuit current amplification factor for the CB configuration

α_{fc} Forward short-circuit current amplification factor for the CC configuration

α_{fe} Forward short-circuit current amplification factor for the CE configuration

Δ Delta; incremental change

∞ Infinity

μA Microampere

μ_{rb} Reverse open-circuited voltage amplification factor for CB configuration

μ_{rc} Reverse open-circuited voltage amplification factor for CC configuration

μ_{re} Reverse open-circuited voltage amplification factor for CE configuration

Microwave Symbols

α	Attenuation
α_o	Attenuation of air filled copper transmission line = 0.35 x 10^{-9} nepers/meter = 0.3 x 10^{-5} db/kilometer
Γ	coefficient of reflection = −1 for short circuit = +1 for open circuit = 0 for matched load
δ	skin depth
ϵ	Dielectric constant
ϵ_o	Dielectric constant for air
λ	Wavelength
λ_g	Guide wavelength
λ_c	Cutoff wavelength
μ	Permeability
μ_o	Permeability of air = 4π x 10^{-7} henries/meter
K	Coupling coefficient
ρ	Resistivity = 1.74 x 10^{-8} ohm-meters for copper
σ	Electrical conductivity
κ_o	Permittivity of air −8.854 x 10^{-12} farad/meter
v	Velocity of propagation
v_o	Velocity of propagation in air = 2.998 x 10^{8} centimeter/second = 2.998 x 10^{8} meters/second = 186,280 miles/second = 11.808 x 10^{9} inches/second
ϕ	Phase angle
β	Phase constant
Z_o	Characteristic impedance = 376.7 ohms for free space = 120π
D	Directivity
P	Power
V	Voltage
I	Current
R	Resistance
C	Capacitance
G	Conductance
B	Susceptance
X	Reactance
Y	Admittance
L	Inductance
f	Frequency
ω	Angular frequency = $2\pi f$
	Figure of merit of a resonator = $2\pi \dfrac{\text{energy stored}}{\text{energy dissipated per cycle}} = \dfrac{\Delta f}{f_o}$
H	Magnetic vector
E	Electric vector
a	Broad waveguide dimension
b	Narrow waveguide dimension
z	Direction of propagation
n	Mode designation (for TE_{10} m = 1, n = 0)
m	Mode designation (for TE_{10} m = 1, n = 0)
VSWR	Voltage Standing Wave Ratio
PSWR	Power Standing Wave Ratio

H Physical and Electrical Conversion Factors

To convert	Into	Multiply by	Conversely multiply by
Ampere-hours	Coulombs	3,600	2.778×10^{-4}
Amperes per sq. cm	Amperes per sq. inch	6.452	.155
Ampere turns	Gilberts	1.257	.7958
Ampere turns per cm	Ampere turns per inch	2.54	.3937
Btu (British thermal unit)	Foot-pounds	778.3	1.285×10^{-3}
Btu	Joules	1,054.8	9.48×10^{-4}
Btu	Kilogram-calories	.252	3.969
Btu	Horsepower-hours	3.929×10^{-4}	2,545
Centigrade	Fahrenheit	$(C^\circ \times 9/5) + 32$	$(F^\circ - 32) \times 5/9.$
Circular mils	Square centimeters	5.067×10^{-6}	1.973×10^{5}
Circular mils	Square mils	.7854	1.273.
Cubic inches	Cubic centimeters	16.39	6.102×10^{-2}
Cubic inches	Cubic feet	5.785×10^{-4}	1,728
Cubic inches	Cubic meters	1.639×10^{-5}	6.102×10^{4}
Cubic meters	Cubic feet	35.31	2.832×10^{-2}
Cubic meters	Cubic yards	1.308	.7646.
Degrees (angle)	Radians	1.745×10^{-2}	57.3
Dynes	Pounds	2.248×10^{-6}	4.448×10^{5}
Ergs	Foot-pounds	7.367×10^{-8}	1.356×10^{7}
Feet	Centimeters	30.48	3.281×10^{-2}
Foot-pounds	Horsepower-hours	5.05×10^{-7}	1.98×10^{6}
Foot-pounds	Kilogram-meters	.1383	7.233
Foot-pounds	Kilowatt-hours	3.766×10^{-7}	2.655×10^{6}
Gauss	Lines per sq. inch	6.452	.155
Grams	Dynes	980.7	1.02×10^{-3}
Grams	Ounces (avoirdupois)	3.527×10^{-2}	28.35
Grams per cm	Pounds per inch	5.6×10^{-3}	178.6
Grams per cubic cm	Pounds per cu. inch	3.613×10^{-2}	27.68
Grams per sq. cm	Pounds per sq. foot	2.0481	.4883
Horsepower (550 ft.-lb. per sec.)	Foot-lb. per minute	3.3×10^{4}	3.03×10^{-5}
Horsepower (550 ft.-lb. per sec.)	Btu per minute	42.41	2.357×10^{-2}
Horsepower (550 ft.-lb. per sec.)	Kg-calories per minute	10.69	9.355×10^{-2}
Horsepower (Metric) (542.5 ft.-lb. per sec.)	Horsepower (550 ft.-lb. per sec.)	.9863	1.014
Inches	Centimeters	2.54	.3937
Inches	Mils	1,000	.001
Joules	Foot-pounds	.7376	1.356.
Joules	Ergs	10^{7}	10^{-7}
Kilogram-calories	Kilojoules	4.186	.2389
Kilograms	Pounds (avoirdupois)	2.205	.4536
Kg per sq. meter	Pounds per sq. foot	.2048	4.882

To convert	Into	Multiply by	Conversely multiply by
Kilometers	Feet	3,281	3.048×10^{-4}
Kilowatt-hours	Btu	3,413	2.93×10^{-4}
Kilowatt-hours	Foot-pounds	2.655×10^{6}	3.766×10^{-7}
Kilowatt-hours	Joules	3.6×10^{6}	2.778×10^{-7}
Kilowatt-hours	Kilogram-calories	860	1.163×10^{-3}
Kilowatt-hours	Kilogram-meters	3.671×10^{5}	2.724×10^{-6}
Liters	Cubic meters	.001	1,000
Liters	Cubic inches	61.02	1.639×10^{-2}
Liters	Gallons (liq. US)	.2642	3.785
Liters	Pints (liq. US)	2.113	.4732
Meters	Yards	1.094	.9144
Meters per min	Feet per min	3.281	.3048
Meters per min	Kilometers per hr	.06	16.67
Miles (nautical)	Kilometers	1.853	.5396
Miles (statute)	Kilometers	1.609	.6214
Miles per hr	Kilometers per min	2.682×10^{-2}	37.28
Miles per hr	Feet per minute	88	1.136×10^{-2}
Miles per hr	Kilometers per hr	1.609	.6214
Poundals	Dynes	1.383×10^{4}	7.233×10^{-5}
Poundals	Pounds (avoirdupois)	3.108×10^{-2}	32.17
Sq inches	Circular mils	1.273×10^{6}	7.854×10^{-7}
Sq inches	Sq centimeters	6.452	.155
Sq feet	Sq meters	9.29×10^{-2}	10.76
Sq miles	Sq yards	3.098×10^{6}	3.228×10^{-7}
Sq miles	Sq kilometers	2.59	.3861
Sq millimeters	Circular mils	1,973	5.067×10^{-4}
Tons, short (avoir 2,000 lb.)	Tonnes (1,000 Kg.)	.9072	1.102
Tons, long (avoir 2,240 lb.)	Tonnes (1,000 Kg.)	1.016	.9842
Tons, long (avoir 2,240 lb.)	Tons, short (avoir 2,000 lb)	1.120	.8929
Watts	Btu per min	5.689×10^{-2}	17.58
Watts	Ergs per sec	10^{7}	10^{-7}
Watts	Ft-lb per minute	44.26	2.26×10^{-2}
Watts	Horsepower (550 ft-lb per sec.)	1.341×10^{-3}	745.7
Watts	Horsepower (metric) (542.5 ft-lb per sec.)	1.36×10^{-3}	735.5
Watts	Kg-calories per min	1.433×10^{-2}	69.77

I Physical and Electrical Engineering Units and Constants

ELECTRICAL ENGINEERING SYMBOLS AND UNITS

Quantity	Symbol	Unit	Symbol
Charge	Q	coulomb	C
Current	I	ampere	A
Voltage, potential difference	V	volt	V
Electromotive force	ξ	volt	V
Resistance	R	ohm	Ω
Conductance	G	mho (siemens)	A/V or mho (S)
Reactance	X	ohm	Ω
Susceptance	B	mho	A/V or mho
Impedance	Z	ohm	Ω
Admittance	Y	mho	A/V or mho
Capacitance	C	farad	F
Inductance	L	henry	H
Energy, work	W	joule	J
Power	P	watt	W
Resistivity	ρ	ohm-meter	Ωm
Conductivity	σ	mho per meter	mho/m
Electric displacement	D	coulomb per sq. meter	C/m^2
Electric field strength	E	volt per meter	V/m
Permittivity (absolute)	ϵ	Farad per meter	F/m
relative permittivity	ϵ_r	(numeric)	
magnetic flux	Φ	weber	Wb
magnetomotive force	$\mathscr{F}$	ampere (ampere-turn)	A
reluctance	$\mathscr{R}$	ampere per weber	A/Wb
permeance	$\mathscr{P}$	weber per ampere	Wb/A
magnetic flux density	B	tesla	T
magnetic field strength	H	ampere per meter	A/m
permeability (absolute)	μ	henry per meter	H/m
relative permeability	μ_r	(numeric)	

(CONT.)

Quantity	Symbol	Unit	Symbol
length	l	meter	m
mass	m	kilogram	kg
time	t	second	s
frequency	f	hertz	Hz
angular frequency	ω	radian per second	rad/s
force	F	newton	N
pressure	p	newton per sq. meter	N/m^2
temperature (absolute)	T	degree Kelvin	°K
temperature (International)	t	degree Celsius	°C

PHYSICAL CONSTANTS USED IN ELECTRICAL ENGINEERING

Constant	Symbol	Rounded Value
electronic charge	e	1.602×10^{-19} C
speed of light in vacuum	c	2.9979×10^{8} m/s
permittivity of vacuum, electric constant	ϵ_o, Γ_e	8.8542×10^{-12} F/m
permeability of vacuum, magnetic constant	μ_o, Γ_m	$4\pi \times 10^{-7}$ H/m†
Planck constant	h	6.63×10^{-34} Js
Boltzmann constant	k	1.38×10^{-23} J/°K
Faraday constant	F	9.649×10^{4} C/mol
proton gyromagnetic ratio	γ	2.6752×10^{8} rad/sT
standard gravitational acceleration	g_n	9.80665 m/s^2†
normal atmospheric pressure	atm	101 325 N/m^2†

†Defined value.

J Mismatch Loss Chart

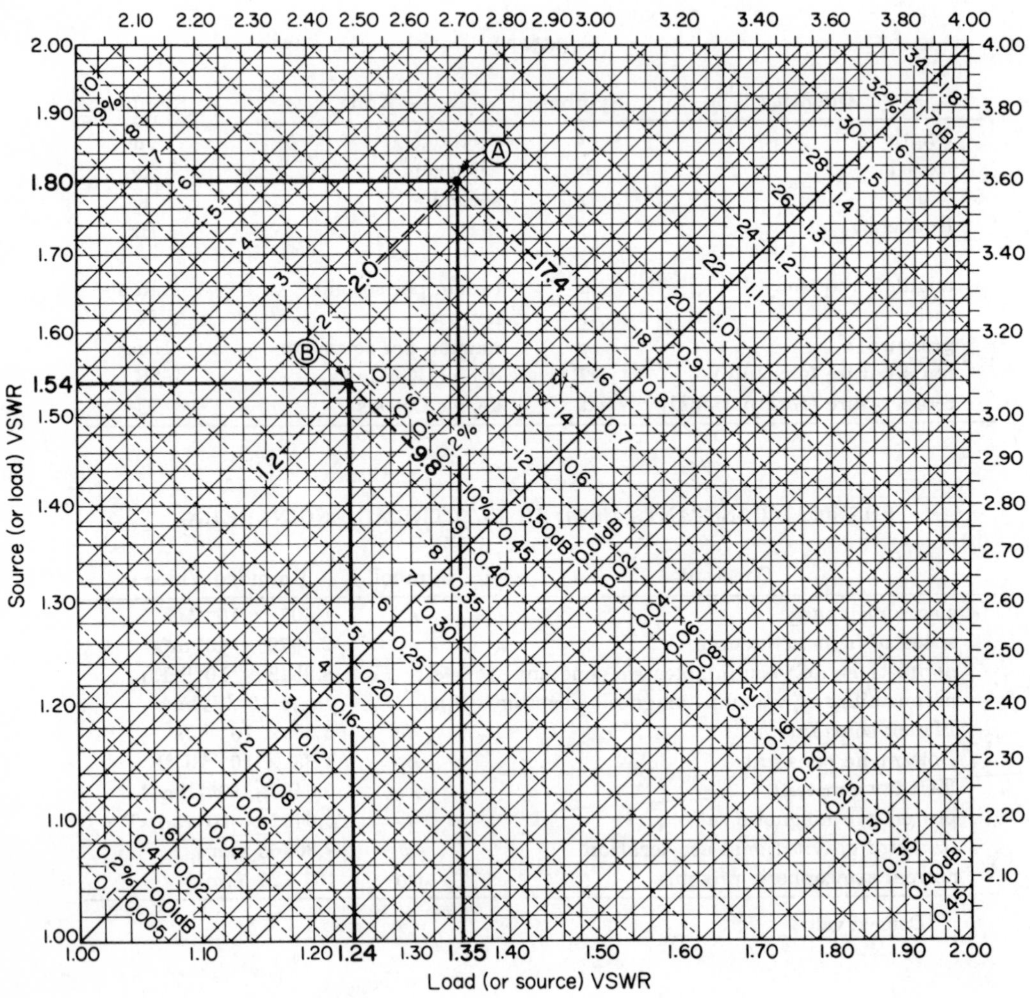

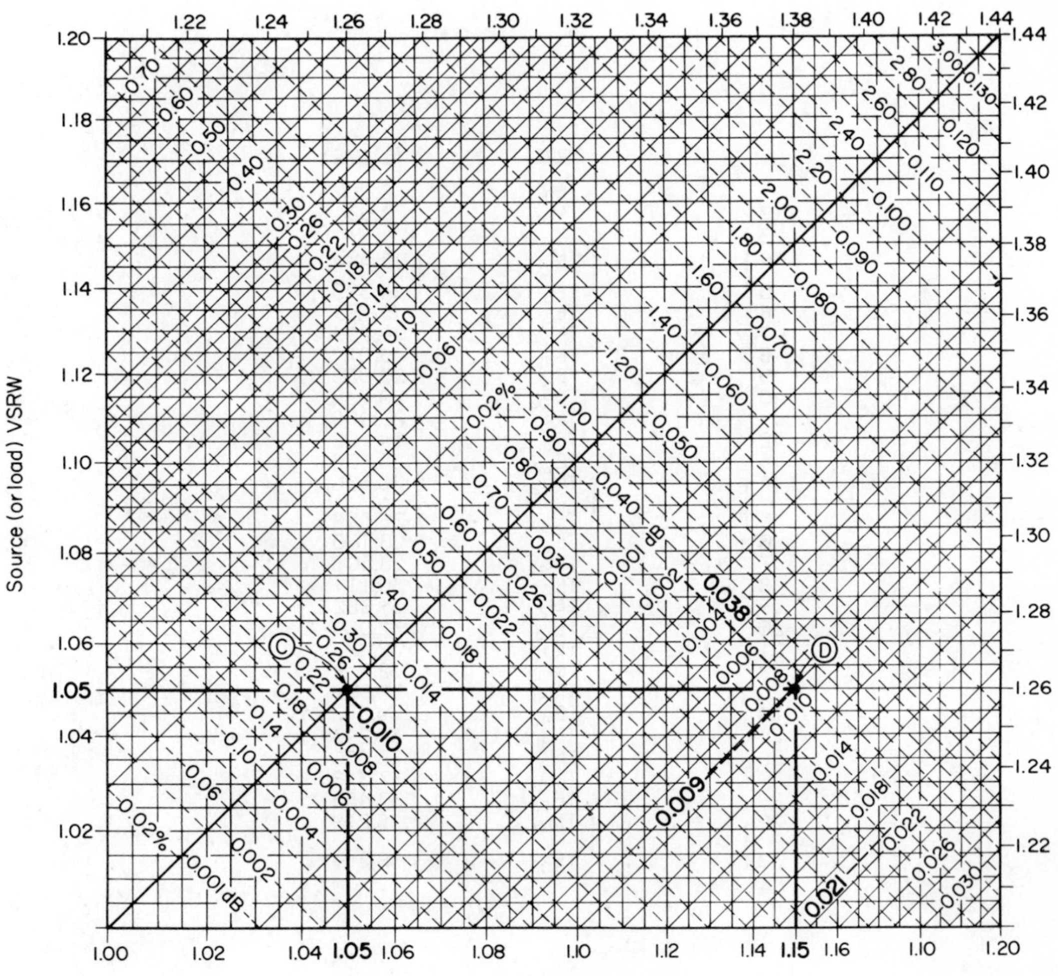
Source (or load) VSRW
Load (or source) VSWR
1.00 1.02 1.04 1.05 1.06 1.08 1.10 1.12 1.14 1.15 1.16 1.10 1.20
1.20 1.18 1.16 1.14 1.12 1.10 1.08 1.06 1.05 1.04 1.02
1.22 1.24 1.26 1.28 1.30 1.32 1.34 1.36 1.38 1.40 1.42 1.44
1.44 1.42 1.40 1.38 1.36 1.34 1.32 1.30 1.28 1.26 1.24 1.22
0.02%
0.001dB
0.001 dB
C
D
0.010
0.009
0.038
0.021

K Cutoff Frequencies for Circular Waveguides

	Mode	$f_c D$ Mc · inch
1	TE_{11}	6,917
2	TM_{01}	9,034
3	TE_{21}	11,474
4	TM_{11}	14,395
5	TE_{01}	14,395
6	TE_{31}	15,783
7	TM_{21}	19,293
8	TE_{41}	19,977
9	TE_{12}	20,029
10	TM_{02}	20,738
11	TM_{31}	23,969
12	TE_{51}	24,102
13	TE_{22}	25,193
14	TM_{12}	26,356
15	TE_{02}	26,356
16	TE_{61}	28,180
17	TM_{41}	28,508
18	TE_{32}	30,111
19	TM_{22}	31,622
20	TE_{13}	32,069
21	TE_{71}	32,225
22	TM_{03}	32,510
23	TM_{51}	32,952
24	TE_{42}	34,872
25	TE_{81}	36,243
26	TM_{32}	36,670
27	TM_{61}	37,328
28	TE_{23}	37,453
29	TM_{13}	38,219
30	TE_{03}	38,219

Radar Frequency Band Code Letters

L

Band P		
	Freq	λ
	0.225 0.390	133.3 76.9

Band L		
Sub	Freq	λ
P	0.390 0.465	76.9 64.5
C	0.465 0.510	64.5 58.8
L	0.510 0.725	58.8 41.4
Y	0.725 0.780	41.4 38.4
T	0.780 0.900	38.4 33.3
S	0.900 0.950	33.3 31.6
X	0.950 1.150	31.6 26.1
K	1.150 1.350	26.1 22.2
F	1.350 1.450	22.2 20.7
Z	1.450 1.550	20.7 19.3

Band S		
Sub	Freq	λ
E	1.55 1.65	19.3 18.2
F	1.65 1.85	18.2 16.2
T	1.85 2.00	16.2 15.0
C	2.00 2.40	15.0 12.5
Q	2.40 2.60	12.5 11.5
Y	2.60 2.70	11.5 11.1
G	2.70 2.90	11.1 10.3
S	2.90 3.10	10.3 9.68
A	3.10 3.40	9.68 8.83
W	3.40 3.70	8.83 8.11
H	3.70 3.90	8.11 7.69
Z	3.90 4.20	7.69 7.15
D	4.20 5.20	7.15 5.77

Band X		
Sub	Freq	λ
A	5.20 5.50	5.77 5.45
Q	5.50 5.75	5.45 5.22
Y	5.75 6.20	5.22 4.84
D	6.20 6.25	4.84 4.80
B	6.25 6.90	4.80 4.35
R	6.90 7.00	4.35 4.29
C	7.00 8.50	4.29 3.53
L	8.50 9.00	3.53 3.33
S	9.00 9.60	3.33 3.13
X	9.60 10.00	3.13 3.00
F	10.00 10.25	3.00 2.93
K	10.25 10.90	2.93 2.75

Band K		
Sub	Freq	λ
P	10.90 12.25	2.75 2.45
S	12.25 13.25	2.45 2.26
E	13.25 14.25	2.26 2.10
C	14.25 15.35	2.10 1.95
U	15.35 17.25	1.95 1.74
T	17.25 20.50	1.74 1.46
Q	20.50 24.50	1.46 1.22
R	24.50 26.50	1.22 1.13
M	26.50 28.50	1.13 1.05
N	28.50 30.70	1.05 0.977
L	30.70 33.00	0.977 0.909
A	33.00 36.00	0.909 0.834

Band Q		
Sub	Freq	λ
A	36.00 38.00	0.834 0.790
B	38.00 40.00	0.790 0.750
C	40.00 42.00	0.750 0.715
D	42.00 44.00	0.715 0.682
E	44.00 46.00	0.685 0.652

Band V		
Sub	Freq	λ
A	46.00 48.00	0.652 0.625
B	48.00 50.00	0.625 0.600
C	50.00 52.00	0.600 0.577
D	52.00 54.00	0.577 0.556
E	54.00 56.00	0.556 0.536

M Microwave Terms and Definitions

Attenuation. Decrease in magnitude of current, voltage, or power of a signal in transmission between points.

Attenuation constant. For a travelling plane wave of a given frequency, the rate of exponential decrease of the amplitude of a field component (or of the voltage or current) in the direction of propagation. Expressed in Nepers or db per unit length.

Attenuator, flap. A device designed to introduce attenuation into a waveguide circuit by means of a resistive material moved into the guide.

Attenuator, rotary vane. A device designed to introduce attenuation into a waveguide circuit by means of varying the angular position of a resistive material in the guide.

Backward-wave tube. A travelling-wave tube in which the electrons travel in a direction opposite to that in which the wave is propagated. A microwave oscillator.

Barretter. A metallic resistor with a positive temperature coefficient of resistivity. Used for detection and power-level measurements.

Bend, E-plane. A bend in a waveguide in the plane of the electric field. ("Easy" bend.)

Bend, H-plane. A bend in a waveguide in the plane of the magnetic field. ("Hard" bend.)

Bolometer. A barretter, a thermistor, or any other device utilizing the temperature coefficient of resistivity of some resistance element.

Choke joint. A type of joint for connecting two sections of waveguide. It is so arranged that there is efficient energy transfer without the necessity of an electrical contact at the insides of the guide.

Coaxial line. A transmission line in which one conductor completely surrounds the other, the two being coaxial and separated by a continuous solid dielectric or by dielectric spacers. Such a line is characterized by no external field and by having no susceptibility to external fields from other sources.

Coupler, directional. A device consisting of two transmission lines coupled together in such a way that a wave travelling in one line in one direction excites a wave in the other guide; ideally, in one direction only.

Coupler, forward. A directional coupler used to sample incident power.

Coupler, reverse. A directional coupler used to sample reflected power.

Coupling coefficient. A ratio between the power entering the main arm of a directional coupler in one direction to the power coupled into the auxiliary arm in the same direction.

Cut-off frquency. The lowest frequency at which lossless waveguide will propagate energy in some particular mode without attenuation.

Cut-off wavelength. The longest wavelength at which lossless waveguide will propagate energy in some particular mode without attenuation.

Demodulator. A device whose output voltage is proportional to the square of its input voltage (i.e., input power).

Detector. An element which reproduces the modulation of an RF wave, usually a semiconductor crystal. Barretters are sometimes used to detect low-frequency modulation.

Directivity. The ratio of (1) power flowing out of the auxiliary arm of a directional coupler when power is flowing in the forward direction in the main arm, to (2) power flowing out of the auxiliary arm of the coupler when power is flowing in the reverse direction in the main arm (both forward and reverse powers in the main arm being equal in magnitude).

Directivity signal. A spurious signal present in the output of a coupler because the directivity of the coupler is not infinite.

Efficiency, bolometer mount. The percentage of net applied power that is absorbed by the RF termination.

EHF. Extremely high frequency. The band of frequencies between 30,000 mc (30 gc) and 300,000 mc (300 gc).

E-H tee. A junction composed of a combination of E and H plane tee junctions having a common point of intersection with the main guide.

E-H tuner. An E-H tee used for impedance transformation, having two arms terminated in adjustable plungers.

Gigacycle. 10^9 cycles (formerly kilomegacycle). Common term for expressing microwave frequencies.

Guide wavelength. The length of waveguide corresponding to one cycle of variation in the axial (transmitted) direction.

Impedance, characteristic (of a rectangular waveguide). For the dominant TE_{10} mode of a lossless rectangular waveguide at a frequency above the cut-off frequency, the ratio of the square of the rms voltage between midpoints of the two conductor faces normal to the electric vector, and the total power flowing when the guide is match-terminated.

Impedance, characteristic (of a two-conductor transmission line). For a travelling, transverse electromagnetic wave, the ratio of the complex voltage between the conductors to the complex current on the conductors.

Impedance, normalized. Any impedance of a system divided by its characteristic impedance.

Incident power or signal. Power flowing from the generator to the load.

Iris. In a waveguide, a conducting plate or plates, of small thickness compared to a wavelength, occupying a part of the cross section of the waveguide. When only a single mode can be supported, an iris acts substantially as a shunt admittance.

Isolator ferrite. A microwave device which allows RF energy to pass through in one direction with very little loss while RF power in the reverse direction is absorbed.

Junction, hybrid. A waveguide arrangement with four branches which, when branches are properly terminated, has the property that energy can be transferred from any one branch into only two of the remaining three. In common usage this energy is equally divided between the two branches.

Magnetron. A high-power microwave oscillator tube with a fixed or limited frequency range. Frequency, efficiency, and power depend on magnetic field strength and anode voltage.

MASER (Microwave Amplification by Stimulated Emission of Radiation). A low-noise, microwave amplifier utilizing a change in energy level of a material to obtain signal amplification. Common materials are gases (ammonia) and crystals (ruby).

Matched termination (waveguide). A termination producing no reflected wave at any transverse section of the waveguide.

Microstrip. A microwave-transmission component utilizing a single conductor supported above a ground plane.

Microwave region. That portion of the electromagnetic spectrum lying between the far infrared and conventional RF portion. Commonly regarded as extending from 1,000 megacycles (30 cm) to 300,000 megacycles (1 mm).

Millimeter waves. The band of frequencies having wavelengths shorter than 1 cm (above 30,000 mc).

Mismatch loss (reflection loss). The ratio, expressed in db, of the incident power to the transmitted power at a discontinuity. A measure of the loss caused by reflection.

Mode (of transmission propagation). A form of propagation of guided waves that is characterized by a particular field pattern in a plane transverse to the direction of propagation. The field pattern is independent of the position along the axis of the waveguide and, for uniconductor waveguide, independent of frequency.

Noise figure. A figure of merit for microwave amplifiers. A ratio in db between actual output noise power, and the output noise power which would come from a noiseless amplifier with identical gain and bandwidth.

Parametric amplifier (MAVAR—Mixer Amplification by Variable Reactance). A microwave amplifier utilizing the nonlinearity of a reactance element to obtain amplification. A low-noise amplifier.

Propagation constant. A transmission characteristic of a line which indicates the effect of the line on the wave being transmitted along the line. It is a complex quantity having a real term, the attenuation constant, and an imaginary term, the phase a constant.

Rat race (hybrid ring). A hybrid junction which consists of a re-entrant line (waveguide) of proper electrical length to sustain standing waves, to which four side arms are connected. Commonly used as an equal power divider.

Reflected power or signal. Power flowing from the load back to the generator.

Reflection coefficient. A numerical ratio between the reflected voltage and the incident voltage.

Reflectometer. A microwave system arranged to measure the incident and reflected voltages and to indicate their ratio (swr).

Reflex klystron. A low-power microwave oscillator tube which depends primarily on the physical size of a cavity resonator for its frequency. Normally has a wider frequency range than a magnetron.

Reike diagram. A polar-coordinate load diagram for microwave oscillators, particularly klystrons and magnetrons.

Return loss. The ratio, expressed in db, between the power incident upon a discontinuity and the power reflected from the discontinuity. (The number of db reflected power is down from incident power.)

Rotator. In waveguides, a means of rotating the plane of polarization. In a rectangular waveguide, rotation is accomplished simply by twisting the guide itself.

SHF. Super high frequency. The band of frequencies between 3,000 and 30,000 mc.

Slotted section. A length of waveguide in the wall of which is cut a nonradiating slot used for standing-wave measurements.

Smith Diagram or Chart. A diagram with polar coordinates; developed to aid in the solution of transmission-line and waveguide problems.

Thermistor. A resistance element made of a semiconducting material which exhibits a high negative temperature coefficient of resistivity.

Travelling-wave tube. A broadband, microwave tube which depends for its characteristics upon the interaction between the field of a wave propagated along a waveguide and the beam of electrons traveling with the wave. A microwave amplifier.

Tuning screw (slide-screw tuner). A screw or probe inserted into the top or bottom of a waveguide (parallel to the E field) to develop susceptance, the magnitude and sign of which is controlled by the depth of penetration of the screw.

UHF. Ultra high frequency, the band of frequencies between 300 and 3,000 mc.

VHF. Very high frequency, the band of frequencies between 30 and 300 mc.

Voltage standing-wave ratio (*VSWR* or *SWR*). The measured ratio of the field strength at a voltage maximum to that at an adjacent minimum.

Wave circuits, slow. A microwave circuit designed to have a phase velocity considerably below the speed of light. The general application for such waves is in travelling-wave tubes.

Wave, dominant. The guided wave having the lowest cut-off frequency. It is the only wave which will carry energy when the excitation is between the lowest cut-off frequency and the next higher frequency of a waveguide.

Waveguide phase shifter. A device for adjusting the phase of a particular field component at the output of the device relative to the phase of that field component at the input.

Waveguide tee. A junction used for the purpose of connecting a branch section of a waveguide in series with or parallel with the main transmission line.

Waveguide tuner. An adjustable device added to a waveguide for the purpose of an impedance transformation.

Waveguide wavelength. For a travelling plane wave at a given frequency, the distance along the waveguide between points at which a field component (or the voltage or current) differs in phase by 2π radians.

Wave, phase velocity. The velocity with which a point of constant phase is propagated in a progressive sinusoidal wave.

Wave, group velocity. The velocity with which the envelope of a group of waves of neighboring frequencies travels in a medium; usually identified with the velocity of energy propagation.

Wave, transverse electric (TE wave). In a homogeneous isotropic medium, an electromagnetic wave in which the electric field vectors are everywhere perpendicular to the direction of propagation.

Wave, transverse electromagnetic (TEM wave). In a homogeneous isotropic medium, an electromagnetic wave in which both the electric and magnetic field vectors are everywhere perpendicular to the direction of propagation. Generally dominant mode of coaxial lines.

Wave, transverse magnetic (TM wave). In a homogeneous istropic medium, an electromagnetic wave in which the magnetic field vector is everywhere perpendicular to the direction of propagation.

Wave, TE_{mn} (in rectangular waveguide). In a hollow, rectangular, metal cylinder, the transverse electric wave for which m is the number of half-period variations of the electric field along the longer transverse dimension, and n is the number of half-period variations of the magnetic field along the shorter transverse dimensions.

Wave, TM_{mn} (in rectangular waveguide). In a hollow, rectangular, metal cylinder, the transverse magnetic wave for which m is the number of half-period variations of the magnetic field along the longer transverse dimension, and n is the number of half-period variations of the magnetic fields along the shorter transverse dimensions.

Wavemeter, absorption. A device which utilizes the characteristics of a resonator, which cause it to absorb maximum energy at its resonant frequency when loosely coupled to a source.

Receiver Noise Figures

N

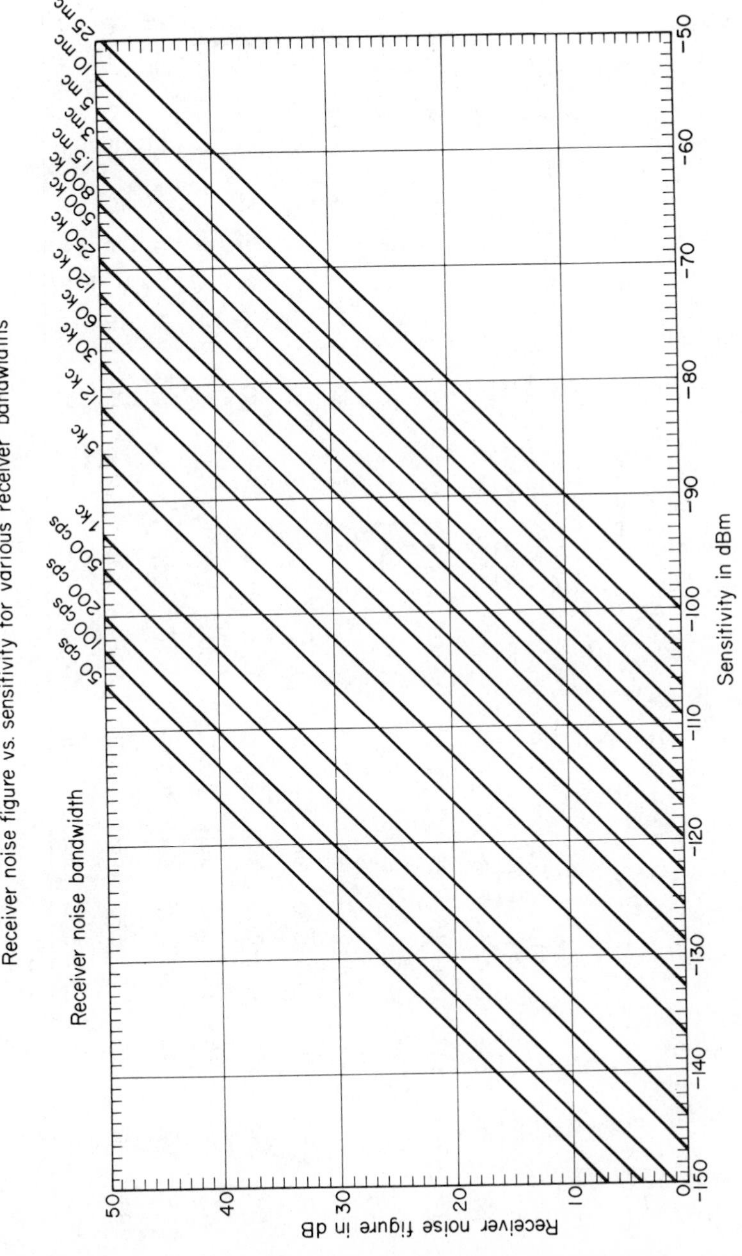

O

Basic Waveguide Lengths

Waveguide size WR-( )	RG Equiv. Brass	RG Equiv. Alum.	Freq. range GHz	Tubing material	Attenuation dB/ft	Typ. max. VSWR Full WG band	Typ. max. VSWR 10% band	Peak power at max. rated press. (megawatts)	Max pressure PSIG	Minimum centerline bending radii: Factory performed E	Factory performed H	Field free-formed E	Field free-formed H	Max. twist °per ft.
62	91	349	12.4–18.0	Brass	0.25	1.10	1.08	0.8	45	3/4	1–1/8	1–1/4	1–3/4	220
75	346	347	10.0–15.0	Brass	0.15	1.10	1.08	0.8	45	1	1–3/8	1–3/8	2	180
90	52	67	8.20–12.4	Brass	0.08	1.10	1.08	1.0	45	1 – 1/8	1–3/8	1–3/8	2–1/4	180
112	51	68	7.05–10.0	Brass	0.06	1.10	1.08	1.0	30	1 –1/4	1–7/8	1–11/16	2–1/2	180
137	50	106	5.85–8.20	Brass	0.05	1.10	1.06	1.2	30	1 –3/8	2–5/16	2–1/2	3–1/2	180
159	343	344	4.90–7.05	Brass	0.04	1.08	1.06	1.7	30	1 –1/2	2–1/2	2–1/2	3–1/2	170
187	49	95	3.95–5.85	Brass	0.025	1.06	1.05	2.4	30	2 –1/2	4	3	4–1/2	150
229	340	341	3.30–4.90	Brass	0.025	1.05	1.02	3.6	30	2 –1/2	4–1/2	4	5–3/4	120
284	48	75	2.60–3.95	Brass	0.017	1.08	1.05	2.0	15	3	5	4–1/2	6–3/8	110
340	112	113	2.20–3.30	Brass	0.015	1.05	1.03	5.0	15	4	6	6	9	100

Twistable, Flexible Waveguide Data

P

This nomograph is used to determine the value of one of the variables: waveguide wavelength λ_g, free space wavelength λ_o (or frequency f), or cutoff wavelength λ_c, when any two are known.

The vertical scale gives waveguide wavelength in centimeters. The horizontal one is for the cutoff wavelength. The points corresponding to the cutoff wavelength in the TE_{10} mode of the three common waveguides are also indicated. The sloping scale is calibrated in free space wavelengths and frequency.

Example I

What is the waveguide wavelength at 6 kmc (5 cm. wavelength) in RG — 50 waveguide?

The straight line between the points f = kmc and RG — 50 on the λ_c scale is extended to intersect the λ_g scale. This insection yields $\lambda_g = 7.17$ cm.

Example II

By measurement on an RG — 51 slotted line, the waveguide wavelength is found to be 6.5 cm. What is the frequency?

The straight line between 6.5 on the λ_g scale and RG - 51 on the λ_c scale intersects the f scale at 7.0 kmc.

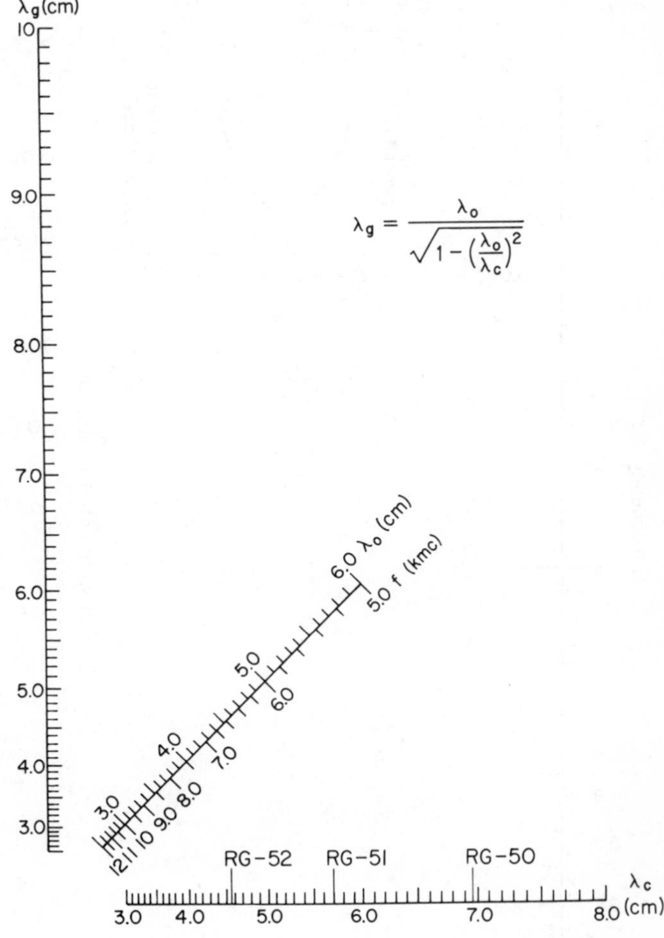

$$\lambda_g = \frac{\lambda_o}{\sqrt{1-\left(\frac{\lambda_o}{\lambda_c}\right)^2}}$$

Q VSWR and Γ Calculation Nomograph

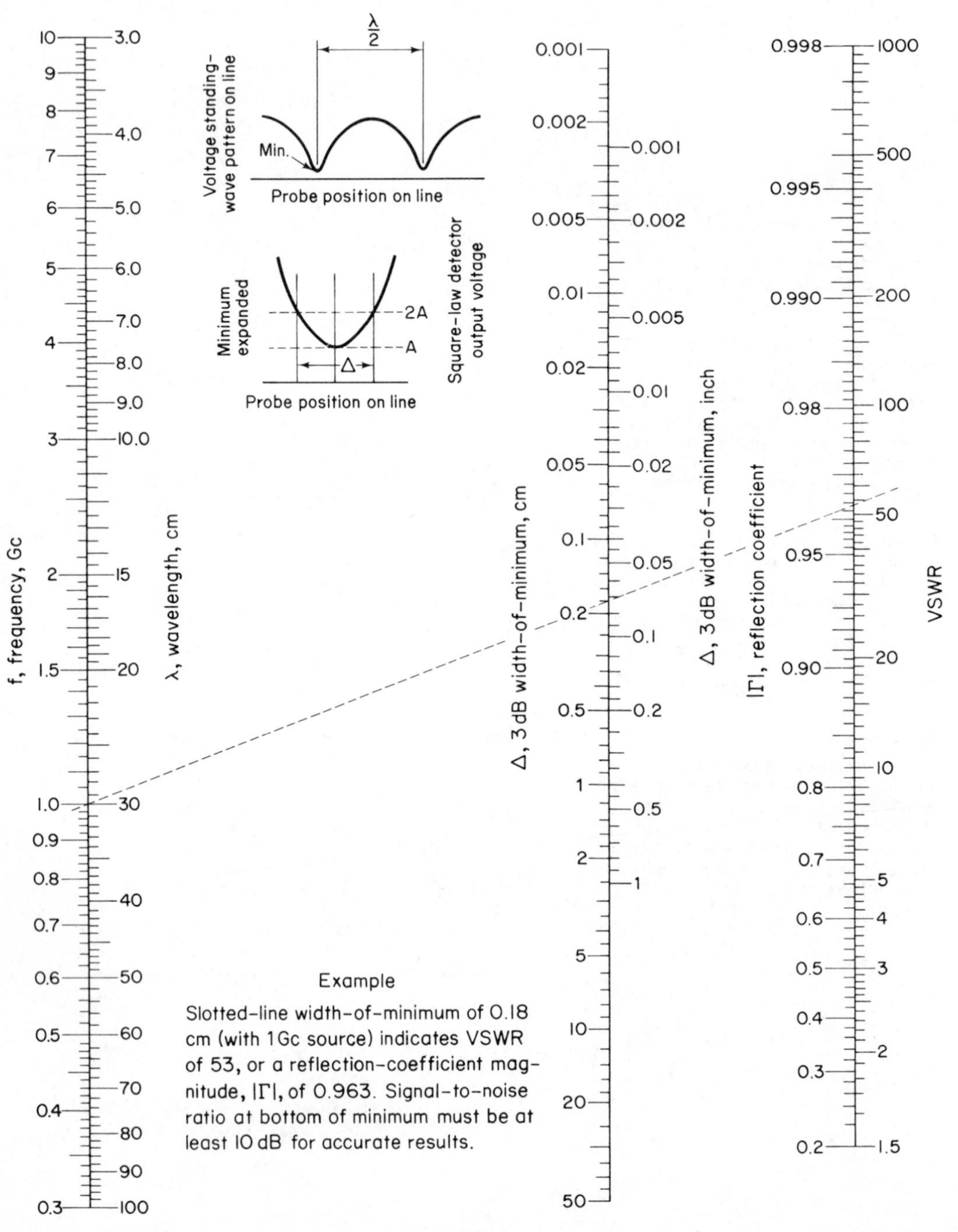

Bolometer Characteristics and Construction

R

Bolometer data and construction

Type B–Barretter T–Thermistor	Max RF power (mw)	Sensitivity (ohms/mw)	Freq range (Gc)	Time Constant (msec)	Ambient temp range (C)
B	3 mw–10 w	4.5–10	dc–40	–0.08 ms–1 sec	–40 to +80
B	1	4.5–12	dc–220	–	–
T	10	–	0.5–220	–	–
B	1–10	4–8	0.01–10	0.01–0.35	–60 to +85
T	10	10	0.01–10	1 sec	–60 to +100
B	10	75	50–90	0.1	0 to 60
B	1–16	approx 10	approx dc–75	approx 0.35	approx –40 to +80
T	approx 300	approx 15	approx dc–18	1 sec	–
B	1.5–40	2.5–10	0.5–18	0.35–0.36	approx –40 to +80
T	150–300	–	–	1 sec	–
B	3–16	4.5–10	0.5–12.4	0.35	–
T	10	–	0.5–12.4	–	–
B	1–100	10–60	0.5–40	–	–
T	10	–	0.5–10	–	–
B	1–10	4.5–11	–	0.07–0.32	–40 to +80
T	10	25	–	75–95	–40 to +80

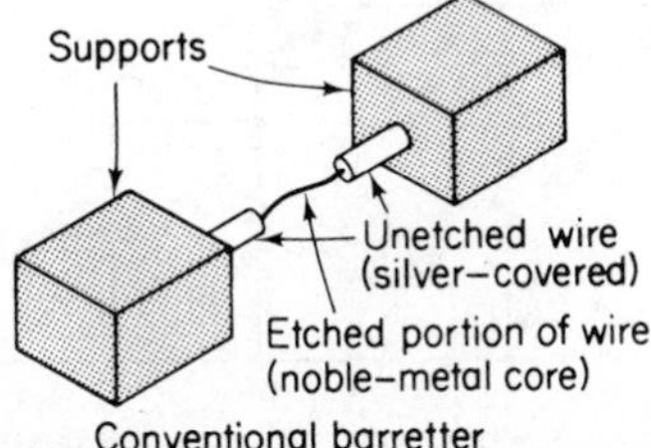

Conventional barretter
(Wollaston wire contributes entire resistance)

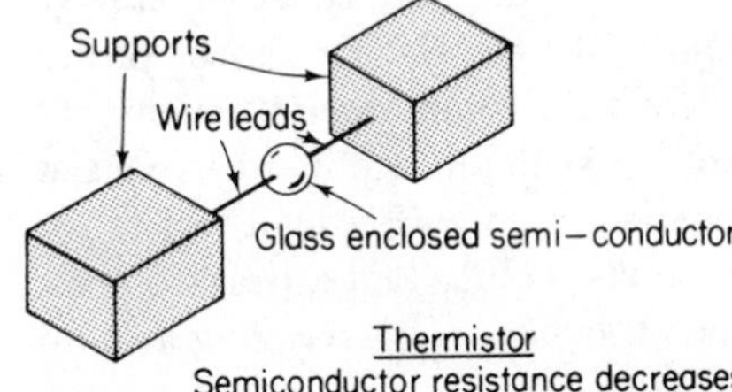

Thermistor
Semiconductor resistance decreases with input microwave energy

S Wide-Band (Video) Amplifiers

Wide-band frequency response in a transistorized amplifier means that the amplifier has uniform gain (within one or two db) from dc (zero frequency) up through several megahertz. In a television video amplifier, the output response is usually flat from zero frequency up to a "roll-off" or cutoff point which lies between 3 and 4 MHz.

At communication, radar, or telemetry frequencies, the term wide-band has another connotation. In these applications, we name receivers wide- or narrow-band according to the width of their signal amplification capabilities centering around the main transmission frequency. Here, the products of detection of the wide- or narrow-band information transmission are amplified again by a wide-band or video amplifier circuit.

TYPICAL TV VIDEO AMPLIFIER

Requirements. The typical, solid-state, TV video output stage must drive the grid of a TV picture tube with constant-amplitude signals from dc to 4 MHz and must deliver an output level of 50-100 V. This calls for a 1- to 2-watt power transistor and collector supply voltages from 130 to 150 V. Since power transistors may have common-emitter current gains of 50 to 100, such an amplifier can consist of a single output stage with a load resistance of from 3000 to 6000 ohms and a peak-to-peak collector current swing (I_{CE}) averaging around 10 mA (2-20 mA).

Figure S-1 shows a typical output stage being driven by an average signal, I_C, of 10 mA ($V_{BE} = 2$ V) and delivering 100 V peak-to-peak across the 5000-ohm collector load resistor.

Response. The gain, power, voltage output, etc. are the basic design considerations; the main problem is (frequencywise) to keep the output voltages uniform in amplitude from d-c to 4 MHz, as well as distortionless and transient-free.

Impedance variations with frequency are the main cause of these output changes, so we must tailor and arrange the output and interstage coupling circuit constants for compensation against inherent frequency drop-off. This must be done at each end of the frequency

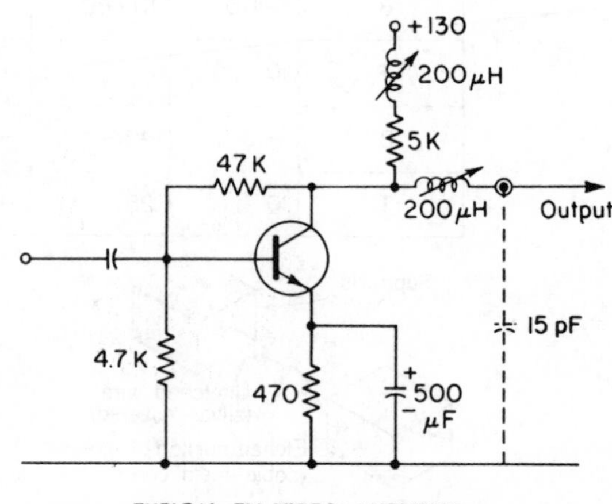

TYPICAL TV VIDEO AMPLIFIER

spectrum where particular treatment and circuit devices apply for low and for high frequencies. Interstage and output coupling thus depend upon the characteristics of the power transistor being used.

The collector-to-base coupling circuit in a two-stage video amplifier in Fig. S-2(a) illustrates the procedure when we include transistor parameters, and particularly, it shows how frequency response falls off at both ends of the spectrum, as follows:

1. at low frequencies, due to the relationship of the amplifier coupling capacitances and their associated load resistances.

2. at high frequencies, due to the fall-off

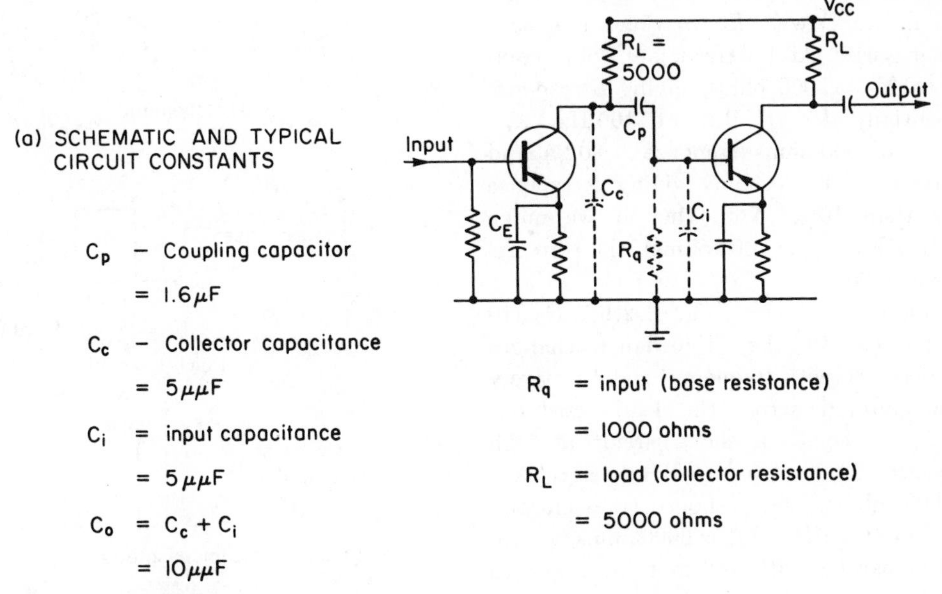

(b) LOW FREQUENCY RESPONSE with → R_q = 1000 ohms, C_p = 1.6 μF

at:	at:	at:
f = 1 MHz	f = 100 Hz	f = 10 Hz
X_{c_p} = 0.1 ohm	X_{c_p} = 1000 ohms	X_{c_p} = 10,000 ohms
Response = 100%	Response = 70%	Response = 10%

Now, if:

C_p = 16μF at f = 10 Hz, X_{c_p} = 1000 ohms and

R_q = 1000 ohms Response = 70%

(c) HIGH FREQUENCY RESPONSE with → R_q = 1000 ohms, C_o = 10 pF

at:	at:
f = 1 MHz	f = 3 MHz
X_{c_o} = 17,000 ohms	X_{c_o} = 6000 ohms
Response = 100%	Response = 60%

FREQUENCY RESPONSE IN A VIDEO AMPLIFIER

caused by the parallel capacitive impedances of the amplifier transistor itself when compared to load and input resistances.

From the tabulation in Fig. H-2(b), we find the low-frequency fall-off from 1-MHz, 100% response by noting that the key factor is the impedance of the coupling capacitor, X_{CP}. Here, the output response depends upon comparison of X_{CP} (when part of a voltage-divider network) with R_G, to which it is connected in series. At 1 MHz it is 0.1 ohm compared to $R_G = 1000$ ohms, giving a response of essentially 100%. But at 100 Hz, X_{CP} increases to 1000 ohms (response = 80%), and at 10 Hz it is 10,000 ohms, giving a response of less than 10%. Note that if we make $C_P = 16$ MF ($X_{CP} = 1000$ ohms), the response increases to 80%.

At high frequencies (Fig. H-2(b)), C_O (10 pF) produces the key impedance changes which decrease with frequency and effectively become shunted across the load resistance since X_{CP} becomes a short circuit at high frequencies. At midband (1 MHz), since X_{CO} is 17,000 ohms, the response is essentially 100%; but at 3 MHz, X_{CO} is 6000 ohms, which, placed in parallel with and at right angles to the load resistance, gives a response of around 60%.

These capacitances are built into the transistor and cannot be altered. Without compensation, their shunting effect on the response can be reduced only by use of a lower R_L which, of course, decreases gain. Stray and wiring capacitances have minor effects; C_E, the emitter bypass capacitor, must be several microfarads (see low-frequency compensation discussed below) and intimately located with respect to the emitter terminal in order to reduce C_{BE}.

S-1.1 COMPENSATING CIRCUITS

Poor low-frequency response is caused by time consumed in discharging the coupling capacitor C_P through R_G—or in other words, the time constant $R_G \times C_P$. Thus for best low-frequency response, both R_G and C_P should be kept as large as practicable. Large physical

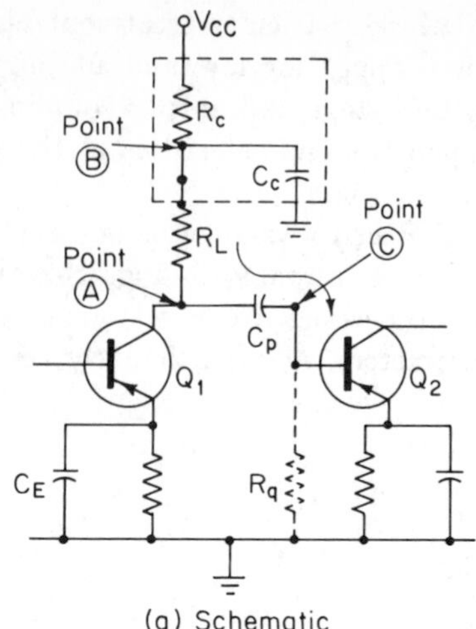

(a) Schematic

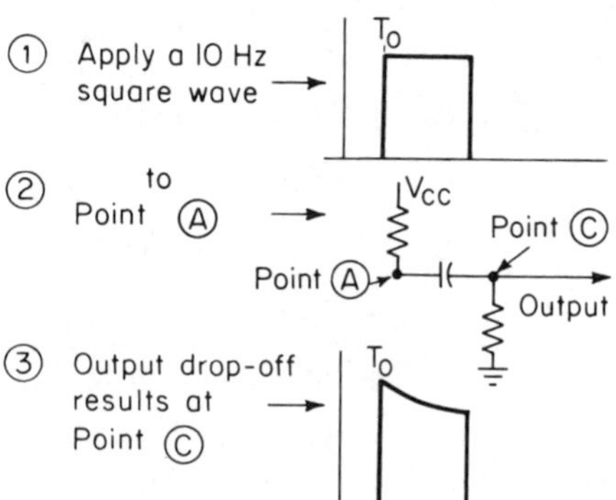

(b) Action in uncompensated amplifier

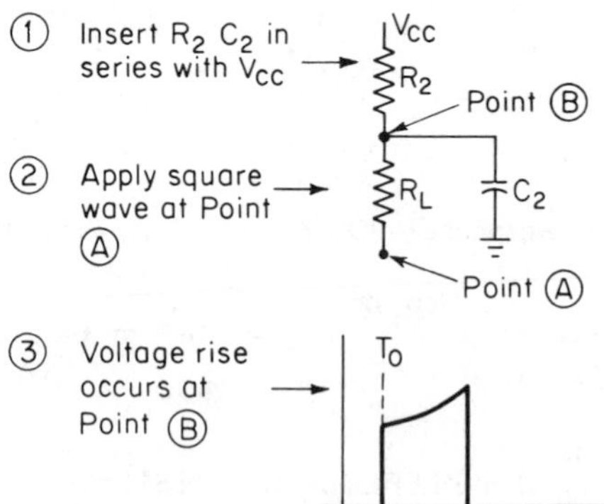

(c) Action in compensated amplifier

Since C_c starts fully charged at T_0, placing the networks in series allows the rise in voltage across C_c to compensate for the decrease in voltage caused by C_p's charging.

LOW-FREQUENCY COMPENSATION IN A VIDEO AMPLIFIER

size is a handicap in using electrolytic capacitors, and any series resistance inserted in the base circuit reduces gain; therefore, other compensating methods are sought.

The best and simplest circuit contains, in series with R_L, a network, R_C, C_C, whose time constant equals that of $R_L \times C_P$. See Fig. H-3(a). Dynamically, the compensation is illustrated by charge and discharge action when, say, a 10-Hz square wave appears across point A—the collector of Q_1. C_P immediately starts to charge, drawing current through R_L, which would cause the voltage at point A to fall exponentially (see Fig. H-3(b)). At T_0, however, C_C, which has been charged at the supply voltage, is given a 10-V boost, which reflects itself on point B as an exponential rise. See Fig. H-3(c). This rise is added in series through R_L to the exponential decrease brought about by the charging of C_P. If the rates of rise and fall of these voltages are equal, the square top of the pulse appearing at B will be preserved, and compensation achieved, while using a relatively small coupling capacitor for C_P.

Note also when an emitter series bias resistor is used that it must be bypassed with a heavy capacitor since feedback voltages developed across a small capacitor will also reduce low-frequency gain.

At high frequencies, we compensate for the shunting effects of parallel capacitances by resonating them with a so-called shunt peaking inductance, shown in Fig. H-4(a). When the dynamic impedance of this resonant circuit equals R_L, the response at 3.5 MHz may increase 20–30%. C_P becomes a short circuit at high frequencies and can be neglected in all calculations.

Series peaking is another method which uses series resonance introduced with coil L_S in series with C_1. See Fig. H-4(b). Impedance pickoff is taken across C_1, giving improved response which can be equal to or better than the shunt peaking method of Fig. H-4(a); however, some stage gain is lost.

Combination shunt and series peaking (see Fig. H-4(c)) is sometimes used with improvement over the series or shunt method but at the cost of greater circuit complexity and additional components.

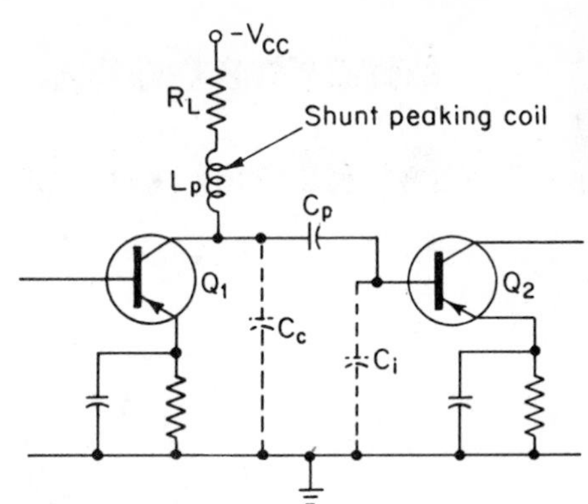

(a) Shunt peaking:

$C_o = C_c + C_i$ (C_p neglected)

If C_o = 10 pF
L_p = 200 μH
(and resonate at 3.5 MHz)

Capacitive shunting effects are reduced and response may go as high as 85%

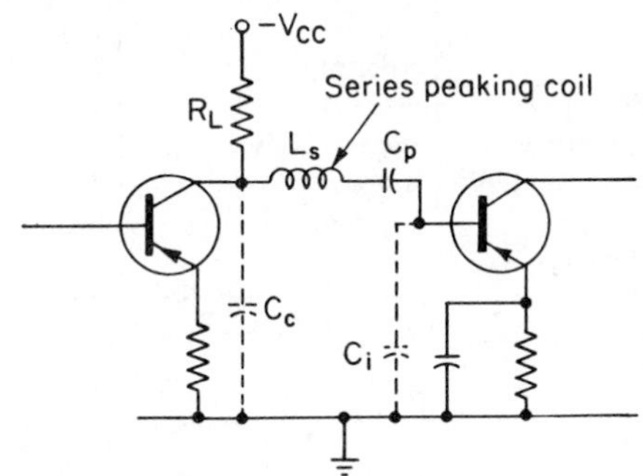

(b) Series peaking:

L_s is made resonant with C_i at 3.5 MHz

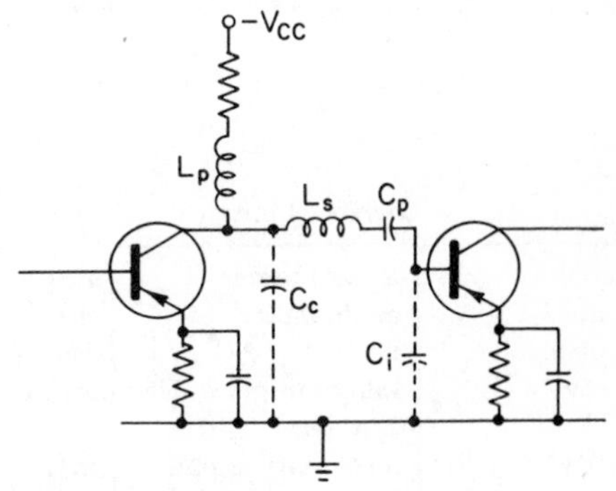

(c) Combination peaking:

Combined effects of series and shunt resonances are used to improve response above either of systems (a) or (b)

HIGH-FREQUENCY COMPENSATION IN A VIDEO AMPLIFIER

T International System of Units

INTERNATIONAL SYSTEM OF UNITS

Basic Units

Quantity	*Unit*	*Symbol*
Length	meter	m
Mass	kilogram	kg
Time	second	s
Temperature	degree Kelvin	°K
Electric current	ampere	A
Luminous intensity	candela	cd

Supplementary Units

Quantity	*Unit*	*Symbol*
Plane angle	radian	rad
Solid angle	steradian	sr

Derived Units

Quantity	*Unit*	*Symbol*
Area	square meter	m^2
Volume	cubic meter	m^3
Frequency	hertz	Hz (s^{-1})
Density	kilogram per cubic meter	kg/m^3
Velocity	meter per second	m/s
Angular velocity	radian per second	rad/s
Acceleration	meter per sec squared	m/s^2
Angular acceleration	radian per sec squared	rad/s^2
Force	newton	N ($kg \cdot m/s^2$)
Pressure	newton per sq meter	N/m^2
Kinematic viscosity	sq meter per second	m^2/s
Dynamic viscosity	newton-second per sq meter	$N \cdot s/m^2$
Work, energy, quantity of heat	joule	J ($N \cdot m$)
Power	watt	W (J/s)
Electric charge	coulomb	C ($A \cdot s$)
Voltage, potential difference, electromotive force	volt	V (W/A)
Electric field strength	volt per meter	V/m
Electric resistance	ohm	Ω (V/A)
Electric capacitance	farad	F ($A \cdot s/V$)
Magnetic flux	weber	Wb ($V \cdot s$)
Inductance	henry	H ($V \cdot s/A$)
Magnetic flux density	tesla	T (Wb/m^2)
Magnetic field strength	ampere per meter	A/m
Magnetomotive force	ampere	A
Flux of light	lumen	lm ($cd \cdot sr$)
Luminance	candela per sq meter	cd/m^2
Illumination	lux	lx (lm/m^2)

MULTIPLE AND SUBMULTIPLE PREFIXES

Multiple	*Prefix*	*Symbol*	*Pronunciation*	*Submultiple*	*Prefix*	*Symbol*	*Pronunciation*
10^{12}	tera	T	tĕr′ ȧ	10^{-1}	deci	d	dĕs′ ĭ
10^{9}	giga	G	jĭ′ gȧ	10^{-2}	centi	c	sĕn′ tĭ
10^{6}	mega	M	mĕg′ ȧ	10^{-3}	milli	m	mĭl′ ĭ
10^{3}	kilo	k	kĭl′ ō	10^{-6}	micro	μ	mī′ krō
10^{2}	hecto	h	hĕk′ tō	10^{-9}	nano	n	năn′ ō
10	deka	da	dĕk′ ȧ	10^{-12}	pico	p	pē′ cō
				10^{-15}	femto	f	fĕm′ tō
				10^{-18}	atto	a	ăt′ tō

U Integrated Circuits in Microwave Equipment and Measurement

Four characteristics typify the movement toward the use of integrated circuit (IC) semiconductors in the 1 GHz range and beyond:

1. lowered noise
2. higher output power
3. shf frequency generation
4. distributed impedance devices

Generally, all of these must also be inherent in the test equipment necessary to design and to tailor microwave hardware, although test equipment design always lags a little behind commercial IC development.

A summary of microwave measurement equipment given in Table M-1 serves as a guide line for the performance areas in which semiconductor devices must be handled. Most of the listed pieces of measuring equipment have been transistorized.

M-1 LOW-NOISE OPERATION

The inherent physical minuteness of IC structures automatically permit high-frequency operation. Tunnel diodes, in particular, and modern transistors, switching diodes, and varactors have been slowly closing the gap toward 100 GHz operation. Some of this improvement has been in germanium transistors which outperform silicon units in this region. Some improvement is due to the use of gallium arsenide.

M-2 POWER OUTPUT

Multistripe construction and increased operating efficiency allow power capability upwards of 1 watt in the 1–10 GHz region. An exploded view illustrating the construction of an X-band mixer is shown is Fig. M-1. Ten-watt transistor outputs, stemming from higher efficiency operation, have been produced at 1 GHz and beyond.

M-3 SHF FREQUENCY GENERATORS

With Gunn effect, IMPATT, and avalanche transit-time oscillators (and efficiency improvements), 100-mW, cw oscillations have been produced at 50 GHz. Special varactors exist with cutoff frequencies at hundreds of gigahertz.

M-4 DISTRIBUTED-IMPEDANCE DEVICES

Transmission lines at centimeter wavelengths can be correctly tailored for optimum wavelength, attenuation, and characteristic impedance by means of microstrip construction. See Fig. M-2.

TABLE M-1 SUMMARY OF MICROWAVE EQUIPMENT AND MEASUREMENT

Type of measurement	Input source and signal	Output measuring instrument	Accessories	Measuring techniques, ranges, comments
Power	Transmitter or equipt. tested	(1) Calorimeter } Power (2) Bolometer } meter	Attenuators, bolometers, mounts, bridges	(1) High power only—generally a complete set-up (2) Milliwatt power range
Frequency	Equipt. tested, signal generator	Cavity wavemeter, slotted line, and crystal frequency standard	Attenuators, directional couplers, stub tuners, mounts, frequency standard	Slotted line positioning less accurate than cavity reading, in turn less than crystal reference
Impedance or admittance	Variable-frequency signal generator	Slotted line, probe, and VSWR meter	Attenuator, frequency meter	Obtain reflection coefficient and read impedance from Smith Chart
Reflection coefficient and VSWR	Signal generator	Slotted line and VSWR indicator	Termination and calibrated attenuator	Calculation from slotted line positioning
Circuit Q	(1) Signal generator	(1) Slotted line, probe, VSWR meter	Attenuator, frequency meter	Reflection coeff. plus Smith chart
	(2) Sweep-frequency sig. generator	(2) Wavemeter, crystal, oscilloscope	Attenuator, directional coupler.	Frequency response at 3-db points
Attenuation	Signal generator	Bolometer and power meter	Variable attenuator	Substitution procedures used with low accuracy due to poor resolution
Frequency spectrum analysis	Spectrum analyzer	Crystal detector and oscilloscope display	Attenuators, directional coupler	Visual observation of oscilloscope pattern
Noise	Equipt. on test and noise generator	Power meter Broad-band voltmeter	Directional coupler, precision attenuator	Comparison of inherent noise signal to added noise signal
Antenna gain	Radiating auxiliary antenna	Antenna under test, receiver, and indicator	Standard gain horn, directional couplers, attenuators, crystal, meter	Pattern plotting from received and detected signal
Radar	Pulsed source equipment itself	Radar display	Echo box, couplers, attenuators	Inspection of reflected signal power, recovery, and other quantities

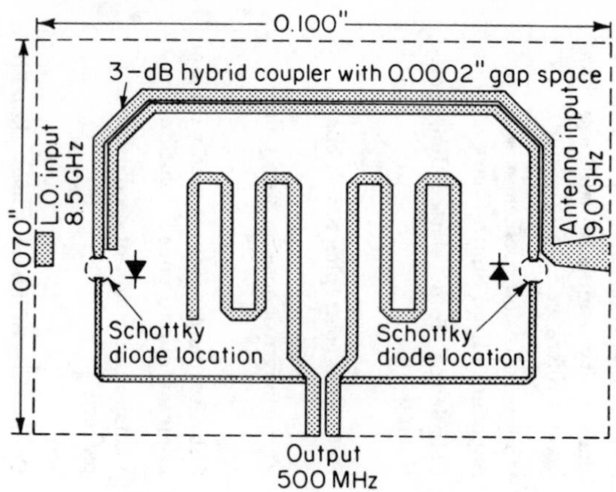

Fig. M-1 X-BAND HYBRID MIXER

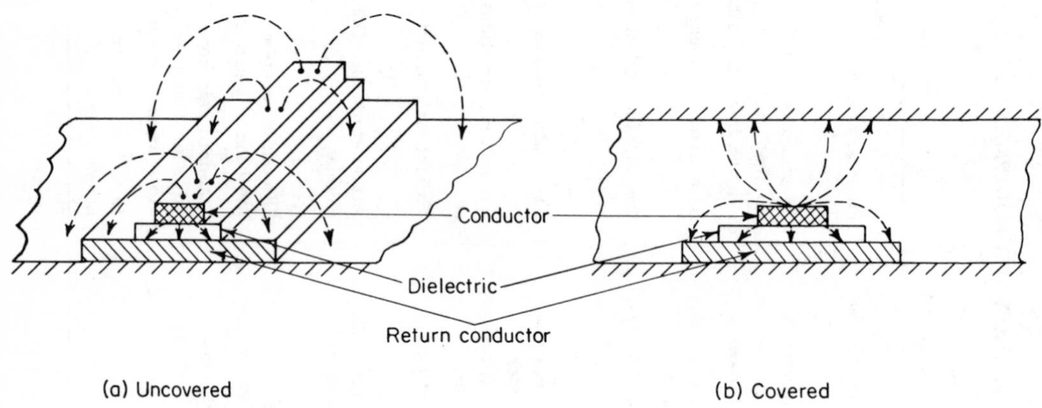

Fig. M-2 MICROSTRIP CONSTRUCTION

Microstrip dimensions allow for monolithic IC construction of a number of circuit components: TR switches, attenuator pads, tuning stubs, hybrid rings, directional couplers, and ferrite circulators. The last component mentioned is constructed by the insertion of ferrite material into the microstrip material.

V

Reactance Chart

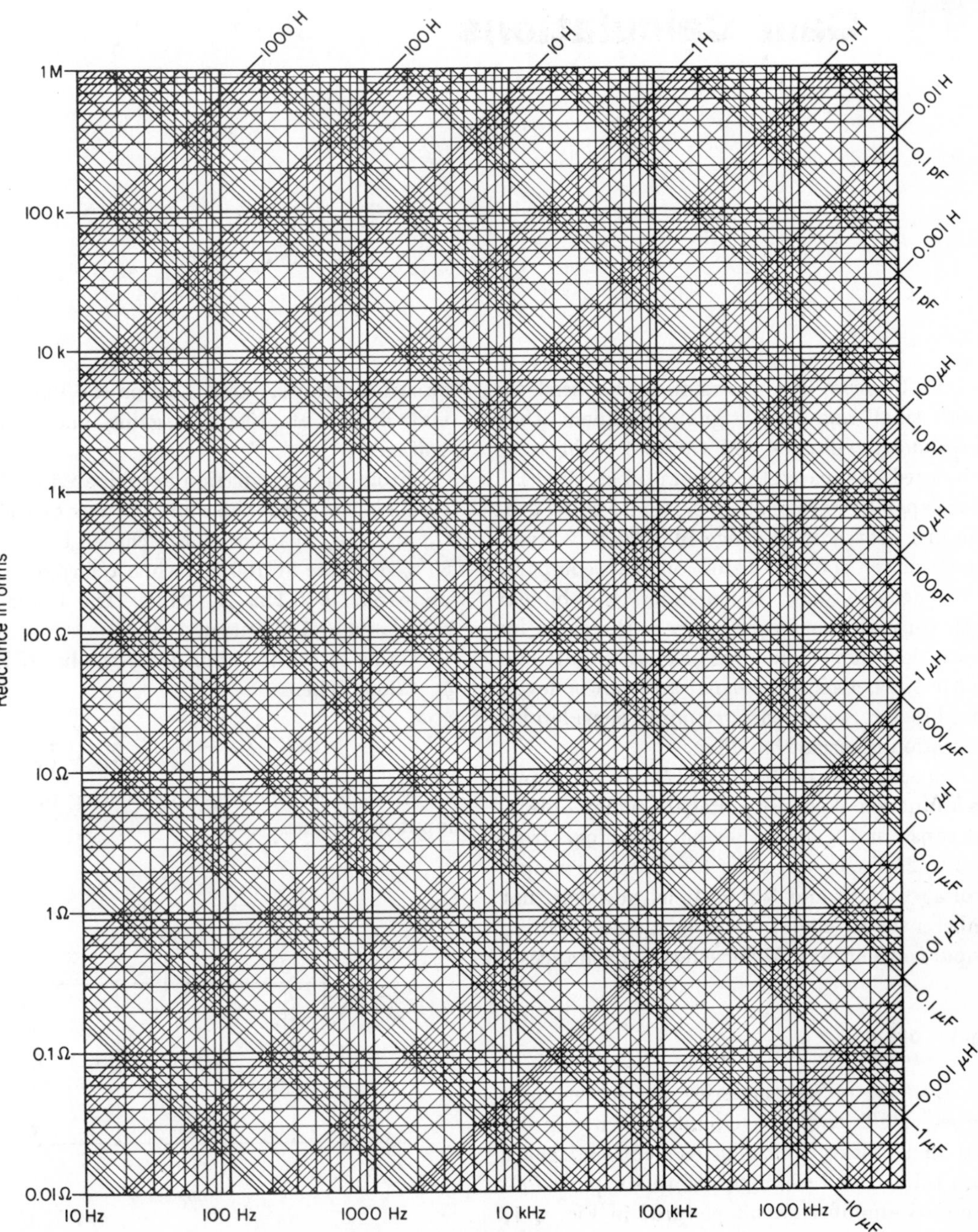

W FET Amplifiers and Gain Calculations

FET versatility allows it to be used for both small and large signal amplifiers when breakdown voltages and thermal ratings are observed. Compared to vacuum tubes, the FET has capabilities which deliver, under equal conditions over a wide range of supply voltage, more power and less distortion than comparable vacuum tubes. It outshines the bipolar transistor in that its circuitry is the same for both low-signal and high-signal input and output, whereas wide signal and power levels in bipolar transistors require considerable variation in circuit handling.

Specifically, maximum FET output signal is a function of supply voltage; we may obtain, at constant 5% distortion, output signals from 1.2 to 25 V corresponding to respective supply voltages of 10 and 300 V. It is, however, not inherently a high-frequency device having input, output, and Miller effect capacitances.

Q-1 DESIGN

Because of its similarity to the vacuum tubes, FET amplifier circuit design can be handled quickly following conventional chart and tabulation techniques. With a typical *RC*-coupled amplifier circuit pictured in Fig. Q-1, we may, with our typical FET (2N3684), select circuit constants, enter them on the Table Q-1, and arrive at 1000-Hz gains and output voltages.

From the explanation accompanying chart symbols, note that the source bypass capacitor, C_s, and the capacitor C_c are chosen to give an 80% response at 100 Hz when referred to the midband, 1000-Hz, A_v value. The latter are calculated at 1-V output.

Gains calculated stem from the listed FET g_m values. Then

$$A_v = g_m \times R_L = g_m \times \frac{R_D \times R_g}{R_D + R_g}$$

E_o is determined by the starting point for input distortion.

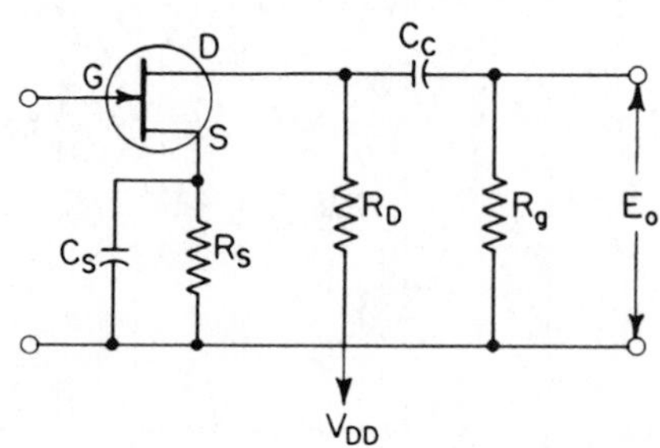

Fig. Q-1 CIRCUIT FOR FET RC-COUPLED AMPLIFIER

TABLE Q-1 FET RC-COUPLED AMPLIFIER GAIN AND OUTPUT CHART

V_{DD}	R_D	R_g	R_s	C_s (μF)	C (μF)	$E_{o(max)}$ (V)	A_v	I_D (μA)
14 V	0.05 M	0.05 M	8500	2.0	0.068	1.5	13.25	190
		0.1 M	9000	1.3	0.036	2.0	15.75	185
		0.22 M	9500	1.2	0.02	2.0	17.50	175
	0.1 M	0.1 M	25 k	0.75	0.036	1.0	15.3*	75
		0.22 M	27 k	0.68	0.02	1.0	18.75*	70
		0.47 M	30 k	0.62	0.006	1.0	20.6*	60
	0.22 M	0.22 M	56 k	0.43	0.02	0.5	16.3**	35
		0.47 M	62 k	0.36	0.006	0.5	18.7**	31.5
		1.0 M	62 k	0.33	0.004	0.5	20.6**	31.5
28 V	0.1 M	0.1 M	11 k	1.9	0.032	2.0	27.0*	160
		0.22 M	11 k	1.4	0.015	2.0	33.5	160
		0.47 M	11 k	1.25	0.005	3.0	37.5	160
	0.22 M	0.22 M	25 k	0.68	0.015	2.0	30.5	79
		0.47 M	26.5 k	0.62	0.005	2.0	37.0	75
		1.0 M	27 k	0.49	0.0032	2.0	42.0	73
	0.47 M	0.47 M	51 k	0.43	0.006	1.0	32.5	40
		1.0 M	55 k	0.33	0.004	2.0	37.5	37.5
		2.2 M	56 k	0.2	0.002	2.0	40.5	37
40 V	0.1 M	0.1 M	7500	2.1	0.032	4.0	33.4	260
		0.22 M	8000	1.95	0.015	4.0	40.0	220
		0.47 M	8500	1.8	0.005	5.0	45.5	210
	0.22 M	0.22 M	17 k	1.5	0.015	4.0	39.5	116
		0.47 M	18 k	0.8	0.005	5.0	47.0	106
		1.0 M	19 k	0.6	0.0032	5.0	51.3	101
	0.47 M	0.47 M	36 k	0.6	0.006	4.0	42.6	55
		1.0 M	39 k	0.47	0.004	5.0	48.8	52
		2.2 M	43 k	0.36	0.002	5.0	52.5	50

*at 500 millivolts (rms)
**at 300 millivolts (rms)

V_{DD} = drain supply voltage (volts)
R_D = drain resistor (megohms)
R_g = gate resistor of the following stage (megohms)
R_s = source self-bias resistor (ohms)
C_s = source resistor bypass capacitor (μF)
C_c = blocking capacitor (μF)
E_o = maximum output voltage (rms volts), where input level is of sufficient magnitude to cause the gate-source diode to be forward-biased and distortion to occur at input (also the point where drain current increases)
A_v = voltage gain (1000 Hz) at 1-V rms output (unless otherwise specified)
I_D = operating drain current ($\mu\alpha$)
C, C_s chosen for $A'_v = 0.8\ A_v$ at 100 Hz

X Waveguide Dimensions versus Operating Frequency Bands

Inside dimensions (inches)	Usual wall thickness (inches)	Outside dimensions (inches)	Frequency band (megacycles)
15 x 7.5	0.125	15.25 x 7.75	470 – 750
11.5 x 5.75		11.75 x 6	640 – 960
9.75 x 4.87		10 x 5.12	750 – 1,120
7.7 x 3.85		7.95 x 4.1	960 – 1,450
6.5 x 3.25 (L band)	0.080	6.66 x 3.41	1,120 – 1,700
5.1 x 2.55		5.26 x 2.71	1,450 – 2,200
4.3 x 2.15		4.46 x 2.31	1,700 – 2,600
3.4 x 1.7		3.56 x 1.86	2,200 – 3,300
2.84 x 1.34 (S band)		3.00 x 1.50	2,600 – 3,950
2.29 x 1.145	0.064	2.418 x 1.273	3,300 – 4,900
1.872 x 0.872 (C band)		2.0 x 1.0	3,950 – 5,850
1.590 x 0.795		1.718 x 0.923	4,900 – 7,050
1.372 x 0.622 (X_b band)		1.5 x 0.75	5,850 – 8,200
1.122 x 0.497		1.25 x 0.625	7,050 – 10,000
0.9 x 0.4 (X band)	0.050	1.0 x 0.5	8,200 – 12,400
0.75 0.375		0.85 x 0.475	10,000 – 15,000
0.622 x 0.311 (K_u band)	0.040	0.702 x 0.391	12,400 – 18,000
0.510 x 0.255		0.590 x 0.335	15,000 – 22,000
0.42 x 0.17 (K band)		0.5 x 0.25	18,000 – 26,500
0.34 x 0.17		0.42 x 0.25	22,000 – 33,000
0.28 x 0.14		0.36 x 0.22	26,500 – 40,000
0.224 x 0.112		0.304 x 0.192	33,000 – 50,000
0.188 x 0.094		0.268 x 0.174	40,000 – 60,000
0.148 x 0.074		0.228 x 0.154	50,000 – 75,000
0.122 x 0.061		0.202 x 0.141	60,000 – 90,000
0.100 x 0.050		0.180 x 0.130	75,000 – 110,000

Glossary of Semiconductor Terms and Definitions

Y

Acceptor Impurity—A substance with three electrons in the outer orbit of its atom which, when added to a semiconductor crystal, provides one hole in the lattice structure of the crystal.

Amplifier, Class A—An amplifier in which the swing of the input signal is always on the linear portion of the characteristic curves of the amplifying device.

Amplifier, Class AB—An amplifier which has the collector current or voltage at zero for less than half of a cycle of input signal.

Amplifier, Class B—An amplifier which operates at collector current cutoff or at zero collector voltage and remains in this condition for $\frac{1}{2}$ cycle of the input signal.

Amplifier, Class C—An amplifier in which the collector voltage or current is zero for more than $\frac{1}{2}$ cycle of the input signal.

AND Circuit (*AND Gate*)—A coincidence circuit that functions as a gate so that, when all the inputs are applied simultaneously, a prescribed output condition exists.

AND-OR Circuit (*AND-OR Gate*)—A gating circuit that produces a prescribed output condition when several possible combined input signals are applied; exhibits the characteristics of the *AND* gate and the *OR* gate.

Astable Multivibrator—A multivibrator that can function in either of two semistable states, switching rapidly from one to the other; referred to as free-running.

Barrier—In a semiconductor, the electric field between the acceptor ions and the donor ions at a junction. (See *Depletion Layer*.)

Barrier Height—In a semiconductor, the difference in potential from one side of a barrier to the other.

Base (*junction transistor*)—The center semiconductor region of a double junction (NPN or PNP) transistor. The base is comparable to the grid of an electron tube.

Base Spreading Resistance—In a transistor, the resistance of the base region caused by the resistance of the bulk material of the base region.

Beat Frequency Oscillator—An oscillator that produces a signal which mixes with another signal to provide frequencies equal to the sum and difference of the combined frequencies.

Bistable Multivibrator—A circuit with two stable states requiring two input pulses to complete a cycle.

Blocking Oscillator—A relaxation type oscillator that conducts for a short period of time and is cut off for a relatively long period of time.

Clamping Circuit—A circuit that maintains either or both amplitude extremities of a waveform at a certain level or potential.

Collector—The end semiconductor material of a double junction (NPN or PNP) transistor which is normally reverse-biased with respect to the base. The collector is comparable to the plate of an electron tube.

Common-Base (*CB*) *Amplifier*—A transistor amplifier in which the base element is common to the input and the output circuit. This configuration is comparable to the grounded-grid triode electron tube.

Common-Collector (*CC*) *Amplifier*—A transistor amplifier in which the collector element is common to the input and the output circuit. This configuration is comparable to the electron tube cathode follower.

Common-Emitter (*CE*) *Amplifier*—A transistor amplifier in which the emitter element is common to the input and the output circuit. This configuration is comparable to the conventional electron tube amplifier.

Complementary Symmetry Circuit—An arrangement of PNP-type and NPN-type transistors that provides push-pull operation from one input signal.

Compound-Connected Transistor—A combination of two transistors to increase the current amplification factor at high emitter currents. This combination is generally employed in power amplifier circuits.

Configuration—The relative arrangement of parts (or components) in a circuit.

Constant Power Dissipation Line—A line (superimposed on the output static characteristic curves) representing the points of collector voltage and current, the products of which represent the maximum collector power rating of a particular transistor.

Cross-Over Distortion—Distortion that occurs at the points of operation in a push-pull amplifier where the input signals cross (go through) the zero reference points.

Current Stability Factor—In a transistor, the ratio of a change in emitter current to a change in reverse-bias current flow between the collector and the base.

Cutoff Frequency—The frequency at which the gain of an amplifier falls below 0.707 times the maximum gain.

Dependent Variable—In a transistor, one of four variable currents and voltages that is arbitrarily chosen and considered to vary in accordance with other currents and voltages (independent variable).

Depletion Region (*or Layer*)—The region in a semiconductor containing the uncompensated acceptor and donor ions; also referred to as the space-charge region or barrier region.

Differentiating Circuit—A circuit that produces an output voltage proportional to the rate of change of the input voltage.

Donor Impurity—A substance with electrons in the outer orbit of its atom; added to a semiconductor crystal, it provides one free electron.

Double-Junction Photosensitive Semiconductor—Three layers of semiconductor material with an electrode connection to each end layer. Light energy is used to control current flow.

Dynamic Transfer Characteristic Curve—In transistors, a curve that shows the variation of output current (dependent variable) with variation of input current under load conditions.

Electron-Pair Bond—A valence bond formed by two electrons, one from each of two adjacent atoms.

Elemental Charge—The electrical charge on a single electron (megatron or positron).

Emitter-Follower Amplifier—See *Common-Collector Amplifier*.

Emitter (*junction transistor*)—The end semiconductor material of a double junction (PNP or NPN) transistor that is forward-biased with respect to the base. The emitter is comparable to the cathode of an electron tube.

Equivalent Circuit—A diagrammatic circuit representation of any device exhibiting two or more electrical parameters.

Fall Time—The length of time during which the amplitude of a pulse is decreasing from 90 percent to 10 percent of its maximum value.

Forward Bias—In a transistor, an external potential applied to a PN junction so that the depletion region is narrowed and relatively high current flows through the junction.

Forward Short-Circuit Current Amplification Factor—In a transistor, the ratio of incremental values of output to input current when the output circuit is a-c short-circuited.

Four-Layer Diode—A diode constructed of semiconductor materials resulting in three PN junctions. Electrode connections are made to each end layer.

Gating Circuit—A circuit operating as a switch, making use of a short or open circuit to apply or eliminate a signal.

Grounded Base Amplifier—See *Common-Base Amplifier.*

Hole—A mobile vacancy in the electronic valence structure of a semiconductor. The hole acts similarly to a positive electronic charge having a positive mass.

Hybrid Parameter—The parameters of an equivalent circuit of a transistor which are the result of selecting the input current and the output voltage as independent variables.

Increment—A small change in value.

Independent Variable—In a transistor, one of several voltages and currents chosen arbitrarily and considered to vary independently.

Inhibition Gate—A gate circuit used as a switch and placed in parallel with the circuit it is controlling.

Interelement Capacitance—The capacitance caused by the PN junctions between the regions of a transistor; measured between the external leads of the transistor.

Junction Transistor—A device having three alternate sections of P-type or N-type semiconductor material. See *PNP Transistor* and *NPN Transistor.*

Lattice Structure—In a crystal, a stable arrangement of atoms and their electron-pair bonds.

Majority Carriers—The holes or free electrons in P-type or N-type semiconductors, respectively.

Minority Carriers—The holes or excess electrons found in the N-type or P-type semiconductors, respectively.

Monostable Multivibrator—A multivibrator having one stable and one semistable condition. A trigger is used to drive the unit into the semistable state where it remains for a predetermined time before returning to the stable condition.

Multivibrator—A type of relaxation oscillator for the generation of nonsinusoidal waves in which the output of each of two stages is coupled to the input of the other to sustain oscillations. See *Astable Multivibrator, Bistable Multivibrator, and Monostable Multivibrator.*

Neutralization—The prevention of oscillation of an amplifier by canceling possible changes in the reactive component of the input circuit caused by positive feedback.

NOR Circuit—An *OR* gating circuit that provides pulse phase inversion.

NOT-AND Circuit—An *AND* gating circuit that provides pulse phase inversion.

NPN Transistor—A device consisting of a P-type section and two N-type sections of semiconductor material, with the P-type in the center.

N-Type Semiconductor—A semiconductor into which a donor impurity has been introduced. It contains free electrons.

Open-Circuit Parameters—The parameters of an equivalent circuit of a transistor which are the result of selecting the input current and output current as independent variables.

OR Circuit (OR Gate)—A gate circuit that produces the desired output with only one of several possible input signals applied.

Parameter—A derived or measured value which conveniently expresses performance; for use in calculations.

Photosensitive Semiconductor—A semiconductor material in which light energy controls current carrier movement.

PN Junction—The area of contact between N-type and P-type semiconductor materials.

PNP Transistor—A device consisting of an N-type section and two P-type sections of semiconductor material, with the N-type in the center.

Point-Contact—In transistors, a physical connection made by a metallic wire on the surface of a semiconductor.

Polycrystalline Structure—The granular structure of crystals which are nonuniform in shape and irregularly arranged.

Preamplifier—A low-level stage of amplification usually following a transducer.

P-Type Semiconductor—A semiconductor crystal into which an acceptor impurity has been introduced. It provides holes in the crystal lattice structure.

Pulse Amplifier—A wide-band amplifier used to amplify square waveforms.

Pulse Repetition Frequency—The number of nonsinusoidal cycles (square waves) that occur in 1 second.

Pulse Time—The length of time a pulse remains at its maximum value.

Quiescence—The operating condition that exists in a circuit when no input signal is applied to the circuit.

Reverse Bias—An external potential applied to a PN junction such as to widen the depletion region and prevent the movement of majority current carriers.

Reverse Open-Circuit Voltage Amplification Factor—In a transistor, the ratio of incremental values of input voltage to output voltage measured with the input a-c open-circuited.

Rise Time—The length of time during which the leading edge of a pulse increases from 10 percent to 90 percent of its maximum value.

Saturation (Leakage) Current—The current flow between the base and collector or between the emitter and collector, measured with the emitter lead or the base lead, respectively, open.

Semiconductor—A conductor whose resistivity is between that of metals and insulators and in which electrical charge carrier concentration increases with increasing temperature over a specific temperature range.

Short-Circuit Parameters—The parameters of an equivalent circuit of a transistor which are the result of selecting the input and output voltages as independent variables.

Single-Junction Photosensitive Semiconductor—Two layers of semiconductor materials with an electrode connection to each material. Light energy controls current flow.

Spacistor—A semiconductor device consisting of one PN junction and four electrode connections, characterized by a low transient time for carriers to flow from the input element to the output element.

Stabilization—The reduction of variations in voltage or current not due to prescribed conditions.

Storage Time—The time during which the output current or voltage of a pulse is falling from maximum to zero after the input current or voltage is removed.

Stray Capacitance—The capacitance introduced into a circuit by the leads and wires used to connect circuit components.

Surge Voltage (or Current)—A large sudden change of voltage (or current) usually caused by the collapsing of a magnetic field or the shorting or opening of circuit elements.

Swamping Resistor—In transistor circuits, a resistor placed in the emitter lead to mask (or minimize the effects of) variations in emitter-base junction resistance caused by variations in temperature.

Tetrode Transistor—A junction transistor with two electrode connections to the base (one to the emitter and one to the collector) to reduce the interelement capacitance.

Thermal Agitation—In a semiconductor, the random movement of holes and electrons within a crystal due to the thermal (heat) energy.

Thyratron—A gas-filled triode electron tube that is used as an electronic switch.

Transducer—A device that converts one type of power to another, such as acoustical power to electrical power.

Transistor—A semiconductor device capable of transferring a signal from one circuit to another and producing amplification. See *Junction Transistor*.

Triggered Circuit—A circuit that requires an input signal (trigger) to produce a desired output determined by the characteristics of the circuit.

Trigger Pulse Steering—In transistors, the routing or directing of trigger signals (usually pulses) through diodes or transistors (called steering diodes or steering transistors) so that the trigger signals affect only one circuit of several associated circuits.

Tuned-Base Oscillator—A transistor oscillator with the frequency-determining device (resonant circuit) located in the base circuit. It is comparable to the tuned-grid electron tube oscillator.

Tuned-Collector Oscillator—A transistor oscillator with the frequency-determining device located in the collector circuit. It is comparable to the tuned-plate electron tube oscillator.

Turnoff Time—The time that it takes a switching circuit (gate) to completely stop the flow of current in the circuit it is controlling.

Unijunction Transistor—A PN junction transistor with one electrode connection to one of the semiconductor materials and two connections to the other semiconductor material.

Unilateralization—The process by which an amplifier is prevented from going into oscillation by canceling the resistive and reactive component changes in the input circuit of an amplifier caused by positive feedback.

Unit Step Current (*or Voltage*)—A current (or voltage) which undergoes an instantaneous change in magnitude from one constant level to another.

Voltage Gain—The ratio of incremental values of output voltage to input voltage of an amplifier under load conditions.

Wide-Band Amplifier—An amplifier capable of passing a wide range of frequencies with equal gain.

Zener Diode—A PN junction diode reverse-biased into the breakdown region; used for voltage stabilization.

Z Commonly Used RF Letter-Band Designations

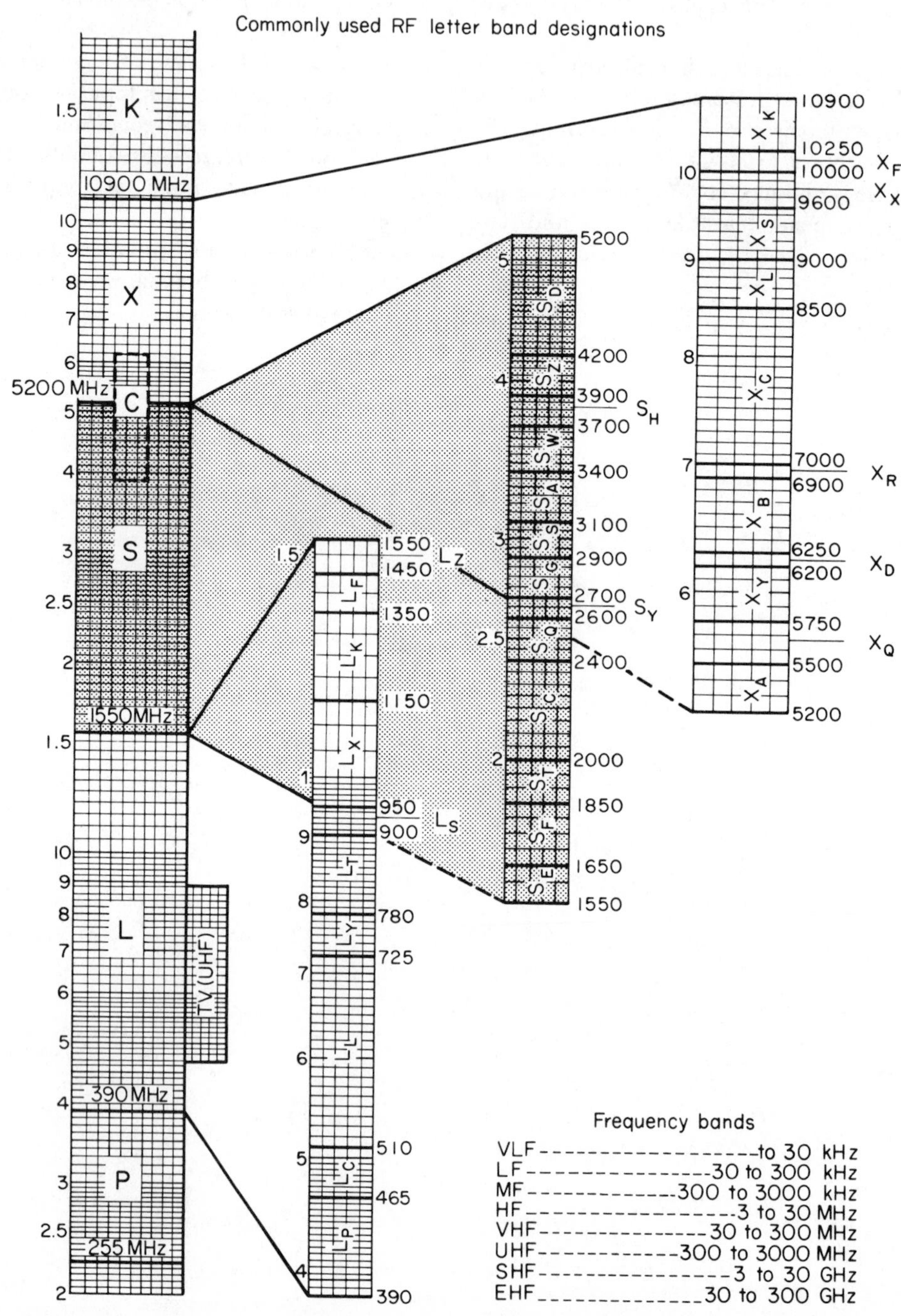

Temperature Conversion Tables ZZ

FAHRENHEIT TO CENTIGRADE*

Temperature Fahrenheit	0	1	2	3	4	5	6	7	8	9
100	73.33	73.89	74.44	75.00	75.56	76.11	76.67	77.22	77.78	78.33
90	67.78	68.33	68.89	69.44	70.00	70.56	71.11	71.67	72.22	72.78
80	62.22	62.78	63.33	63.89	64.44	65.00	65.56	66.11	66.67	67.22
70	56.67	57.22	57.78	58.33	58.89	59.44	60.00	60.56	61.11	61.67
60	51.11	51.67	52.22	52.78	53.33	53.89	54.44	55.00	55.56	56.11
50	45.56	46.11	46.67	47.22	47.78	48.33	48.89	49.44	50.00	50.56
40	40.00	40.56	41.11	41.67	42.22	42.78	43.33	43.89	44.44	45.00
30	34.44	35.00	35.56	36.11	36.67	37.22	37.78	38.33	38.89	39.44
20	28.89	29.44	30.00	30.56	31.11	31.67	32.22	32.78	33.33	33.89
10	23.33	23.89	24.44	25.00	25.56	26.11	26.67	27.22	27.78	28.33
0	17.78	18.33	18.89	19.44	20.00	20.56	21.11	21.67	22.22	22.78
0	17.78	17.22	16.67	16.11	15.56	15.00	14.44	13.89	13.33	12.78
10	12.22	11.67	11.11	10.56	10.00	9.44	8.89	8.33	7.78	7.22
20	6.67	6.11	5.56	5.00	4.44	3.89	3.33	2.78	2.22	1.67
30	1.11	0.56	0.00	0.56	1.11	1.67	2.22	2.78	3.33	3.89
40	4.44	5.00	5.56	6.11	6.67	7.22	7.78	8.33	8.89	9.44
50	10.00	10.56	11.11	11.67	12.22	12.78	13.33	13.89	14.44	15.00
60	15.56	16.11	16.67	17.22	17.78	18.33	18.89	19.44	20.00	20.56
70	21.11	21.67	22.22	22.78	23.33	23.89	24.44	25.00	25.56	26.11
80	26.67	27.22	27.78	28.33	28.89	29.44	30.00	30.56	31.11	31.67
90	32.22	32.78	33.33	33.89	34.44	35.00	35.56	36.11	36.67	37.22
100	37.78	38.33	38.89	39.44	40.00	40.56	41.11	41.67	42.22	42.78

*Basic formula: $°C = (5/9)(°F - 32)$

CENTIGRADE TO FAHRENHEIT*

Temperature Centigrade	0	1	2	3	4	5	6	7	8	9
100	148.0	149.8	151.6	153.4	155.2	157.0	158.8	160.6	162.4	164.2
90	130.0	131.8	133.6	135.4	137.2	139.0	140.8	142.6	144.4	146.2
80	112.0	113.8	115.6	117.4	119.2	121.0	122.8	124.6	126.4	128.2
70	94.0	95.8	97.6	99.4	101.2	103.0	104.8	106.6	108.4	110.2
60	76.0	77.8	79.6	81.4	83.2	85.0	86.8	88.6	90.4	92.2
50	58.0	59.8	61.6	63.4	65.2	67.0	68.8	70.6	72.4	74.2
40	40.0	41.8	43.6	45.4	47.2	49.0	50.8	52.6	54.4	56.2
30	22.0	23.8	25.6	27.4	29.2	31.0	32.8	34.6	36.4	38.2
20	4.0	5.8	7.6	9.4	11.2	13.0	14.8	16.6	18.4	20.2
10	14.0	12.2	10.4	8.6	6.8	5.0	3.2	1.4	0.4	2.2
0	32.0	30.2	28.4	26.6	24.8	23.0	21.2	19.4	17.6	15.8
0	32.0	33.8	35.6	37.4	39.2	41.0	42.8	44.6	46.4	48.2
10	50.0	51.8	53.6	55.4	57.2	59.0	60.8	62.6	64.4	66.2
20	68.0	69.8	71.6	73.4	75.2	77.0	78.8	80.6	82.4	84.2
30	86.0	87.8	89.6	91.4	93.2	95.0	96.8	98.6	100.4	102.2
40	104.0	105.8	107.6	109.4	111.2	113.0	114.8	116.6	118.4	120.2
50	122.0	123.8	125.6	127.4	129.2	131.0	132.8	134.6	136.4	138.2
60	140.0	141.8	143.6	145.4	147.2	149.0	150.8	152.6	154.4	156.2
70	158.0	159.8	161.6	163.4	165.2	167.0	168.8	170.6	172.4	174.2
80	176.0	177.8	179.6	181.4	183.2	185.0	186.8	188.6	190.4	192.2
90	194.0	195.8	197.6	199.4	201.2	203.0	204.8	206.6	208.4	210.2
100	212.0	213.8	215.6	217.4	219.2	221.0	222.8	224.6	226.4	228.2

*Basic formula: $°F = (9/5)°C + 32$

Index

A

B

C

D

T

U

V

W

Y